Modern Biocatalysis

Edited by
Wolf-Dieter Fessner and
Thorleif Anthonsen

Related Titles

Sheldon, R. A., Arends, I:, Hanefeld, U

Green Chemistry and Catalysis

2007
Hardcover
ISBN: 978-3-527-30715-9

Aehle, W. (ed.)

Enzymes in Industry
Production and Applications

2007
Hardcover
ISBN: 978-3-527-31689-2

Liese, A., Seelbach, K., Wandrey, C. (eds.)

Industrial Biotransformations
Second, Completely Revised and Extended Edition

2006
Hardcover
ISBN: 978-3-527-31001-2

Garcia-Junceda, E. (ed.)

Multi-Step Enzyme Catalysis
Biotransformations and Chemoenzymatic Synthesis

2008
Hardcover
ISBN: 978-3-527-31921-3

Gotor, V., Alfonso, I., García-Urdiales, E. (eds.)

Asymmetric Organic Synthesis with Enzymes

2008
Hardcover
ISBN: 978-3-527-31825-4

Rothenberg, G.

Catalysis
Concepts and Green Applications

2008
Hardcover
ISBN: 978-3-527-31824-7

Modern Biocatalysis

Stereoselective and Environmentally Friendly Reactions

Edited by
Wolf-Dieter Fessner and Thorleif Anthonsen

WILEY-VCH

WILEY-VCH Verlag GmbH & Co. KGaA

The Editors

Prof. Dr. Wolf-Dieter Fessner
TU Darmstadt, Inst. f. Organische Chemie und Biochemie
Petersenstr. 22
64287 Darmstadt

Prof. Dr. Thorleif Anthonsen
Norwegian Unvi. of Science & Technology, Dept. of Chemistry
Hogskoleringen 5
7491 Trondheim
Norwegen

Cover Illustration:
Background: Modified part of the "Biochemical Pathways Chart", 4th edition, by courtesy of F. Hoffmann-La Roche Ltd.

Protein structure taken from the RSCB Protein Data Base and modified:
PDB ID: 1ADO
N. Blom, J. Sygusch, Product binding and role of the C-terminal region in class I D-fructose 1,6-bisphosphate aldolase, *Nat. Struct. Biol.* **4** (1997), 36–39

Library of Congress Card No.:
applied for

British Library Cataloguing-in-Publication Data
A catalogue record for this book is available from the British Library.

Bibliographic information published by the Deutsche Nationalbibliothek
The Deutsche Nationalbibliothek lists this publication in the Deutsche Nationalbibliografie; detailed bibliographic data are available on the Internet at http://dnb.d-nb.de.

Typesetting SNP Best-set Typesetter Ltd., Hong Kong
Printing Strauss GmbH, Mörlenbach
Binding Litges & Dopf GmbH, Heppenheim

Printed in the Federal Republic of Germany
Printed on acid-free paper

ISBN: 978-3-527-32071-4

Contents

Modern Biocatalysis: Stereoselective and Environmentally Friendly Reactions.
Edited by Wolf-Dieter Fessner and Thorleif Anthonsen

ISBN: 978-3-527-32071-4

Preface

Asymmetric compounds cover a steadily increasing market share, e.g. as fine chemicals, chiral intermediates or pharmaceutical ingredients. The worldwide commercial volume of single isomers of chiral drugs currently approaches US$200 billion. Chirality dictates how stereoisomers interact with biological systems. Because of the often much improved biological specificity profiles of pure enantiomers as compared to their racemates, methods for resolving racemates and for preparing single enantiomers are in high demand. For the challenge of asymmetric synthesis to generate complex chiral compounds in high enantiomeric purity and yield, biocatalysis offers a tremendous advantage due to the homochiral nature of protein catalysts, which offer unparalleled levels of stereoselectivity and reaction specificity, in addition to their remarkable catalytic efficiency under mild conditions and regenerative production from biological materials. Breakthrough technologies in modern molecular biotechnology research, development and application have recently paved the way for the rapid discovery and engineering of novel enzymes, and today allow the generation of biocatalysts that are optimally adaptable to even demanding industrial process parameters. Such features form the fundaments of so-called white biotechnology – the use of microorganisms and enzymes for industrial chemical production, which currently strongly benefits from political support because of its impact on sustainable development, lower energy consumption and independence from fossil raw materials. Rising oil prices from a shortage of crude oil resources and political uncertainty in oil-producing countries; global warming due to greenhouse gases; increasing population; and environmental pollution are strong drivers for this trend.

Utilization of biocatalysis for chiral chemical synthesis and pharmaceutical manufacturing has come a long way, from highly specific niche applications a century ago, through hundreds of small to medium scale industrial processes in the fine chemical sector, e.g. including antibiotics, to even bulk processes for non-chiral commodity chemicals such as acrylamide. The conversion of traditional industrial processes to biotechnological alternatives is still at an early stage, because high costs in the development and adaptation of new processes have slowed down a substitution; development cycles for such biocatalytic processes used to be longer than anticipated and considerably longer than those for comparable chemical process alternatives. In the long run, however, biocatalytic processes have often

Modern Biocatalysis: Stereoselective and Environmentally Friendly Reactions.
Edited by Wolf-Dieter Fessner and Thorleif Anthonsen

ISBN: 978-3-527-32071-4

proven to be economically feasible, ecologically advantageous and more sustainable than current chemical technologies because of the intrinsic advantages of biocatalysts in higher reaction selectivity, milder reaction conditions and potential use of inexpensive regenerable resources. After all, the tools of the trade have changed dramatically over the past decade, and with current advanced technology in protein design and engineering, further developments are expected to strongly gain momentum for the immediate future. As one of the most impressive examples, the newly developed one-step fermentation process for vitamin B_2 at BASF has cut CO_2 emissions by 30%, production costs by 40%, consumption of resources by 60% and waste generation by 95% when compared to the conventional eight-step chemical synthesis.

Still, the field of biocatalysis research for preparative synthesis poses a range of intellectual frontiers and needs further developments to broaden the range of applicable reaction types, addressable target structures and advanced method integration. There is both general and industrial interest in further research to extend the scope of biocatalysis for asymmetric synthesis, and a collaboration of organic chemists, biochemists, molecular geneticists and biotechnical engineers is needed for success. The development of biocatalytic methodologies undoubtedly requires strong interdisciplinary and transdisciplinary research cooperation, and meeting the challenges for an environmentally friendly, sustainable process design adds another dimension, from the supply and efficient use of raw materials to the minimization and recycling of enzymes, by-products and waste under economical constraints. This book summarizes the efforts and current state of the art in several important arenas of biocatalysis research that has been coordinated within the activities of the European Union-funded COST network D25, entitled 'Applied Biocatalysis: Stereoselective and Environmentally Friendly Reactions Catalysed by Enzymes', over the last five years. The topics of the chapters span from modern assay technologies for enzyme screening over different factors influencing enzyme selectivities, including the consequences from enzyme formulation and various solvent effects, to the manifold of preparative applications. More than half of the book chapters deal with the various conceptual strategies and synthetic opportunities available for the rational synthesis or targeted modification of different compound classes, such as phenolic natural products, nucleoside analogs, monosaccharides and oligosaccharides, iminosugars, proteinogenic and non-proteinogenic amino acids, nitriles, hydroxy acids, and various oxidation products, including lactones from enzymatic Baeyer–Villiger reactions.

We are grateful to all those friends and colleagues who helped to start, and then participated in, the D25 action and who kindly contributed to this project as authors of dedicated and informative chapters in order to share their expertise with you. It is our hope that this volume will encourage scientific discussion and foster imaginative new developments in applied biocatalysis for tomorrow's novel applications.

Wolf-Dieter Fessner and Thorleif Anthonsen
Darmstadt and Trondheim, October 2008

List of Contributors

Miguel Alcalde
Departamento de Biocatálisis
Instituto de Catálisis y
Petroleoquímica
CSIC
Cantoblanco
28049 Madrid
Spain

Veronique Alphand
Université Paul Cézanne
Biosciences – FRE CNRS 3005
Case 432
Av. Escadrille Normandie-Niemen
13397 Marseille cedex 20
France

Thorleif Anthonsen
Norwegian University of Science
and Technology
Department of Chemistry
7491 Trondheim
Norway

Antonio Ballesteros
Departamento de Biocatálisis
Instituto de Catálisis y
Petroleoquímica
CSIC
Cantoblanco
28049 Madrid
Spain

Karel Bezouška
Charles University Prague
Faculty of Science
Department of Biochemistry
128 40 Prague
Czech Republic

Laura Cantarella
Department of Industrial Engineering
University of Cassino
Via di Biasio 43
03043 Cassino (FR)
Italy

Maria Cantarella
University of L'Aquila
Department of Chemistry
Chemical Engineering and Materials
67040 Monteluco di Roio – L'Aquila
Italy

Franck Charmantray
Université Blaise Pascal
CNRS
UMR 6504 – SEESIB
63177 Aubière Cedex
France

Modern Biocatalysis: Stereoselective and Environmentally Friendly Reactions.
Edited by Wolf-Dieter Fessner and Thorleif Anthonsen

ISBN: 978-3-527-32071-4

Andrzej Chmura
Delft University of Technology
Department of Biotechnology
Laboratory of Biocatalysis and
Organic Chemistry
Julianalaan 136
2628 BL Delft
The Netherlands

Pere Clapés
Catalonia Institute for Advonced
Chemistry (IQAC)-CSIC. Group
of Biotransformation and
Bioactive Molecules
Jordi Girona 18-26
08034 Barcelona
Spain

Josefa María Clemente-Jiménez
Departamento de Química-Física
Bioquímica y Química Inorgánica
Edificio CITE I
Carretera de Sacramento S/N
La Cañada de San Urbano
04120 Almería
Spain

Attilio Converti
University of Genoa
Department of Chemical and
Process Engineering 'G.B.
Bonino'
Via Opera Pia 15
16145 Genoa
Italy

Bruno C. M. Fernandes
Delft University of Technology
Department of Biotechnology
Laboratory of Biocatalysis and
Organic Chemistry
Julianalaan 136
2628 BL Delft
The Netherlands

Lucía Fernández-Arrojo
Departamento de Biocatálisis
Instituto de Catálisis y Petroleoquímica
CSIC
Cantoblanco
28049 Madrid
Spain

Marco W. Fraaije
University of Groningen
Groningen Biomolecular Sciences and
Biotechnology Institute
Biochemical Laboratory
Nijenborgh 4
9747 AG Groningen
The Netherlands

Alberto Gallifuoco
University of L'Aquila
Department of Chemistry
Chemical Engineering and Materials
67040 L'Aquila
Italy

Raffaella Gandolfi
University of Milan
Institute of Organic Chemistry
'Alessandro Marchesini'
Via Venezian 21
20133 Milan
Italy

Lucia Gardossi
Università degli Studi di Trieste
Dipartimento di Scienze Farmaceutiche
Laboratory of Applied and
Computational Biocatalysis
Piazzale Europa 1
34127 Trieste
Italia

Iraj Ghazi
Departamento de Biocatálisis
Instituto de Catálisis y Petroleoquímica
CSIC
Cantoblanco
28049 Madrid
Spain

Vicente Gotor
Universidad de Oviedo
Departamento de Química Orgánica e Inorgánica
Instituto de Biotecnología de Asturias
33006 Oviedo
Spain

Maja Habulin
University of Maribor
Faculty of Chemistry and Chemical Engineering
Laboratory for Separation Processes and Product Design
Smetanova 17
2000 Maribor
Slovenia

Ulf Hanefeld
Delft University of Technology
Department of Biotechnology
Biocatalysis and Organic Chemistry
Julianalaan 136
2628 BL Delft
The Netherlands

Laurence Hecquet
Université Blaise Pascal
Synthèse et Etudes de Systèmes à Intérêt Biologique
UMR 6504
24 avenue des Landais
63177 Aubiere Cedex
France

Virgil Hélaine
Université Blaise Pascal
CNRS
UMR 6504–SEESIB
63177 Aubière Cedex
France

Francisco Javier Las Heras-Vázquez
Departamento de Química-Física
Bioquímica y Química Inorgánica
Edificio CITE I
Carretera de Sacramento S/N
La Cañada de San Urbano
04120 Almería
Spain

Elisabeth Egholm Jacobsen
Norwegian University of Science and Technology
Department of Chemistry
7491 Trondheim
Norway

Jesús Joglar
Catalonia Institute for Advanced Chemistry (IQAC)-CSIC.
Group of Biotransformation and Bioactive Molecules
Jordi Girona 18-26
08034 Barcelona
Spain

Ondrej Kaplan
Academy of Sciences of the Czech Republic
Institute of Microbiology
Center of Biocatalysis and Biotransformation
142 20 Prague
Czech Republic

Maria H. Katsoura
University of Ioannina
Department of Biological
Applications and Technologies
Laboratory of Biotechnology
45110 Ioannina
Greece

Norbert Klempier
Technische Universität Graz
Institut für Organische Chemie
Stremayrgasse 16
8010 Graz
Austria

Željko Knez
University of Maribor
Faculty of Chemistry and
Chemical Engineering
Laboratory for Separation
Processes and Product Design
Smetanova 17
2000 Maribor
Slovenia

Fragiskos N. Kolisis
National Technical University
of Athens
Chemical Engineering
Department
Biotechnology Laboratory
5 Iroon Polytechniou Str.
Zografou Campus
15700 Athens
Greece

Vladimír Kren
Academy of Sciences of the Czech
Republic
Institute of Microbiology
Center of Biocatalysis and
Biotransformation
142 20 Prague
Czech Republic

Morten Kristensen
Arinco Arla Foods Amba
Mælkevejen 4
6920 Videbæk
Denmark

Marielle Lemaire
Université Blaise Pascal
CNRS
UMR 6504–SEESIB
63177 Aubière Cedex
France

Sergio Martínez-Rodríguez
Departamento de Química-Física
Bioquímica y Química Inorgánica
Edificio CITE I
Carretera de Sacramento S/N
La Cañada de San Urbano
04120 Almería
Spain

Ludmila Martínková
Academy of Sciences of the Czech
Republic
Institute of Microbiology
Center of Biocatalysis and
Biotransformation
142 20 Prague
Czech Republic

Cesar Mateo
Delft University of Technology
Department of Biotechnology
Laboratory of Biocatalysis and Organic
Chemistry
Julianalaan 136
2628 BL Delft
The Netherlands

Marko D. Mihovilovic
Vienna University of Technology
Institute for Applied Synthetic Chemistry
Getreidemarkt 9
1060 Vienna
Austria

Francesco Molinari
University of Milan
Department of Food Science and Microbiology
Via Celoria 2
20133 Milan
Italy

Gianluca Molla
Università degli Studi dell'Insubria
Dipartimento di Biotecnologie e Scienze Molecolari
Via J.H. Dunant 3
21100 Varese
Italy

Bernd Nidetzky
Graz University of Technology
Institute of Biotechnology and Biochemical Engineering
Petersgasse 12
8010 Graz
Austria

Gianluca Ottolina
Istituto di Chimica del Riconoscimento Molecolare CNR
Via Mario Bianco 9
20131 Milano
Italy

Lars Haastrup Pedersen
Aalborg University
Department of Biotechnology
Chemistry and Environmental Engineering
Sohngårdsholmvej 49
9000 Aalborg
Denmark

Francisco J. Plou
Departamento de Biocatálisis
Instituto de Catálisis y Petroleoquímica
CSIC
Cantoblanco
28049 Madrid
Spain

Loredano Pollegioni
Università degli Studi dell'Insubria
Dipartimento di Biotecnologie e Scienze Molecolari
Via J.H. Dunant 3
21100 Varese
Italy

Mateja Primožič
University of Maribor
Faculty of Chemistry and Chemical Engineering
Laboratory for Separation Processes and Product Design
Smetanova 17
2000 Maribor
Slovenia

Fred van Rantwijk
Delft University of Technology
Department of Biotechnology
Laboratory of Biocatalysis and Organic Chemistry
Julianalaan 136
2628 BL Delft
The Netherlands

Jean-Louis Reymond
University of Berne
Department of Chemistry and Biochemistry
Freiestraße 3
3012 Berne
Switzerland

Sinthuwat Ritthitham
Aalborg University
Department of Biotechnology
Chemistry and Environmental Engineering
Sohngårdsholmvej 49
9000 Aalborg
Denmark

Felipe Rodríguez-Vico
Departamento de Química-Física
Bioquímica y Química Inorgánica
Edificio CITE I
Carretera de Sacramento S/N
La Cañada de San Urbano
04120 Almería
Spain

Diego Romano
University of Milan
Department of Food Science and Microbiology
Via Celoria 2
20133 Milan
Italy

Francesco Secundo
Istituto di Chimica del Riconoscimento Molecolare
CNR
Via Mario Bianco 9
20131 Milano
Italy

Stefano Servi
Politecnico di Milano
Dipartimento CMIC 'G. Natta'
Via Mancinelli 7
20131 Milano
Italy

Roger A. Sheldon
Delft University of Technology
Department of Biotechnology
Laboratory of Biocatalysis and Organic Chemistry
Julianalaan 136
2628 BL Delft
The Netherlands

Agata Spera
Department of Chemistry
Chemical Engineering and Materials
University of L'Aquila
67040 L'Aquila
Italy

Patrizia Spizzo
Università degli Studi di Trieste
Dipartimento di Scienze Farmaceutiche
Laboratory of Applied and Computational Biocatalysis
Piazzale Europa 1
34127 Trieste
Italia

Georg A. Sprenger
Universität Stuttgart
Institute of Microbiology
Allmandring 31
70569 Stuttgart
Germany

Haralambos Stamatis
University of Ioannina
Department of Biological Applications and Technologies
Laboratory of Biotechnology
45110 Ioannina
Greece

Davide Tessaro
Politecnico di Milano
Dipartimento CMIC 'G. Natta'
Via Mancinelli 7
20131 Milano
Italy

Eleni Theodosiou
National Technical University of Athens
Chemical Engineering Department
Biotechnology Laboratory
5 Iroon Polytechniou Str.
Zografou Campus
15700 Athens
Greece

Malene S. Thomsen
Research Centre Applied Biocatalysis
Petersgasse 14
8010 Graz
Austria.

Vojtech Vejvoda
Academy of Sciences of the Czech Republic
Institute of Microbiology
Center of Biocatalysis and Biotransformation
142 20 Prague
Czech Republic

Margit Winkler
Technische Universität Graz
Institut für Organische Chemie
Stremayrgasse 16
8010 Graz
Austria

Roland Wohlgemuth
Sigma-Aldrich
Research Specialities
Industriestrasse 25
9470 Buchs
Switzerland

1
Fluorescence Assays for Biotransformations

Jean-Louis Reymond

1.1
Introduction

High-throughput activity screening in microbial collections, expression libraries of enzyme mutants, or metagenomic libraries using enzyme activity assays is one of the key steps in the development of new or improved biocatalysts. Ideally, screening is carried out with the authentic reaction of interest under its operating conditions and reaction progress is monitored using analytical instruments such as high-performance liquid chromatography, gas chromatography, nuclear magnetic resonance, or mass spectrometry, or simply thin-layer chromatography. In most laboratories, however, such analyses are not realizable in high-throughput or in the format imposed by the expression conditions of the library under study. In such cases the enzyme discovery process can be assisted by a fluorescence assay. Fluorescence can be recorded with high sensitivity in a variety of settings, in particular in a high-throughput screening format such as microtiter plates, micro-arrays, or fluorescence-activated cell sorting. Fluorescence is also usually directly visible by the eye under illumination with an ultraviolet (UV) lamp.

Fluorescence monitoring of a biotransformation is possible either by coupling the reaction of interest to a fluorogenic sensor system, for example an enzyme-coupled assay linking the reaction to the production or consumption of NAD^+ or by using a fluorogenic model substrate instead of the authentic substrate. Fluorogenic substrates are particularly useful for high-throughput screening and routine identification of an enzyme during cloning when it has been established that this enzyme catalyzes the desired reaction. In recent years the author and others have reported on a variety of fluorescence sensors and substrates that expand the chemistry of classical assays to a broad diversity of reactions and substrates. The present chapter reviews the development of fluorogenic substrates based on umbelliferone and related fluorescent phenols by the author's research group and related reports by others. A more general presentation of recent advances in high-throughput screening assays for biotransformations has been recently published [1].

Modern Biocatalysis: Stereoselective and Environmentally Friendly Reactions.
Edited by Wolf-Dieter Fessner and Thorleif Anthonsen

ISBN: 978-3-527-32071-4

1.2
Alcohol Dehydrogenases (ADHs) and Aldolases

Alcohol dehydrogenases and aldolases are some of the most synthetically significant enzymes in biocatalysis because they carry out classical and highly useful reactions with ketones and aldehydes that generate new asymmetric centers. Although only a limited number of these enzymes are currently in use, their potential for complex asymmetric synthesis is very high and efficient fluorogenic substrates are particularly important for facilitating research in this area.

1.2.1
Chiral Fluorogenic ADH Substrates

Alcohol dehydrogenases can be assayed by following the conversion of $NAD(P)^+$ to NAD(P)H at 340 nm, either by UV or by fluorescence. However, the signal is often obscured by contaminants or other assay components if the assay is done in whole cells or in the presence of other chromophores and the reduced forms of the co-factors are not very stable. Signal stability is also a problem in indirect assays where the reduced co-factors are detected in a colorimetric secondary reaction such as the formazane test [2]. As an alternative to these methods, one can consider using a fluorogenic or chromogenic alcohol substrate as a reporter for the reaction. Umbelliferyl ethers of chiral 1,3-diols such as **1**, **2**, or **3** that serve as fluorogenic substrates for alcohol dehydrogenases have been developed (Scheme 1.1) [3–5]. The alcohols are obtained by alkylation of an umbelliferyl anion with activated precursors and this allows various substitution patterns and stereochemistries to be installed around the enzyme reactive functional group. The enzyme

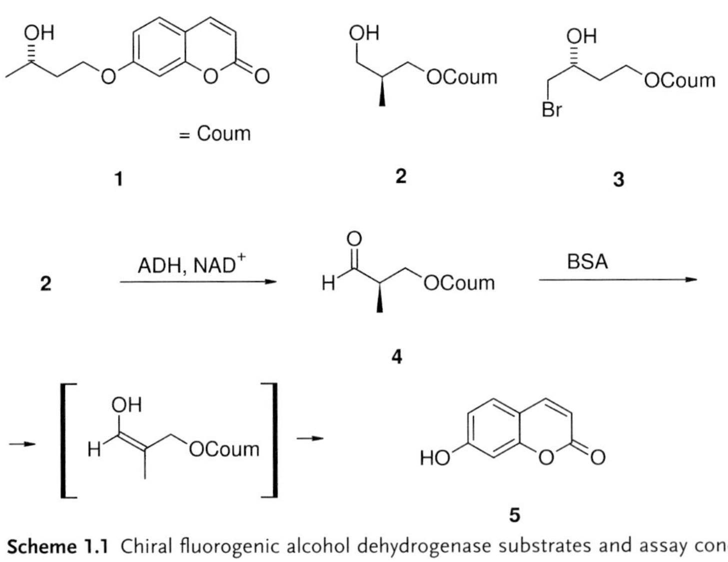

Scheme 1.1 Chiral fluorogenic alcohol dehydrogenase substrates and assay conditions.

to the corresponding carbonyl (e.g. **4**) oxidizes the alcohol function and this spontaneously liberates the fluorescent product umbelliferone **5** by β-elimination. This secondary reaction occurs above pH 7 and can be catalyzed by the addition of bovine serum albumin (BSA). The alkaline pH is also necessary because the fluorescent umbelliferone anion only exists above pH 7. The blue fluorescent signal produced around 440 nm increases over 20-fold upon conversion of the alcohol substrates to umbelliferone and is visible by the naked eye under illumination with UV light, which is particularly convenient during screening. On the other hand the β-elimination process has a half-life of 3–15 min depending on the substrate, which limits the reaction kinetics if the alcohol oxidation itself is very fast. Therefore the assay is only suitable for kinetic studies under slow reaction conditions, for example using low enzyme concentrations.

1.2.2 Fluorogenic Aldolase Probes

The discovery of aldolase catalytic antibodies in the 1990s [6–8] led to an interest in screening biocatalytic aldol reactions of non-natural aldolase-type substrates. The ADH assay above can be modified in the form of aldol substrates such as **6** to assay retro-aldolization reactions via a retro-aldol/β-elimination sequence. The aldolase substrate **6** and analogs are prepared by direct synthesis from aldehyde **4** (Scheme 1.2) [9, 10]. Aldehyde **4** is unstable towards β-elimination in water, but can be handled in organic solvent. For instance, the reaction of aldehyde with silyl enolethers under catalysis by $SnCl_4$ at low temperature in dichloromethane proceeds in excellent yields without β-elimination to deliver a stereoisomeric mixture of aldol **6**, which can then be separated by high-performance liquid chromatography for individual assays. The retro-aldol reaction can also be assayed with aldol **7**, which follows the same mechanism as **6**, as well as with aldols **8** and **9**, which directly yield a fluorescent aldehyde product upon retro-aldolization, as shown by List *et al.* [11]. The advantage of assaying retro-aldol reactions as opposed to the

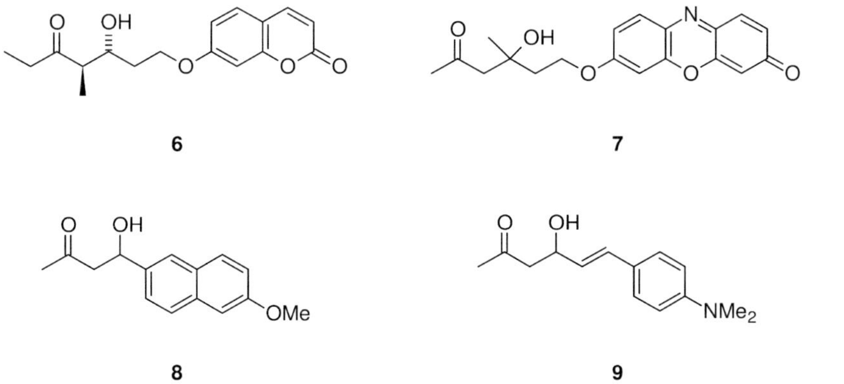

Scheme 1.2 Fluorogenic substrates for aldolase catalytic antibodies and related biocatalysts.

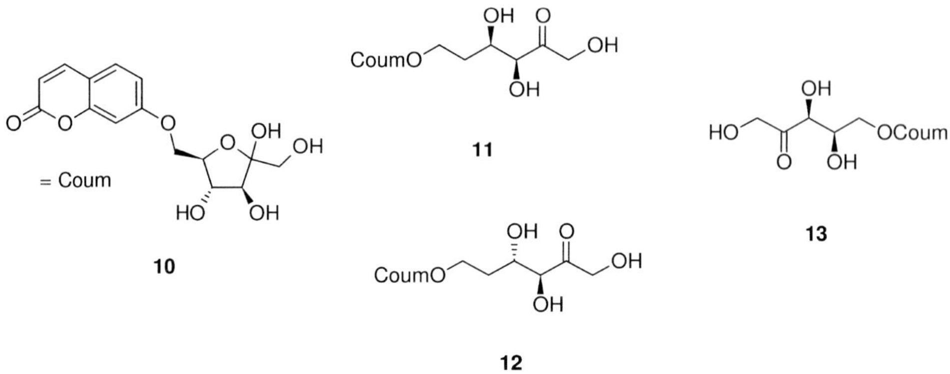

Scheme 1.3 Fluorogenic substrates for transaldolase and transketolase.

forward reaction lies in the possibility of testing individual stereoisomers of the aldol product for reactivity, leading to a prediction of stereoselectivity, as was analyzed using all eight diastereoisomers of aldol **6** [10].

1.2.3
Transaldolases and Transketolases

Related systems were later developed for transaldolases **10–12** (Scheme 1.3) [12]. The fructo/tagato stereoselectivity of various transaldolases was determined by fluorescence for the stereoisomeric substrate pair **11/12**. However, the reactivity of the substrates towards transaldolases is much lower than with the natural substrate due to the replacement of the phosphate group at position 6 of the natural fructose-6-phosphate substrate with the neutral, aromatic coumarin ether, which is not well recognized by the enzyme. Sevestre *et al.* [13] reported substrate **13** as a fluorogenic substrate for transketolases, based on a similar fluorescence release mechanism.

It should be mentioned that most natural aldolase enzymes can also be assayed using enzyme-coupled systems relaying the reaction to a redox process with NAD^+. The formation of NADH by active microbial colonies in expression libraries of mutant enzymes was detected colorimetrically in agar plates using phenazine methosulfate and nitroblue tetrazolium, which forms an insoluble precipitate. The assay was used by Williams *et al.* [14] and Woodhall *et al.* [15] for evolving sialic acid aldolases to accept non-natural aldehyde acceptors.

1.2.4
Enolase Probe

Aldolase-type biocatalysts can generally also be expected to catalyze the enolization of the ketone donor. However, directly detecting enolization by fluorescence is not possible. It was recently found that dihydroxyacetone coumarin ether **14** functions

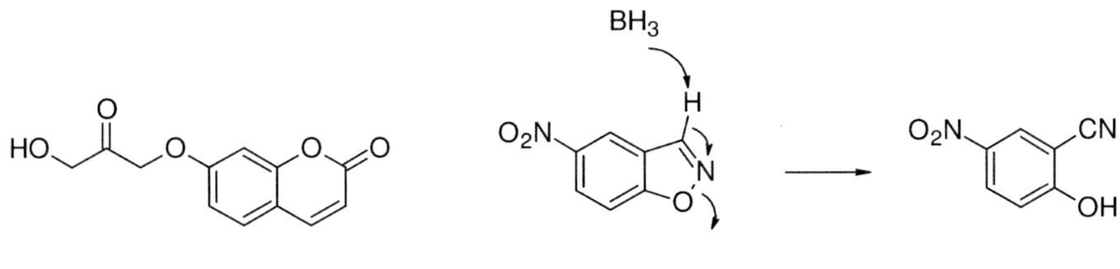

14 15 16

Scheme 1.4 Substrates to assay enolization and deprotonation at carbon.

as a fluorogenic enolization probe in aqueous buffer (Scheme 1.4) [16]. Enol formation is followed by β-elimination of umbelliferone and the corresponding fluorescence signal. The probe detects the activity of a variety of small molecule enolization catalysts and is generally useful for screening aldolase type biocatalysts [17]. Alkylation of umbelliferone with allyl bromide, dihydroxylation of the allyl ether double bond, silylation of the primary alcohol, oxidation of the secondary alcohol to the ketone, and acidic deprotection provide excellent yields of the probe.

Enolization involves a deprotonation step at carbon, which is a fundamental reaction step of interest in studying the principles of enzyme action. Investigations of biocatalytic carbon deprotonation have focused on the chromogenic ring opening of nitrobenzoxazole **15** to form the yellow product 2-cyano-4-nitrophenol **16**. The reaction is promoted by weak bases such as carboxylate side chains in a hydrophobic environment, in particular with albumins and catalytic antibodies [18–21]. The unstable aldehyde **4** (Scheme 1.1) and related carbonyls also provide interesting fluorogenic substrates for such reactions and the β-elimination process is indeed catalyzed by BSA in an enzyme-like manner [3].

1.3
Lipases and Esterases

Lipases and esterases can be assayed using pH indicators. A particularly elegant implementation of this principle was reported by the group of Kazlauskas for screening enantioselectivities using the Quick E assay [22, 23]. A variety of other high-throughput screening assays have been reported based on the chemoselective detection of the reaction products from untagged esters. These include fluorogenic hydrazone formation with acetaldehyde produced by lipase/esterase-catalyzed transesterification of vinyl esters in organic solvents [24], the detection of acetic acid produced from acetate esters using an enzyme kit [25], or the back titration of periodate with adrenaline after oxidation of the glycerol release from triglyceride hydrolysis [26, 27].

Various chromogenic and fluorogenic esters for lipases and esterase are also available commercially in the form of simple esters of umbelliferone or *p*-nitrophenol. Unfortunately such substrates are often relatively unreactive towards the enzyme while displaying a high level of non-specific hydrolysis in the reaction medium. In order to overcome these limitations various assays and substrates that significantly improve the fluorogenic detection of lipases and esterases have now been developed.

1.3.1
Assays on Solid Support

The disadvantage of esters of umbelliferone or nitrophenol esters as lipase substrates lies in their low aqueous solubility and high non-specific reactivity in water. Because lipases are interfacial enzymes, there has been a search for an assay format where these substrates would be presented to the enzyme solution in stable, insoluble form over a high surface material. The best implementation of this principle consists of impregnating silica gel plates (used for analytical thin-layer chromatography) with a solution of the umbelliferone esters in dichloromethane [28]. Evaporation of the solvent leaves the substrate homogeneously adsorbed on the surface. The silica gel plates can then be used for assaying enzymes by simply adding a drop of test solution to the surface. The buffer impregnates the plate and the substrates are dissolved and hydrolyzed if an active enzyme in present. This format is extremely practical since it allows the assay volume to be reduced to only 1 μl per assay. Its use has been demonstrated in high-throughput by profiling 35 different enzymes across 20 different ester substrates, realizing over 7000 individual assays in a matter of hours. The cyclohexane carboxylic ester **17**, which releases 4-methylumbelliferone **18** as a fluorescent product, appeared as the most uniformly reactive substrate across all lipases and esterases tested (Scheme 1.5). It should be mentioned that silica gel plates can be impregnated with fluorogenic substrates for other enzymes, for example umbelliferyl glycosides for testing glycosidases.

The photograph insert shows a typical plate impregnated with substrate 17 under UV illumination after reaction of 1-μl reactions containing various lipases (columns 1–10) or controls (BSA) (columns 11 and 12) under various conditions (lanes A–L) (see the original publication for details).

Substrate micro-arrays for testing lipases have also been investigated (Scheme 1.6) [29]. The assay is based on an indirect tagging strategy involving periodate cleavage of a primary 1,2-diol product followed by chemoselective tagging with rhodamine sulfohydrazide. The tag becomes covalently attached to the surface whenever an aldehyde product has been formed and its red fluorescence can be quantified using a standard micro-array scanner. The 1,2-diol esters showed the desired chemical stability on the micro-array and the 1,2-diol could be successfully tagged chemoselectively using the hydrazide reagent without non-specific staining under appropriate washing conditions. However, most lipases proved unreactive in the assay and the most reactive enzymes on the micro-array were not those most active in solution, highlighting strong differences between the two reaction formats.

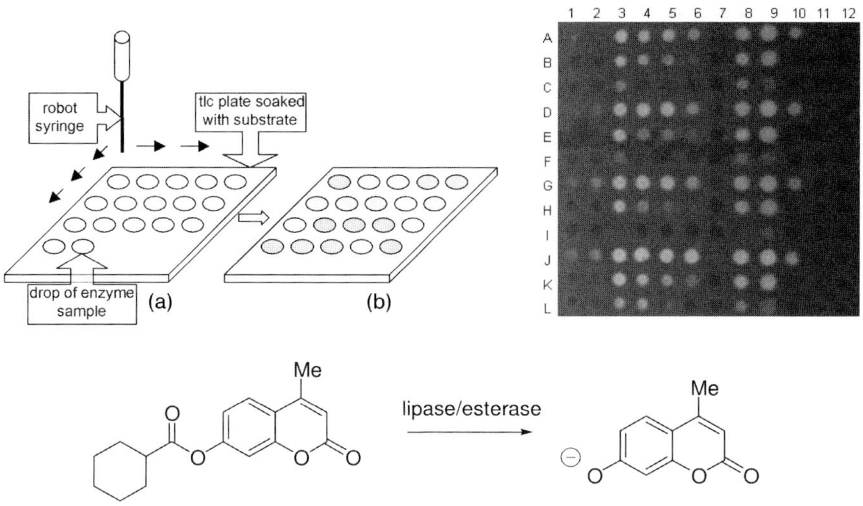

17 **18**

Scheme 1.5 Fluorescence lipase assay on a solid support.

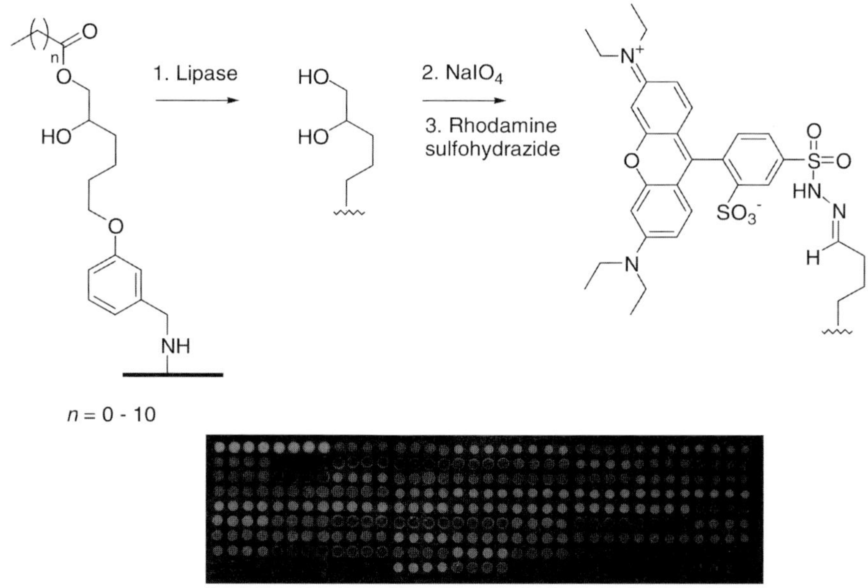

Scheme 1.6 A substrate micro-array for testing lipases.

1.3.2 The Clips-O Substrates with Periodate

The Clips-O substrates form one of the most reliable classes of fluorogenic and chromogenic substrates for hydrolytic enzymes, in particular for lipases (e.g. **19**) [30–32]. These substrates are based on a double indirect release mechanism for the fluorescence signal exploiting sequential cleavage of a primary 1,2-diol product (e.g. **20**) with sodium periodate to form an unstable aldehyde or ketone (e.g. **21**), which undergoes β-elimination to release umbelliferone **5** or nitrophenol (Scheme 1.7). Sodium periodate oxidizes 1,2-diols and 1,2-aminoalcohols selectively, but does not interact with most of the other functional groups, in particular the substrates of the assay and the enzymes. This indirect release strategy separates the enzyme reactive functional group from the fluorescent label. In addition, the functional group reacting with the enzyme is non-activated and therefore much less prone to uncatalyzed spontaneous hydrolysis in the assay medium. The Clips-O substrate **19** is particularly reactive with lipases because the enzyme reactive ester is part of a 1,2-diol monoester function resembling the natural glyceride ester substrates of these enzymes [33].

Interestingly, it has been found that **19** not only reacts well with lipases, but also shows a high reactivity with esterases, such that the substrate serves as a general

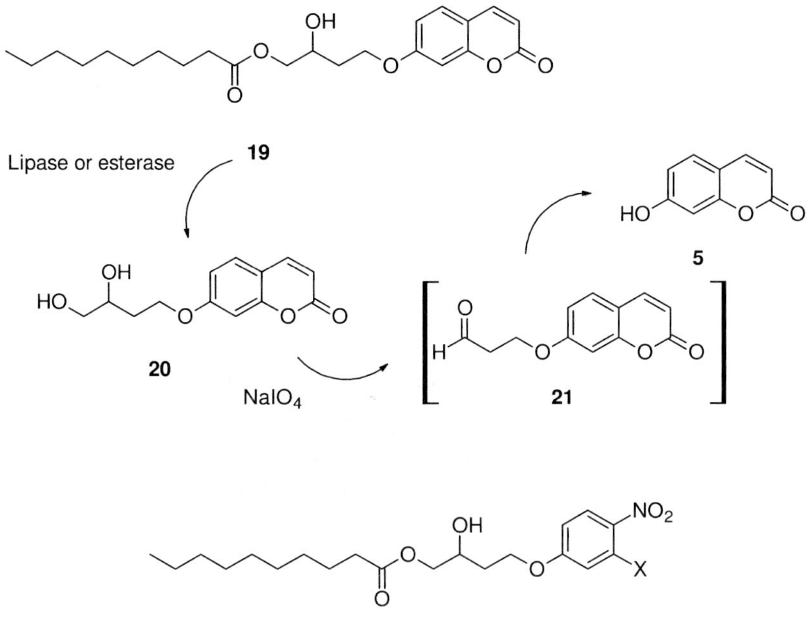

22 (X = H)
23 (X = NO_2)

Scheme 1.7 Clips-O substrates for lipases and esterases.

reference fluorogenic substrate for esterases and lipases. An enzyme fingerprinting study involving eight enantiomeric pairs of analogs of **19** with various acyl chain length showed that lipases and esterases can be distinguished by their chain length selectivity [34]. The corresponding nitrophenyl and dinitrophenyl derivatives **22** and **23** display similar reactivities [35]. As for the ADH and aldolase assays discussed above (Scheme 1.1), the secondary decomposition of the 1,2-diol product leading to umbelliferone may be kinetically limiting in the assay, implying that the substrates are not suitable for following the kinetics of very fast enzymes.

1.3.3
Esters of Fluorogenic Cyanohydrins and Hydroxyketones

Using aliphatic alcohols rather than acidic phenols in the ester function of lipase and esterase substrates is the key to obtaining selective probes for these enzymes. In the examples above the problem was solved by using an oxidative decomposition of the primary aliphatic alcohol product. Fluorogenic cyanohydrins and hydroxyketones that spontaneously react to form a fluorescent phenol following the β-elimination sequence presented above have been investigated as alternative reagent-free versions of fluorogenic aliphatic alcohols (Scheme 1.8) [36]. Reaction of the unstable aldehyde **21** with cyanotrimethyl silane (TMSCN) in dichloromethane and acidic hydrolysis of the trimethylsilyl (TMS) group affords cyanohydrin **25** without β-elimination. The cyanohydrin **25** can then be esterified with various acyl chlorides to afford the corresponding esters such as **24**. These substrates undergo a fluorogenic reaction with lipases and esterases. The enzyme-catalyzed

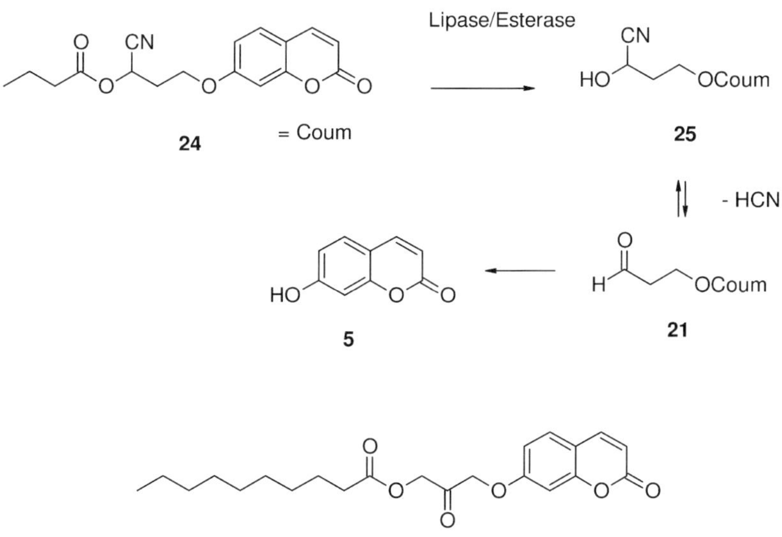

26

Scheme 1.8 Fluorogenic cyanohydrins and hydroxyketone esters.

ester hydrolysis liberates the cyanohydrin **25**, which then undergoes a rapid and spontaneous decomposition in aqueous buffer to form umbelliferone as a fluorescent product. Interestingly, the equilibration of the cyanohydrin **25** to the aldehyde **21** is faster than the β-elimination.

The fluorogenic hydroxyketone **14** discussed above (Scheme 1.4) can also be used to form esters such as **26** which can be used as fluorogenic substrates for lipases [37]. In this case, however, the esters are quite unstable despite being aliphatic alcohol esters. The relatively rapid spontaneous hydrolysis in this case is probably due to an assisted mechanism involving acyl transfer in the hydrated form of the ketone.

1.3.4 Fluorogenic Acyloxymethyl Ethers

Alkylation of umbelliferone with chloromethyl esters of various aliphatic acids affords the corresponding acyloxymethyl ethers (e.g. **27**) in good yields [38]. These substrates undergo lipase/esterase-catalyzed hydrolysis in buffer to form umbelliferone (Scheme 1.9) [39]. The mechanism probably involves an unstable hemiacetal intermediate, which spontaneously reacts to formaldehyde and umbelliferone **5**. The substrates can also be prepared in a two-step sequence involving alkylation of umbelliferone with 2-chloroketones followed by Baeyer–Villiger oxidation. The sequence is particularly interesting for obtaining the fluorogenic lactones **28** and **29**, which only react with esterases [40].

The alkylation chemistry described above can also be applied with fluorescein to yield the corresponding fluorescein mono-ethers such as **30** as green fluorescent probes for lipases [41]. These substrates are water soluble due to the anionic carboxylate on the fluorescein group and can be used in pure aqueous buffer without any co-solvent. This latter class of substrates reacts extremely fast and specifically with the enzymes, with assay times under 1 min. An activity fingerprinting study

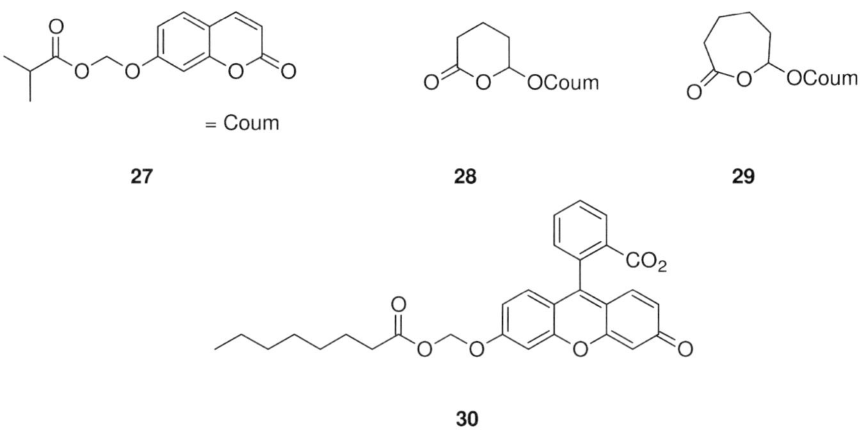

27 **28** **29** **30**

Scheme 1.9 Fluorogenic acyloxymethyl ethers for lipases and esterases.

using various fluorescein mono-ether substrates and enzymes either in pure buffer or in buffer containing 20% v/v dimethylsulfoxide co-solvent showed that enzymes classified as esterases react faster in pure aqueous buffer, while those classified as lipases show a stronger reactivity in the presence of the co-solvent, which might be related to the fact that lipases are activated at interfaces.

1.3.5
FRET-Lipase Probes

FRET (fluorescence or Förster resonance energy transfer) is one of the most general principles for designing fluorogenic substrates for bond-cleaving reactions and is well known for protease and lipase substrates. Despite the strong influence of the double labels on reactivity, the use of FRET substrates for lipases has been investigated. A survey of various variations around the 1,2-diol monoester function led to the discovery that most lipases react strongly with the pyrene carboxylic ester **31**, which is non-fluorescent due to intramolecular fluorescence quenching by the dinitrophenylamino group, to liberate the fluorescent pyrene butyrate product **32** [42]. The assay is carried out at a low substrate concentration due to the limited solubility and very low K_M values for this substrate. Most interestingly, the substrate is extremely resistant towards non-specific hydrolysis, in particular at alkaline pH. This allows one to screen enzymes under strongly basic conditions, as illustrated for the Roche lipase L8 (Scheme 1.10). A survey of various lipases and esterases at pH 11 showed that most enzymes in fact lose their activity under these conditions.

1.4
Other Hydrolases

Most of the reactivity principles used for fluorogenic lipase substrates are also applicable for other types of hydrolases. Such developments in the case of the epoxide hydrolases, amidases, esterases, and phosphatases have been investigated. Glycosidases were not investigated because the well-known aryl glycosides function perfectly well for their corresponding enzymes, with negligible non-specific reactivity in the absence of enzyme. On the other hand, there is still a need for developing fluorogenic substrates for nitrilases, a further important class of enzymes for biocatalysis.

1.4.1
Epoxide Hydrolases

The 1,2-diol product **20** in the Clips-O lipase substrate **19** (Scheme 1.7) can also form upon hydrolysis of the parent epoxide **33**, leading to a convenient periodate coupled assay for epoxide hydrolases with a variety of epoxides (Scheme 1.11) [5, 32, 43]. These substrates are readily prepared in racemic form from umbelliferone by alkylation

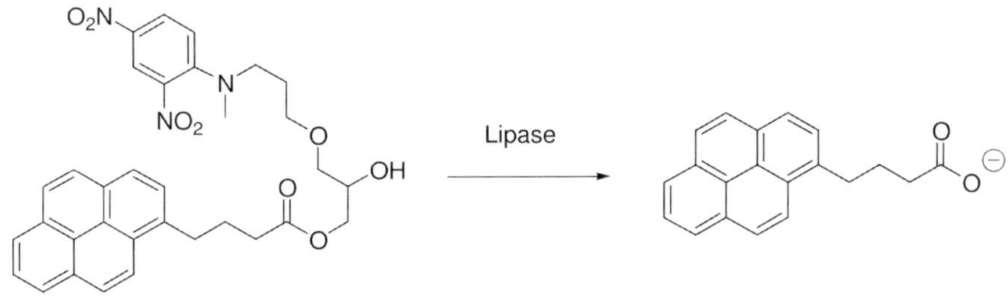

31 **32** (λ_{em} = 376 nm)

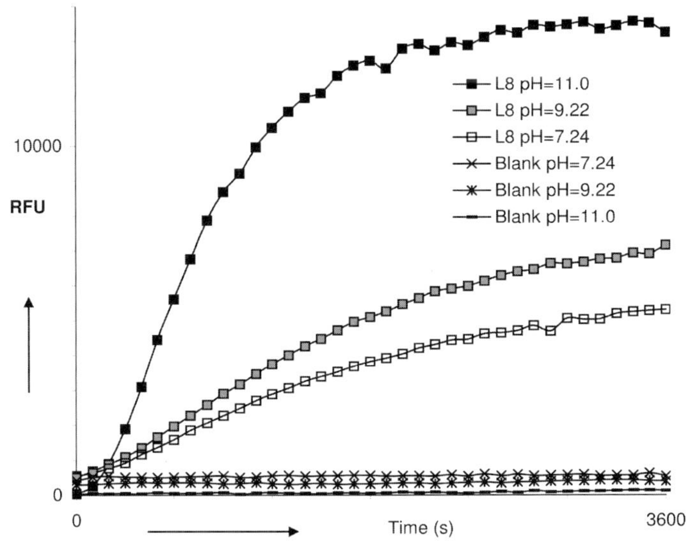

Scheme 1.10 FRET lipase substrate for alkaline assay of lipases.

with an allyl halide and epoxidation of the allyl ether with a peracid. The optically pure fluorogenic epoxides are obtained by asymmetric Sharpless dihydroxylation of the allyl ether, followed by regioselective ring closure. Such enantiomerically pure epoxides may be used for screening enantioselective epoxide hydrolases, although the problem of enantioselectivity is more complex. Indeed epoxide hydrolases enzymes may catalyze epoxide opening at both sides of the epoxide, leading to enantiomeric products [44]. Alternatively, the enzyme reaction may also lead to an enantioconvergent hydrolysis where a racemic epoxide is converted to a single enantiomer of the 1,2-diol [45].

The Clips-O epoxide **33–41** and related substrates are currently the only available selective fluorogenic substrates for EH. On the other hand various indirect assays have been reported to detect either the unreacted epoxide [46] or the carbonyl product resulting from periodate cleavage of the 1,2-diol product [26, 47, 48]. These assays are suitable for fluorescence or colorimetric assays for the hydrolysis of any epoxide of interest.

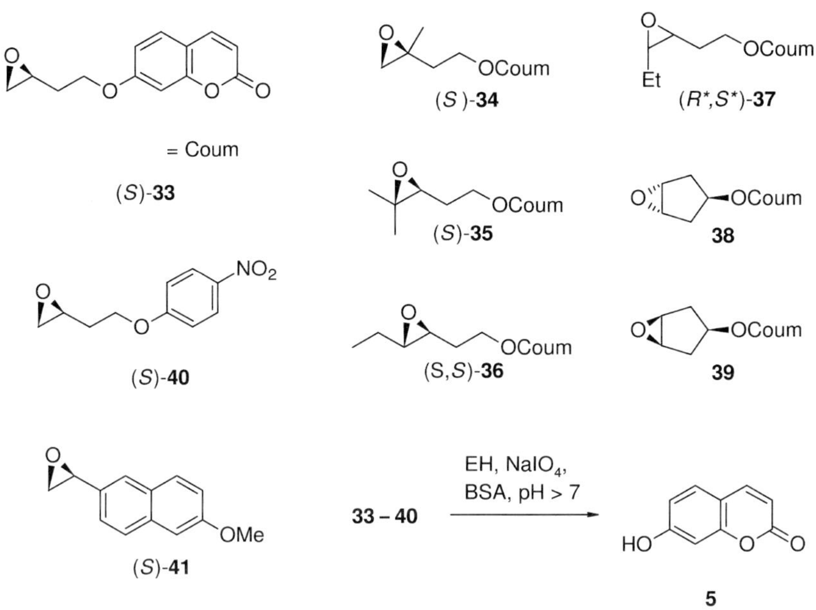

Scheme 1.11 Clips-O fluorogenic substrates for epoxide hydrolases.

1.4.2
Amidases and Proteases

1,2-Aminoalcohols are oxidized by sodium periodate to form the corresponding carbonyl products. The reaction is much faster than for 1,2-diols in a surprisingly pH-independent manner. The reaction can be used for designing non-volatile precursors of fragrant ketones and aldehydes by cyanide addition to form the cyanohydrins and reduction to the corresponding amino alcohol [49]. The oxidative cleavage of 1,2-aminoalcohol for designing selective substrates for amidases such as the phenacetylamide **42** suitable for screening penicillin G amidase has been exploited (Scheme 1.12) [32, 43]. A related fluorogenic substrate for ceramidase was also recently reported by others [50]. The phenacetyl derivative **43** also serves a fluorogenic substrate for amidases [37]. This substrate is however completely unstable towards alkaline conditions, probably due to hydrolysis by an assisted mechanism with an intramolecular acyl shift to the hydrated diol as discussed above for the corresponding lipase substrate. A chromogenic threonine derivative to form a selective HIV-protease substrate **44** has also been prepared [51].

A particularly elegant solution for the assay of proteases and acylases is offered by the fluorogenic detection of free amino acids by decomplexation of copper from calcein, which removes its quenching effect. This principle has been used for assays of acylases, amidases, and proteases (Scheme 1.13) [52, 53]. For the case of proteases combinatorial assays are particularly in demand for testing multiple peptides in parallel and determining the cleavage specificity [54]. New solutions

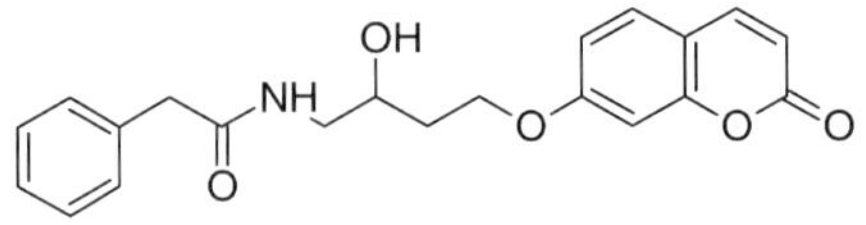

42

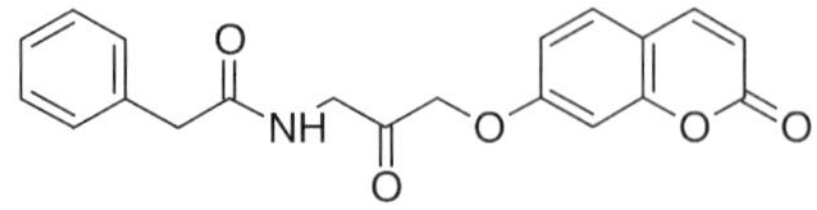

43

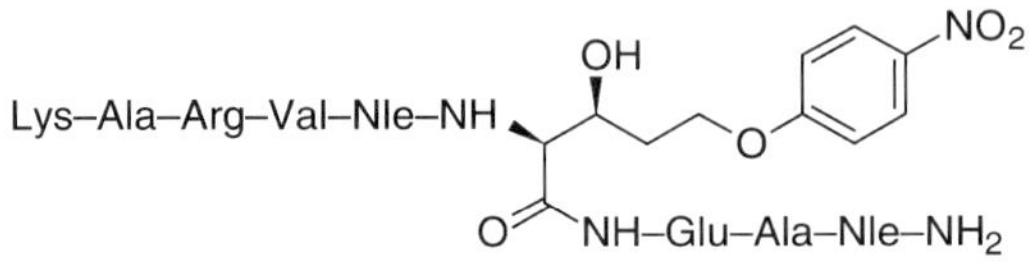

44

Scheme 1.12 Clips-O acylase and amidase substrates.

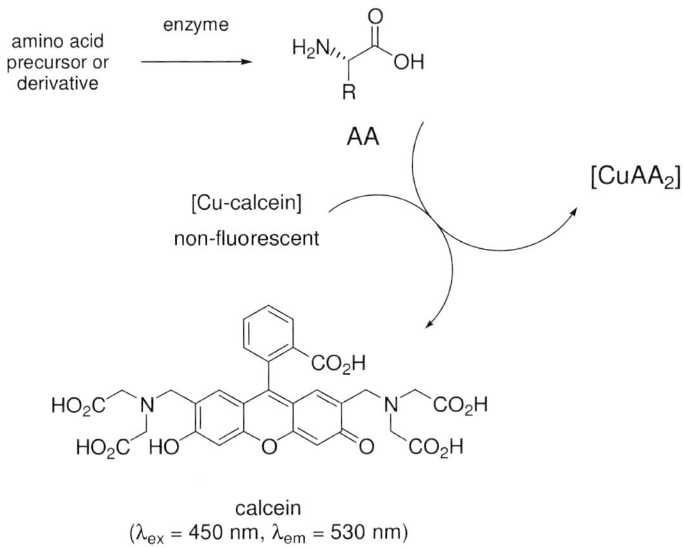

Scheme 1.13 Copper calcein as a fluorogenic sensor for amino acids.

to this problem using cocktails of fluorescence-labeled peptides [55] or combinatorial peptide libraries were recently reported [56, 57].

1.4.3
Phosphatases

Although phosphatases are only of moderate interest in organic synthesis, phosphatase enzymes and particularly phytases that hydrolyze phytic acids bearing multiple phosphate groups are of biotechnological interest. The standard substrates for phosphatases are nitrophenyl phosphates and related monophosphates of aromatic phenols. Di- and triphosphates **45** and **46** related to the Clips-O principle discussed above have been investigated (Scheme 1.14) [58]. The substrates

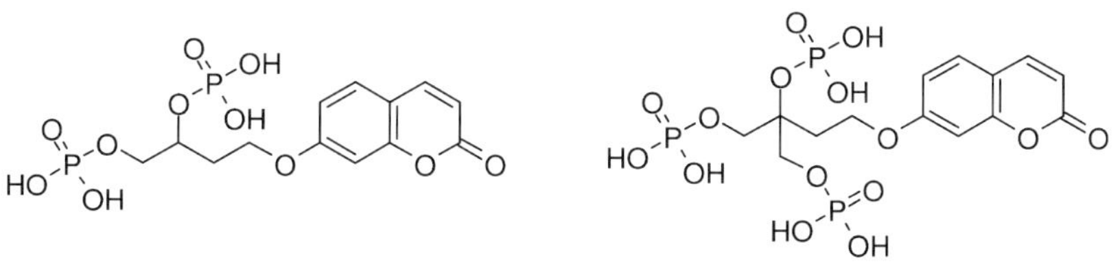

45 46

Scheme 1.14 Clips-O substrates for phosphatases.

are obtained from the diols by phosphorylation with dibenzyl phosphoramidite, followed by oxidation to the phosphate and debenzylation by hydrogenation. These phosphates display the expected reactivity towards phosphatases, however without selectivity for phytases against alkaline phosphatases, which can be differentiated rather on the basis of their pH rate profile.

1.5 Baeyer–Villigerases

Baeyer–Villiger monoxygenases (BVMOs) have been recently recognized as preparatively useful enzymes, in particular in relation to their enantioselectivity with prochiral ketones. It has recently been shown that 2-aryloxyketones **47–50** act as fluorogenic substrates for these enzymes (Scheme 1.15) [40]. The natural selectivity of oxidative cleavage leads to insertion of the oxygen atom on the side of the aryl ether to form an unstable lactone or acetal ester, which is rapidly hydrolyzed to form umbelliferone. The assay is useful for the detection of BVMOs in whole cells. However, in some cases oxidative activity at the ether methylene group not involving a BVMO reaction was detected as a side reaction with these substrates. The substrates are particularly attractive because they can be prepared in a single step in high yield from commercially available starting materials. They represent to date the only available fluorogenic assay for these enzymes.

1.6 Conclusion

Fluorescence assays for biotransformations are an indispensable tool for enzyme engineering and the daily practice of enzyme studies. Fluorogenic substrates are particularly useful as general probes for enzyme classes that can be used in routine screening and activity checking. However, they cannot replace the authentic substrate in cases where an optimization towards a particular biotransformation is desired. In such cases an indirect fluorogenic assay or an instrumental assay may be required in order to follow the reaction. A variety of fluorescence assays for enzymes still remain to be discovered and the development of new enzyme assays

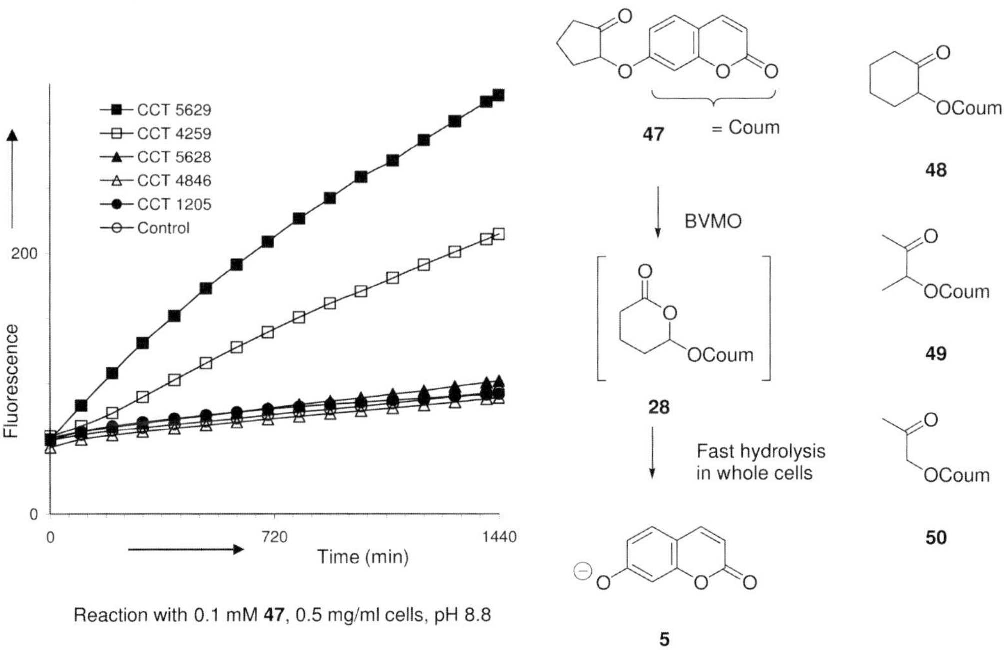

Scheme 1.15 Fluorogenic assay for BVMOs in whole cells.

continues to provide fertile ground for innovative developments. One area of recent progress concerns the use of multiple substrates in parallel assays or as mixtures to record activity profiles, also called fingerprints, which enable the characterization of enzyme activities in great detail [59, 60].

Acknowledgment

This work was financially supported by the University of Berne, Swiss National Science Foundation, Swiss Federal Office of Education, COST Program Action D25 on Biocatalysis, and Protéus SA, Nîmes, France.

References

1 Reymond, J.L. (2005) *Enzyme Assays: High-throughput Screening, Genetic Selection and Fingerprinting*, Wiley-VCH Verlag GmbH, Weiheim.

2 Ravot, G., Wahler, D., Favre-Bulle, O., Cilia, V. and Lefevre, F. (2003) High throughput discovery of alcohol dehydrogenases for industrial biocatalysis. *Advanced Synthesis & Catalysis*, **345**, 691–694.

3 Klein, G. and Reymond, J.L. (1998) An enantioselective fluorimetric assay for alcohol dehydrogenases using albumin-catalyzed beta-elimination of

umbelliferone. *Bioorganic & Medicinal Chemistry Letters*, **8**, 1113–1116.
4 Klein, G. and Reymond, J.L. (1999) Enantioselective fluorogenic assay of acetate hydrolysis for detecting lipase catalytic antibodies. *Helvetica Chimica Acta*, **82**, 400–407.
5 Badalassi, F., Klein, G., Crotti, P. and Reymond, J.L. (2004) Fluorescence assay and screening of epoxide opening by nucleophiles. *European Journal of Organic Chemistry*, 2557–2566.
6 Reymond, J.L. and Chen, Y.W. (1995) Catalytic, enantioselective aldol reaction using antibodies against a quaternary ammonium ion with a primary amine cofactor. *Tetrahedron Letters*, **36**, 2575–2578.
7 Reymond, J.L. and Chen, Y.W. (1995) Catalytic, enantioselective aldol reaction with an artificial aldolase assembled from a primary amine and an antibody. *Journal of Organic Chemistry*, **60**, 6970–6979.
8 Wagner, J., Lerner, R.A. and Barbas III, C.F. (1995) Efficient aldolase catalytic antibodies that use the enamine mechanism of natural enzymes. *Science*, **270**, 1797–1800.
9 Jourdain, N., Carlon, R.P. and Reymond, J.L. (1998) A stereoselective fluorogenic assay for aldolases: detection of an anti-selective aldolase catalytic antibody. *Tetrahedron Letters*, **39**, 9415–9418.
10 Perez Carlon, R., Jourdain, N. and Reymond, J.L. (2000) Fluorogenic polypropionate fragments for detecting stereoselective aldolases. *Chemistry*, **6**, 4154–4162.
11 List, B., Barbas, C.F. and Lerner, R.A. (1998) Aldol sensors for the rapid generation of tunable fluorescence by antibody catalysis. *Proceedings of the National Academy of Sciences of the United States of America*, **95**, 15351–15355.
12 Gonzalez-Garcia, E., Helaine, V., Klein, G., Schuermann, M., Sprenger, G.A., Fessner, W.D. and Reymond, J.L. (2003) Fluorogenic stereochemical probes for transaldolases. *Chemistry–A European Journal*, **9**, 893–899.
13 Sevestre, A., Helaine, V., Guyot, G., Martin, C. and Hecquet, L. (2003) A fluorogenic assay for transketolase from *Saccharomyces cerevisiae*. *Tetrahedron Letters*, **44**, 827–830.
14 Williams, G.J., Domann, S., Nelson, A. and Berry, A. (2003) Modifying the stereochemistry of an enzyme-catalyzed reaction by directed evolution. *Proceedings of the National Academy of Sciences of the United States of America*, **100**, 3143–3148.
15 Woodhall, T., Williams, G., Berry, A. and Nelson, A. (2005) Creation of a tailored aldolase for the parallel synthesis of sialic acid mimetics. *Angewandte Chemie–International Edition in English*, **44**, 2109–2112.
16 Kofoed, J., Darbre, T. and Reymond, J.L. (2006) Dual mechanism of zinc-proline catalyzed aldol reactions in water. *Chemical Communications (Cambridge, England)*, 1482–1484.
17 Kofoed, J., Darbre, T. and Reymond, J.L. (2006) Artificial aldolases from peptide dendrimer combinatorial libraries. *Organic & Biomolecular Chemistry*, **4**, 3268–3281.
18 Kikuchi, K., Hannak, R.B., Guo, M.J., Kirby, A.J. and Hilvert, D. (2006) Toward bi-functional antibody catalysis. *Bioorganic & Medicinal Chemistry*, **14**, 6189–6196.
19 Debler, E.W., Ito, S., Seebeck, F.P., Heine, A., Hilvert, D. and Wilson, I.A. (2005) Structural origins of efficient proton abstraction from carbon by a catalytic antibody. *Proceedings of the National Academy of Sciences of the United States of America*, **102**, 4984–4989.
20 Manetsch, R., Zheng, L., Reymond, M.T., Woggon, W.D. and Reymond, J.L. (2004) A catalytic antibody against a tocopherol cyclase inhibitor. *Chemistry*, **10**, 2487–2506.
21 Zheng, L., Manetsch, R., Woggon, W.D., Baumann, U. and Reymond, J.L. (2005) Mechanistic study of proton transfer and hysteresis in catalytic antibody 16E7 by site-directed mutagenesis and homology modeling. *Bioorganic & Medicinal Chemistry*, **13**, 1021–1029.
22 Janes, L.E. and Kazlauskas, R.J. (1997) Quick E. A fast spectrophotometric method to measure the enantioselectivity of hydrolases. *Journal of Organic Chemistry*, **62**, 4560–4561.
23 Janes, L.E., Lowendahl, A.C. and Kazlauskas, R.J. (1998) Quantitative screening of hydrolase libraries using pH

indicators: identifying active and enantioselective hydrolases. *Chemistry – A European Journal*, **4**, 2324–2331.

24 Konarzycka-Bessler, M. and Bornscheuer, U.T. (2003) A high-throughput-screening method for determining the synthetic activity of hydrolases. *Angewandte Chemie – International Edition in English*, **42**, 1418–1420.

25 Baumann, M., Sturmer, R. and Bornscheuer, U.T. (2001) A high-throughput-screening method for the identification of active and enantioselective hydrolases. *Angewandte Chemie – International Edition*, **40**, 4201–4204.

26 Wahler, D. and Reymond, J.L. (2002) The adrenaline test for enzymes. *Angewandte Chemie – International Edition in English*, **41**, 1229–1232.

27 Wahler, D., Boujard, O., Lefevre, F. and Reymond, J.L. (2004) Adrenaline profiling of lipases and esterases with 1,2-diol and carbohydrate acetates. *Tetrahedron*, **60**, 703–710.

28 Babiak, P. and Reymond, J.L. (2005) A high-throughput, low-volume enzyme assay on solid support. *Analytical Chemistry*, **77**, 373–377.

29 Grognux, J. and Reymond, J.L. (2006) A red-fluorescent substrate microarray for lipase fingerprinting. *Molecular Biosystems*, **2**, 492–498.

30 Badalassi, F., Wahler, D., Klein, G., Crotti, P. and Reymond, J.L. (2000) A versatile periodate-coupled fluorogenic assay for hydrolytic enzymes. *Angewandte Chemie – International Edition in English*, **39**, 4067–4070.

31 Wahler, D., Badalassi, F., Crotti, P. and Reymond, J.L. (2001) Enzyme fingerprints by fluorogenic and chromogenic substrate arrays. *Angewandte Chemie – International Edition in English*, **40**, 4457–4460.

32 Wahler, D., Badalassi, F., Crotti, P. and Reymond, J.L. (2002) Enzyme fingerprints of activity, and stereo- and enantioselectivity from fluorogenic and chromogenic substrate arrays. *Chemistry*, **8**, 3211–3228.

33 Nyfeler, E., Grognux, J., Wahler, D. and Reymond, J.L. (2003) A sensitive and selective high-throughput screening fluorescence assay for lipases and esterases. *Helvetica Chimica Acta*, **86**, 2919–2927.

34 Grognux, J. and Reymond, J.L. (2004) Classifying enzymes from selectivity fingerprints. *Chembiochem*, **5**, 826–831.

35 Grognux, J., Wahler, D., Nyfeler, E. and Reymond, J.L. (2004) Universal chromogenic substrates for lipases and esterases, *Tetrahedron, Asymmetry*, **15**, 2981–2989.

36 Leroy, E., Bensel, N. and Reymond, J.L. (2003) Fluorogenic cyanohydrin esters as chiral probes for esterase and lipase activity. *Advanced Synthesis & Catalysis*, **345**, 859–865.

37 Sicart, R., Collin, M.P. and Reymond, J.L. (2007) Fluorogenic substrates for lipases, esterases, and acylases using a TIM-mechanism for signal release. *Biotechnol Journal*, **2**, 221–231.

38 Bensel, N., Reymond, M.T. and Reymond, J.L. (2001) Pivalase catalytic antibodies: towards abzymatic activation of prodrugs. *Chemistry*, **7**, 4604–4612.

39 Leroy, E., Bensel, N. and Reymond, J.L. (2003) A low background high-throughput screening (HTS) fluorescence assay for lipases and esterases using acyloxymethylethers of umbelliferone. *Bioorganic & Medicinal Chemistry Letters*, **13**, 2105–2108.

40 Sicard, R., Chen, L.S., Marsaioli, A.J. and Reymond, J.L. (2005) A fluorescence-based assay for Baeyer–Villiger monooxygenases, hydroxylases and lactonases. *Advanced Synthesis & Catalysis*, **347**, 1041–1050.

41 Yang, Y.Z., Babiak, P. and Reymond, J.L. (2006) New monofunctionalized fluorescein derivatives for the efficient high-throughput screening of lipases and esterases in aqueous media. *Helvetica Chimica Acta*, **89**, 404–415.

42 Yang, Y.Z., Babiak, P. and Reymond, J.L. (2006) Low background FRET-substrates for lipases and esterases suitable for high-throughput screening under basic (pH 11) conditions. *Organic & Biomolecular Chemistry*, **4**, 1746–1754.

43 Badalassi, F., Wahler, D., Klein, G., Crotti, P. and Reymond, J.L. (2000) A versatile periodate-coupled fluorogenic assay for hydrolytic enzymes. *Angewandte Chemie – International Edition*, **39**, 4067–4070.

44 Archelas, A. and Furstoss, R. (2001) Synthetic applications of epoxide hydrolases. *Current Opinion in Chemical Biology*, **5**, 112–119.

45 Monterde, M.I., Lombard, M., Archelas, A., Cronin, A., Arand, M. and Furstoss, R. (2004) Enzymatic transformations. Part 58: enantioconvergent biohydrolysis of styrene oxide derivatives catalysed by the *Solanum tuberosum* epoxide hydrolase, *Tetrahedron, Asymmetry*, **15**, 2801–2805.

46 Zocher, F., Enzelberger, M.M., Bornscheuer, U.T., Hauer, B. and Schmid, R.D. (1999) A colorimetric assay suitable for screening epoxide hydrolase activity. *Analytica Chimica Acta*, **391**, 345–351.

47 Doderer, K., Lutz-Wahl, S., Hauer, B. and Schmid, R.D. (2003) Spectrophotometric assay for epoxide hydrolase activity toward any epoxide. *Analytical Biochemistry*, **321**, 131–134.

48 Mateo, C., Archelas, A. and Furstoss, R. (2003) A spectrophotometric assay for measuring and detecting an epoxide hydrolase activity. *Analytical Biochemistry*, **314**, 135–141.

49 Yang, Y.Z., Wahler, D. and Reymond, J.L. (2003) Beta-amino alcohol properfumes. *Helvetica Chimica Acta*, **86**, 2928–2936.

50 Bedia, C., Casas, J., Garcia, V., Levade, T. and Fabrias, G. (2007) Synthesis of a novel ceramide analogue and its use in a high-throughput fluorogenic assay for ceramidases. *Chembiochem*, **8**, 642–648.

51 Badalassi, F., Nguyen, H.K., Crotti, P. and Reymond, J.L. (2002) A selective HIV-protease assay based on a chromogenic amino acid. *Helvetica Chimica Acta*, **85**, 3090–3098.

52 Klein, G. and Reymond, J.L. (2001) An enzyme assay using pM. *Angewandte Chemie – International Edition*, **40**, 1771–1773.

53 Dean, K.E.S., Klein, G., Renaudet, O. and Reymond, J.L. (2003) A green fluorescent chemosensor for amino acids provides a versatile high-throughput screening (HTS) assay for proteases. *Bioorganic & Medicinal Chemistry Letters*, **13**, 1653–1656.

54 Diamond, S.L. (2007) Methods for mapping protease specificity. *Current Opinion in Chemical Biology*, **11**, 46–51.

55 Yongzheng, Y. and Reymond, J.L. (2005) Protease profiling using a fluorescent domino peptide cocktail. *Molecular Biosystems*, **1**, 57–63.

56 Kofoed, J. and Reymond, J.L. (2007) A general method for designing combinatorial peptide libraries decodable by amino acid analysis. *Journal of Combinatorial Chemistry*, **9**, 1046–1052.

57 Kofoed, J. and Reymond, J.L. (2007) Identification of protease substrates by combinatorial profiling on TentaGel beads. *Chemical Communications (Cambridge, England)*, 4453–4455.

58 Gonzalez-Garcia, E.M., Grognux, J., Wahler, D. and Reymond, J.L. (2003) Synthesis and evaluation of chromogenic and fluorogenic analogs of glycerol for enzyme assays. *Helvetica Chimica Acta*, **86**, 2458–2470.

59 Reymond, J.L. and Wahler, D. (2002) Substrate arrays as enzyme fingerprinting tools. *Chembiochem*, **3**, 701–708.

60 Goddard, J.P. and Reymond, J.L. (2004) Enzyme activity fingerprinting with substrate cocktails. *Journal of the American Chemical Society*, **126**, 11116–11117.

2
Immobilization as a Tool for Improving Enzymes

Ulf Hanefeld

2.1
Introduction

The application of enzymes as catalysts in organic chemistry is closely linked to their immobilization. Indeed, many enzymes are only available in an immobilized form. The immobilized enzymes can be used as received, greatly easing their application. Numerous of these readily available immobilized enzymes are now the working horses of biocatalysis. This has even led to the incorrect use of the abbreviation of an enzyme name for a specific enzyme preparation, that is CALB for the immobilized form of *Candida antarctica* lipase B on cross-linked polymethacrylate (also known as Novozym 435). Vice versa the commercial name of an enzyme preparation – Amano PS – has taken the place of the enzyme (*Burkholderia cepacia* lipase on dextrin or diatomaceous earth). Surprisingly, often no attention is paid to the fact that the enzyme is immobilized [1].

Why then is it important whether the enzyme is immobilized and what are the advantages of immobilizing an enzyme? In general, homogeneous catalysts and enzymes in solution have a number of advantages over heterogeneous catalysts (Table 2.1). However, homogeneous catalysts also have drawbacks: in particular it is difficult to separate them from the reaction mixture [1–3]. In addition recycling is problematic and continuous processes are not possible. By immobilizing enzymes several of the advantages of heterogeneous catalysts can be added to the list of advantages of using an enzyme.

For enzymes another point can be important. Not all pure enzymes are stable in solution. In particular in organic solvents enzymes can be very unstable [4]. The stability of enzymes can be improved by genetic modifications, by chemical modification, and most easily and without specialist equipment: by immobilization. Furthermore, immobilization opens up the possibility of improving other properties of the enzyme, such as its activity, substrate specificity, and enantioselectivity. Although the underlying principles are not always understood enzyme immobilization is a powerful tool for the improvement of activity and enzyme stability, specificity, and selectivity.

Modern Biocatalysis: Stereoselective and Environmentally Friendly Reactions.
Edited by Wolf-Dieter Fessner and Thorleif Anthonsen

ISBN: 978-3-527-32071-4

Table 2.1 Comparison of homogeneous catalysts and enzymes with heterogeneous catalysts.

	Homogeneous catalysts and enzymes	Heterogeneous catalysts
Advantages	Mild reaction conditions High activity and selectivity Efficient heat transfer	Excellent separation Straightforward recycling of the catalyst Continuous processes
Disadvantages	Difficult separations Product contamination No continuous process	Heat transfer problems Low activity and selectivity

When a homogeneous catalyst, that is an enzyme, is immobilized there are in principle three different approaches [3]. The enzyme can be adsorbed onto a carrier, it can be encapsulated in the carrier, or it can be covalently attached to the carrier. In an extreme case of covalent attachment the enzyme can be cross-linked, making a carrier unnecessary. The choice of immobilization method and carrier can greatly influence the properties of the immobilized enzyme, ideally improving them. To date many excellent reviews [5–13] and even comprehensive books [14] on the topic of enzyme immobilization have been written. Many of these reviews and books use their own division of approaches: they range from three to eight and can also include membrane techniques. In order to ease the comparison with chemical catalysts, a division according to reference [3] was chosen. This chapter aims to demonstrate, with examples from the research performed within the COST D25 Action, the importance of the area under discussion.

2.2 Adsorption/Electrostatic Interactions

The adsorption of an enzyme can proceed via different types of interactions. If the enzyme has a large lipophilic surface area, Van der Waals interactions with a hydrophobic carrier lead to the fixation of the enzyme on the carrier. Hydrophilic surface areas or sugar residues attached to the protein can ensure adsorption via hydrogen bonds. In the most extreme case ionic interactions, where the enzyme acts as the counter ion to the carrier, guarantee the binding of the enzyme to the carrier. In all cases the enzyme does not have to be chemically modified and can be employed without pre-treatment. By varying the immobilization conditions the results can be greatly influenced, allowing a straightforward manipulation of the enzyme properties. A significant disadvantage of immobilizing by adsorption is that the enzyme may leach from the carrier under reaction conditions.

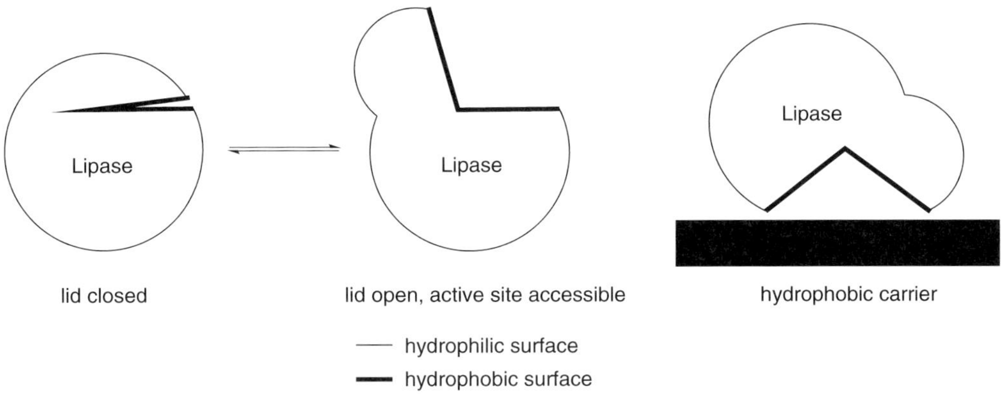

Figure 2.1 Most lipases are interfacially activated. When immobilized on hydrophobic carriers they are assumed to be in their active conformation.

2.2.1 Van der Waals Interactions

The enzymes that are almost by definition most suited for immobilization via Van der Waals interactions are the lipases. Lipases are active at the interface: indeed, many lipases show interfacial activation [15]. It is commonly assumed that those lipases that have a lid covering their active site will only change into the active conformation (lid open and active site accessible) when they are in contact with a lipophilic interface. By adsorbing them onto lipophilic surfaces the lid should ideally be fixed in an open conformation (Figure 2.1). Moreover, as interfacially active enzymes, part of their surface is lipophilic, ensuring the right orientation of the active site during the reaction [6, 9]. Consequently many lipases have been immobilized on hydrophobic carriers. Increases in activity have been reported for several lipases that were immobilized on hydrophobic carriers such as EP-100 polypropylene, Accurel MP1004 polypropylene, octyl-silica, and octyl-agarose [9, 14, 16–20].

This has led to the development of many different applications. In the fermentation broth of lipase producing micro-organisms the lipase is often the only lipophilic component. It was demonstrated that lipases can be extracted from the fermentation broth and purified by adsorption of hydrophobic carriers in one single step [21, 22]. However, this isolation method also reveals one of the disadvantages of immobilization on hydrophobic carriers. After immobilization and removal from the fermentation broth the lipases could be washed off the carrier rather readily. Indeed, when utilizing lipases immobilized on hydrophobic carriers in aqueous media, leaching can be a major problem. If the immobilized enzyme is, however, used in hydrophobic organic solvents, no leaching occurs since the enzyme is insoluble in the solvents.

When the lidless CALB is immobilized on a hydrophobic macroporous polymer that is based on methyl and butyl methacrylic esters and cross-linked with divinyl-

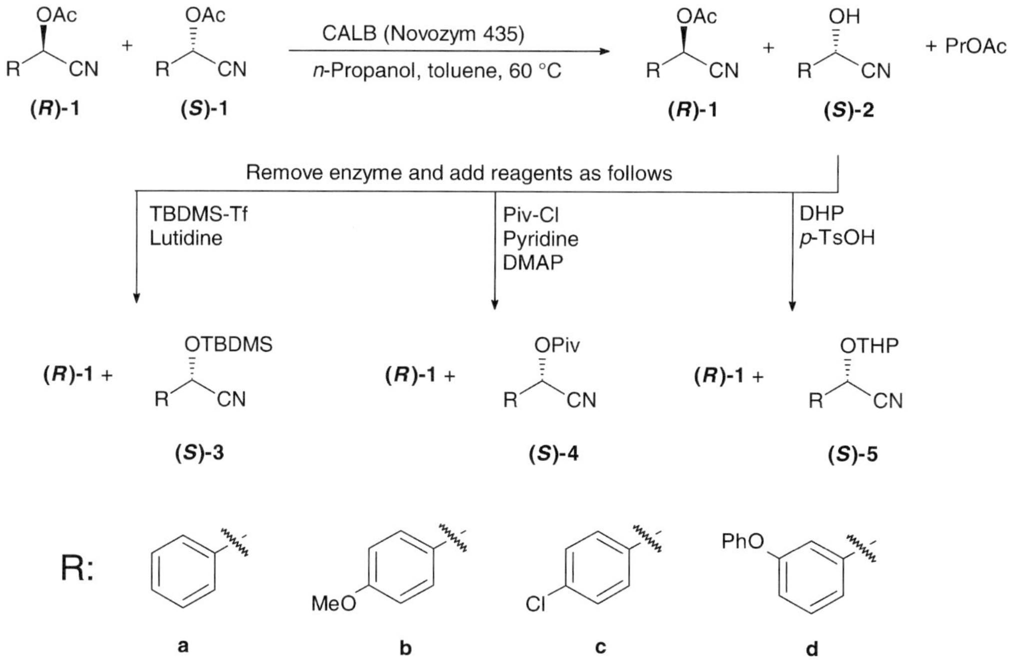

Scheme 2.1 Immobilized CALB can be used in organic solvents: it is readily removable and recyclable.

Table 2.2 Recycling of Novozym 435 (CALB immobilized on a hydrophobic carrier) in the kinetic resolution of **1c** in toluene at 60 °C.

Ratio (% *ee*)	Cycle				
	1	2	3	4	5
R-**1c**	98 (99)	96 (99)	98 (97)	98 (98)	98 (98)
S-**5c**	86 (90)	86 (90)	88 (93)	90 (93)	91 (93)
Corresponding aldehyde	16	18	14	12	10

benzene it is called Novozym 435 [23]. This preparation of CALB is extremely successful and popular. Unlike most other lipases CALB and, hence, Novozym 435 is not activated at the interface [15, 24]. When used in toluene for the kinetic resolution of aromatic cyanohydrin acetates (**1**) the advantage of using an immobilized enzyme could be clearly demonstrated [25]. At the end of the enzyme-catalyzed reaction the enzyme could be filtered off and the reaction mixture could be submitted to a second transformation, the chemical protection of the (*S*)-cyanohydrins (**2**). This yielded both enantiomers of the cyanohydrins, protected with different protection groups and readily separable (Scheme 2.1). Straightforward removal of the enzyme allowed its recycling. After five cycles no loss of

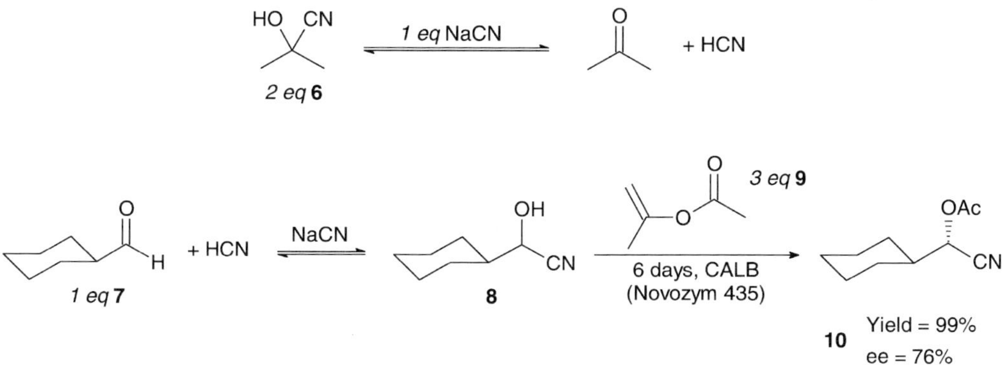

Scheme 2.2 Novozym 435 catalyzes the enantioselective synthesis of aliphatic cyanohydrins via a DKR.

activity or enantioselectivity was observed (Table 2.2), clearly demonstrating the value of immobilization.

The same CALB preparation was applied in many dynamic kinetic resolutions; combining two types of catalysts with each other. In the presence of homogeneous transition metal catalysts that catalyze the racemization and heterogeneous acids or bases or immobilized transition metals Novozym 435 was not deactivated [1, 26–28]. This is all the more remarkable since the reactions catalyzed by these catalysts include redox reactions at elevated temperatures (>60 °C). When Novozym 435 was applied for the enantioselective synthesis of cyanohydrin acetates (**10**) from aliphatic aldehydes (**7**), good results were achieved (Scheme 2.2) for this dynamic kinetic resolution (DKR) [29]. Here NaCN is used as the base for the dynamic racemic formation and degradation of the cyanohydrins (**6** and **8**).

It has to be noted that CALB has also been immobilized on Accurel. On this carrier its stereoselectivity and activity is even better than when immobilized as Novozym 435. However, it is also stated that diffusion effects need to be taken into account, emphasizing that there can always be more than one reason why an enzyme performs better/worse when modifying the carrier and that the improvements seen might actually have nothing to do with the carrier [30]. Indeed, Novozym 435 and free CALB were more recently shown to have identical stereoselectivity [31]. When looking at the surface of CALB it can be concluded that a large part of it, including the area around the active site, is lipophilic (Figure 2.2) [32]. One might intuitively say that, when immobilized on a hydrophobic carrier, this enzyme should be inactive, since the active site might be inaccessible. It seems that the enzyme instead orients itself in such a way that the active site is still readily approachable.

2.2.2 Hydrogen Bonds

Most enzymes have hydrophilic surfaces, which are often glycosylated. Consequently they can readily form hydrogen bonds with polar surfaces. Therefore they

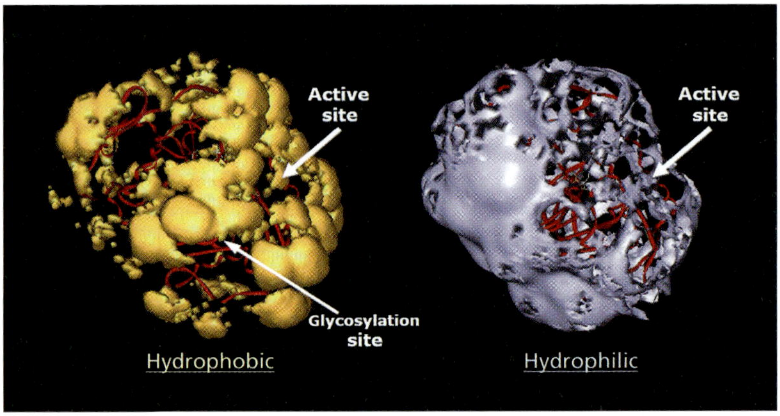

Figure 2.2 Hydrophobic and hydrophilic surface areas of CALB. Reproduced with permission from [32]. Copyright (2007) Wiley-VCH Verlag GmbH & Co. KGaA.

can easily be immobilized on hydrophilic carriers (cellulose, lignine, Avicel, Celite, porous glass, clay, silica gel) [9, 12, 14]. One such carrier is Celite (diatomaceous earth), the silicate skeletons of diatoms [33]. There are many different types of Celite and care has to be taken to employ the right type. A particularly interesting characteristic of Celite is that it can be used for controlling the water activity in organic solvents, thereby helping to maintain the correct reaction conditions for the enzyme [34]. Celite is a carrier for many commercial lipase preparations. Given the great ease of immobilizing on Celite, it is also used for the immobilization of many different enzymes in the laboratory [35–37]. Often sugars, but also polyethylene glycol (PEG) or albumin, are added in order to stabilize the enzyme and protect it against the negative influence of organic solvents.

Hydroxynitrile lyase (Oxynitrilase) from *Hevea brasiliensis* (HbHNL) immobilized on Celite displayed higher activity and slightly higher enantioselectivity than when immobilized on Avicel (microcrystalline cellulose) or hydrophobic polyamide Accurel EP 700. HbHNL on Celite showed good to excellent stability when stored in dry organic solvents, but was not active when used in dry solvents [37–39]. HNLs from *Prunus amygdalus, Manihot esculanta,* and *Sorghum bicolor* were included in the study and it was shown that, for all four HNLs, they are active and selective in organic solvents in the presence of 1–1.5% water, that is with a water layer.

When lipases are immobilized on Celite they can readily be used in dry organic solvents. *Pseudomonas cepacia* (also named *B. cepacia*) lipase was immobilized in the presence of sucrose on Hyflo super-cel Celite [36, 40, 41] and used in the first enantioselective synthesis of cyanohydrin acetates via a DKR. Similarly, other successful syntheses of cyanohydrin acetates via DKR were catalyzed by lipases immobilized on Celite [42]. This is in contrast to Novozym 435, which had successfully been used for the DKR of aliphatic cyanohydrin acetates (see Section 2.2.1 and Scheme 2.2) [29]. When Novozym 435 was used for the enantioselective synthesis of mandelonitrile acetate (**1a**) via DKR, the reaction did not proceed [43]. It was

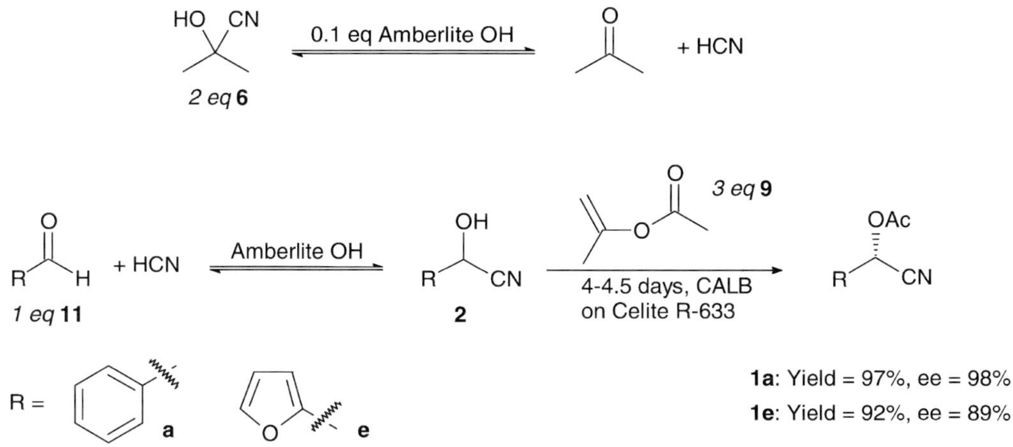

Scheme 2.3 CALB immobilized on Celite R-633 catalyzes the enantioselective synthesis of aromatic cyanohydrins via a DKR.

identified that acetic acid formed during the reaction via hydrolysis of part of the acyldonor (**9**) neutralized the base that catalyzed the racemic dynamic formation of the cyanohydrin (**2**). This hydrolysis was caused by residual water that was released from the hydrophobic carrier of Novozym 435. When the carrier was replaced by Celite R-633 CALB proved to be an excellent catalyst for this DKR starting from aromatic aldehydes (Scheme 2.3) [44, 45]. Obviously the Celite maintains a very low water activity in the reaction medium, thereby preventing any hydrolysis and acetic acid formation. As mentioned above, Celite can control water activities in organic solvents, keeping them very low and constant. In a direct comparison under optimized conditions, both enzyme preparations catalyzed this DKR, but still CALB immobilized on Celite was the catalyst of choice (Figure 2.3). The difference in performance can easily be explained by the different orientation of the enzyme on the hydrophobic and hydrophilic carrier (Figure 2.2) [32]. Furthermore, it was recently demonstrated that proteins undergo conformational changes when immobilized and that these changes are dependent on the carrier [46]. Consequently, it would actually be surprising if an enzyme immobilized on different carriers would behave the same on each of these different carriers.

2.2.3 Ionic Interactions

The strongest electrostatic interactions are ionic interactions. The surface areas of enzymes contain amino and acid groups, due to lysine (ε-amino group), aspartic acid, or glutamic acid (second acid group) [13]. Depending on the pH of the solution these groups can bear a charge and can thus interact with an ionic surface. Essentially this type of immobilization can take place on any ion exchanger. The ion exchanger can be positively or negatively charged, depending on the predomi-

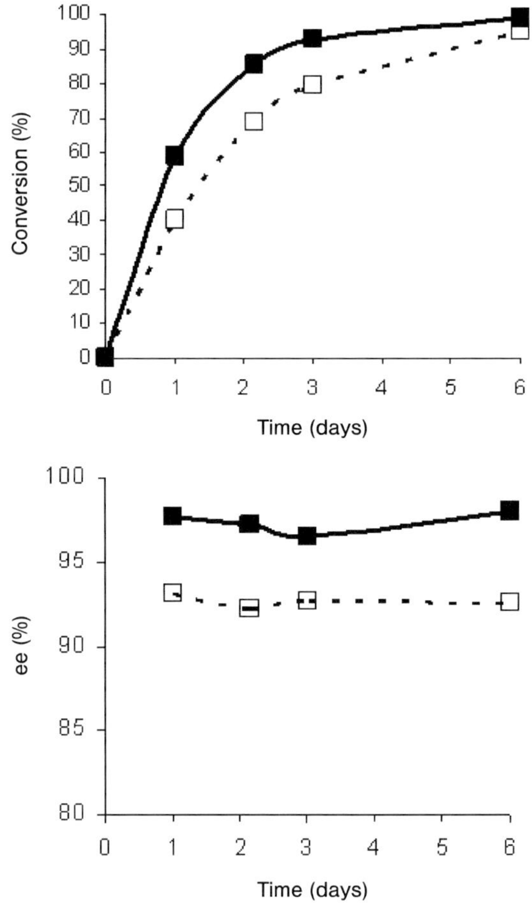

Figure 2.3 Enantioselective synthesis of **1a** via DKR catalyzed by CALB on Celite R-633 (■) and Novozym 435 (□).

nate charge on the enzyme (Figure 2.4) [14]. CALB was immobilized on the anion exchanger polyethylenimine (PEI, containing many amino groups, which are positively charged). It was shown that the pH value and temperature at which the enzyme was immobilized greatly influenced the activity and enantioselectivity of the enzyme [47]. It had earlier already been demonstrated that PEI can also activate lipases that were first bound onto hydrophobic carriers. Obviously PEI interacts with the hydrophilic surface of the lipases and influences their conformation [48].

In addition to anion and cation exchangers as enzyme carriers it has been demonstrated that mixed ion exchange supports can be used for binding enzymes with both acid and amino groups at pH values close to the isoelectric point, such as penicillin G acylase from *Escherichia coli* (Figure 2.4) [49].

Tailored mesoporous siliceous carriers were prepared in a systematic study of the immobilization of chloroperoxidase from *Caldariomyces fumago*, an enzyme

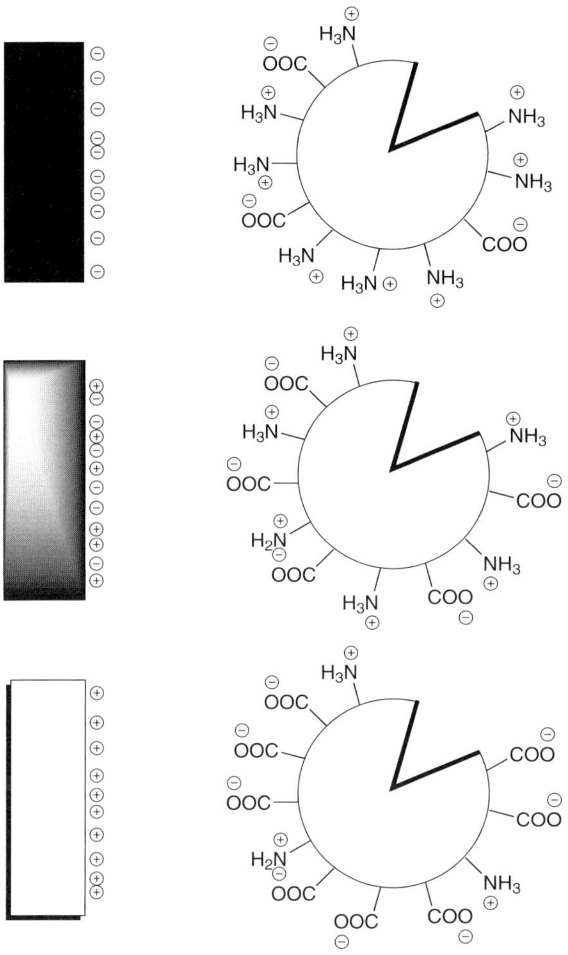

Figure 2.4 Different ionic interactions between ion exchangers and enzymes.

that nature did not develop for chlorinations [50, 51]. Based on the size of the enzyme (diameter 6.2 nm), the electrostatic surface potential (Figure 2.5a), and its isoelectric point of 4 (Figure 2.5b) the pore size of the carrier was tuned and amino groups were incorporated into the structure of the carrier. The best carrier was a mesoporous material with an average pore diameter of 15 nm with positively charged amino propyl groups to match the negative charge of the chloroperoxidase. At pH 6 virtually all the enzyme was adsorbed on and into the carrier and the immobilized enzyme could be recycled five times before the activity decreased.

That ionic immobilization is strongly dependent on the pH value and salt concentrations during immobilization and of the reaction mixture in which the enzyme is used cannot be overemphasized. If these parameters are not taken into

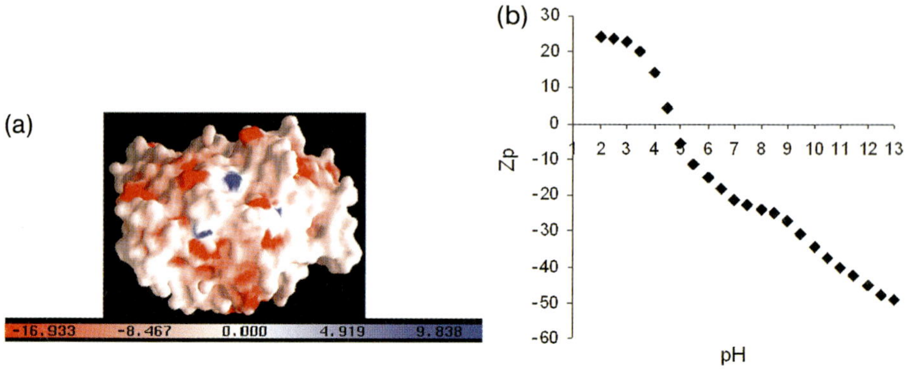

Figure 2.5 (a) Poisson–Boltzmann electrostatic surface potential of chloroperoxidase at pH 7.0 (blue represents areas of positive charge and red, areas of negative charge) and (b) calculated charge on chloroperoxidase as a function of pH. Reprinted with permission from [51]. Copyright (2007) American Chemical Society.

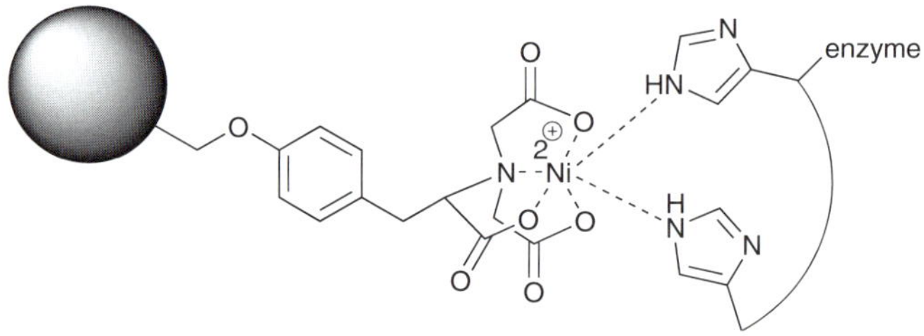

Figure 2.6 A carrier fixed Ni^{2+} acts as anchor for an enzyme with a His_6 tag.

account or if in the case of porous materials the wrong pore size is chosen the enzyme will not be immobilized or will leach immediately [52].

Another approach towards immobilization via ionic interactions is the use of tethered metal ions, such as Cu^{2+}, Co^{2+}, or Ni^{2+}, in order to bind the enzyme [53, 54]. This is particularly so when the enzyme contains an easily accessible imidazole residue from histidine [55] or a His_6 tag (Figure 2.6) [56–59], that is a short tag with six histidines. This tag can be readily introduced by genetically modifying the enzyme. Little influence of the tag on the catalytic performance has been noticed. When benzaldehyde lyase [60] was immobilized on an Ni^{2+}-containing polyvinylpyrrolidinone-based matrix, it could be reused several times for the formation of benzoin (**12**) (Scheme 2.4) [58].

Protein-coated micro-crystals (PCMC) are straightforwardly prepared forms of immobilized enzymes. An aqueous solution of the enzyme and a salt such as potassium sulfate is made and a water miscible solvent is added. The salt precipitates and the enzyme forms a layer on top of the micro-crystals. It is thus readily accessible. Given the direct interaction of the protein with the growing crystal of

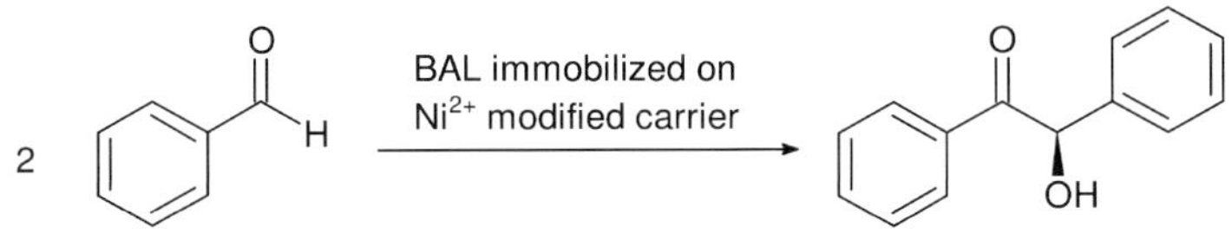

11a 12

Scheme 2.4 Immobilized BAL with a His_6 tag catalyzes the formation of benzoin **12**.

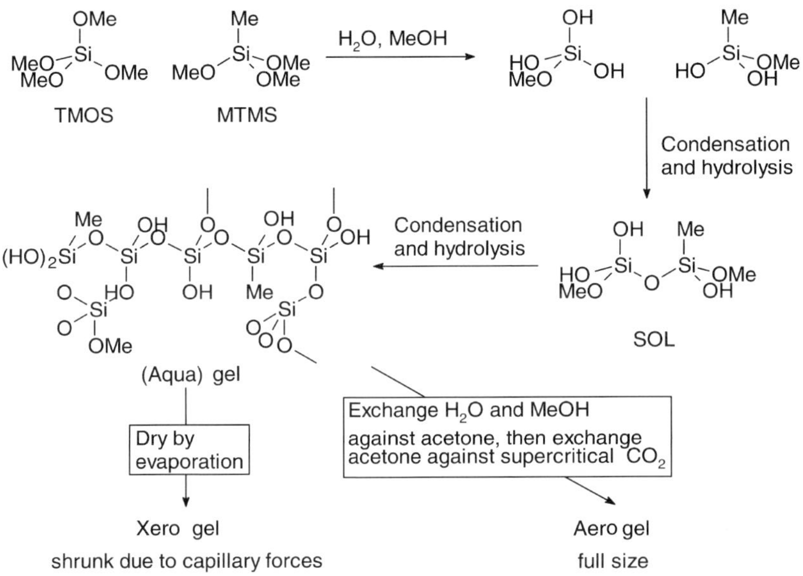

Scheme 2.5 Synthesis of sol gels.

the salt the immobilization will most likely proceed via ionic interactions. The PCMCs can only be used in organic solvents and are stable to storage [61].

2.3 Encapsulation

The best way to immobilize an enzyme while affecting its structure as little as possible is to encapsulate it. Out of all the different encapsulation methods the most prominent and widely used is the sol gel technique [5, 7, 62]. Sol gels are highly porous silica materials that are readily prepared and modified (Scheme 2.5). The sol gel obtained is a chemically inert glass that be formed into any desired shape and can be designed to be thermally and mechanically very stable. Most importantly the synthesis proceeds under conditions that are relatively benign for many enzymes. In the first step a tetra-alkoxysilane such as tetramethoxylsilane (TMOS) is hydrolysed by acid catalysis. Hydrolysis is followed by condensation and the sol is formed, which is a mixture of partially hydrolyzed and partially condensed monomers. Further condensation leads to the gel. Here all the pores are filled with water and it is therefore also known as aqua gel. This aqua gel can then be dried by evaporation of the water and alcohol. Under these conditions capillary forces will cause the sol gel to shrink significantly, losing some of its structure. The dry gel thus obtained is called xero gel. If the water and alcohol is exchanged against acetone and then against supercritical carbon dioxide the aqua gel can be dried avoiding the capillary forces and, thus, shrinkage of the gel. The aero gel obtained in this manner has a huge pore volume but is rather brittle [5]. The different sol gels obtained in this manner are all hydrophilic. If an alkyltrialkoxysilane such as methyltrimethoxysilane (MTMS) is added to the synthesis mixture, the sol gel prepared will have a hydrophobic surface. Thus, a straightforward change in the synthesis allows a major modification of the gel. The modifications of the surface ensure that the sol gel can either be used as a hydrophilic carrier, which forms hydrogen bonds with the enzyme (see also Section 2.2.2) or as hydrophobic carrier (see also Section 2.2.1). Sol gels with hydrophobic surfaces have a very positive influence on the reactivity of lipases, possibly because they are immobilized in their active conformation (lid open) (Figure 2.1) [63, 64]. Similar to the adsorption approach it is possible to stabilize and activate the enzymes even further by adding surfactants, polyvinylalcohol, or crown ethers. In addition, the sol gel can be mechanically strengthened by including porous glass beads or silica glass fibers at different stages of the sol gel synthesis [64–66].

Given the huge increase in activity that is often observed for lipases in hydrophobic sol gels, HbHNL was also immobilized in a hydrophobic sol gel. HbHNL is structurally closely related to the α/β hydrolases and, thus, the lipases. However, HbHNL was already denatured in the sol gel (Scheme 2.5). When the methanol in the sol gel was removed under vacuum before adding the enzyme, 65% of the HbHNL activity was retained in the aqua gel. Drying to obtain xero gels or aero gels caused complete loss of activity. However, the aqua gels could be used in organic solvents and catalyzed the formation of cyanohydrins with high enantioselectivity (Scheme 2.6) [67].

Based on these results sol gels of PaHNL and MeHNL were prepared and the three HNLs were compared as aqua gel and as free enzyme (Table 2.3) [68]. Strickly

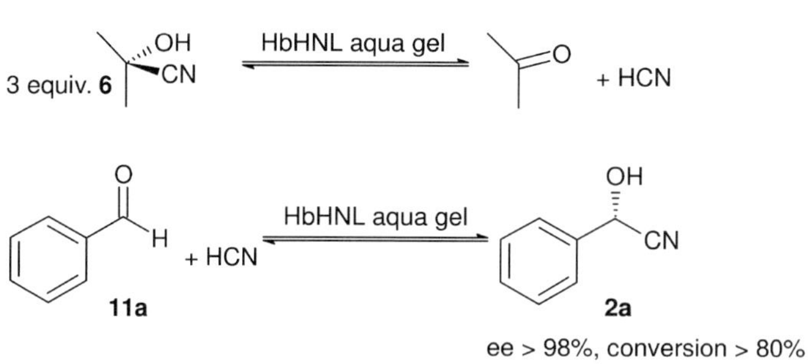

Scheme 2.6 HbHNL encapsulated in an aqua gel catalyzes the enantioselective formation of S-cyanohydrins.

Table 2.3 Conversion ratios and *ee*'s (parentheses) and reaction times in the synthesis of mandelonitrile (**2a**) catalyzed by free HNLs and their corresponding aquagels.

(*S*)-HbHNL		(*S*)-MeHNL		(*R*)-PaHNL	
Free[a]	Aqua gel[b]	Free[a]	Aqua gel[b]	Free[a]	Aqua gel[b]
4 h: 97(97)	0.5 h: 97(99)	4 h: 97(98)	0.5 h: 96(99)	4 h: 98(97)	2 h: 97(97)

Reaction conditions: benzaldehyde **11a** (0.5 mmol/ml diisopropylether), HCN (3 eq), and HNL (6 U/mmol) were shaken at room temperature.
a The HNL stock solution was diluted with citrate/phosphate buffer (50 mM, pH = 5.0) to a DIPE:aqueous media ratio of 5 : 1.
b DIPE saturated with citrate/phosphate buffer (50 mM, pH = 5.0).

speaking, both reactions are two-phase reactions: the free enzymes are in a buffer layer and so are the immobilized enzymes. However, this time the buffer layer is inside the aqua gel and the system therefore at first glance looks like a reaction performed in organic solvent. The immobilized enzymes perform better than the free enzymes. This might be due to an improvement of the enzymes. However, it is more likely that the large surface area of the buffer layer in the aqua gels enhances phase transfer of benzaldehyde (**11a**) and mandelonitrile (**2a**). As already observed earlier, enhancement of the performance of an enzyme upon immobilization is not always due to an improved action of the enzyme [30].

The lipase from *B. cepacia* (BCL) has successfully been immobilized in xero gels [66] and aero gels [65]. Recently, BCL was immobilized in a xero gel prepared from MTMS and TMOS (Scheme 2.5) and lyophilized. This BCL preparation showed excellent properties for acylations in dry organic solvents [69]. As mentioned above (Section 2.2.2 and Scheme 2.3) residual water in the reaction mixture can cause acyl donors and esters to hydrolyze, thus releasing acid. This in its turn can significantly disturb the desired reaction. When the xero gel-immobilized BCL was compared with Amano PS (a commercial preparation of BCL) it hydrolyzed significantly less of the esters (Table 2.4). Consequently this xero gel is the BCL preparation of choice when reactions are performed in dry media. In addition, the xero gel could be recycled eight times in the synthesis of **14** from the racemic alcohol and vinylacetate. Only modest loss of activity or selectivity was observed (Figure 2.7).

2.4 Covalent Binding/Cross-linking

Covalent binding of an enzyme to a carrier has the advantage that the enzyme is tightly fixed. Thus deactivation due to leaching is no problem and the immobilized enzymes can often be reused. A disadvantage is, however, that the enzyme is

Table 2.4 Conversion (%) after 24 h of enantiopure esters (0.05 M) by hydrolysis due to residual water in the enzyme preparation in dry organic solvent at room temperature.

Entry	Substrate	Solvent	log *P*	Conversion (%) by Amano PS[a]	Conversion (%) by xero gel[b]
1	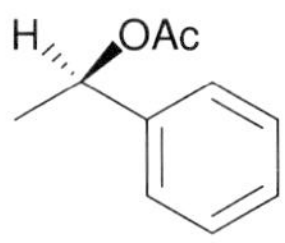 13	Toluene	2.8	4	1
2		TBME	1.35	18	3
3	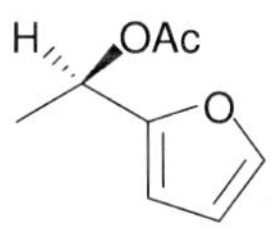 14	DIPE	1.9	20	10
4		TBME	1.35	12	2
5[c]	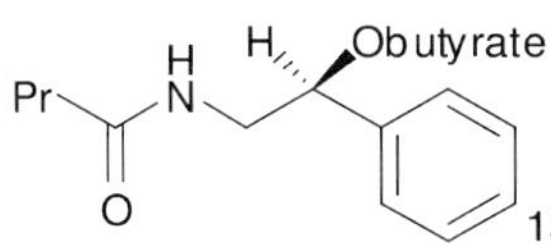 15	TBME	1.35	86	16

a 100 mg/ml of lipase PS powder.
b Lyo sol gel based on 100 mg/ml of lipase PS powder.
c Reaction temperature 48 °C.

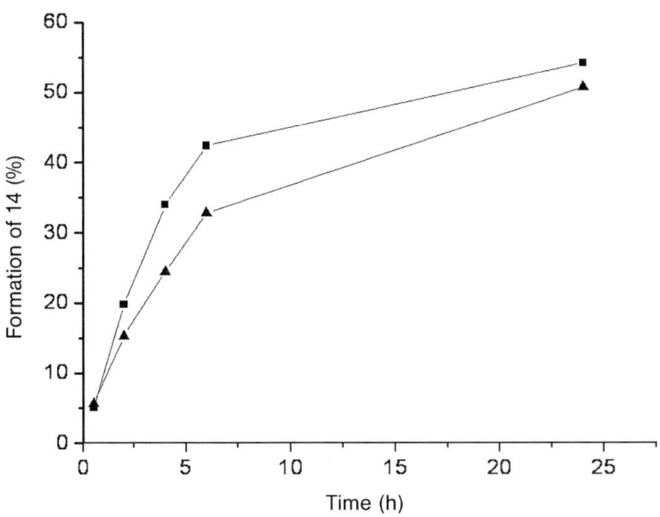

Figure 2.7 Lyophilized xero gel of BCL maintains activity and enantioselectivity in the synthesis of **14** over eight cycles. (■) cycle 1, $E = 139$, (▲) cycle 8, $E = 90$.

chemically modified. These modifications are not always easy to direct and in most cases the immobilization is not uniform for all enzyme molecules in one batch. However, enzymes can be bound stably by multipoint attachment and even multimeric enzymes have been efficiently bound on a range of carriers [8, 9, 13].

An extreme case of covalent binding is cross-linking of enzymes. Instead of fixing the enzyme to a carrier, the enzyme acts as a carrier itself. Enzyme aggregates or crystals, enzymes in a spray-dried form, or even enzymes in solution can be cross-linked. The immobilized enzyme is carrier free, that is the material is virtually pure enzyme and the negative effects of carriers can thus be avoided [10, 70].

For the covalent immobilization or cross-linking of enzymes all the functional groups on the enzyme surface can be utilized, including sugar residues [13, 71]. Most commonly the amino groups of the enzyme are employed, either as nucleophile, to attack for instance an epoxide or an aldehyde. In the case of the aldehyde the Schiff base formed can be reduced with $NaBH_4$, thus making the immobilization irreversible. Alternatively carbodiimide can be used for activating an acid group on the enzyme to couple with an amino group of the carrier. The carrier can have hydrophobic or hydrophilic surfaces and the reactive groups can be attached via short or long spacers to the carrier (Figure 2.8). The carriers Eupergit C and Eupergit C 250 L have a reactive epoxy group and are porous acrylic beads [72]. Sepabeads are methacrylic carriers bearing either epoxy or amino groups. The amino groups of the carrier can be linked to the enzyme via glutaraldehyde [32]. Agarose, glyoxyl agarose [73, 74], and the aminated glyoxyl agarose (MANA–the primary amino group is particularly nucleophilic) [75], as well as glutaraldehyde-modified agarose or silica [76] have also proven their value as carriers.

A bifunctional molecule, such as glutaraldehyde, has to be used for cross-linking, but otherwise the same basic principles apply [10, 11, 70, 71]. An additional possibility that has recently been explored is to first immobilize an enzyme on a carrier and cross-link it afterwards. In this manner particularly stable preparations can be obtained [77].

When *Candida rugosa* lipase (formerly named *Candida cylindracea* lipase) was immobilized on an epoxy-activated resin it became resistant against acetaldehyde. Due to this immunization it could be repeatedly employed for the enantioselective acylation of secondary alcohols with vinyl acetate in dry organic solvents (Scheme 2.7) [78].

In a comparative study CALB was immobilized on epoxy-activated Sepabeads and amino Sepabeads with long and short spacers (glutaraldehyde was used as the coupling reagent) (see Figure 2.8). Lyophilized CALB and Novozym 435 were also included in the test. The specific activity (U/g dry immobilized CALB) of Novozym 435 was much lower than for CALB immobilized on the different Sepabeads. In a thermal stability screening, Novozym 435 and CALB immobilized on amino Sepabeads with short spacers displayed equal stability, while all the other CALB preparations were less stable [32].

Cross-linked enzyme aggregates (CLEA) are prepared by first aggregating enzymes via the addition of precipitants such as ammonium sulfate, acetone,

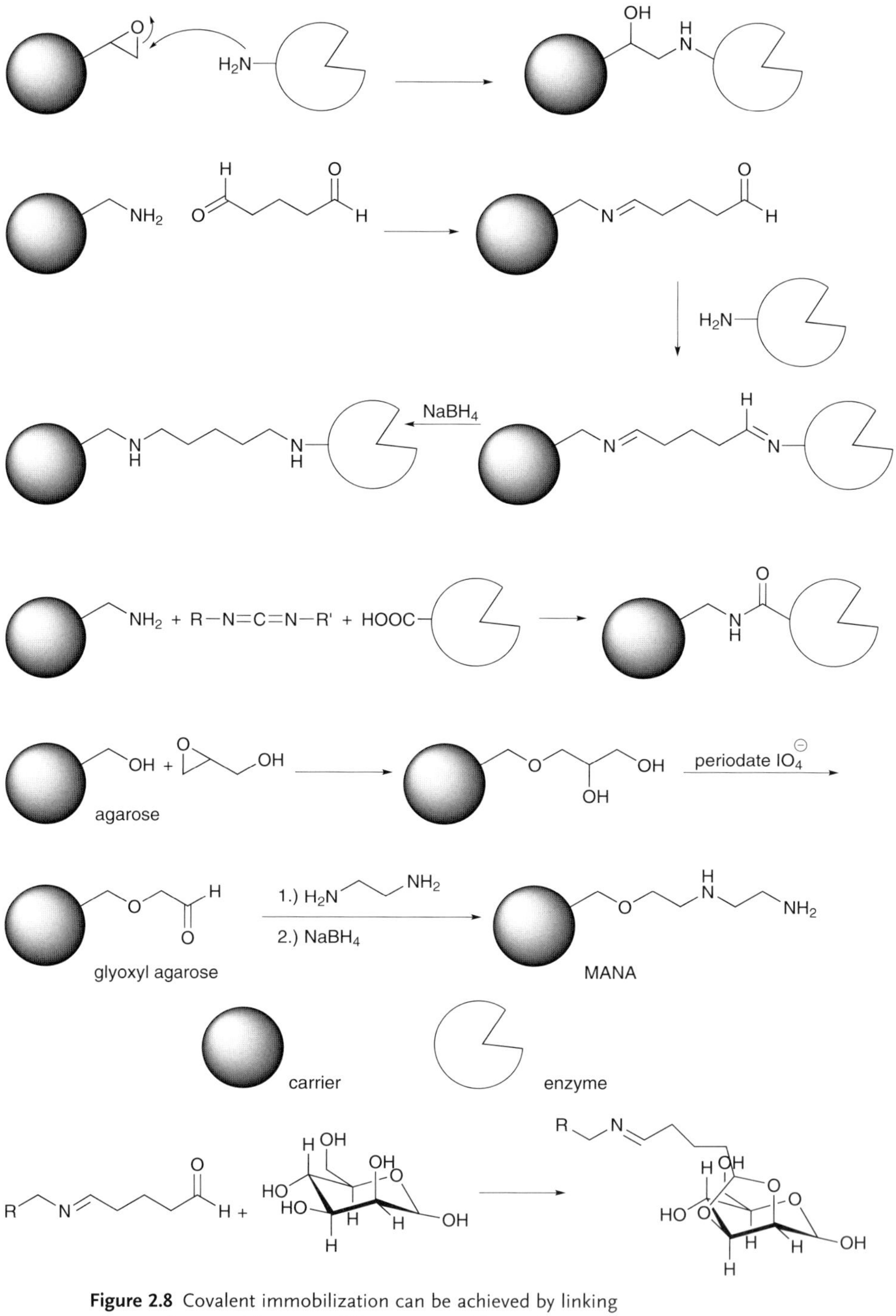

Figure 2.8 Covalent immobilization can be achieved by linking either amino or acid groups of the enzyme with epoxides, amines, or aldehyde groups on the carrier. In addition, sugar residues can be coupled to the carrier.

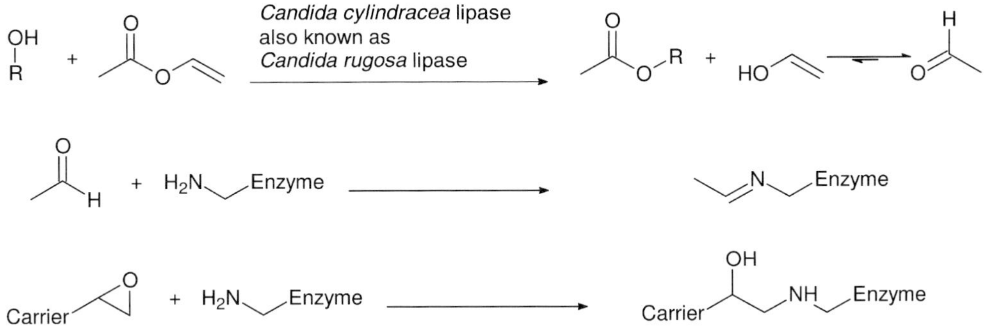

Scheme 2.7 Immobilization on an epoxy-activated resin protects *C. rugosa* lipase against acetaldehyde-induced deactivation.

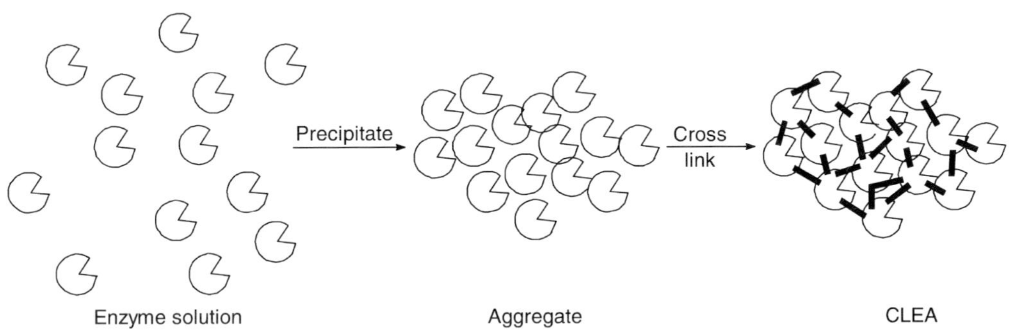

Scheme 2.8 Preparation of CLEAs.

ethanol, or 1,2-dimethoxyethane. Subsequently a cross-linker, commonly glutaraldehyde, is added (Scheme 2.8). CLEAs of BCL were prepared with bovine serum albumin as the protecting reagent [79] and in the presence of dextrin [69]. The BCL CLEA containing dextrin was significantly more active than the commercial Amano PS preparation of BCL and also more active than the xero gel of BCL (see also Section 2.3) [69]. In addition, the enantioselectivity of BCL is improved when immobilized as a CLEA. However, when the BCL CLEA is submitted to recycling experiments (acylation of **14** with vinyl acetate in dry diisopropylether (DIPE)) the CLEA rapidly loses activity (28% in the second cycle). In comparison the BCL xero gel (Figure 2.7) shows much higher stability and recyclability [69].

For an assessment of different immobilization techniques, CLEAs of HbHNL, PaHNL, and MeHNL were compared with aqua gels of these enzymes [68]. When the CLEAs were used in the presence of a buffer layer; that is under conditions comparable to the aqua gels (see also Section 2.3), all of them were stable [68, 80] and PaHNL can be recycled up to 10 times without loss of activity [81]. The real advantage of the immobilization as CLEAs only becomes obvious when they are employed in pure organic solvents. Here MeHNL displays remarkable stability

Table 2.5 Conversion ratios and *ee*'s (parentheses) in the synthesis of mandelonitrile (**2a**) catalyzed by CLEAs of different HNLs.

(*S*)-HbHNL CLEA	(*S*)-MeHNL CLEA	(*R*)-PaHNL CLEA
72 h: 55(67)	2 h: 96(97)	72 h: 97(99)

Reaction conditions: benzaldehyde **11a** (0.5 mmol/ml DIPE containing traces of water from the HCN solution), HCN (3eq), and the respective CLEA (6 U/mmol) were shaken at room temperature.

and reactivity (Table 2.5, compare with Table 2.3). Immobilization of MeHNL as a CLEA enables its use in pure organic solvents, thus expanding the scope of the enzyme significantly [68, 82].

2.5 Conclusion

The immobilization of enzymes has been extensively studied during the last decades [14]. Many approaches have been investigated and it is now possible to improve the stability of enzymes in such a way that they can be used under harsh, non-natural conditions. Organic solvents, supercritical fluids, and ionic liquids (Chapters 5 and 8) [83] do not constitute an insurmountable problem any longer. At the same time, enzyme immobilization has helped to improve the activity of enzymes, in particular that of lipases. In addition to the activity the selectivity and enantioselectivity of the enzymes could be modified and improved. However, although all these steps have been undertaken the approach is still an iterative experimental improvement, not a theory-based choice of 'the right method'. Given the huge variety in structure between the enzymes, indeed already between the different isoenzymes of one enzyme, the vast choice of carriers and the different binding methods there is – to date – no general solution to all problems.

Acknowledgments

The author would like to thank all the students, PhD students, and postdoctorals, past and present, for their work that led quite unintended into the field of immobilization. The contacts stimulated, established, and strengthened during the COST D25 Action, significantly helped in undertaking large parts of the research described here. In particular L. Gardossi (Trieste), P. J. Halling (Strathclyde), L. T. Kanerva (Turku), E. Magner (Limerick), F. Molinari (Milan), and A. Pierre (Lyon) are gratefully acknowledged for introducing the author to many different aspects of enzyme chemistry.

References

1 Veum, L. and Hanefeld, U. (2006) *Chemical Communications*, 825–831.

2 Sheldon, R.A., Arends, I.W.C.E. and Hanefeld, U. (2007) *Green Chemistry and Catalysis*, Wiley-VCH Verlag GmbH, Weinheim.

3 McMorn, P. and Hutchings, G.J. (2004) *Chemical Society Reviews*, **33**, 108–122.

4 Faber, K. (2004) *Biotransformations in Organic Chemistry*, 5th edn, Springer-Verlag, Berlin.

5 Pierre, A.C. (2004) *Biocatalysis and Biotransformation*, **22**, 145–170.

6 Bornscheuer, U.T. (2003) *Angewandte Chemie – International Edition*, **42**, 3336–3337.

7 Avnir, D., Coradin, T., Lev, O. and Livage, J. (2006) *Journal of Materials Chemistry*, **16**, 1013–1030.

8 Cao, L. (2005) *Current Opinion in Chemical Biology*, **9**, 217–226.

9 Mateo, C., Palomo, J.M., Fernandez-Lorente, G., Guisan, J.M. and Fernandez-Lafuente, R. (2007) *Enzyme and Microbial Technology*, **40**, 1451–1463.

10 Cao, L., van Langen, L. and Sheldon, R.A. (2003) *Current Opinion in Biotechnology*, **14**, 387–394.

11 Sheldon, R.A. (2007) *Advanced Synthesis Catalysis*, **349**, 1289–1307.

12 End, N. and Schöning, K.-U. (2004) *Topics in Current Chemistry*, **242**, 273–317.

13 Tischer, W. and Wedekind, F. (1999) *Topics in Current Chemistry*, **200**, 95–126.

14 (a) Cao, L. (2005) *Carrier-bound Immobilized Enzymes*, Wiley-VCH Verlag GmbH, Weinheim.
(b) Hartmeier, W. (1988) *Immobilised Biocatalysts; an Introduction*, Springer Verlag, Berlin.

15 Bornscheuer, U.T. and Kazlauskas, R.J. (2006) *Hydrolases in Organic Synthesis*, 2nd edn, Wiley-VCH Verlag GmbH, Weinheim.

16 Persson, M., Mladenoska, I., Wehtje, E. and Adlercreutz, P. (2002) *Enzyme and Microbial Technology*, **31**, 833–841.

17 Salis, A., Sanjust, E., Solinas, V. and Monduzzi, M. (2003) *Journal of Molecular Catalysis B: Enzymatic*, **24–25**, 75–82.

18 Gitlesen, T., Bauer, M. and Adlercreutz, P. (1997) *Biochimica et Biophysica Acta*, **1345**, 188–196.

19 Blanco, R.M., Terreros, P., Munoz, N. and Serra, E. (2007) *Journal of Molecular Catalysis B: Enzymatic*, **47**, 13–20.

20 Fernandez-Lorente, G., Terreni, M., Mateo, C., Bastida, A., Fernandez-Lafuente, R., Dalmases, P., Huguet, J. and Guisan, J.M. (2001) *Enzyme and Microbial Technology*, **28**, 389–396.

21 Gupta, N., Rathi, P., Singh, R., Goswami, V.K. and Gupta, R. (2005) *Applied Microbiology and Biotechnology*, **67**, 648–653.

22 Bastida, A., Sabuquillo, P., Armisen, P., Fernandez-Lafuente, R., Huguet, J. and Guisan, J.M. (1998) *Biotechnology and Bioengineering*, **58**, 486–493.

23 Heinsman, N.W.J.T., Schroën, C.G.P.H., van der Padt, A., Franssen, M.C.R., Boom, R.M. and van't Riet, K. (2003) *Tetrahedron, Asymmetry*, **14**, 2699–2704.

24 Kirk, O. and Christensen, M.W. (2002) *Organic Process Research & Development*, **6**, 446–451.

25 Veum, L., Kuster, M., Telalovic, S., Hanefeld, U. and Maschmeyer, T. (2002) *European Journal of Organic Chemistry*, **9**, 1516–1522.

26 Wuyts, S., Wahlen, J., Jacobs, P.A., and De Vos, D.E. (2007) *Green Chemistry*, **9**, 1104–1108.

27 Ko, S.-B., Baburaj, B., Kim, M.-J. and Park, J. (2007) *Journal of Organic Chemistry*, **72**, 6860–6864.

28 Kim, M.-J., Kim, W.-H., Han, K., Choi, Y.K. and Park, J. (2007) *Organic Letters*, **9**, 1157–1159.

29 Veum, L. and Hanefeld, U. (2005) *Synlett*, **15**, 2382–2384.

30 Rotticci, D., Norin, T. and Hult, K. (2000) *Organic Letters*, **2**, 1373–1376.

31 Egholm Jacobsen, E., Andresen, L.S. and Anthonsen, T. (2005) *Tetrahedron Asymmetry*, **16**, 847–850.

32 Basso, A., Braiuca, P., Cantone, S., Ebert, C., Linda, P., Spizzo, P., Caimi, P., Hanefeld, U., Degrassi, G. and Gardossi, L. (2007) *Advanced Synthesis Catalysis*, **349**, 877–886.

33 (a) Gebeshuber, I.C., Kindt, J.H., Thompson, J.B., Del Amo, Y., Stachelberger, H., Brzezinski, M.A., Stucky, G.D., Morse, D.E. and Hansma, P.K. (2003) *Journal of Microscopy.*, **212**, 292–299.
(b) Gebeshuber, I.C., Kindt, J.H., Thompson, J.B., Del Amo, Y., Stachelberger, H., Brzezinski, M.A., Stucky, G.D., Morse, D.E. and Hansma, P.K. (2004) *Journal of Microscopy*, **214**, 101.

34 Basso, A., De Martin, L., Ebert, C., Gardossi, L. and Linda, P. (2000) *Journal of Molecular Catalysis B: Enzymatic*, **8**, 245–253.

35 Kanerva, L.T. and Sundholm, O. (1993) *Journal of the Chemical Society – Perkin Transactions*, **1**, 2407–2410.

36 Inagaki, M., Hiratake, J., Nishioka, T. and Oda, J. (1992) *Journal of Organic Chemistry*, **57**, 5643–5649.

37 Costes, D., Rotcenkovs, G., Wehtje, E. and Adlercreutz, P. (2001) *Biocatalysis and Biotransformation*, **19**, 119–130.

38 Costes, D., Wehtje, E. and Adlercreutz, P. (1999) *Enzyme and Microbial Technology*, **25**, 384–391.

39 Persson, M., Costes, D., Wehtje, E. and Adlercreutz, P. (2002) *Enzyme and Microbial Technology*, **30**, 916–923.

40 Inagaki, M., Hiratake, J., Nishioka, T. and Oda, J. (1991) *Journal of the American Chemical Society*, **113**, 9360–9361.

41 Inagaki, M., Hatanaka, A., Mimura, M., Hiratake, J., Nishioka, T. and Oda, J. (1992) *Bulletin of the Chemical Society of Japan*, **65**, 111–120.

42 Paizs, C., Tähtinen, P., Tosa, M., Majdik, C., Irimie, F.-D. and Kanerva, L.T. (2004) *Tetrahedron*, **60**, 10533–10540.

43 Li, Y.-X., Straathof, A.J.J., and Hanefeld, U. (2002) *Tetrahedron, Asymmetry*, **13**, 739–743.

44 Veum, L., Kanerva, L.T., Halling, P.J., Maschmeyer, T. and Hanefeld, U. (2005) *Advanced Synthesis Catalysis*, **347**, 1015–1021.

45 Veum, L. and Hanefeld, U. (2004) *Tetrahedron Asymmetry*, **15**, 3707–3709.

46 Roach, P., Farrar, D. and Perry, C.C. (2005) *Journal of the American Chemical Society*, **127**, 8168–8173.

47 Torres, R., Ortiz, C., Pessela, B.C.C., Palomo, J.M., Mateo, C., Guisan, J.M. and Fernandez-Lafuente, R. (2006) *Enzyme and Microbial Technology*, **39**, 167–171.

48 Guisan, J.M., Sabuquillo, P., Fernandez-Lafuente, R., Fernandez-Lorente, G., Mateo, C., Halling, P.J., Kennedy, D., Miyata, E. and Re, D. (2001) *Journal of Molecular Catalysis B: Enzymatic*, **11**, 817–824.

49 Fuentes, M., Batalla, P., Grazu, V., Pessela, B.C.C., Mateo, C., Montes, T., Hermoso, J. A., Guisan, J.M. and Fernandez-Lafuente, R. (2007) *Biomacromolecules*, **8**, 703–707.

50 Murphy, C.D. (2006) *Natural Product Reports*, **23**, 147–152.

51 Hudson, S., Cooney, J., Hodnett, B.K. and Magner, E. (2007) *Chemistry of Materials*, **19**, 2049–2055.

52 Essa, H., Magner, E., Cooney, J. and Hodnett, B.K. (2007) *Journal of Molecular Catalysis B: Enzymatic*, **49**, 61–68.

53 Wang, F., Guo, C., Liu, H.-Z. and Liu, C.-Z. (2007) *Journal of Molecular Catalysis B: Enzymatic*, **48**, 1–7.

54 Osman, B., Kara, A., Uzun, L., Besirli, N. and Denizli, A. (2005) *Journal of Molecular Catalysis B: Enzymatic*, **37**, 88–94.

55 Hochuli, E., Döbeli, H. and Schacher, A. (1987) *Journal of Chromatography*, **411**, 177–184.

56 Augé, C., Malleron, A., Tahrat, H., Marc, A., Goergen, J.-L., Cerutti, M., Steelant, W.F.A., Delannoy, P. and Lubineau, A. (2000) *Chemical Communications*, 2017–2018.

57 Nahalka, J., Liu, Z., Chen, X. and Wang, P.G. (2003) *Chemistry – A European Journal*, **9**, 373–377.

58 Dräger, G., Kiss, C., Kunz, U. and Kirschning, A. (2007) *Organic and Biomolecular Chemistry*, **5**, 3657–3664.

59 Cassimjee, K.E., Trummer, M., Branneby, C. and Berglund, P. (2008) *Biotechnology and Bioengineering*, **99**, 712–716.

60 Sukumaran, J. and Hanefeld, U. (2005) *Chemical Society Reviews*, **34**, 530–542.

61 Kreiner, M., Moore, B.D. and Parker, M.C. (2001) *Chemical Communications*, 1096–1097.

62 Avnir, D., Braun, S., Lev, O. and Ottolenghi, M. (1994) *Chemistry of Materials*, **6**, 1605–1614.

63 Reetz, M.T., Zonta, A. and Simpelkamp, J. (1995) *Angewandte Chemie – International Edition in English*, **34**, 301–303.

64 Reetz, M.T., Wenkel, R. and Avnir, D. (2000) *Synthesis*, **6**, 781–783.

65 Orcaire, O., Buisson, P. and Pierre, A.C. (2006) *Journal of Molecular Catalysis B: Enzymatic*, **42**, 106–113.

66 Reetz, M.T., Tielmann, P., Wiesenhöfer, W., Könen, W. and Zonta, A. (2003) *Advanced Synthesis Catalysis*, **345**, 717–728.

67 Veum, L., Hanefeld, U. and Pierre, A. (2004) *Tetrahedron*, **60**, 10419–10425.

68 Cabirol, F.L., Hanefeld, U. and Sheldon, R.A. (2006) *Advanced Synthesis Catalysis*, **348**, 1645–1654.

69 Hara, P., Hanefeld, U. and Kanerva, L.T. (2008) *Journal of Molecular Catalysis B: Enzymatic*, **50**, 80–86.

70 Cao, L., van Rantwijk, F. and Sheldon, R.A. (2000) *Organic Letters*, **2**, 1361–1364.

71 Schoevaart, R., Siebum, A., van Rantwijk, F., Sheldon, R. and Kieboom, T. (2005) *Starch*, **57**, 161–165.

72 Boller, T., Meier, C. and Menzler, S. (2002) *Organic Process Research & Development*, **6**, 509–519.

73 Guisan, J.M. (1988) *Enzyme and Microbial Technology*, **10**, 375–382.

74 Mateo, C., Palomo, J.M., Fuentes, M., Betancor, L., Grazu, V., Lopez-Gallego, F., Pessela, B.C.C., Hidalgo, A., Fernandez-Lorente, G., Fernandez-Lafuente, R. and Guisan, J.M. (2006) *Enzyme and Microbial Technology*, **39**, 274–280.

75 Fernandez-Lafuente, R., Rosell, C.M., Rodriguez, V., Santana, C., Soler, G., Bastida, A. and Guisan, J.M. (1993) *Enzyme and Microbial Technology*, **15**, 546–550.

76 Betancor, L., Lopez-Gallego, F., Hidalgo, A., Alonso-Morales, N., Dellamora-Ortiz, G., Mateo, C., Fernandez-Lafuente, R. and Guisan, J.M. (2006) *Enzyme and Microbial Technology*, **39**, 877–882.

77 Lopez-Gallego, F., Betancor, L., Hidalgo, A., Alonso, N., Fernandez-Lorente, G., Guisan, J.M. and Fernandez-Lafuente, R. (2005) *Enzyme and Microbial Technology*, **37**, 750–756.

78 Berger, B. and Faber, K. (1991) *Journal of the Chemical Society D – Chemical Communications*, 1198–1200.

79 Shah, S., Sharma, A. and Gupta, M.N. (2006) *Analytical Biochemistry*, **351**, 207–213.

80 Roberge, C., Fleitz, F., Pollard, D. and Devine, P. (2007) *Tetrahedron Letters*, **48**, 1473–1477.

81 van Langen, L.M., Selassa, R.P., van Rantwijk, F. and Sheldon, R.A. (2005) *Organic Letters*, **7**, 327–329.

82 Chmura, A., van der Kraan, G.M., Kielar, F., van Langen, L.M., van Rantwijk, F. and Sheldon, R.A. (2006) *Advanced Synthesis Catalysis*, **348**, 1655–1661.

83 Cantone, S., Hanefeld, U. and Basso, A. (2007) *Green Chemistry*, **9**, 954–971.

3
Continuous-flow Microchannel Reactors with Surface-immobilized Biocatalysts

Malene S. Thomsen and Bernd Nidetzky

3.1
Introduction

Microstructured reaction technologies offer fundamental advantages to synthetic chemistry [1]. Besides the benefits of miniaturization for high-throughput reaction optimization [1b], the atypical fluid behavior in microchannels with inner dimensions of between several tens to hundreds of micrometers provides a uniquely useful and defining feature of microstructured reactors, in comparison with conventional macroscale reactors [1c, 2]. Fluid flow is mostly laminar, directed, and symmetric, thus facilitating continuous reaction engineering and reactor design. Improved control of key process parameters such as flow rate or temperature can result in enhanced yield and improved chemical selectivity of chemical conversions. Although mass transport occurs by diffusion, the high surface area-to-volume ratio of microchannels ($\approx$10 000 m^2/m^3 or greater) often allows conversions to be performed in the chemical regime where the rate of transport is high compared with the reaction rate. The use of microstructured reactors opens up possibilities for new production concepts, most notably continuous processing and flexible scale-up on demand via parallelization (numbering-up) [2, 3].

While microstructured reactors have gained widespread acceptance as tools of synthetic chemistry [1], biocatalytic transformations in microfluidic systems are not advancing as quickly ([4] and references therein). Although enzymes are often integrated into microfluidic 'lab-on-a-chip' systems for analytical applications, the development of scalable microstructured reactors for enzyme-catalyzed conversions has attracted comparably little attention so far. This is surprising considering that microwell systems, featuring parallelized shaken vessels for batch and fed-batch operation in the microliter scale, are now widely in use for miniaturized bioprocess studies [5]. Not only would a microstructured flow reactor facilitate screening of candidate enzymes from natural biodiversity or mutant libraries, but it should also provide an excellent system for biocatalytic process analysis and design. Experimentation in a suitable microstructured reactor under flow conditions could allow one to collect process information early during the

Modern Biocatalysis: Stereoselective and Environmentally Friendly Reactions.
Edited by Wolf-Dieter Fessner and Thorleif Anthonsen

ISBN: 978-3-527-32071-4

development which in turn would have a direct impact on reducing the time to market [5a].

To be useful as a technology, continuous flow processing in microstructured biocatalytic reactors must integrate recycling of the enzyme used most conveniently and efficiently via immobilization ([4] and references therein). One possibility is to attach the enzyme onto microparticles and assemble the functionalized beads in the microchannel as a miniature fixed bed. The complex liquid flow pattern and the high differential pressure required for obtaining practical flows are possible disadvantages. A second option is to immobilize the enzyme directly on the microchannel walls (for a review, see reference [4]; for examples of coated-wall microreactors using chemical catalysts, see references [1b, 2, 6]). Although the specific surface area available for the reaction may be relatively small in this case, a microfluidic reactor featuring enzyme-coated channels would seem to be a useful tool for biocatalytic microprocess technology. After briefly reviewing the state of the art, this chapter will summarize the results of microreactor development for enzymatic hydrolysis of lactose by the hyperthermostable β-glycoside hydrolase CelB from *Pyrococcus furiosus* [7].

3.2
Biocatalytic Synthesis Using Microreaction Technology with Free and Immobilized Enzymes

Recent reviews have provided systematic coverage of the enzymatic microreactors used in chemical analysis [4]. Considering that the focus of this chapter is biocatalytic synthesis, it does not consider the analytical applications and the reader is referred to the cited literature ([4] and references given therein). The use of microreactors for high-throughput kinetic characterization of enzymes is another very interesting application of the technology [8], which, for reasons of limited space, is not discussed herein.

Kanno *et al.* [9] reported on on-chip synthesis of oligosaccharides using glycosidase-catalyzed transgalactosylation. The authors did not use immobilized enzyme. However, a remarkable result of their study was that the product yield could be enhanced significantly in comparison with the conventional batch conversion. Other hydrolytic enzymes, namely lipase [10] and protease [11] were also tested in microchannel reactors. Different approaches of entrapment were used for enzyme immobilization. The results showed that the yields of product could be improved and that the requirements for reactant were reduced through the use of microreactors. Jones *et al.* [12] designed an enzyme reactor in which urease was immobilized on the microreactor wall by incorporating the enzyme directly into the polydimethylsiloxane (PDMS) material used for microstructure fabrication. Belder *et al.* [13] performed enantioselective catalysis using a soluble epoxide hydrolase and analysis on a microfluidic chip. NAD(H)-dependent enzymatic redox reactions were also carried out in microfluidic systems [14]. Methods for electrochemical regeneration of NADH were described [14b]. Oxidative enzymatic reactions examined in microchannel systems include degradation of *p*-chlorophenol catalyzed by laccase [15] and hydroxylation of an alkaloid catalyzed by a bacterial P450 enzyme [16]. More recently, Ku *et al.* [17] reported polyketide synthesis and functionalization using two microchannel reactors in series. The first reactor contained type III polyketide synthase immobilized onto agarose microbeads via interactions of the protein 'his tag' and Ni-nitrilotriacetic acid groups displayed on the surface of the particles. The second reactor contained soybean peroxidase attached on the surface of the microchannel. The study showed the rapid evaluation of different reaction conditions for the two-step enzymatic conversion and the effect of the varied conditions on the final product structure. Microstructured multi-enzyme reactors may become more widely used in the future for combinatorial synthesis. Honda *et al.* [18] introduced the technology of cross-linked enzyme aggregates to microstructured reactors. They deposited an aggregate formed between an aminoacylase and poly-L-lysine in the presence of glutardialdehyde on the inner wall of a silica capillary. The enzyme reactor thus obtained showed good operational stability and the activity of the immobilized enzyme was more resistant to thermal stress and organic solvents than that of the free enzyme. In a more recent paper Honda *et al.* [19] integrated an immobilized aminoacylase microreactor with a microdevice for two-phase liquid extraction to perform continuous enzymatic resolution of racemic amino acid derivatives with *in situ* product removal. Luckarift *et al.* [20] reported on multistep chemo-enzymatic synthesis of 2-aminophenoxazin-3-one using a continuous flow microdevice in which several microfluidic chips were connected in series. Silica-bound immobilized enzymes were used in the reaction.

3.3 Novel Microfluidic Immobilized Enzyme Reactors

3.3.1 Microreactor Design

Two different microstructured reactors for use with immobilized enzymes have been developed.

The first device integrated a microstructured multichannel plate fabricated by micro-injection molding from a two-component liquid silicon rubber material (Silopren LSR 4070) with an appropriately interfaced and temperature-controlled housing, as shown in Figure 3.1a.

The plate was designed to contain nine linear microchannels, an inlet zone that ensured equal division of liquid into the channels, and an outlet zone (Figure 3.1b). Each microchannel was characterized by a length of 64 mm, a width of 350 μm, and a height of 250 μm. Bas-relief microstructures were incorporated every 2.5 mm along each microchannel, alternating between the left and right walls (Figure 3.1c). They were included as flow obstacles to improve mass transfer to and from the microchannel surface through a passive mixing effect [1b, c, 2, 3].

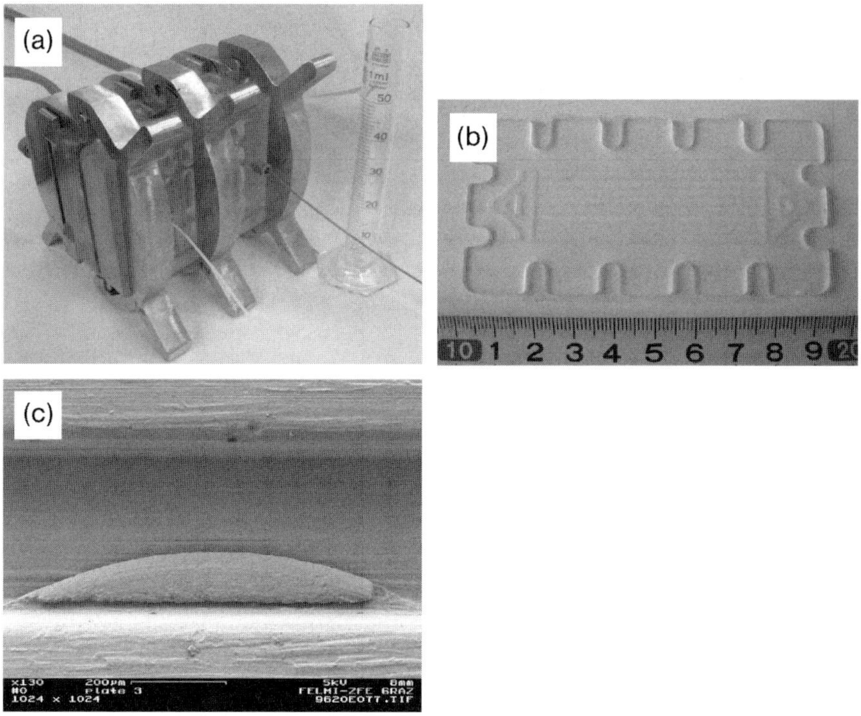

Figure 3.1 Microreactor featuring a multichannel microfluidic element fabricated from PDMS [21]. Panel a: the fully assembled microreactor. Panel b: microstructured multichannel plate. Panel c: electron micrograph of a segment of a microfluidic channel that shows a passive mixing element.

The liquid volume accommodated by the PDMS plate was 167 μl. The microfluidic element was placed in a housing designed as a sandwich-like construction consisting of two plates made of poly(methyl methacrylate), as can be seen in Figure 3.1a. One plate was transparent and contained inlet and outlet for fluid flow. The other element consisted partly of poly(methyl methacrylate) and an embedded metal plate. It was used for temperature control from an external water bath. The microreactor was designed for use under conditions of pressure-driven continuous fluid flow, which was delivered from a high-pressure liquid chromatography (HPLC) pump (Knauer Smartline 1000). Optionally, the outlet of the device was connected to a flow-through absorbance detector [21].

The second device was a multiplate-stacked microfluidic reactor, originally developed by Hessel, Löwe and co-workers for gas-phase transformations [1b] (Figure 3.2) and now adapted for biocatalytic processing with immobilized enzymes.

It is termed GPMR (gas phase microreactor) throughout. The microfluidic element of the microreactor was a stainless steel plate possessing 34 linear channels (Figure 3.2b). Its inlet was interfaced with the multichannel domain in order

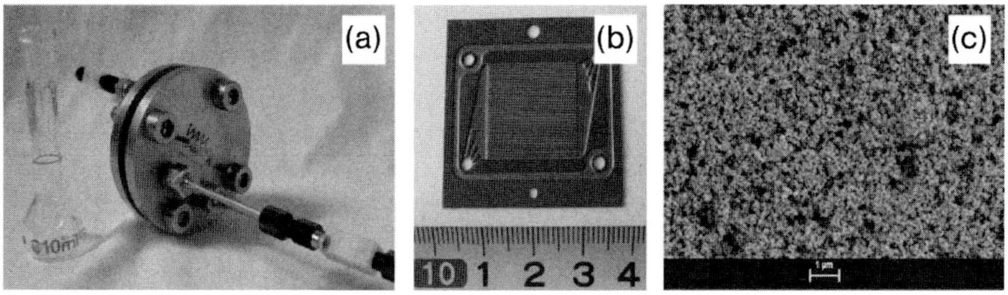

Figure 3.2 GPMR used for biocatalytic transformations with immobilized enzymes [22]. (a) the fully assembled microreactor, (b) microstructured multichannel plate, and (c) electron micrograph of the wash-coat layer of γ-aluminum oxide covering the microchannel walls.

to achieve proper fluid equipartition over the microchannels, each of which had a length of 20 mm and was characterized by an aspect ratio of 1.5 (width 300 μm and height 200 μm). The microreactor could be assembled with up to 10 microreaction plates, each accommodating a liquid volume of 24.5 μl and stacked alternating with microheat-exchanger plates. The walls of each channel were wash coated with a 10–20 μm layer of γ-aluminum oxide (Figure 3.2c), which allowed random tethering of the enzyme via chemical coupling [23]. Note that, according to literature reports [24], the γ-aluminum oxide coating of the microchannels does not influence the residence time distribution of the continuously operated microreactor. The inlet of the microreactor was connected to the HPLC pump mentioned above. The outlet was connected with an Upchurch Micro-Splitter Valve from Postnova Analytics (Landsberg, Germany) to enable manual back pressure regulation, typically maintained at 0.5–1.0 MPa.

3.3.2
Enzyme Immobilization

A useful compilation of methods employed for the immobilization of enzymes on microchannel surfaces is given by Miyazaki and Maeda [4a]. We used glutardialdehyde-mediated cross-linking of amino groups on the microstructured plates with amino groups of the enzyme. The surface area available for attachment of protein on the PDMS plate was estimated to be 793 mm^2. Due to the presence of the γ-aluminum oxide coating it is difficult to know the exact value for the relevant surface area on the GPMR plate. From the number and the geometry of the microchannels on the microstructured plate, a minimum surface area of 340 mm^2 was calculated.

The surface of PDMS is hydrophobic which results in poor wettability with aqueous solvents and promotes non-specific protein adsorption. It is also relatively inert to chemical modification [25]. The liquid silicon rubber chosen for fabrication of the reaction plate contained pyrogenic silicic acid as a filler. Aside from its effect on elastomer properties the silicic acid can be expected to provide additional silanol groups for surface chemistry. The microfluidic plate was silanized using in-flow treatment with an aqueous solution of (3-aminopropyl)triethoxysilane at pH 3.5–4.0. The silanized plate was then derivatized with glutardialdehyde solution. In the final step, the pre-activated plate was incubated with enzyme solution, typically containing about 0.1 mg protein/ml. A full description of the protocol used has been published elsewhere [21]. Enzyme immobilization on the GPMR plate took place by an analogous procedure with the exception that a pH of 7.0 was used during the silanization step to prevent destruction of the wash-coat layer of γ-aluminum oxide [22].

Immobilization experiments involved recombinant β-glycosidase CelB from *P. furiosus* produced in *Escherichia coli*. A partially purified preparation obtained from the soluble protein fraction of a heat-treated *E. coli* cell was employed [7]. The specific activity of CelB towards lactose was about 800 units/mg protein at 80 °C. We were able to immobilize about 60 μg and up to 100 μg of CelB on the PDMS and GPMR plates, respectively.

3.4 Enzymatic Hydrolysis of Lactose

3.4.1 Catalytic Effectiveness of Immobilized CelB

The enzyme activity bound on the reaction plate was measured under continuous flow conditions at 80 °C. Substrate solution typically containing 600 mM of lactose in 20 mM sodium citrate buffer at pH 5.5 was delivered at different flow rates and the concentration of glucose in samples collected at the outlet of the microreactor operated in the steady state was measured off-line. Figure 3.3 shows typical results. The enzyme activity corresponds to the slope of the linear relationship between the lactose converted and mean residence time τ_{av}, which is the ratio of the total reactor volume (24.5 μl per plate, a volume determined by the supplier (Institut für Mikrotechnik Mainz GmbH) and the flow rate (Figure 3.3)). CelB tethered onto GPMR plates retained about 50% of the specific activity (= activity bound/protein bound) of the free enzyme. By contrast, the specific activity of CelB immobilized on the PDMS plate was only about 3% that of the free enzyme. By way of comparison, CelB attached onto silanized macroporous glass beads using conditions otherwise exactly identical to the ones employed for immobilization of the enzyme on microstructured plates, showed about 35% of the original specific activity.

3.4.2 Continuous Conversion of Lactose

Figure 3.4 summarizes the results of microreactor experiments in which the effect of a varied flow rate on the enzymatic conversion of lactose was analyzed. The

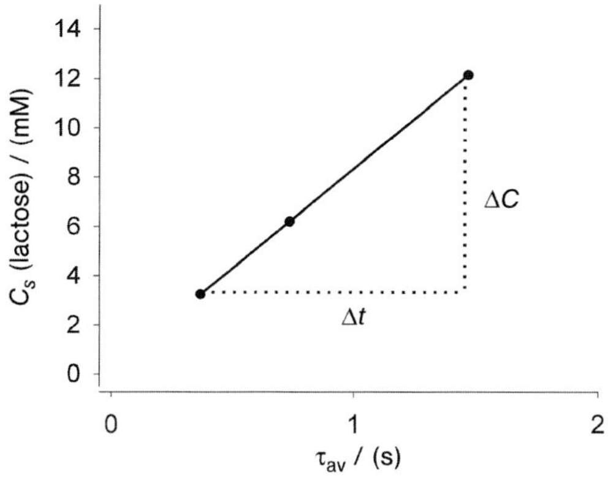

Figure 3.3 Determination of the activity of immobilized CelB under conditions of continuous flow [22]. The slope of the linear dependence of product concentration on residence time (τ_{av}) was used to calculate the activity $\Delta CV/\Delta\tau_{\alpha\varpi}$ where V is the reactor volume. Reactions were performed in the GPMR at 80 °C using 600 mM of lactose (pH 5.5) as the substrate. A single microreaction plate was used ($V = 24.5\,\mu l$). It contained a total amount of 13 U of enzyme activity.

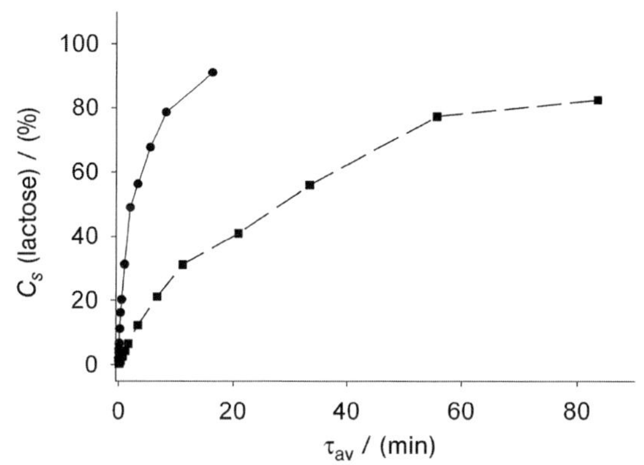

Figure 3.4 Comparison of the performance of the immobilized enzyme microreactors during continuous conversion of 100 mM of lactose (80 °C and pH 5.5) [21, 22]. GPMR (full circles); PDMS microreactor (full squares). Reactions were performed using a volumetric enzyme activity of 82 U/ml (GPMR, two stacked plates, and total volume of 49 µl) and 6 U/ml (PDMS microreactor and total volume of 167 µl).

average residence time required to hydrolyze ≥60% of the initial concentration of lactose (100 mM) was 4 min in the GPMR and about eight times lower than in the PDMS reactor. However, the comparison of the two microreactors must take into account that the concentration of active enzyme per milliliter of reactor working volume used in the GPMR (≈500 U/ml) was 56-fold higher than that in the PDMS reactor (≈12 U/ml). Therefore, this implies that, after correction for the effect of enzyme loading, the catalytic performance of the PDMS reactor was about 7 (=56/8) times more efficient than that of the GPMR.

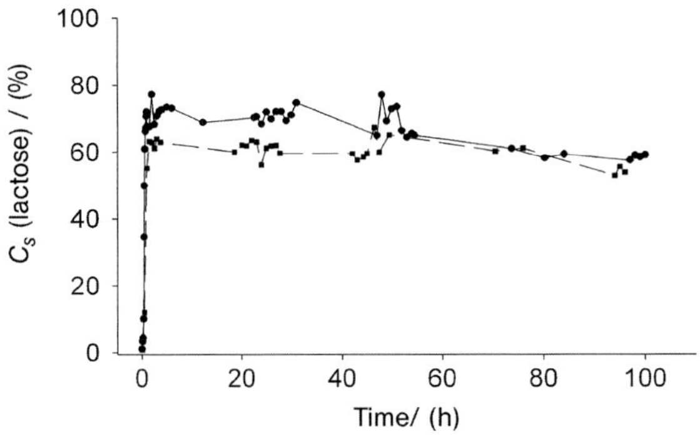

Figure 3.5 Operational stability of the immobilized enzyme microreactors during continuous conversion of 100 mM of lactose (80 °C and pH 5.5) [21, 22]. GPMR (full circles); PDMS microreactor (full squares). Reactions were performed using a τ_{av} value of 3.2 min (GPMR) and 33 min (PDMS microreactor). The volumetric enzyme activity was 184 U/ml in the GPMR (two stacked plates and total volume of 49 μl) and 12 U/ml in the PDMS microreactor (total volume of 167 μl).

In order to examine the operational stability of CelB in each of the two microreactors, we carried out time course experiments, the results of which are shown in Figure 3.5. The average residence time for lactose hydrolysis in the GPMR reactor (3.2 min) and the PDMS reactor (33 min) was constant and chosen to enable conversion of 60–70% of the initial concentration of substrate present, assuming a fully stable enzyme. Under these conditions, a decrease in the steady-state level of released glucose during the time span of each experiment, typically 100 h, would indicate gradual loss of activity. Both enzyme reactors showed useful stability.

3.5
Biocatalytic Process Intensification Using Microreaction Technology

The term process intensification encompasses a multilevel design approach aiming at the improvement of key features of chemical processing, such as size reduction of equipment, a decrease in energy consumption, and in safety of operation while achieving a given production objective [26]. Microreaction technology is a promising tool with which to achieve process intensification through matching of the fluid dynamics of a process with the physicochemical requirements of a reaction [2, 27]. According to an analysis performed for heterogeneously catalyzed gas-phase transformations [2, 27a], coated-wall microchannel reactors hold great potential for eventually outperforming conventional fixed-bed technology with respect to the above-mentioned criteria of intensified chemical processing. In particular, the characteristic time scale for a chemical transformation is expected to vary with microchannel diameter (d_M), from about $d_M^{\ 1}$ to $d_M^{\ 2}$ for reaction rate-limited to diffusion-limited conversions, respectively. Therefore, chemical processing with immobilized catalysts appears to be quite suitable for miniaturization to the microscale and the general idea is strongly supported by an increasing number of case studies of catalytic gas- or liquid-phase reactions [1b, 2]. A rigorous chemical engineering analysis that compares the performance of the microstructured device with that of a relevant macro fixed bed reactor is presently lacking for immobilized biocatalytic systems.

The potential benefit of equipment size reduction when using microreaction technology can be exemplified through space–time yield (STY) comparison. A STY of about 62 g/(l h) was reported for a continuously operated fixed bed reactor that utilized processing at 70 °C with CelB immobilized onto Eupergit C particles (90 units/ml of reactor) [7a]. The linear flow rate used with the fixed bed reactor was 0.64 cm/min, being in the range of linear flow rates (0.125–400 cm/min) employed with the microreactor. The STY obtained with the immobilized CelB microreactor (530 units/ml of reactor) operated at 80 °C was 500 g/(l h). Therefore, when taking into consideration the 25% decrease in enzyme activity caused by lowering the temperature from 80 to 70 °C (data not shown), the comparison of the two reactors reveals that, after normalization in the used enzyme loading to 1 kU CelB/l, the performance of the microfluidic device and the packed bed reactor was almost identical with 0.71 g glucose/lh and 0.69 g glucose/lh, respectively. This shows that the miniaturization to the microfluidic format concomitant with a change from catalytic particle to catalytic wall geometry does not enhance the intrinsic productivity of lactose conversion into glucose and galactose catalyzed by immobilized CelB. The benefits of introducing the enzyme into the microfluidic system was in this case the higher volumetric productivity obtained. Critical analysis of the performance of the different CelB reactors provides a note of caution in showing that reactor miniaturization to a microfluidic format alone is clearly not sufficient to improve the productivity of enzymatic lactose conversion. The STY of the continuously operated multichannel microreactor fabricated from PDMS was only 20 glucose/lh in spite of the fact that the available microstructured surface area was close to that of a two-plate stacked steel microreactor. Retention of just 3% of the specific activity of soluble CelB in the enzyme preparation immobilized onto PDMS explains the relatively much lower value for STY. The importance of selection of a suitable surface for enzyme immobilization is stressed and a wash-coat layer of γ-aluminum oxide seems to be a promising candidate.

3.6
Conclusions and Outlook

Two types of bioreactors are commonly used in industry for performing enzyme-catalyzed transformations: the stirred tank reactor and the packed or fluidized bed reactor [28]. Substitution of these well-established reactors by a system based on microfluidic technology will happen only if there are clear and compelling advantages in so doing. It is therefore necessary to consider, in a rigorous and case-specific manner, the potential assets of microstructured reactors for biocatalytic synthesis other than serving as a tool during initial process development in the laboratory. No clear and generally valid answers can be given at the present time. However, the following general benefits could be anticipated based on the results summarized herein and obtained in other groups [1–4]: intensification of processing, especially for multiphase reactions, easier integration of unit operations like biocatalytic reaction and product removal, by liquid–liquid extraction for example [19], ability to carry out multistep organic syntheses in the absence of intermediate purification, easier implementation of kinetically controlled reactions, change of product properties, faster transfer of research results into production and therefore earlier start of production at lower costs, and, finally, easier scale-up of production capacity.

Acknowledgements

We gratefully acknowledge the contributions from A. P. Zurk (dbs – daily business solutions GmbH) in the fabrication of the microstructured PDMS reactor and from Dr D. Kirschneck (Microinnova GmbH) in the assembly and set-up of the GPMR. Dr P. Pölt and C. Elis from the Institute for Electron Microscopy at Graz University of Technology kindly performed measurements by scanning electron microscopy. Work from the senior author's laboratory was supported by a grant from the Austrian Federal Ministry of Transport, Innovation, and Technology (Program 'Factory of Tomorrow', Project 807947). Dr D. Kirschneck, Dr M. Koncar (VTU Engineering), and Mag. H. Reichl (Hämosan) were industry partners in the project. They are thanked for their encouragement and for financial support.

References

1 (a) Baxendale, I.R., Deeley, J., Griffiths-Jones, C.M., Ley, S.V., Saaby, S. and Tranmer, G.K. (2006) *Chemical Communications*, 2566–2568.
(b) Jähnisch, K., Hessel, V., Löwe, H. and Baerns, M. (2004) *Angewandte Chemie – International Edition*, **43**, 406–446.
(c) deMello, A.J. (2006) *Nature*, **442**, 394–402.
(d) Watts, P. and Haswell, S.J. (2005) *Chemical Engineering & Technology*, **28**, 290–301.
(e) Jensen, K.F. (2001) *Chemical Engineering Science*, **56**, 293–303.
(f) Watts, P. and Wiles, C. (2007) *Chemical Communications*, 443–467.

2 Hessel, V., Hardt, S., Löwe, H., Müller, A. and Kolb, G. (2005) *Chemical Micro Process Engineering*, Vol. **1** and **2**, Wiley-VCH Verlag GmbH, Weinheim.

3 (a) Ehrfeld, W., Hessel, V. and Löwe, H. (2000) *Microreactors*, Wiley-VCH Verlag GmbH, Weinheim.
(b) Brand, O., Fedder, G.K., Hierold, C., Korvink, J.G. and Tabata, O. (series eds) (2006) *Micro Process Engineering* (N. Kockmann, volume ed), Wiley-VCH Verlag GmbH, Weinheim.

4 (a) Miyazaki, M. and Maeda, H. (2006) *Trends in Biotechnology*, **24**, 463–470.
(b) Urban, P.L., Goodall, D.M. and Bruce, N. C. (2006) *Biotechnology Advances*, **24**, 42–57.
(c) Krenkova, J. and Foret, F. (2004) *Electrophoresis*, **25**, 3550–3563.

5 (a) Micheletti, M. and Lye, G.J. (2006) *Current Opinion in Biotechnology*, **17**, 611-618.

(b) Kumar, S., Wittmann, C. and Heinzle, E. (2004) *Biotechnology Letters*, **26**, 1–10.
(c) Hermann, R., Lehmann, M. and Büchs, J. (2003) *Biotechnology and Bioengineering*, **81**, 178-186.
(d) Weuster-Botz, D., Puskeiler, R., Kusterer, A., Kaufmann, K., John, G.T. and Arnold, M. (2005) *Bioprocess and Biosystems Engineering*, **28**, 109–119.
(e) Thomsen, M. and Nidetzky, B. (2008) *Principles and Applications of Chemical Microreactors* (ed. T.R. Dietrich), Blackwell Publishing, Berlin, Germany (in press).

6 (a) Berger, R.J. and Kapteijn, F. (2007) *Industrial and Engineering Chemistry Research*, **46**, 3863–3870.
(b) Schouten, J.C., Rebrov, E.V. and de Croon, M.H.J.M. (2002) *Chimia*, **56**, 627–635.
(c) Kreutzer, M.T., Kapteijn, F., Moulijn, J.A., Ebrahimi, S., Kleerebezem, R. and van Loosdrecht, M.C.M. (2005) *Industrial and Engineering Chemistry Research*, **44**, 9646–9652.

7 (a) Petzelbauer, I., Kuhn, B., Splechtna, B., Kulbe, K.D. and Nidetzky, B. (2002) *Biotechnology and Bioengineering*, **77**, 619–631.
(b) Kamrat, T. and Nidetzky, B. (2007) *Journal of Biotechnology*, **129**, 69–76.
(c) Petzelbauer, I., Nidetzky, B., Haltrich, D. and Kulbe, K.D. (1999) *Biotechnology and Bioengineering*, **64**, 322–332.
(d) Splechtna, B., Petzelbauer, I., Kuhn, B., Kulbe, K.D. and Nidetzky, B. (2002) *Applied Biochemistry and Biotechnology*, **99**, 473–488.
(e) Lang, M., Kamrat, T. and Nidetzky, B. (2006) *Biotechnology and Bioengineering*, **95**, 1093–1100.

8 (a) Kerby, M.B., Legge, R.S. and Tripathi, A. (2006) *Analytical Chemistry*, **78**, 8273–8280.
(b) Mao, H., Yang, T. and Cremer, P.S. (2002) *Analytical Chemistry*, **74**, 379–385.
(c) Mao, H., Yang, T. and Cremer, P.S. (2002) *Journal of the American Chemical Society*, **124**, 4432–4435.
(d) Seong, G.H., Heo, J. and Crooks, R.M. (2003) *Analytical Chemistry*, **75**, 3161–3167.
(e) Koh, W.-G. and Pishko, M. (2005) *Sensors and Actuators. B, Chemical*, **106**, 335–342.
(f) Liu, A.L., Zhou, T., He, F.Y., Xu, J.J., Lu, Y., Chen, H.Y. and Xia, X.H. (2006) *Lab on a Chip*, **6**, 811–818.
(g) DeLouise, L.A. and Miller, B.L. (2005) *Analytical Chemistry*, **77**, 1950–1956.
(h) Jiang, H., Zou, H., Wang, H., Ni, J., Zhang, Q. and Zhang, Y. (2000) *Journal of Chromatography A*, **903**, 77–84.
(i) Gleason, N.J. and Carbeck, J.D. (2004) *Langmuir*, **20**, 6374–6381.
(j) Garcia, E., Hasenbank, M.S., Finlayson, B. and Yager, P. (2007) *Lab on a Chip*, **7**, 249–255.

9 (a) Kanno, K., Maeda, H., Izumo, S., Ikuno, M., Takeshita, K., Tashiro, A. and Fujii, M. (2002) *Lab on a Chip*, **2**, 15–18.
(b) Hisamoto, H., Shimizu, Y., Uchiyama, K., Tokeshi, M., Kikutani, Y., Hibara, A. and Kitamori, T. (2003) *Analytical Chemistry*, **75**, 350–354.

10 (a) Urban, P.L., Goodall, D.M., Bergström, E.T. and Bruce, N.C. (2006) *Journal of Biotechnology*, **126**, 508–518.
(b) Pijanowska, D.G., Baraniecka, A., Wiater, R., Ginalska, G., Lobarzewski, J. and Torbicz, W. (2001) *Sensors and Actuators. B, Chemical*, **78**, 263–266.
(c) Nakamura, H., Li, X., Wang, H., Uehara, M., Miyazaki, M., Shimizu, H. and Maeda, H. (2004) *Chemistry–Engineering Journal*, **101**, 261–268.
(d) Park, C.B. and Clark, D.S. (2002) *Biotechnology and Bioengineering*, **78**, 229–235.

11 (a) Miyazaki, M., Kaneno, J., Uehara, M., Fujii, M., Shimizu, H. and Macda, H. (2003) *Chemical Communications*, 648–649.
(b) Miyazaki, M., Kaneno, J., Kohama, R., Uehara, M., Kanno, K., Fujii, M., Shimizu, H. and Maeda, H. (2004) *Chemistry–Engineering Journal*, **101**, 277–284.
(c) Miyazaki, M., Kaneno, J., Yamaori, S., Honda, T., Briones, M.P., Uehara, M., Arima, K., Kanno, K., Yamashita, K., Yamaguchi, Y., Nakamura, H., Yonezawa, H., Fujii, M. and Maeda, H. (2005) *Protein and Peptide Letters*, **12**, 207–210.
(d) Kawakami, K., Sera, Y., Sakai, S., Ono, T. and Ijima, H. (2005) *Industrial and Engineering Chemistry Research*, **44**, 236–240.
(e) Wu, H., Tian, Y., Liu, B., Lu, H., Wang, X., Zhai, J., Jin, H., Yang, P., Xu, Y. and

Wang, H. (2004) *Journal of Proteome Research*, **3**, 1201–1209.

12 (a) Jones, F., Lu, Z. and Elmore, B. (2002) *Applied Biochemistry and Biotechnology*, **98**, 627–640.
(b) Jones, F. *et al.* (2004) *Applied Biochemistry and Biotechnology*, **113**, 261–272.

13 Belder, D. et al. (2006) *Angewandte Chemie – International Edition*, **45**, 2463–2466.

14 (a) Zhao, D.S. and Gomez, F.A. (1998) *Electrophoresis*, **19**, 420–426.
(b) Yoon, S.K., Choban, E.R., Kane, C., Tzedakis, T. and Kenis, P.J. (2005) *Journal of the American Chemical Society*, **30**, 10466–10467.

15 Maruyama, T., Uchida, J., Ohkawa, T., Futami, T., Katayama, K., Nishizawa, K., Sotowa, K., Kubota, F., Kamiya, N. and Goto, M. (2003) *Lab on a Chip*, **3**, 308–312.

16 Srinivasan, A., Bach, H., Sherman, D.H. and Dordick, J.S. (2004) *Biotechnology and Bioengineering*, **88**, 528–535.

17 Ku, B., Cha, J., Srinivasan, A., Kwon, S.J., Jeong, J.C., Sherman, D.H. and Dordick, J.S. (2006) *Biotechnology Progress*, **22**, 1102–1107.

18 Honda, T., Miyazaki, M., Nakamura, H. and Maeda, H. (2006) *Advanced Synthesis & Catalysis*, **348**, 2163–2171.

19 Honda, T., Miyazaki, M., Yamaguchi, Y., Nakamura, H. and Maeda, H. (2007) *Lab on a Chip*, **7**, 366–372.

20 Luckarift, H.R., Ku, B.S., Dordick, J.S. and Spain, J.C. (2007) *Biotechnology and Bioengineering*, **98**, 701–705.

21 (a) Thomsen, M.S., Pölt, P. and Nidetzky, B. (2007) *Chemical Communications*, 2527–2529.
(b) Thomsen, M.S. and Nidetzky, B. (2008) *Engineering in Life Sciences*, **8**, 40–48.

22 Thomsen, M.S. and Nidetzky, B. (2008) *Biotechnology Journal*, in press; DOI 10. 1002/biot.200800057.

23 (a) Rouge, A., Spoetzl, B., Gebauer, K., Schenk, R. and Renken, A. (2001) *Chemical Engineering Science*, **56**, 1419–1427.
(b) Zapf, R., Kolb, G., Pennemann, H. and Hessel, V. (2006) *Chemical Engineering & Technology*, **29**, 1509–1512.
(c) Germani, G., Stefanescu, A., Schuurman, Y. and van Veen, A.C. (2007) *Chemical Engineering Science*, **62**, 5084–5091.

24 Rouge, A., Spoetzl, B., Gebauer, K., Schenk, R. and Renken, A. (2001) *Chemical Engineering Science*, **56**, 1419–1427.

25 Makamba, H., Kim, J.H., Lim, K., Park, N. and Hahn, J.H. (2003) *Electrophoresis*, **24**, 3607–3619.

26 (a) Ramshaw, C. (1999) *Green Chemistry*, G15–17.
(b) Stankiewicz, A.I. and Moulijn, J.A. (2000) *Chemical Engineering Progress*, **96**, 22–34.

27 (a) Commenge, J.-M., Falk, L., Corriou, J.-P. and Matlosz, M. (2005) *Chemical Engineering & Technology*, **28**, 446–458.
(b) Becht, S., Franke, R., Geißelmann, A. and Hahn, H. (2007) *Chemical Engineering & Technology*, **3**, 295–299.
(c) Mae, K. (2007) *Chemical Engineering Science*, **62**, 4842–4851.

28 Buchholz, K., Kasche, V., Bornscheuer, U.T. (2005) *Biocatalysis and Enzyme Technology*, Wiley-VCH Verlag GmbH, Weinheim.

4
Activity and Stability of Proteases in Hydrophilic Solvents

Lars Haastrup Pedersen, Sinthuwat Ritthitham and Morten Kristensen

4.1
Introduction

Proteases and lipases have proven to be very efficient biocatalysts in both hydrophobic and hydrophilic solvents. The choice of enzyme and solvent system opens up a wide range of possibilities in process design for manipulating the selectivity and solubility of reactants and products. When it comes to specific transformations and regioselective substitution of complex carbohydrates, lipids, and peptides biocatalysis can be used for producing acceptable yields at relatively mild reaction conditions. Therefore biocatalysis offers a strong alternative to conventional chemistry, avoiding the need for protection and deprotection steps required in chiral and regioselective organic synthesis. In contrast to lipases, proteases do not catalyze direct esterification with fatty acids as acyl donors, but proteases have proven to be efficient catalysts of transesterification reactions with carbohydrates ranging in size from mono- to polysaccharides as acyl acceptors.

Proteases catalyze the esterification of carbohydrates using activated esters as acyl donors. Members of the subtilisin family (S8) have been shown to catalyze specific acylation of sucrose at the C-1′ position (see Figure 4.1), whereas thermolysin (EC 3.4.24.27) catalyzes the acylation of sucrose primarily at the C-2 position as well as regioselective acylation of the glucose moieties of cyclodextrins. This regioselectivity differs from most lipases including the ones from *Humicola lanuginosa* and *Pseudomonas* sp., which catalyze the acylation of sucrose primarily at the C-6 position and to a minor extent both at the C-6, C-6′ positions and the C-6, C-1′ positions. Taking advantage of the differences in regioselectivity allows for esterification at specific positions of the carbohydrate. Thus, by controlling both the acyl chain length and the positional distribution of fatty acids on the sugar molecule, systematical changes of the physico-chemical properties of these amphiphilic biosurfactants are obtained.

In order to achieve an optimal process the solvent must solubilize the reactants, while at the same time maintaining enzyme activity. For synthesis of biosurfactants hydrophilic solvents are good solvents of both hydrophilic carbohydrates and hydrophobic acyl donors, but are known as denaturants of most proteins. Reduced

Modern Biocatalysis: Stereoselective and Environmentally Friendly Reactions.
Edited by Wolf-Dieter Fessner and Thorleif Anthonsen

ISBN: 978-3-527-32071-4

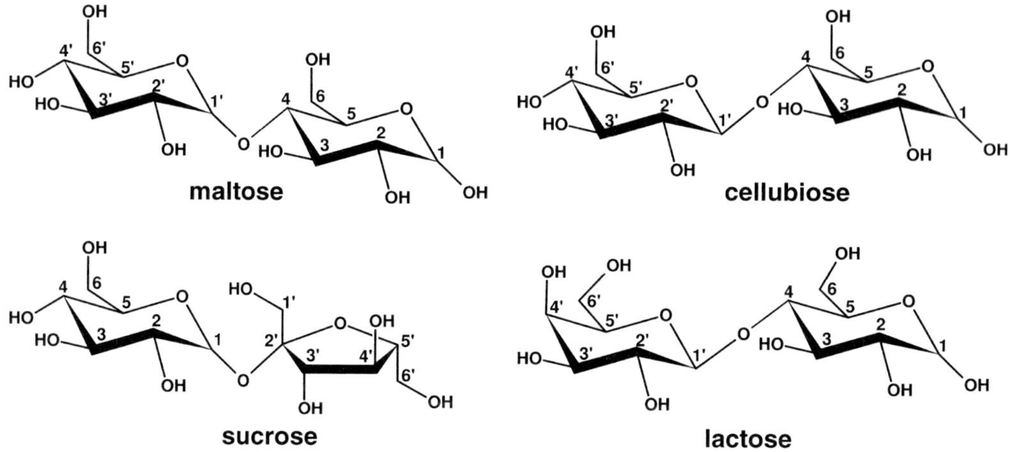

Figure 4.1 Important disaccharides.

activity and complete inactivation of lipases in hydrophilic, aprotic solvents at concentrations of 20–30% (v/v) has been observed [1–3]. In contrast, alkaline proteases have proven active and relatively stable under these conditions. Therefore the stability of these enzymes in organic solvents has been investigated and attempts to relate stability characteristics to protein structure and conformation have been made.

Protease- and lipase-catalyzed sugar ester synthesis has been reviewed by Plou *et al.* [4] and Kennedy *et al.* [5], whereas specific reviews on proteases in organic solvents and industrial utilization of alkaline proteases were compiled by Bordusa [6] and Gupta *et al.* [7], respectively.

The aim of the present chapter is to give a brief overview of protease-catalyzed synthesis of sugar esters in hydrophilic solvents and to present some of the most recent investigations on the effect of these solvents on activity and stability of proteases. Consequently, the perspectives in solvent engineering are outlined with focus on hydrophilic solvents and ionic liquids.

4.2 Activity and Selectivity of Proteases in Synthesis of Carbohydrate Fatty Acid Esters

The catalytic properties and regioselectivity of proteases of the subtilisin family including subtilisin Carlsberg have been thoroughly investigated in transestrification reactions with trihaloethyl esters and vinyl esters of short- to medium-chained fatty acids and a range of mono- and disaccharides as acyl acceptors. Anhydrous pyridine and *N,N*-dimethylformamide (DMF) have been shown to be good solvents for these reactions both in terms of solubility of the substrates and sustained activity of the alkaline protease. Some of the most abundant hexoses were all substituted at the primary hydroxyl group at the C-6 position (see Table 4.1). However, by adding dimethyl sulfoxide (DMSO) to the reaction medium the regioselectivity

Table 4.1 Selectivity of alkaline proteases in the synthesis of monosaccharide-based biosurfactants.

Alkaline protease	Solvent and conditions	Acyl donor	Acyl acceptor	Position of substitution	Yield %	Reference
Subtilisin *Bacillus subtilis*	DMF 45 °C, 18 h	Trichloroethyl butanoate	D-glucose	6	64	[8]
Bacillus sp. (Proleather)	Pyridine 45 °C, 2 days	Trichloroethyl propanoate	D-glucose	6	5.2	[9]
		Vinyl acetate	D-glucose	6	31	
		Vinyl benzoate	D-glucose	6	33	
Streptomyces sp.	**DMF** 35 °C, 7 days	Divinyl adipate	D-glucose	6	66	[10]
			D-galactose	**6**	63	
			D-mannose	6	54	
			α-methyl-D-galactose	6	74	
	DMF : DMSO 4 : 1 35 °C, 7 days	Divinyl adipate	D-glucose	6	30	
			D-mannose	6	46	
			α-methyl-D-galactose	6	21	
			D-galactose	**2**	49	

in a *Streptomyces* sp. protease-catalyzed reaction was shifted from the primary hydroxyl to the secondary hydroxyl group at the C-2 position of galactose [10].

In both pyridine and DMF subtilisin Carlsberg and other subtilisins showed selectivity towards acylation of primary hydroxyl groups of important disaccharides (see Figure 4.1). Hence, monoesters exclusively substituted at the C-6′ position were obtained with cellubiose and maltose, whereas with lactose the major product was the 6′-*O*-monoester, but other monoesters substituted at either the C-3′ or the C-4′ position were detected in smaller amounts. The overall substitution pattern clearly reveals regioselectivity towards the primary hydroxyl group situated at the non-reducing end of these three reducing sugars. Using proteinase N as biocatalyst in either DMF or DMSO and water (7% v/v) as co-solvent, sucrose monoesters substituted at each of the three primary hydroxyl groups were obtained, respectively, with the 1′-*O*-ester as the major product [11, 12]. Sucrose monoesters were also synthesized in anhydrous DMF and pyridine, respectively with different acyl donors and other subtilisins as biocatalysts – in all cases 1′-*O*-monoester was the major product (see Table 4.2). Using vinyl acrylate as the acyl donor Park and Chang [14] also observed the formation of two different acrylic acid esters of sucrose (1′,6′-di-*O*-acryloyl- and 6,1′-di-*O*-acryloyl-sucrose) both substituted at the C-1′ position in combination with one of the two remaining primary hydroxyl groups. With the alkaline protease AL 89 in a mixture of DMF, DMSO, and water (7% v/v) the regioselectivity towards the C-2 position of sucrose was observed.

Table 4.2 Selectivity of alkaline proteases in the synthesis of disaccharide-based biosurfactants.

Alkaline protease	Solvent and conditions	Acyl donor	Acyl acceptor	Sugar ester yield (%)	Substitution pattern		Reference
					position	distribution mol/mol (%)	
Subtilisin *B. subtilis*	DMF at 45 °C for 2–7 days	Trichloroethyl butanoate	Maltose		6′		[8]
			Cellubiose		6′		
			Lactose		6′	75	
					4′	10	
					3′	10	
Subtilisin Carlsberg *Bacillus licheniformis*	DMF at 45 °C	Trifluoroethyl butanoate	Sucrose	50	1′	90	[13]
Subtilisin Carlsberg *B. licheniformis* (Optimase M-440)	Pyridine at 30 °C for 5 days	Vinyl acrylate	Sucrose		1′	70	[14]
					6, 1′	18	
					1′, 6′	12	
Proteinase N *B. subtilis*	DMF : H_2O 13.3 : 1 at 45 °C for 1 day	Trifluoroethyl-methacrylate	Sucrose	15	1′	90	[11]
					6′	10	
					6	10	
	DMF : H_2O 24 : 1 at 45 °C for 3 days	Vinyl esters of fatty acids (C8, C10 and C12)	Sucrose	17 (C8)	1′	75	[12]
				40 (C10)	6′	10–13	
				21 (C12)	6	7–11	
Protease AL 89 *Bacillus pseudofirmus*	DMSO : DMF : H_2O 6.2 : 6.2 : 1 at 45 °C for 1 day	Vinyl-laurate (C12)	Sucrose	57	2		[15]
B. subtilis	Pyridine at 50 °C for 5 days	Divinyl esters of fatty acids (C4, C6 and C10)	Maltose	53 (C4)	6′		[16]
				42 (C6)			
				34 (C10)			
			Lactose	62 (C4)	6′		
				46 (C6)			
				31 (C10)			
			Sucrose	55 (C6)	1′		

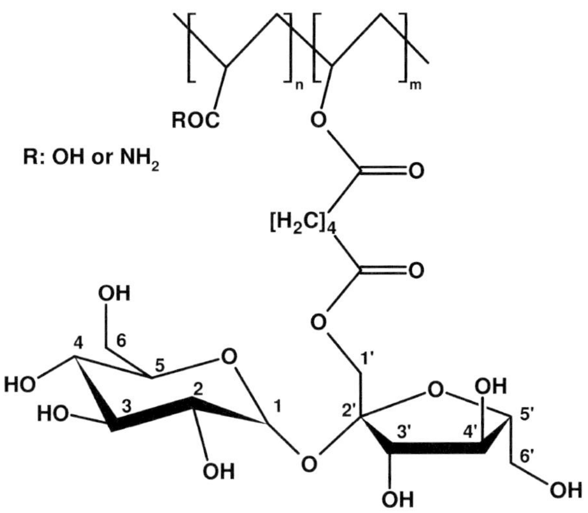

Figure 4.2 Sucrose containing biodegradable polymers. 1′-*O*-vinyl hexanedioyl-sucrose co-polymerized with either acrylic acid or acrylamide [21].

Thus, using vinyl laurate as the acyl donor the enzyme catalyzed the synthesis of 2-*O*-lauroyl sucrose [15].

The neutral protease thermolysin was shown to have the same regioselectivity with sucrose as the acyl acceptor in anhydrous DMSO [17] and, depending on the chain length of the acyl donor, the glucose moiety in cyclodextrins was substituted at the C-2 or both the C-2 and C-3 or C-6 positions [18]. The carbohydrate fatty acid esters described above have a wide range of applications due to their physical and chemical properties. Particularly laurate monoesters of mono- and disaccharides have anti-microbial properties [19] and vinyl esters of disaccharides can be used for polymerization – either homopolymerized or co-polymerized with acrylic acid or arylamide to produce hydrogels [20, 21]. In this way Wang *et al.* [21] synthesized biodegradable homopolymers of 1′-*O*-vinyl hexanedioyl-sucrose and 6′-*O*-vinyl hexanedioyl-lactose respectively, in which the disaccharide is linked to the polyvinyl main chain via the ester linkage to the hexanedioyl spacer. Furthermore 1′-*O*-vinyl hexanedioyl-sucrose was co-polymerized with either acrylic acid or acrylamide (see Figure 4.2). Fatty acid esters of cyclodextrins have been used for producing self-assembling nanoparticles for drug delivery [22].

4.3 Enzyme Stability and Conformation

The conformation and dynamics of secondary structural elements of subtilisin Carlsberg in 100% DMSO was studied by Fourier-transform infrared (FTIR) spectroscopy and suggested, on one hand, random coil formation at room temperature due to complete unfolding [23]. On the other hand, formation of aggregates or intermolecular β-sheets at 30 °C resulting in very low catalytic activity has been reported [24]. At DMSO concentrations of 54–56% only a partial destruction of the tertiary structure of subtilisin was observed by FTIR and circular dichroism. This structural change induced an enhanced conformational flexibility of the enzyme resulting in an increase in both hydrolytic activity and enantioselectivity towards (*S*)-ethyl-2-(4-substituted phenoxy) propanoates. Changing the DMSO concentration from 0 to 55% (v/v) the enantiomeric ratio (*E*-value) increased by a factor 2.4–4.5 depending on the chain length of the alkyl substituent. With a racemic mixture of ethyl-2-(4-hexyl phenoxy) propanoate in 65% (v/v) DMSO an *E*-value of 53 was obtained, whereas in buffer the *E*-value was 5.4 [25]. A similar effect caused by DMF on a papaya-type plant cystein protease was observed by Quiroga *et al.* [26], namely a simultaneous increase in hydrolytic activity and partial unfolding of the enzyme in a mixture of DMF and Tris–HCl buffer pH 8 (1 : 1 v/v). The structural analysis was based on FTIR revealing a decrease in the total content of β-sheet and β-turn secondary structural elements during 4 h of incubation.

At a DMSO concentration of 30% (v/v) Ruiz and De Castro [27] reported that the halo-alkalophilic protease from the archaeon *Natrialba magadii* showed a residual activity of approximately 80% after 7 days of incubation at 30 °C, 1.5 M NaCl, pH 8 and pH 10, respectively. At suboptimal salt concentrations the presence of DMSO had a stabilizing effect on the enzyme.

Ogino and Ishikawa [28] compared the hydrolytic activity half-life of four different proteases as effected by a range of solvents. DMSO (25% v/v) had a stabilizing effect on α-chymotrypsin and the two alkaline proteases, subtilisin Carlsberg and *Pseudomonas aeruginosa* protease PST-01. In fact the latter proved to be extremely stable in DMSO and a range of alcohols showing activity half-lives longer than 50 days. DMF destabilized both α-chymotrypsin and thermolysin, but had a stabilizing effect on the alkaline proteases, subtilisin Carlsberg in particular. The same effect was observed with methanol, but instead of subtilisin Carlsberg it was PST-01 showing the highest stability. In terms of stability of the alkaline proteases, water was found to be the worst solvent. Of the more hydrophobic solvents, toluene had a stabilizing effect on all four proteases (see Table 4.3).

Klibanov [30] and co-workers [31, 32] showed that the tertiary structure of subtilisin Carlsberg was nearly the same in water, acetonitrile, and dioxane and that the differences observed were of the same magnitude as those observed between structures in different aqueous systems. In particular the active site structures were essentially the same in all three solvents, but also the solvents did not substantially affect the exposed surface area of the enzyme, which was prepared from slightly cross-linked crystals.

Recent investigations using circular dichroism have suggested that protein stability was closely related to specific elements of secondary structure with α-helical structures remaining unchanged – whereas β-pleated sheet structures were affected by the addition of methanol. Hence, the relatively high content of α-helical structural elements and low content of β-sheet of PST-01 (see Table 4.4) was seen as

Table 4.3 Activity half-life of proteases in the presence of organic solvent [28, 29]. Values in parentheses are the results of repetitions carried out in 2007. Enzymes were incubated at 30 °C and pH 8 in the presence of 25% organic solvent. Residual hydrolytic activity was measured by a casein assay.

Solvent	Log *P*	Half life (days)			
		α-Chymotrypsin	Thermolysin	Subtilisin	PST-01
Water		13.2 (0.05)	10.8 (17.3)	0.3 (0.4)	9.7 (10.2)
DMSO	−1.35	33.6	2.6	6.4	>50
DMF	−1.01	2.2	0.9	39.8	25.3
Methanol	−0.74	6.0 (0.006)	4.6 (4.1)	26.2 (20.3)	>50 (41.0)
Ethanol	−0.30	27.0	3.0	>50	>100
Acetone	−0.24	0.6	0.7	24.8	23.1
Tert-butanol	0.35	0.5	0.8	41.6	>50
Heptanol	2.62	3.8	13.1	8.6	>50
Toluene	2.73	>100	22.5	5.7	12.0

Table 4.4 Secondary structural elements composition of proteases [29]. The compositions were estimated from circular dichroism spectra of enzyme solutions at 30 °C and pH 8.

Protease	α-helix (%)	β-sheet (%)
α-Chymotrypsin	7	30
Thermolysin	38	17
Subtilisin	30	16
PST-01	37	13

important factors for the increased half-life observed for this protease in 25% methanol (see Table 4.3), but also the presence of two disulfide bonds contributed to the stability characteristics of PST-01 [29].

The stabilizing effect of introducing new disulfide bonds was demonstrated in a thermolysin-like double mutant, where enhanced thermostability was clearly related to the stabilization of a 14 amino acid long flexible loop region (56–69) by the introduction of a disulfide bond [33]. A closely related mutant, boilysin, proved resistant to boiling with a half-life of 170 min at 100 °C [34]. Both enzymes were mutated far from the active site and therefore the flexibility of the amino acid residues involved in catalysis was not affected. Stabilization against thermal inactivation simultaneously led to increased stability in organic solvents. Although the difference in thermal stability between the two mutants was clear (10 K), no significant difference in the stability towards organic solvents was observed [35].

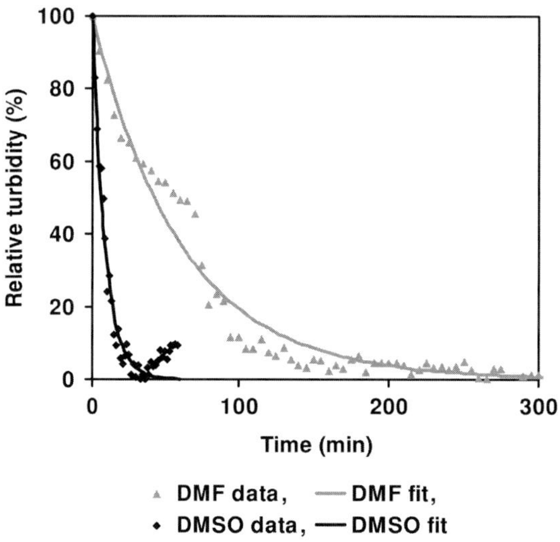

Figure 4.3 Dissolution of alkaline protease AL 89 suspended in hydrophilic solvents (M. Kristensen and L.H. Pedersen, unpublished). DMF $t_{½}$: 29 min; DMSO $t_{½}$: 6 min. A crude preparation of alkaline protease AL 89 (10 g/l) was suspended and incubated in anhydrous DMF or DMSO at 40 °C. OD_{600} was measured in triplicates of 200 μl sample volumes in a Tecan microplate reader with continuous shaking.

The effect of DMSO on the initial reaction rate of a protease PST-01-catalyzed synthesis of *Z*-aspartame in a mixture of DMSO and Tris–HCl buffer at pH 8 was investigated by Tsuchiyama *et al.* [36]. The initial reaction rate decreased with increasing DMSO concentration (20–70% (v/v)). At 50% (v/v) DMSO an 83% yield of aspartame precursor was obtained and at DMSO concentrations higher than 60% a precipitate assumed to be the protease was observed.

We investigated the dissolution behavior of a lyophilized crude preparation of the alkaline protease AL 89 in anhydrous DMSO and DMF, respectively. At first a turbid suspension was obtained in both cases, then the suspension was completely solubilized following an exponential decrease in turbidity (OD_{600}), with the fastest decrease observed in DMSO (see Figure 4.3). During this process the residual hydrolytic activity of the enzyme suspension was analyzed and shown to decrease, correlating with the decrease in turbidity. These observations indicated that the highest activity was maintained as long as the enzyme preparation was in suspension and that solvation caused inactivation. DMSO dissolved the enzyme suspension faster than DMF corresponding to the decrease in residual activity (see Figure 4.4). The optimal balance between substrate solubility and sustained enzyme activity can therefore be found by adjusting the mixture ratio between these two solvents.

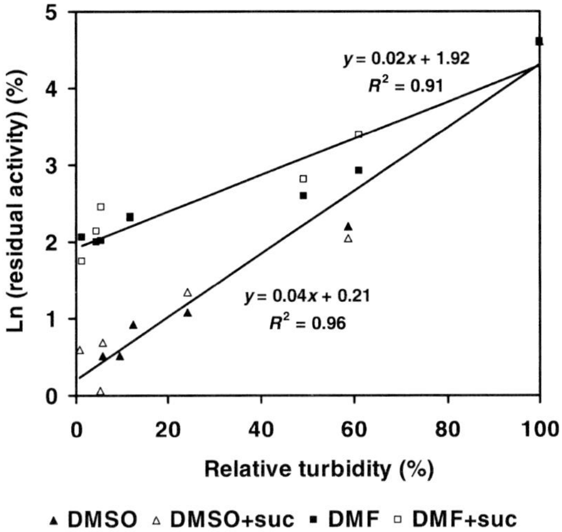

Figure 4.4 Residual activity as effected by dissolution of alkaline protease AL 89 in hydrophilic solvents (M. Kristensen and L.H. Pedersen, unpublished). Enzyme incubated in anhydrous DMF (squares) or DMSO (triangles) in the absence (filled symbols) or presence (open symbols) of 0.2 M sucrose. Relative turbidity (OD_{600}) was used as a measure of the degree of solubilization of suspended alkaline protease AL 89 (see the description and progression curves in Figure 4.1). Residual hydrolytic activity of 20 μl samples diluted in buffer 1 : 50 was measured by an azo-casein assay at pH 10 and 50 °C.

4.4 Solvent Engineering

Enzyme activity is generally significantly lower in anhydrous organic solvents than in water, not just because of reduced conformational mobility, but also because of factors directly related to the physical–chemical properties of the organic solvent, namely unfavorable energetics of substrate desolvation and transition state destabilization [30]. These effects can often be compensated for by an appropriate choice of organic solvent. Therefore solvent engineering and investigation of new types of reaction media will still be important research areas in relation to biocatalysis. In this context ionic liquids have recently received increasing attention. Ionic liquids are low melting point (<100 °C) salts representing a new class of non-aqueous, but polar solvents. Unlike conventional organic solvents, ionic liquids possess no measurable vapour pressure. Because of their polarity they are able to dissolve many hydrophilic compounds and some are even miscible with water [37]. Thus their hydrophilicity can be tuned by the choice of cation and anion and this way the process can be designed from water miscible to water immiscible systems [38] and a wide range of combinations exists: examples of cations are mono-, di-, and tri-substituted imidazoliums, pyridiniums, pyrrolidiniums, phophoniums, ammoniums, guanidiniums, and isouroniums while anions can be halides, sulfates, sulfonates, amides, imides, methyls, borates, phosphates, and antimonates.

Ionic liquids have been shown to be good solvents of carbohydrates including linear polysaccharides as cellulose and amylose, hence it has been reported that 1-butyl-3-methylimidazolium chloride ([BMIM] [Cl]) dissolved 100 g/l cellulose at 100 °C [39]. Ionic liquids containing the dicyanide anion, $(CN)_2N^-$ have been reported to dissolve approximately 200 g/l of glucose, sucrose, and lactose, respectively at 25–75 °C [40] and from the same investigation [BMIM] [$(CN)_2N$] was reported to dissolve 750 g/l of β-cyclodextrin at 75 °C.

In addition to being good solvents of carbohydrates, ionic liquids have shown to sustain enzyme activity and stability. Erbeldinger *et al.* [41] showed that thermolysin catalyzed the synthesis of aspartame precursor (*Z*-aspartame) in a mixture of [BMIM] [PF_6] and water 95 : 5 (v/v) resulting in a 95% yield. In suspension the enzyme was active and stable, but when dissolved in lower concentrations in the ionic liquid, it was inactivated. In addition, subtilisin Carlsberg has been shown to be active in a Hofmeister series of hydrophilic ionic liquids [42] and both α-chymotrypsin and subtilisin Carlsberg catalyzed the esterification of *N*-acetyl-phenylalanine in water miscible 1-ethyl-3-methylimidazolium trifluoromethyl sulfonate ([EMIM] [Tf]) with 0.2% water (v/v) at 30 °C [43]. The transestrification activity half-life of α-chymotrypsin was increased in four different ionic liquids as compared to 1-propanol as reaction medium at 50 °C and 2% water (v/v): the largest increase was obtained with the ionic liquid of lowest polarity, methyltrioctylammonium triflimide and it was 13.5 times higher than that of 1-propanol [44].

4.5
Conclusion

Hydrophilic organic solvents can be used as co-solvents with water to produce one-phase systems in which optimal mass transfer of substrates and products can be obtained. At the same time these solvents can increase the activity, influence the regioselectivity and improve the enantioselectivity of enzymes due to partial denaturation and increased flexibility of the enzyme structure. At concentrations below 50% hydrophilic solvents can even have a stabilizing effect on alkaline proteases, but at higher concentrations and particularly in anhydrous systems most enzymes including alkaline protease will denature and consequently lose activity. Although the structural reasons may not be identical, stabilization against thermal inactivation often leads to increased stability in organic solvents. Using hydrolytic enzymes in reversed hydrolysis reactions by shifting the thermodynamic equilibrium at low water activity offers potential alternatives to conventional organic chemistry. The advantages of biocatalysis have been realized by the chemical and pharmaceutical industry. Alkaline proteases have been shown to be active not only on peptides, but on a wide range of renewable resources for synthesis of biologically active molecules and carriers and in the synthesis of carbohydrate derivatives with designed functional properties for drug delivery and health care systems. Although these enzymes are extremely stable in hydrophilic organic solvents and ionic liquids it is even possible to improve their properties by modern biotechnology and to design highly efficient enzymes for specific industrial processes. Among the challenges in biocatalysis are securing optimal mass transfer and high yields, and recirculation of the reaction medium. Therefore, solvent engineering remains an important element of bioprocess design. An increased understanding of the interactions between the reaction medium and the biocatalyst, its substrates, and products will be essential to the further development of green chemistry.

References

1 Castillo, E., Pezzotti, F., Navarro, A. and López-Munguia, A. (2003) *Journal of Biotechnology*, **102**, 251.

2 Degn, P. and Zimmermann, W. (2001) *Biotechnology and Bioengineering*, **74** (6), 483.

3 Moniruzzaman, M., Hayashi, Y., Talukder, M.R., Kawanishi, T. (2007) *Biocatalysis and Biotransformation*, **25** (1), 51.

4 Plou, F.J., Cruces, M.A., Ferrer, M., Fuentes, G., Pastor, E., Bernabé, M., Christensen, M., Comelles, F., Parra, J.L. and Ballesteros, A. (2002) *Journal of Biotechnology*, **96**, 55.

5 Kennedy, J.F., Kumar, H., Panesar, P.S., Marwaha, S.S., Goyal, R., Parmar, A. and Kaur, S. (2006) *Journal of Chemistry and Technology*, **81**, 866.

6 Bordusa, F. (2002) *Chemical Reviews*, **102**, 4817.

7 Gupta, R., Beg, Q.K. and Lorenz, P. (2002) *Applied Microbiology and Biotechnology*, **59**, 15.

8 Riva, S., Chopineau, J., Kieboom, A.P.G. and Klibanov, A. (1988) *Journal of the American Chemical Society*, **110**, 584.

9 Watanabe, T., Matsue, R., Honda, Y. and Kuwahara, M. (1995) *Carbohydrate Research*, **275**, 215.

10 Kitagawa, M., Fan, H., Raku, T., Shibatani, S., Maekawa, Y., Hiraguri, Y., Kurane, R. and Tokiwa, Y. (1999) *Biotechnology Letters*, **21**, 355.

11 Potier, P., Bouchu, A., Descotes, G., and Queneau, Y. (2000) *Tetrahedron Letters*, **41**, 3597.

12 Potier, P., Bouchu, A., Gagnaire, J., and Queneau, Y. (2001) *Tetrahedron Asymmetry*, **12**, 2409.

13 Riva, S., Nonini, M., Ottolina, G. and Danieli, B. (1998) *Carbohydrate Research*, **314**, 259.

14 Park, H.G. and Chang, H.N. (2000) *Biotechnology Letters*, **22**, 39.

15 Pedersen, N.R., Wimmer, R., Matthiesen, R., Pedersen, L.H. and Gessesse, A. (2003) *Tetrahedron Asymmetry*, **14**, 667.

16 Wu, Q., Wang, N., Xiao, Y.M., Lu, D.S. and Lin, X.F. (2004) *Carbohydrate Research*, **339**, 2059.

17 Pedersen, N.R., Halling, P.J., Pedersen, L.H., Wimmer, R., Matthiesen, R. and Veltman, O.R. (2002) *FEBS Letters*, **519**, 181.

18 Pedersen, N.R., Kristensen, J.B., Bauw, G., Ravoo, B.J., Darcy, R., Larsen, K.L. and Pedersen, L.H. (2005) *Tetrahedron: Asymmetry*, **16**, 615.

19 Watanabe, T., Katayama, S., Matsubara, M., Honda, Y. and Kuwahara, M. (2000) *Current Microbiology*, **41**, 210.

20 Patil, N.S., Yanzi, L., Rethwisch, D.G. and Dordick, J.S. (1997) *Journal of Polymer Science Part A – Polymer Chemistry*, **35** (11), 2221.

21 Wang, X., Wu, Q., Wang, N. and Lin, X.F. (2005) *Carbohydrate Polymers*, **60**, 357.

22 Choisnard, L., Gèze, A., Yaméogo, B.G.J., Putaux, J.L. and Wouessidjewea, D. (2007) *International Journal of Pharmacy*, **344**, 26.

23 Griebenow, K. and Klibanov, A.M. (1997) *Biotechnology and Bioengineering*, **53** (4), 351.

24 Xu, K., Griebenow, K. and Klibanov, A.M. (1997) *Biotechnology and Bioengineering*, **56** (5), 351.

25 Watanabe, K. and Ueji, S. (2000) *Biotechnology Letters*, **22**, 599.

26 Quiroga, E., Cami, G., Marchese, J. and Barberis, S. (2007) *Biochemical Engineering Journal*, **35**, 198.

27 Ruiz, D.M. and De Castro, R.E. (2007) *Journal of Industrial Microbiology & Biotechnology*, **34**, 111.

28 Ogino, H. and Ishikawa, H. (2001) *Journal of Bioscience and Bioengineering*, **91**, 109.

29 Ogino, H., Gemba, Y., Yutori, Y., Doukyu, N., Ishimi, K. and Ishikawa, H. (2007) *Biotechnology Progress*, **23**, 155.

30 Klibanov, A.M. (1997) *Trends in Biotechnology*, **15** (3), 97.

31 Schmitke, J.L., Stern, L.J. and Klibanov, A.M. (1997) *Proceedings of the National Academy of Sciences of the United States of America*, **94**, 4250.

32 Schmitke, J.L., Stern, L.J. and Klibanov, A.M. (1998) *Proceedings of the National Academy of Sciences of the United States of America*, **95**, 12918.

33 Mansfeld, J., Vriend, G., Dijkstra, B.W., Veltman, O.R., Van den Burg, B., Venema, G., Ulbrich-Hofmann, R. and Eijsink, V.G.H. (1997) *Journal of Biological Chemistry*, **272** (17), 11152.

34 Van den Burg, B., Vriend, G., Veltman, O.R., Venema, G. and Eijsink, V.G.H. (1998) *Proceedings of the National Academy of Sciences of the United States of America*, **95** (5), 2056.

35 Mansfeld, J. and Ulbrich-Hofmann, R. (2006) *Biotechnology and Bioengineering*, **97** (4), 672.

36 Tsuchiyama, S., Doukyu, N., Yasuda, M., Ishimi, K. and Ogino, H. (2007) *Biotechnology Progress*, **23**, 820.

37 Kragl, U., Eckstein, M. and Kaftzik, N. (2007) *Current Opinion in Biotechnology*, **13**, 565.

38 Sheldon, R.A., Lau, R.M., Sorgedrager, M.J., van Rantwijk, F. and Seddon, K. (2002) *Green Chemistry*, **4**, 147.

39 Swatloski, R.P., Spear, S.K., Holbrey, J.D. and Rogers, R.D. (2002) *Journal of the American Chemical Society*, **124**, 4974.

40 Liu, Q., Janssen, M.H.A., van Rantwijk, F. and Sheldon, R.A. (2005) *Green Chemistry*, **7**, 39.

41 Erbeldinger, M., Mesiano, A.J. and Russell, A.J. (2000) *Biotechnology Progress*, **16**, 1129.

42 Zhao, H., Campbell, S.M., Jackson, L., Song, Z. and Olubajo, O. (2006) *Tetrahedron Asymmetry*, **17**, 377.

43 Noritomi, H., Nishida, S. and Kato, S. (2007) *Biotechnology Letters*, **29**, 1509.

44 Lozano, P., De Diego, T., Carrié, D., Vaultier, M. and Iborra, J.L. (2003) *Journal of Molecular Catalysis B: Enzymatic*, **21**, 9.

5
Importance of Enzyme Formulation for the Activity and Enantioselectivity of Lipases in Organic Solvents

Francesco Secundo

5.1
Introduction

Lipases are of remarkable practical interest since they have been used in numerous biocatalytic applications, such as kinetic resolution of alcohols and carboxyl esters (both in water and in non-aqueous media) [1], regioselective acylations of polyhydroxylated compounds, and the preparation of enantiopure amino acids and amides [2, 3]. Moreover, lipases are stable in organic solvents, do not require cofactors, possess broad substrate specificity, and exhibit, in general, a high enantioselectivity. All these features have contributed to make lipases the class of enzyme with the highest number of biocatalytic applications carried out in neat organic solvents.

The use of organic solvents is especially advantageous in transforming substrates that are unstable or poorly soluble in water. Furthermore, in the absence of water, the synthesis of ester and amide bonds by hydrolases (mainly lipases and proteases) can be favored over hydrolysis and many side reactions that are water dependent can be prevented. By varying the organic solvent it is also possible to control the substrate specificity and the regio- and enantioselectivity of a given enzyme. However, although enzymes in organic media show numerous advantages, their catalytic efficiency is, in most cases, orders of magnitude lower than in aqueous systems. This behavior can be ascribed to different causes such as high saturating substrate concentrations, diffusional limitations, low stabilization of the transition state of the enzyme–substrate intermediate, restricted protein flexibility, and aggregation and distortions of the enzyme molecules caused by dehydration, which become irreversible in anhydrous solvents [4].

The present chapter briefly examines and discusses the effects of some additives, namely lyoprotectants (sugars, crown ethers, and methoxypoly(ethylene glycol)), on the activity and enantioselectivity of several lipases. A possible rationale for the increment usually observed when using lyoprotectants, on the basis of conformational data obtained by means of Fourier-transform infrared (FT-IR) spectroscopy is also suggested.

Modern Biocatalysis: Stereoselective and Environmentally Friendly Reactions.
Edited by Wolf-Dieter Fessner and Thorleif Anthonsen

ISBN: 978-3-527-32071-4

5.2
Lipase Formulations and their Activity and Enantioselectivity in Neat Organic Solvent

It is generally stated that biocatalysis in organic solvents refers to those systems in which the enzymes are suspended (or, sometimes, dissolved) in neat organic solvents in the presence of enough aqueous buffer (less than 5%) to ensure enzymatic activity. However, in the case of hydrolases water is also a substrate and it might be critical to find the water activity (a_w) value to which the synthetic reaction (e.g. ester formation) can be optimized. Valivety *et al.* [5] found that, in some cases, the activity of *Candida rugosa* lipase immobilized on different supports showed the same activity profile versus a_w but a different absolute rate. With lipase from *Burkholderia cepacia* (lipase BC), previously known as lipase from *Pseudomonas cepacia*, and *Candida antarctica* lipase B (CALB) it was found that the enzyme activity profile versus a_w and even more the specific activity were dependent on the way the enzyme was freeze dried or immobilized [6, 7]. A comparison of the transesterification activity of different forms of lipase BC or CALB can be observed in Tables 5.1 and 5.2, respectively.

It can be noted that the way in which the enzyme is prepared in the dry form for catalysis in organic solvent is responsible for striking differences (up to two orders of magnitude) in the enzyme-specific activity. Furthermore, it is worth mentioning that the transesterification activity of lipase from *B. cepacia* entrapped in sol gel (sol gel-AK-lipase BC) was 83% of the activity in water measured using tributyrin as a substrate [6]. Analogously, in the case of CALB lyophilized with methoxypoly(ethylene glycol) (CALB + PEG) the activity was 51% of the activity in water in the hydrolysis of vinyl acetate [7]. It is important to note that, for both

Table 5.1 Transesterification activity in carbon tetrachloride ($a_w < 0.1$) of the various lipase BC formulations.

Enzyme form	Relative transesterification activity[a,b]
Purified and lyophilized with PEG[c]	7.8
Purified and covalently linked to PEG[c]	3.5
Crude[d]	1.0
CLEC of lipase BC[e]	1.4
Sol gel-AK-lipase BC[f]	76.4

a Relative to crude lipase BC taken as 1.
b The transesterification between 1-octanol (0.19 M) and vinyl butyrate (0.79 M) to give 1-octyl butyrate and acetaldehyde was used as a model reaction. An amount of enzyme form containing 10 μg of protein was used in a reaction volume of 1 ml. The activity of crude lipase BC was 0.5 μmol/min × 10 μg of protein.
c PEG, methoxy poly(ethylene glycol) (molecular mass 5000 Da).
d Lipase BC as commercialized by AMANO company (commercial name lipase PS).
e CLEC, cross-linked enzyme crystals. Sample obtained as a kind gift of Altus.
f Lipase BC entrapped in sol gel-AK. Sample purchased from Fluka.

lipase BC and CALB, the comparison of activity between the various enzyme forms and that with the enzyme in water was performed referring to the same amount of lipase. A 6.4-fold increase in the transesterification activity was also observed in the case of wild type *Bacillus subtilis* lipase A (wtBSLA) when it was previously lyophilized with PEG (Table 5.3) [8].

Some carbohydrates are other additives that have been used by different research groups for increasing the catalytic activity of several hydrolases in organic solvents. In particular, Kanerva and Sundholm [9] with lipase BC and Pu *et al.* [10] with lipases from *C. rugosa* and *Pseudomonas* sp. found that co-lyophilization of these

Table 5.2 Transesterification activity in toluene ($a_w < 0.1$) of the various CALB formulations.

Enzyme form	Relative transesterification activity[a,b]
Crude CALB[c]	0.05
Purified CALB[d]	1.0
CALB + PEG[e]	13.0
CALB + OA[f]	6.7
Novozym 435[g]	7.3

a Relative to purified CALB taken as 1.

b The transesterification between 1-octanol (0.19 M) and vinyl acetate (1.1 M) to give 1-octyl acetate and acetaldehyde was used as a model reaction. An amount of enzyme form containing 10 μg of protein was used in a reaction volume of 1 ml. The activity of purified CALB was 6 μmol/min × 10μg of protein.

c Crude CALB was a gift from Novo-Nordisk (experimental product SP 525).

d Purified CALB was purified from crude CALB as reported in [7].

e Purified CALB was lyophilized with PEG (molecular mass 5000 Da).

f Purified CALB was lyophilized with oleic acid (OA).

g Novozym 435 was purchased from Novo-Nordisk and it is CALB immobilized on macroporous acrylic resin.

Table 5.3 Transesterification rate in petroleum ether of wtBSLA and BSLA lid mutants.

Enzyme	Rate (nmol/min)[a]
BSLA[b]	122 (19)
HPlip[c]	14 (3)
AXElip[d]	16 (3)

a Rate of transesterification at $a_w = 0.11$ of sulcatol (0.013 M) with vinyl acetate (0.11 M) as acyl donor. Forty milligrams of enzyme were lyophilized with 5 mg of PEG and added to a final reaction volume of 1 ml. The reaction mixture was shaken at 150 rpm and at 25 °C. The reaction progress was monitored by GLC using the conditions described in [8]. The rate obtained with the same amount of enzyme lyophilized without PEG is given in parentheses.

b wtBSLA.

c, d Mutated BSLA in which the lid of acetylxylanesterase from *Penicillium purpurogenum* ([c]) or the lid of the human pancreatic lipase ([d]) was inserted. The lids were chosen from these latter two enzymes because of their structural homology to BSLA. Moreover, the lids were engineered into the wild-type enzyme without disturbing the core of the α/β-hydrolase fold [8].

enzymes with sucrose had a beneficial effect on the catalytic activity in hexane. An increment in the activity was also observed for other serine hydrolases. Adlercreutz [11] reported an increment in the alcoholysis rate when α-chymotrypsin was adsorbed on Celite in the presence of sorbitol. De Paz *et al.* [12] also observed an increase in activity of subtilisin co-lyophilized with trehalose and sucrose. Analogously, an increment in the transesterification activity in dioxane of subtilisin lyophilized or adsorbed on Celite in the presence of sorbitol was found [13]. The activity and enantioselectivity of lipase BC lyophilized in the presence of different kinds of sugars has also been studied.

Figure 5.1 shows the activity of lipase BC as a function of the sugar/lipase ratio (w/w). It can be observed that lyophilizing the lipase in the presence of sugars and

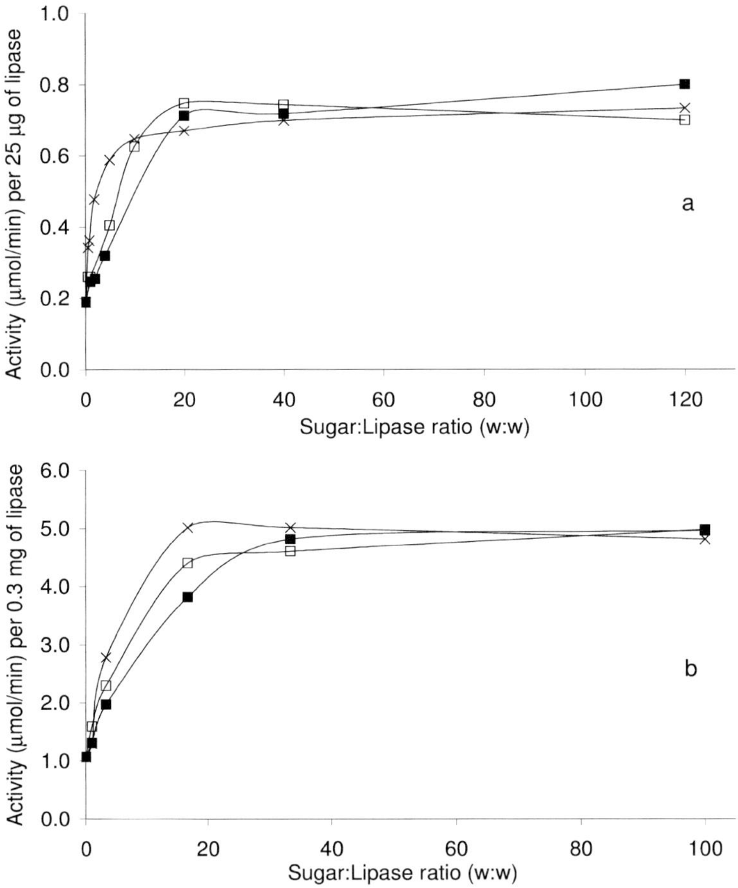

Figure 5.1 Transesterification activity of (a) 0.025 or (b) 0.3 mg of lipase BC co-lyophilized with different amounts of sucrose (□), trehalose (×) and mannitol (■). Activity was tested in toluene by measuring the rate of 1-octyl acetate formation starting from 1-octanol (0.19 M) and vinyl acetate (1.1 M) as substrates. The reaction volume was 1 ml. All the reagents, the solvent, and the enzyme were equilibrated at a water activity of 0.33.

Table 5.4 Effects of sugars on the enantiomeric ratio (*E*) and transesterification rate of lipase BC with 6-methyl-5-hepten-2-ol as substrate and toluene as the reaction medium.

Sugar	E^a	Rate^a (nmol/min)	E^b	Rate^b (nmol/min)
None	14	14	21	162
Sucrose	39	89	32	533
Trehalose	26	95	33	235
Mannitol	31	151	37	491

a Data obtained with 0.025 mg lipase BC lyophilized with 0.5 mg of sugar. **b** Data obtained with 0.1 mg lipase BC lyophilized with 1 mg of sugar. For reaction conditions see [14].

with a sugar/lipase ratio ≥20 it was possible to increase the transesterification activity of lipase BC up to 4.7-fold. Analogously, lipase enantioselectivity, expressed as the enantiomeric ratio, increased up to 2.8-fold in the presence of sugars (Table 5.4) [14].

Several research groups using crown ethers as additives in the lyophilization buffer [15–19] observed a beneficial effect on the activity and enantioselectivity of hydrolases in organic solvents. Conditions in which the use of 18-crown-6 has a beneficial effect on lipase activity in organic solvent were also found. In particular, the transesterification activity in toluene increased to 2.5- and 1.4-fold for lipases BC and CALB, respectively (Figure 5.2a and b). Interestingly, 18-crown-6 increases the activity in spite of the fact that, in water, it decreases lipase activity. Indeed, in water with 4% (w/w) of additive, the activity left was about 12% (CALB) and 0.25% (lipase BC) of that displayed by the enzymes in the absence of additive. This might also explain why the activity increase was obtained only using 18-crown-6/lipase molar ratios ≤100, while higher molar ratios caused a reduction in enzyme activity to values lower than that of the enzyme lyophilized without additive. Instead, the possibility that this decrease was due to the formation of a complex between the additive and the substrate (which would cause a decrease in the apparent substrate concentration) was excluded because the addition of 18-crown-6 to the reaction after enzyme lyophilization did not cause significant variation in the enzyme activity (Figure 5.2a and b). It can be concluded that the loss of activity is due to the interaction of 18-crown-6 with the enzyme in water and during lyophilization. With regard to the use of other organic solvents, it was found that the increase in transesterification activity in 1,4-dioxane as reaction medium was less pronounced for lipase BC (1.7-fold) and for CALB (1.5-fold).

The effect of 18-crown-6 on the enantioselectivity depends on the lipase. It was noted that, with lipase BC, the *E*-value (about 33) was scarcely affected by co-lyophilization with the crown ether (inset in Figure 5.2a). This finding is in agreement with those reported by Mine *et al.* [17], who did not observe variation in the *E*-value for lipase BC lyophilized with 18-crown-6 in the transesterification between 2,2-dimethyl-1,3-dioxolane-4-methanol and vinyl butyrate. Contrary to lipase BC, CALB co-lyophilized with 18-crown-6 shows an increase in its *E*-value (inset Figure

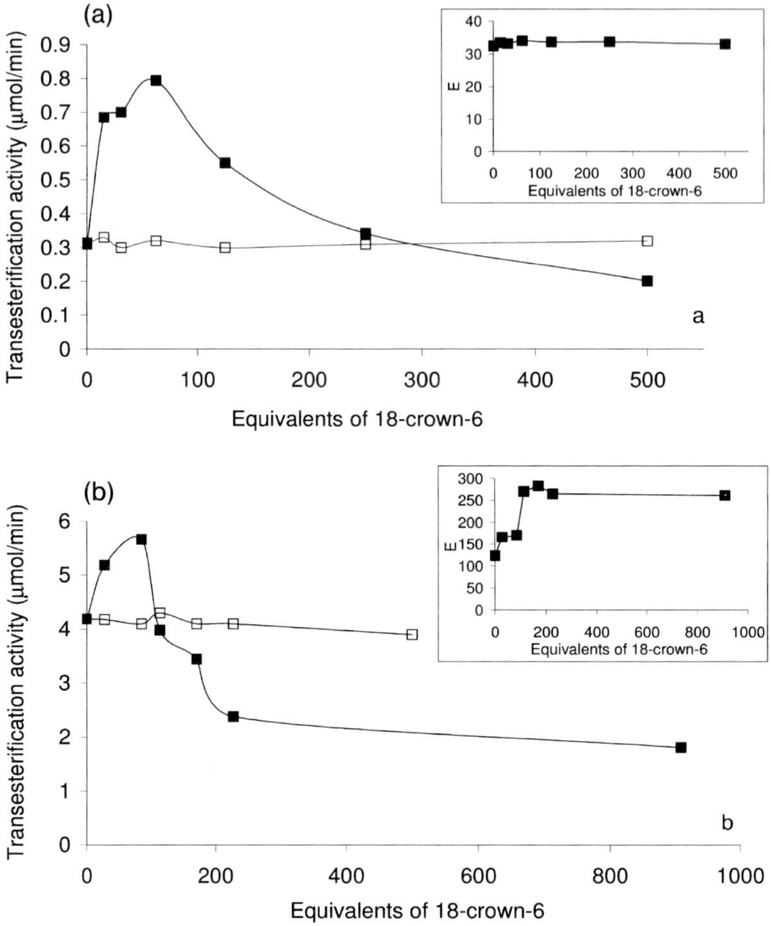

Figure 5.2 Transesterification activity of 0.1 mg of (a) lipase BC and (b) CALB in toluene as a function of molar equivalents of 18-crown-6 added to the enzyme in aqueous solution before lyophilization (■) or to the reaction medium (□). The reaction was carried out with sulcatol as nucleophile and vinyl acetate as acyl donor in a reaction volume of 1 ml. Inset: effect of the crown ether on lipase BC enantioselectivity. The data were averages of at least three determinations.

5.3) from about 120 up to 280. The different behaviors of the two enzymes might be due to different small conformational changes in lipase BC and CALB induced by the additive during lyophilization and/or to its different interaction with the catalytic active site of the two enzymes. It is worth pointing out that the E-value of CALB increases as a function of 18-crown-6 up to about 200 equivalents of the additive (inset in Figure 5.2b). Instead the increase of activity was observed up to about 100 equivalents of additive and then it decreased to about 200 equivalents, where it remained almost constant at around 50% of the activity in the absence of the crown ether (Figure 5.2b). These data suggest that higher concentrations of

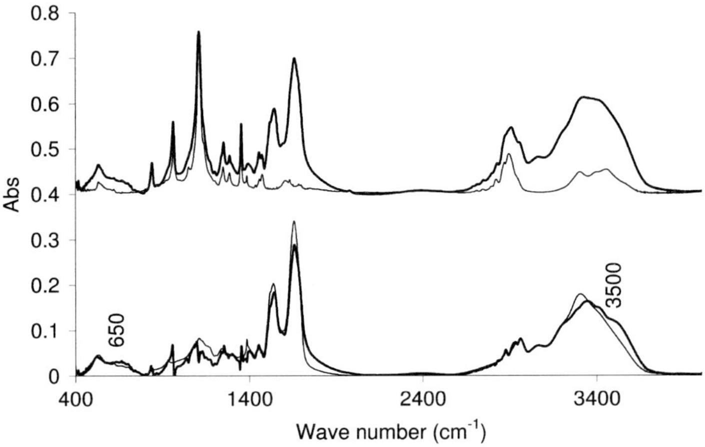

Figure 5.3 Upper part: FT-IR spectra of lipase BC lyophilized with 62 molar equivalents of 18-crown-6 (thick line) and of lyophilized 18-crown-6 only (thin line). Lower part: lipase BC lyophilized without 18-crown-6 (thin line) is compared with the difference (thick line) between the spectra reported in the upper part (lipase BC lyophilized with 18-crown-6 minus 18-crown-6). The spectra are offset along the *y*-axis for clarity.

the additive cause the formation of a CALB intermediate that is less active but more enantioselective [20].

5.3 Why do Additives Affect the Activity and Enantioselectivity of Lipases in Organic Solvent?

In order to shed light on the mechanisms by which a given additive can increase the enzyme activity in organic solvent, we have been mainly focusing on how the additive affects the enzyme conformation and the amount of water adsorbed on the enzyme after lyophilization. We monitored these effects on the protein by means of FT-IR spectroscopy, a spectroscopic technique that allows conformational comparisons between enzyme samples in different physical states and in the presence of additives. In particular, the band that proteins show in the 1600–1700/cm region (amide I band) is composed of several overlapping components that can be assigned to different secondary structure conformations [21].

In the case of sugars, which are largely employed in the pharmaceutical field as lyoprotectants for the formulation of proteins used as drugs, it was found that FT-IR spectra of lipase BC co-lyophilized with these additives are more similar to that of lipase BC lyophilized without sugars than to that of the enzyme in water (the comparison of spectra was done on the basis of correlation coefficients between the infrared amide I band) [14]. An analysis of the relative area of the components that form the amide I band is shown in Table 5.5. In the case of lipase BC lyophilized with sugars, no correlation was found between activity and the relative intensity of the amide I band at 1654–1657/cm, which is usually assigned to α-helix secondary structures (Table 5.5). This band resulted always lower in the lyophilized sample compared to that in water. The fact that this band was higher in the case of lipase BC lyophilized with trehalose compared to that of the sample lyophilized with the other sugars could not be related to a more evident increase in activity. Instead, the sum of the relative area of the bands at around 1680–1683/cm with that at 1693–1695/cm (both assigned to intermolecular β-sheets with a total relative area not higher than 20.8%) is always lower than that obtained with lipase BC lyophilized without sugars (total area 24.0%) (Table 5.5). An analogous trend was observed for the band at around 1614/cm, which was lower for the sample of lipase BC lyophilized in the presence of sugars than in the absence of additive. These data suggest that the presence of sugars reduces the formation of intermolecular β-sheets, a situation that is typical of protein aggregates. Thus, sugars might have a role breaking protein–protein interactions and this would favor the access of substrates to the enzyme catalytic active sites and of water to those enzyme molecules positioned in the internal part of the beads of the enzyme powder. Better access to the enzyme molecules could also be the reason for the increment of enantioselectivity induced by all the sugars tested (Table 5.3). In fact, Rottici *et al.* [22] showed that immobilized CALB (e.g. Novozym 435), in which the substrate should be less hindered in reaching the catalytic active site, has both higher activity and enantioselectivity with respect to other enzyme preparations (e.g. free enzyme).

That additives might play an important role also favoring the hydration of the enzyme was evinced by the comparison of the spectra of lipase BC lyophilized in the presence and in the absence of 18-crown-6 (Figure 5.3).

When observing the bands due to water in the spectra at 3500/cm (O H stretching) and at 650/cm (libration band) [23] in the differential spectrum obtained by subtracting the spectrum of 18-crown-6 from the spectrum of lipase BC lyophilized with 18-crown ether (Figure 5.3, lower part and thick line), it can be noted that their intensities are more pronounced in lipase BC lyophilized with 18-crown ether (0.030 and 0.128 instead of 0.022 and 0.091 Abs, at 650/cm and 3500/cm, respectively) than for lipase BC lyophilized without 18-crown-6. It was concluded that 18-crown-6 favored lipase activity (especially in the case of lipase BC) allowing a better hydration of the enzyme. The activating effect was less pronounced in the case of CALB, because this latter enzyme is *N*-glycosylated and the sugar moieties could themselves be responsible for hydrogen bond formation with CALB or for better hydration.

Another possible role of the additive concerns the peculiar structure of some lipases endowed with a mobile element (named lid) that regulates the entrance to the catalytic active site. It has been suggested that the lower enzyme activity in organic solvent could be due to the fact that, after lyophilization, the enzyme is in a close conformation (less active) [24]. Then, if an additive (e.g. *n*-octyl-β-glucopyranoside) also favors the open conformation after lyophilization, the

Table 5.5 Infrared band positions, band areas, and assignments in the amide I spectral region of lipase BC in aqueous solution or lyophilized in the absence or in the presence of sugars.

Band position (cm^{-1}) and assignment[a]	Band area (%)[a]							
	In water	Without sugar	Sucrose 0.3 mg	Sucrose 15 mg	Trehalose 0.3 mg	Trehalose 15 mg	Mannitol 0.3 mg	Mannitol 15 mg
1614–1620 β-sheets[b]	–	4.6	3.8	3.1	2.9	2.0	6.3	2.8
1625 Hydrated amide C=O groups	6.6	–	–	–	–	–	–	–
1628–1631 β-sheets	–	12.1	8.4	12.3	9.3	8.5	9.5	9.7
1638–1640 β-sheets	21.5	8.3	13.3	14.1	11.8	13.9	11.1	13.6
1645–1649 random coil	–	11.6	7.7	8.5	10.9	5.0	5.0	7.1
1654–1657 α-helix	32.4	18.3	19.5	18.4	26.5	27.8	20.8	20.6
1666–1668 β-turns	20.3	21.1	28.3	24.1	20.1	23.0	26.5	27.2
1680–1683 β-turns/ β-sheets[b]	12.2	15.2	12.1	14.5	12.7	14.4	17.0	12.7
1693–1695 β-sheets[b]	–	8.8 (24.0)[c]	6.9 (19.0)[c]	5.0 (19.5)[c]	5.7 (18.4)[c]	5.3 (19.7)[c]	3.8 (20.8)[c]	6.2 (18.9)[c]
1698 (not-assigned)	7.0	–	–	–	–	–	–	–

a The position and the relative contribution of each band component of the amide I band was determined by curve fitting of the deconvoluted spectra following the procedure of the GRAMS/32 program. For details see [14].
b Intermolecular β-sheets.
c Relative area of the band at 1693–1695 cm^{-1} plus that at 1680–1683 cm^{-1}.

enzyme might be more active in organic solvent [25]. However, this possibility was excluded with PEG because, when used with BSLA (a lipase that does not have the lid), an increase of activity was also observed. Furthermore, mutated BSLA (in which a lid was inserted) freeze dried with PEG showed an increase in activity in organic solvent similar to that of wtBSLA (Table 5.3). Instead, the fact that PEG allows protein dissolution in some organic solvents (e.g. 1,4-dioxane) [26] suggests that, analogously to sugars, it also acts by breaking protein–protein interactions and decreasing diffusional limitations of the substrate to the enzyme catalytic active sites.

5.4 Conclusions

The present chapter has highlighted the way in which an enzyme prepared in a dry form is extremely important for optimal activity and enantioselectivity of lipases in organic solvent. The addition of some additives before (freeze) drying appears beneficial for lipases. It protects the enzyme from deleterious interactions with supports (e.g. inorganic materials) and/or prevents strong protein–protein interactions (e.g. such as those in protein aggregates). These actions appear responsible for better accessibility of water and the substrates to the enzyme molecule resulting in higher enzyme activity. On the basis of the numerous bodies of evidence provided within this chapter, we think that the enzyme formulation represents a factor that cannot be eluded in a biocatalytic process carried out with a lipase in organic solvent that aims to scale up to industrial level.

References

1 Drauz, K. and Waldmann, H. (2002) *Enzyme Catalysis in Organic Synthesis: A Comprehensive Handbook*, 2nd edn, **I–III**, Wiley-VCH Verlag GmbH, Weinheim.

2 Kanerva, L.T., Csomos, P., Sundholm, O., Bernath, G. and Fulop, F. (1996) *Tetrahedron, Asymmetry*, **7**, 1705–1716.

3 Gotor, V. (1999) *Bioorganic & Medicinal Chemistry*, **7**, 2189–2197.

4 Klibanov, A.M. (1997) *Trends in Biotechnology*, **15**, 97–101.

5 Valivety, R.H., Halling, P.J., Peilow, A.D. and Macrae, A.R. (1994) *European Journal of Biochemistry*, **222**, 461–466.

6 Secundo, F., Spadaro, S., Carrea, G. and Overbeeke, P.L.A. (1999) *Biotechnology and Bioengineering*, **62**, 554–561.

7 Secundo, F., Carrea, G., Varinelli, D. and Soregaroli, C. (2001) *Biotechnology and Bioengineering*, **73**, 157–163.

8 Secundo, F., Carrea, G., Tarabiono, C., Gatti-Lafranconi, P., Brocca, S., Lotti, M., Jaeger, K.-E., Puls, M. and Eggert, T. (2006) *Journal of Molecular Catalysis B: Enzymatic*, **39**, 166–170.

9 Kanerva, L.T. and Sundholm, O. (1993) *Journal of the Chemical Society – Perkin Transactions*, **1** (1), 2407–2410.

10 Pu, W., Li-rong, Y. and Jian-ping, W. (2001) *Biotechnology Letters*, **23**, 1429–1433.

11 Adlercreutz, P. (1993) *Biochimica et Biophysica Acta*, **1163**, 144–148.

12 Paz, R. A. De, Dale, D.A., Barnett, C.C., Carpenter, J.F., Gaertner, A.L. and Randolph, T.W. (2002) *Enzyme and Microbial Technology*, **31**, 765–774.

13 Bovara, R., Carrea, G., Gioacchini, A.M., Riva, S. and Secundo, F. (1997) *Biotechnology and Bioengineering*, **54**, 50–57.

14 Secundo, F. and Carrea, G. (2005) *Biotechnology and Bioengineering*, **92**, 438–446.

15 van Unen, D.J., Engbersen, J.F.J. and Reinhoud, D.N. (1998) *Biotechnology and Bioengineering*, **59**, 553–556.

16 Persson, M., Mladenoska, I., Wehtje, E. and Adlercreutz, P. (2002) *Enzyme and Microbial Technology*, **31**, 833–841.

17 Santos, A.M., Vidal, M., Pacheco, Y., Frontera, Y., Baez, C., Ornellas, O., Barletta, G. and Griebenow, K. (2001) *Biotechnology and Bioengineering*, **74**, 295–308.

18 Mine, Y., Fukunaga, K., Itoh, K., Yoshimoto, M., Nakao, K. and Sugimura, Y. (2003) *Journal of Bioscience and Bioengineering*, **95**, 441–447.

19 Tsukube, H., Yamada, T. and Shinoda, S. (2001) *Journal of Heterocyclic Chemistry*, **38**, 1401–1408.

20 Secundo, F., Barletta, G.L., Dumitriu, E. and Carrea, G. (2007) *Biotechnology and Bioengineering*, **97**, 12–18.

21 Byler, D.M. and Susi, H. (1986) *Biopolymers*, **25**, 469–487.

22 Roticci, D., Norin, T. and Hult, K. (2000) *Organic Letters*, **2**, 1373–1376.

23 Grdadolnik, J. and Maréchal, Y. (2001) *Biopolymers*, **62**, 40–53.

24 Louwrier, A., Drtina, G.J. and Klibanov, A.M. (1996) *Biotechnology and Bioengineering*, **50**, 1–5.

25 Gonzalez-Navarro, H., Bano, M.C. and Abad, C. (2001) *Biochemistry*, **40**, 3174–3183.

26 Secundo, F., Carrea, G., Vecchio, G. and Zambianchi, F. (1999) *Biotechnology and Bioengineering*, **64**, 624–629.

6
Direct Esterification with Dry Mycelia of Molds: a (Stereo)selective, Mild and Efficient Method for Obtaining Structurally Diverse Esters

Francesco Molinari, Diego Romano, Raffaella Gandolfi, Lucia Gardossi, Ulf Hanefeld, Attilio Converti and Patrizia Spizzo

6.1
Mycelia and Biotransformations in Organic Media

Enzymatic reactions in organic media have been a major issue in the field of biocatalysis over the last two decades. Carboxylesterases (mostly lipases) have been used in monophasic organic solution under controlled values of water activity (a_w) for catalyzing ester formation: the reaction equilibrium can be shifted towards ester formation by interesterification or transesterification [1]. Direct esterification is often hampered by water formation, which may increase a_w, thus negatively influencing the equilibrium.

Cell-bound enzymes are difficult to purify and are often not very stable outside their natural environment: as a consequence, only few data are available on cell-associated enzymes used as biocatalysts. However, by using dry whole cells, cell-bound enzymes can be directly used as biocatalysts in organic solvents.

Whole cell catalysts do not need immobilization, especially when mycelial micro-organisms are involved, since their morphological structure allows for easy filtration and re-utilization. Carboxylesterases bound to the mycelia of molds have been advantageously employed as biocatalysts in water and/or organic solvents: the first report of the use of fungal mycelia in organic solvent dates back to 1978 [2], followed by the work of Gancet and co-workers [3]. Here this chapter reports on the properties of mycelium-bound fungal carboxylesterases and their applications in flavor ester production as well as in stereoselective esterification.

6.2
Screening and Microbiological Aspects

Different lyophilized micro-organisms, ranging from bacteria (46 strains) to yeasts (42 strains) and molds (15 strains), were screened for the hydrolysis of a number of esters (acetate, butyrate, and caprylate) of geraniol and *n*-hexanol [4].

Modern Biocatalysis: Stereoselective and Environmentally Friendly Reactions.
Edited by Wolf-Dieter Fessner and Thorleif Anthonsen

ISBN: 978-3-527-32071-4

During the screening, various strains of *Aspergillus oryzae* and *Rhizopus oryzae* showed remarkable mycelium-bound activity, being able to hydrolyze all the esters completely with excellent rates. *Aspergillus oryzae* MIM and *R. oryzae* CBS 112.07 were also employed as dry mycelia in the study of the direct esterification of geraniol with butyric acid in organic solvent. This investigation allowed determination of the crucial role played by the growth conditions in inducing extracellular and/or mycelium-bound carboxylesterase activity. It must be underlined that lipases isolated from *Rhizopus delemar, Rhizopus javanicus* and *Rhizopus niveus* have identical amino acid sequences and that lipase from *R. oryzae* differs by two amino acids [5]. *Rhizopus delemar, Rhizopus liquefaciens, R. javanicus* and *R. niveus* are now ascribed to the species *R. oryzae* [6]. Lipases from *R. oryzae* (ROL) are normally secreted extracellularly after cleavage of the prolipase and it has been suggested that the various lipase forms found in these micro-organisms are due to the degree of the proteolytic action involved in the post-translational cleavage, rather than to the involvement of different genes [7].

Optimization of the biotransformation showed that the best conditions for achieving high rates and molar conversions included the use of *n*-hexane and *n*-heptane as solvents, a temperature of 50 °C, an equimolar concentration of the substrates around 50 mM, and a concentration of the lyophilized biocatalyst around 25–30 g/l. As a matter of fact, Table 6.1 shows how the carbon source affects the expressed activity of *A. oryzae* MIM and *R. oryzae* CBS 112.07.

Tween 80 proved to be the best carbon source for supporting the activity of lyophilized mycelia, while the use of more conventional carbon sources (glucose, sucrose, and glycerol) gave mostly extracellular lipolytic activities. On the basis of these data, growth conditions could be optimized for mycelium-bound activity and, consequently, fermentations were scaled-up to 50 l.

Table 6.1 Molar conversion (%) after 24 h of geranyl butyrate starting from 50 mM of geraniol and butyric acid in *n*-hexane at 50 °C using 30 g/l of dry mycelia grown on different carbon sources.

Carbon source	*A. oryzae* MIM	*R. oryzae* CBS 112.07
Malt extract	10	75
Glucose	<5	10
Glycerol	<5	20
Sucrose	<5	15
Oleic acid	<5	15
Olive oil	35	90
Soybean oil	20	60
Triacetin	35	45
Tween 80	>95	>95
Glucose + Tween 80	<5	<5
Oleic acid + Tween 80	70	10

The lipolytic activity associated with the mycelia was stable for more than 8 months when the lyophilized biocatalyst was maintained under dry conditions over P_2O_5.

The carboxylesterases involved in the esterification of geraniol and butyric acid have been partially purified: they are tightly membrane bound and only the use of a surfactant such as 3-[(3-cholamidopropyl)dimethylammonium]-1-propanesulfonate (CHAPS) allowed their removal from cellular membranes. From a biocatalytic point of view, this situation can be seen as if the enzymes were 'immobilized' into a hydrophobic carrier.

6.3 Production of Acetates

Many acetate esters (such as those of isoamyl, benzyl, citronellyl, and geranyl alcohols) are components of natural flavors. They can be obtained by lipase-catalyzed esterification in organic solvents, but the major problem with enzymatic acetylations is deactivation of lipases by acetic acid [8, 9]. Most of the lipase-catalyzed syntheses of esters have been carried out by transesterification to avoid free acid toxicity and water formation. Claon and Akoh [10] found that immobilized lipases from *Candida antarctica* promote highly effective direct esterification of geraniol and citronellol with acetic acid.

Dry mycelia of *A. oryzae* MIM efficiently catalyzed the esterification between free acetic acid and primary alcohols in organic solvent [11] (Table 6.2).

The most relevant features of this biotransformation are the following.

- In most of the cases very high molar conversions were observed although no means of withdrawing water formed during the direct esterification were adopted

Table 6.2 Molar conversion after 12 h of the synthesis of different acetates using a starting concentration of alcohol and acetic acid of 65 mM in *n*-heptane at 50 °C and 30 g/l of dry mycelia of *A. oryzae* MIM.

Alcohol	Molar conversion (%)
Ethanol	88
n-Propanol	90
n-Butanol	>95
n-Pentanol	>95
n-Hexanol	95
2-Methyl-1-butanol	85
3-Methyl-1-butanol	80
Geraniol	>95
Benzyl alcohol	65
2-Phenylethanol	95

(see also the final section of this chapter concerning partition phenomena and esterification reaction equilibria).

- Only 30 g/l of lyophilized mycelia was used, which means that the specific activity of the enzymes involved in the transformation was very high.
- The enzymes involved showed a notable thermostability when suspended in heptane: the mycelia lost only 10–30% of activity after 14 days at temperatures between 30 and 50 °C [11]; this also means that the organic solvent did not remove the enzymes from the mycelia.
- The mycelia can be easily filtered off through macroporous filters.
- Different strains of *R. oryzae* and *A. oryzae* were able to perform efficient esterification under optimized conditions [12].

These observations indicate that direct acetylation of primary alcohols catalyzed by lyophilized mycelia of *A. oryzae* MIM can be a preparative method and continuous processes may be feasible.

Kinetic and thermodynamic studies of geranyl acetate production by direct geraniol acetylation with lyophilized cells of *A. oryzae* MIM were carried out in *n*-heptane [13–15]. Batch tests were performed varying the starting substrates equimolar level from 25 to 150 mM, cell concentration from 5 to 30 g/l, and temperature from 30 to 95 °C. The initial rates at different initial substrate concentrations were measured and an apparent Michaelis constant (K_m) of 62 mM and a k_{cat} value of 0.88 mmol/g/h were estimated by direct fitting of the initial rates against the initial substrate concentrations ≤75 mM [15].

The very high K_m values calculated for both the ethanol and geraniol acetylations, which are consistent with the well-known remarkable increase observed for this parameter in organic solvents, could be ascribed to a competitive inhibition between solvent and substrate for binding at the active site [13, 15].

The progressive increase in the starting product formation rate observed with increasing the temperature up to 80 °C and the successive decrease beyond this value confirmed the occurrence of reversible biocatalyst inactivation. The Arrhenius model was used for estimating the apparent activation enthalpies of the acetylation of geraniol ($\Delta H^{\#}$ = 35 kJ/mol) and the reversible inactivation of the biocatalyst ($\Delta H_i^{\#} = 150$ kJ/mol) [15]. The thermodynamic data were compared with those of ethanol acetylation (Table 6.3).

These thermodynamic data confirmed that the phenomenon responsible for the reversible thermal inactivation of the biocatalyst required higher activation energy for the formation of the transition state than the acetylation reaction. Besides, since the activation enthalpies of ethanol and geraniol acetylations were almost the same, the transition state of the limiting step should scarcely be influenced by steric hindrance. The activation enthalpy of reversible inactivation was much higher using geraniol (150 kJ/mol) rather than ethanol (63 kJ/mol), which is in agreement with the well-known greater inactivation of enzymes in organic solvents by shorter chain alcohols [16, 17].

Table 6.3 Apparent thermodynamic parameters of geraniol and ethanol acetylations by lyophilized cells of *A. oryzae* MIM estimated by the Arrhenius model under different conditions, with reference temperature 50 °C.

	Ethyl acetate			Geranyl acetate		
	$\Delta H^{\#}$ (kJ/mol)	$\Delta S^{\#}$ (kJ mol/K)	r^2	$\Delta H^{\#}$ (kJ/mol)	$\Delta S^{\#}$ (kJ/mol/K)	r^2
Alcohol acetylation	31	–	0.98	35	–	0.94
Reversible inactivation	63	–	0.96	150	–	0.99
Irreversible denaturation	22	–0.29	0.99	28	–0.28	0.99

6.4 Stereoselective Esterifications of Racemic Alcohols

Lipases have been extensively used for the kinetic resolution of racemic alcohols or carboxylic acids in organic solvents. Chiral alcohols are usually reacted with achiral activated esters (such as vinyl, isopropenyl, and trichloroethyl esters) for shifting the equilibrium to the desired products and avoiding problems of reversibility. For the same reasons, chiral acids are often resolved by using acidolysis of esters. In both cases, the overall stereoselectivity is affected by the thermodynamic activity of water (a_w): water favors hydrolytic reactions leading to a decrease in the optical purity of the desired ester. Direct esterifications are therefore difficult to apply since water formed during the reaction may increase the a_w of the system, favors reversibility, and diminishes the overall stereoselectivity.

The resolution of the racemic mixture of 2-octanol and of other secondary alcohols (2-butanol, 2-pentanol, 2-hexanol, and 2-heptanol) by direct esterification in organic solvent was studied by using lyophilized mycelia of *R. oryzae* CBS112.07 as catalysts [18]. The profile of the resolution of (*R,S*)-2-octanol under optimized conditions (1 g/l of alcohol and equimolar butanoic acid as acylating agent in *n*-heptane, 30 g/l of dry biocatalyst, and 30 °C) is shown in Figure 6.1.

Maximum molar conversion (42–43%) was reached after 7 days with a corresponding enantiomeric excess (e.e.) of the *R*-product in the range of 97–98%: the optical purity of the ester slightly diminished (95–96%) after 21 days, while the molar conversion remained the same. This stereochemical behavior indicates that the esterification is not reversible under these conditions, despite formation of water during the esterification.

A very different situation was observed when the resolution of racemic 1,2-*O*-isopropylideneglycerol (IPG or solketal) was studied [19]. In this case, the kinetics of the esterification with butyric acid catalyzed by *A. oryzae* MIM and *R. oryzae* CBS 112.07 were investigated by carrying out independent batch tests on the commercially available *R*- and *S*-IPG (Table 6.4).

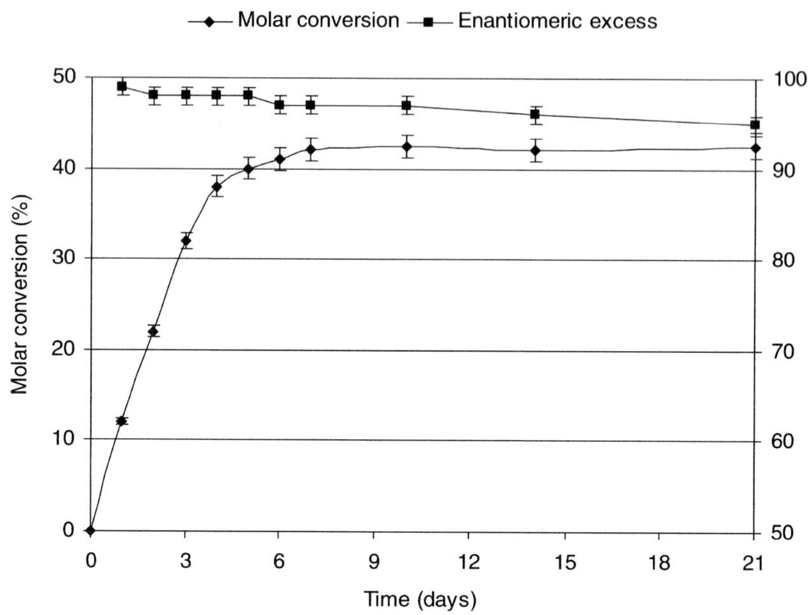

Figure 6.1 Time course of (*R*)-2-octylbutanoate production catalyzed by lyophilized cells (30 g/l) of *R. oryzae* CBS 112.07 in *n*-heptane at 30 °C. Alcohol concentration = 1.0 g/l, with an equimolar concentration of butanoic acid as the acylating agent.

Table 6.4 Kinetic parameters of *R*- and *S*-IPG esterification with butyric acid catalyzed by *R. oryzae* CBS 112.07 and *A. oryzae* MIM.

	R. oryzae		***A. oryzae***	
	***R*-IPG**	***S*-IPG**	***R*-IPG**	***S*-IPG**
K_m (mM)	87.2	284.4	125.2	3277.0
k_{cat} (mmol$_P$/g/h)	1.31	1.26	4.29	14.45
Enantiomeric ratio	3.4		8.0	

Aspergillus oryzae MIM was more enantioselective than *R. oryzae* CBS 112.07: therefore, it was used for studying the time course of the esterification with butyric acid of (*R,S*)-IPG under optimal conditions (Figure 6.2).

The esterification of IPG was largely reversible, with the e.e. of the substrate reaching a maximum after 1 h and progressively diminishing as the reaction progressed further, while the best resolution was obtained after 15 min (e.e. of the product = 68–70%). Biotransformations were performed at different initial water

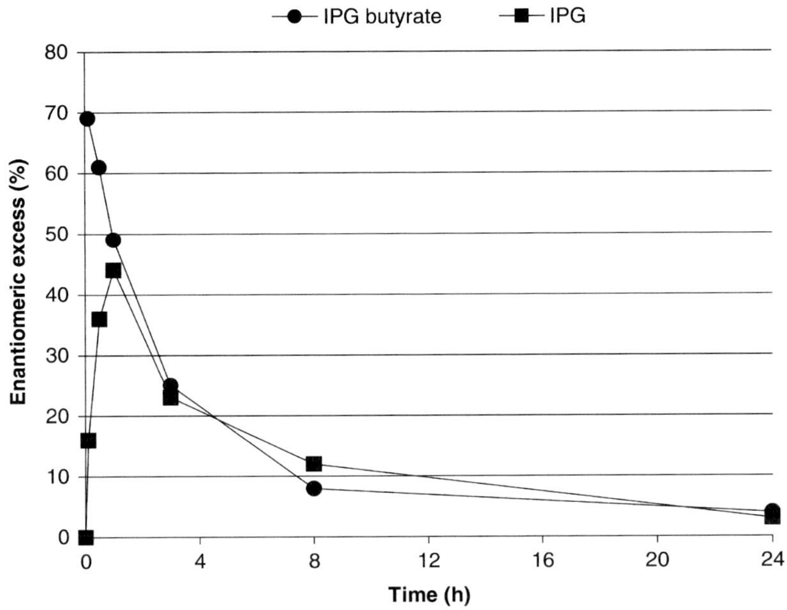

Figure 6.2 Profile of the e.e. of the esterification of (*R*,*S*)-IPG with butyric acid under optimized conditions: $T = 30\,°C$, the concentration of the mycelia = 30 g/l, the concentration of (*R*, *S*)-IPG = 3 g/l, an equimolar concentration of butyric acid, solvent *n*-heptane, $a_w = 0.75$, and biocatalyst lyophilized cells of *A. oryzae* MIM.

activities, ranging from 0.15 to 0.95. No significant differences in the initial rates and the overall profile of the kinetic resolution were observed, suggesting that even at very low a_w values mycelia display such a high activity that they establish a fast equilibrium.

6.5 Stereoselective Esterifications of Racemic Carboxylic Acids

The resolution of racemic 2-arylpropionic acids by enzymatic esterification in organic solvents is a straightforward way of obtaining enantiomerically pure products that can be used as biologically active compounds or as chiral resolving agents [20]. Different chiral aromatic acids (2-phenylpropanoic acid, naproxen, ibuprofen, flurbiprofen, and tropic acid) were tested using various strains of *A. oryzae* and *R. oryzae* as lyophilized catalysts in organic solvents. Racemic 2-arylpropionic acids were esterified with good enantioselectivity, while tropic acid did not react under any condition tested.

The enantioselective esterification of (*R,S*)-2-phenylpropanoic acid in organic solvent was studied by using dry mycelia of *A. oryzae* MIM and *R. oryzae* CBS

112.07. The two micro-organisms were enantiocomplementary since *A. oryzae* MIM showed preference for the formation of the *S*-ester, while *R. oryzae* CBS 112.07 preferentially gave the *R*-ester. As shown in Table 6.5, the highest enantioselectivity was obtained with both strains when the esterification was carried out with ethanol: a suitable combination of solvent and temperature allows for high enantioselectivity.

Flurbiprofen is a cyclooxygenase (COX)-inhibiting non-steroidal anti-inflammatory drug (NSAID). The COX-inhibiting activity resides primarily in the (*S*)-enantiomer whereas the (*R*)-enantiomer has scarce COX activity. (*R*)-Flurbiprofen has been found to inhibit tumor growth in various animal models. *In vitro* experiments have shown that this effect is based on the induction of a cell cycle block and apoptosis [21].

Mycelia were used at 10 g/l (12.3×10^{-6} U/ml, *p*-nitrophenyl palmitate hydrolysis) with a 50 mM concentration of (*R,S*)-flurbiprofen and 27.5 mM of 1-octanol in toluene (Scheme 6.1). Both intact mycelia and lysate of the mycelia were tested

Table 6.5 Esterification of 2-phenylpropanoic acid with ethanol catalyzed by dry mycelia of *R. oryzae*CBS 112.07 at different temperatures: molar conversion and e.e. of the (*R*)-ester after 6 days.

Solvent	Temperature (°C)	Molar conversion (%)	e.e. (%)	*E*
Heptane	20	9	>98R	>100
	30	16	>98R	>100
	40	24	95R	52
Pentadecane	20	8	>98R	>100
	30	15	>98R	>100
	40	18	97R	80

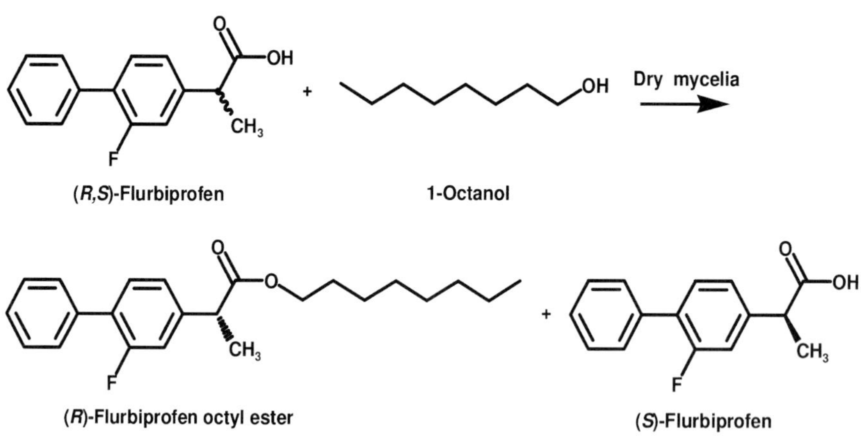

Scheme 6.1 Enantioselective esterification of (*R,S*)-flurbiprofen with 1-octanol.

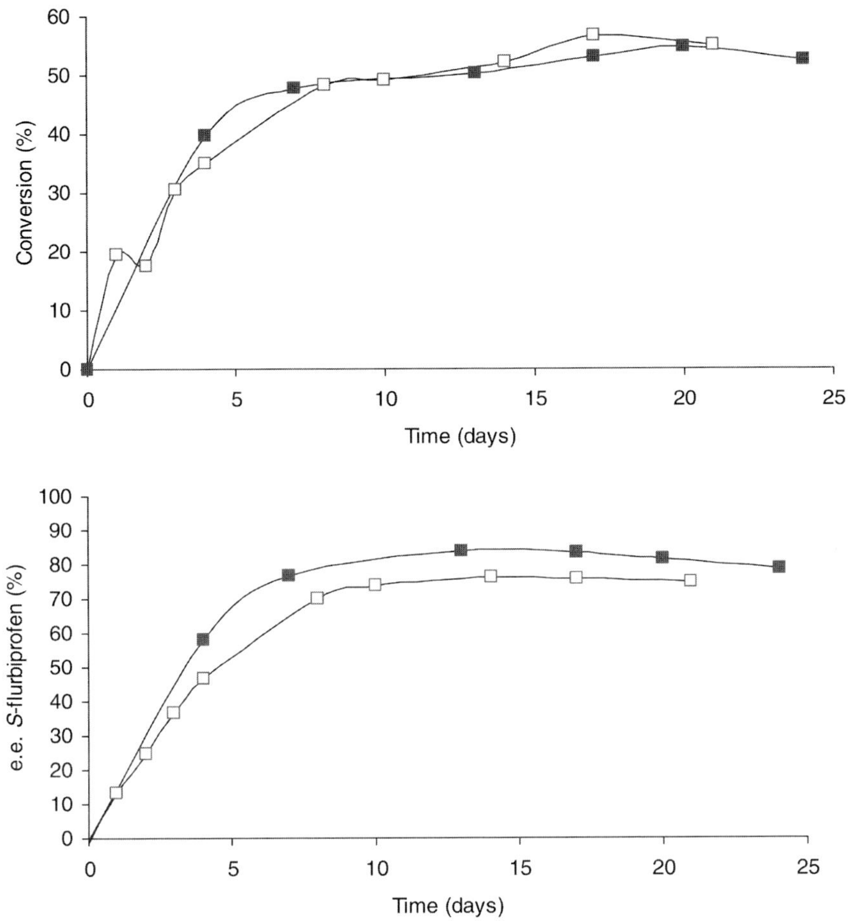

Figure 6.3 Molar conversion and enantioselectivity of the esterification of (*R*,*S*)-flurbiprofen with 1-octanol catalyzed by dry mycelia (10 g/l) of *A. oryzae* MIM in toluene at 50 °C. Empty squares, lysate of mycelia; black squares, intact mycelia.

(Figure 6.3). No relevant variations were observed in terms of the reaction kinetics, whereas a slight decrease in the maximum enantioselectivity was detected in the lysate.

At 50 °C the maximum conversion (about 55%) was achieved with 84% e.e. of the unreacted (*S*)-flurbiprofen. It is noteworthy that molar conversion was complete, although during the direct esterification some water was produced.

By removing biomass by centrifugation and selectively extracting the acid from the organic solvent with a saturated solution of $NaHCO_3$, recovery of the unreacted flurbiprofen and its separation from the ester were easily accomplished.

The results obtained with dry mycelia represent a significant improvement as compared to the data obtained using two commercial preparations of lipase B from *Candida antarctica* (CAlB) as biocatalysts [22, 23].

6.6
Partition Phenomena and Equilibrium of Esterification Reactions

The high molar conversion obtainable in esterifications catalyzed by dry mycelia encouraged us to investigate the partition of water in these heterogeneous systems further. The objective was to verify whether mycelia were able to affect the thermodynamic equilibrium of the reactions by modifying the partitioning of the water formed during the esterification. The synthesis of two esters by lyophilized mycelia of *R. oryzae* CBS 112.07 was studied and the results were compared with those obtained with a commercial immobilized CALB (Novozym 435).

Table 6.6 reports on the molar conversions of the esterifications of geraniol with hexanoic acid and of cinnamyl alcohol with butyric acid obtained using these two biocatalysts under different conditions. In order to exclude possible adsorption of reagents and/or products on the biocatalysts, a procedure of extensive rinsing of the biocatalysts was applied and the liquid phases were analyzed by high-performance liquid chromatography. No appreciable presence of products nor reagents was observed. The effective achievement of the equilibrium was confirmed in all cases by the addition of further fresh biocatalyst after the reactions stopped.

In all cases the conversions achieved by using *R. oryzae* CBS 112.07 were slightly higher, although both systems led to >90% conversions.

In order to understand whether the observed differences were ascribable to any effect of the mycelia on the reaction equilibrium, a series of measurements was aimed at monitoring the variations in the water vapour pressure in the gas phase, expressed as relative humidity (RH). A hygrometer specifically conceived for providing fast measurements in systems involving organic solvents was used for this.

Table 6.6 Molar conversions at equilibrium in the esterifications catalyzed by *R. oryzae* CBS 112.07 and Novozym 435®.

Solvent	Temperature (°C)	Molar conversion at equilibrium (%)	
		***R. oryzae* CBS 112.07**	**Novozym 435®**
Esterification of geraniol with hexanoic acid			
Toluene	50	93	91
Hexane	30	97	90
Esterification of cinnamyl alcohol with butyric acid			
Toluene	50	95	91

It must be underlined that, in those cases where the system had reached an apparent equilibrium (constant reading for at least 1 h), such values were considered as an acceptable approximation of the a_w value of the system. Firstly, the water content of the two dry biocatalysts was evaluated and it was found that 11% of the water was still present in the mycelia after lyophilization (weight variation after 6 h in an oven at 110 °C). The two biocatalysts were then equilibrated in toluene for 24 h and the amounts were chosen so as to obtain the same water content in both systems. The results of Table 6.7 indicate that the mycelia led to a lower water activity, hence suggesting that water might partition inside the mycelia.

In order to verify whether water produced during the esterification was released in the bulk solvent or sequestrated inside the cell, samples were prepared where the mycelia wre incubated in toluene in the presence of amounts of water equal to the amount that would be formed in corresponding 50 and 100% of molar conversion of the esterifications studied. The variation in RH in the gas phase was measured for 7 days as reported in Figure 6.4a and b. Moreover, to exclude any possible interference of the procedure in opening the vials upon the insertion of the hygrometer for the reading (empty squares in Figure 6.4a and b), similar experiments were performed where the measurements were done only at the start of the incubation and then after 1 week.

A long-lasting process of water partitioning and equilibration leads to values of RH of about 0.3, whereas the theoretical water activity at the end of the reaction for 100% molar conversion would be 0.12 (as calculated by the UNIFAC method). Therefore, these observations suggest that the mycelia contribute marginally to the water take-up. Rather than sequestrating water inside the cell wall, it seems that the mycelia provide a micro-environment where the water produced during the esterification is promptly removed. The profiles of Figures 6.4 and 6.5 indicate that the water formed during the esterification is promptly released into the bulk solvent (RH of about 0.75). The esterification of butyric acid with three different alcohols was studied: cinnamyl alcohol, 1-phenyl-1-butanol, and *Z*-L-Ser-OBz (Figure 6.5).

Table 6.7 Initial water content of the dry biocatalysts and water activity measured in a closed system obtained after incubating 60 mg of mycelia and 200 mg of Novozym 435® for 24 h in 6 ml of toluene.

Biocatalyst	Water content of the dry biocatalyst (%, w/w)[a]	a_w in toluene[b]
R. oryzae CBS 112.07	11	0.15
Novozym 435®	3	0.22

a Difference in weight between the wet and the dried sample after drying samples at 110 °C for 6 h.

b Measured using hygrometer DARAI (Trieste, Italy).

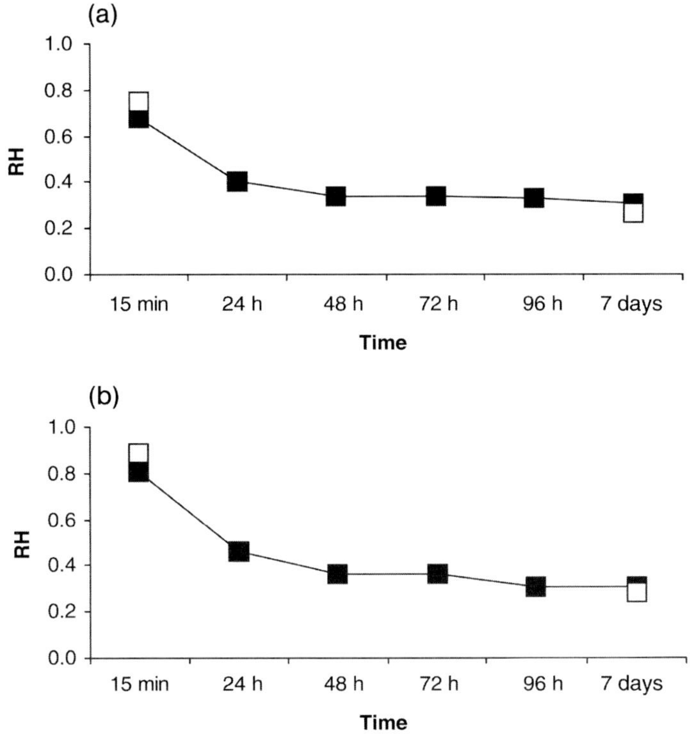

Figure 6.4 Variation in the RH in systems constituted of toluene (2 ml), mycelia (10 g/l), and water corresponding to (a) 50% and (b) 100% molar conversion of direct esterifications performed starting from equimolar substrate concentrations of 25 mM and 50 mM, respectively. Empty squares indicate the values measured in identical experiments where no intermediate reading was performed.

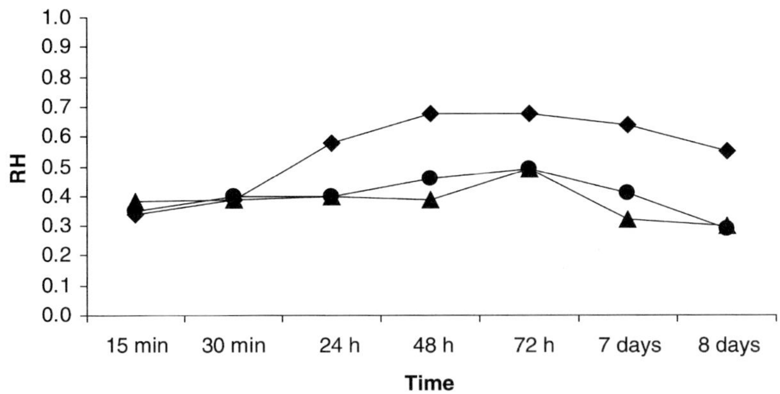

Figure 6.5 Variation in the RH measured during the esterification reaction of butyric acid with different alcohols: cinnamyl alcohol (◆), 1-phenyl-1-butanol (●), and Z-L-Ser-OBzl (▲). Experimental conditions: 2 ml of toluene, 20 mg of lyophilized mycelia, and 50 mM equimolar substrates in a 6 ml sealed vial.

After 7 days the conversion was nearly quantitative only in the case of cinnamyl alcohol, whereas in the other two cases the conversion was below 5%. By monitoring the variation in the RH in the gas phase of the three systems it was observed that a significant increase in the RH values was associated only with the reaction corresponding to cinnamyl alcohol: the RH value then decreased with equilibration of the system, which appeared to be very slow.

6.7 Conclusions

Dry mycelia of molds can be effectively used for (stereo) selective esterification of different alcohols and carboxylic acids in organic solvent with notable advantages, such as the following.

- High stability in organic solvents and high resistance to the inactivation due to free carboxylic acids, including acetic acid.
- High molar conversions also enabled by favorable partition of water.
- Good enantioselectivity also permitted by favorable partition of water.
- Easy set-up of continuous membrane-bioreactor.

References

1 Faber, K. (2004) *Biotransformations in Organic Chemistry*, 5th edn, Springer Verlag, Berlin.
2 Bell, G., Blain, J.A., Paterson, J.D.E., Shaw, C.E.L. and Todd, R.J. (1978) *FEMS Microbiology Letters*, **3**, 223–228.
3 Gancet, C. and Guignard, C. (1986) *Biocatalysis in Organic Media*, Elsevier, Amsterdam.
4 Molinari, F., Marianelli, G. and Aragozzini, F. (1995) *Applied Microbiology and Biotechnology*, **43**, 967–973.
5 Bornscheuer, U.T. and Kazlauskas, R.J. (1999) *Hydrolases in Organic Synthesis*, Wiley-VCH Verlag GmbH, Weinheim.
6 Schipper, M.A.A. and Stalpers, J.A. (1984) *Studies in Micology*, CBS, Baar.
7 Beer, H.D., Bornscheuer, U.T., McCarthy, J.E.G. and Schmid, R.D. (1998) *Biochimica et Biophysica Acta*, **1399**, 173–180.
8 Langrand, G., Rondot, N., Triantaphylides, C. and Baratti, J. (1990) *Biotechnology Letters*, **12**, 581–586.
9 De Castro, H.F., De Oliveira, P.C. and Pereira, E.B. (1997) *Biotechnology Letters*, **19**, 229–232.
10 Claon, A.C. and Akoh, C.C. (1993) *Biotechnology Letters*, **15**, 1211–1216.
11 Molinari, F., Gandolfi, R., Zilli, M. and Converti, A. (2000) *Enzyme and Microbial Technology*, **27** (8), 626–630.
12 Gandolfi, R., Converti, A., Pirozzi, D. and Molinari, F. (2001) *Journal of Biotechnology*, **92**, 21–26.
13 Converti, A., Del Borghi, A., Lodi, A., Palazzi, E., Gandolfi, R. and Molinari, F. (2002) *Biotechnology and Bioengineering*, **77**, 232–237.
14 Converti, A., Del Borghi, M., Gandolfi, R., Molinari, F., Palazzi, E. and Zilli, M. (2002) *World Journal of Microbiology & Biotechnology*, **18**, 409–416.
15 Converti, A., Del Borghi, A., Gandolfi, R., Molinari, F., Palazzi, E., Perego, P. and Zilli, M. (2002) *Enzyme and Microbial Technology*, **30**, 216–223.

16 Garcia, H.S., Malcata, F.X., Hill, C.G.Jr, and Amundson, C.H. (1992) *Enzyme and Microbial Technology*, **14**, 535–545.
17 Garcia-Alles, L.F. and Gotor, V. (1998) *Biotechnology and Bioengineering*, **59**, 163–170.
18 Molinari, F., Mantegazza, L., Villa, R. and Aragozzini, F. (1998) *Journal of Fermentation and Bioengineering*, **86**, 62–64.
19 Romano, D., Ferrario, V., Molinari, F., Gardossi, L., Sanchez Montero, J.M., Torre, P. and Converti, A. (2006) *Journal of Molecular Catalysis B: Enzymatic*, **41**, 71–74.
20 Alcantara, A.R., Sanchez Montero, J. M. and Sinisterra, J.V. (2000) *Stereoselective Biocatalysis*, Marcel Dekker, New York-Basel.
21 Grösch, S., Schilling, K., Janssen, A., Maier, T.J., Niederberger, E. and Geisslinger, G. (2005) *Biochemical Pharmacology*, **69**, 831–839.
22 Spizzo, P., Basso, A., Ebert, C., Gardossi, L., Ferrario, V., Romano, D. and Molinari, F. (2007) *Tetrahedron*, **63**, 11005–11010.
23 Morrone, G., Nicolosi, G., Patti, A. and Piattelli, M. (1995) *Tetrahedon: Asymmetry*, **6**, 1178–1773.

7
Factors Affecting Enantioselectivity: Allosteric Effects

Elisabeth Egholm Jacobsen and Thorleif Anthonsen

7.1
How to Provide Enantiopure Compounds

Many naturally occurring compounds are chiral and often enantiopure. They may be carbohydrates, amino acids, terpenoids, hydroxy carboxylic acids, or alkaloids. They can serve as building blocks for the synthesis of chiral compounds. In these compounds nature has already provided the chirality since natural compounds are synthesized by enzyme catalysis.

Selectivity is an important issue in most organic syntheses. It is common to distinguish between selectivity on three levels: chemoselectivity is to discern between reactions at chemically different groups of a molecule such as between a hydroxy group and an amino group, regioselectivity is the discrimination of reactions at sites with chemically equivalent groups but in chemically different environments such as the various hydroxy groups of a carbohydrate, And, thirdly; discrimination of reactions at chemical groups that only differ in stereochemical environment. Enzyme catalysis is able to discern between all of these three: however, stereo discrimination is the most demanding one.

The goal of enzyme-catalyzed stereo discriminating synthesis is to produce enantiomerically pure compounds. This target can be reached by two different routes, either by asymmetric synthesis or by resolution of a racemic mixture. These two methods have their particular characteristics. By asymmetric synthesis, apart from reactions based on chiral natural products, the substrate is achiral and the stereochemical principle is inherent in the catalyst, the enzyme. The enzyme is able to catalyze a reaction with an achiral substrate to yield excess of one single enantiomer of a product. The theoretical yield is 100% of product with 100% enantiomeric excess (e.e.). The e.e. of the product is in theory constant throughout the reaction, that is it is independent of the degree of conversion. On the contrary, kinetic resolution gives at the most 50% of each enantiomer and the e.e. of the two enantiomers depends on the degree of conversion (see Figure 7.1). However, by using a deracemization technique such as dynamic kinetic resolution or *in situ* inversion, the yield and e.e. may be 100%.

Modern Biocatalysis: Stereoselective and Environmentally Friendly Reactions.
Edited by Wolf-Dieter Fessner and Thorleif Anthonsen

ISBN: 978-3-527-32071-4

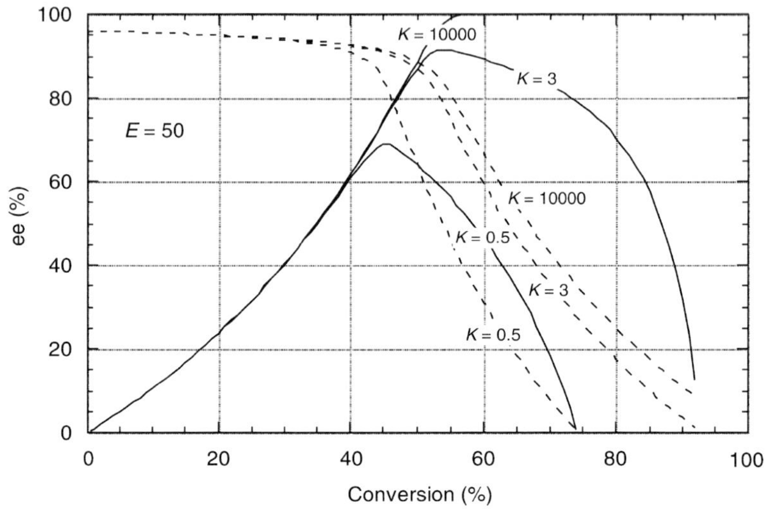

Figure 7.1 The graph shows how the e.e. values of the substrate (solid lines) and the product (dotted lines) change during the conversion of a racemic mixture. The *E*-value is set to 50 and calculations have been performed for three different equilibrium constants, where $K = 10000$ corresponds to an irreversible reaction. In this situation the substrate curve touches the 100% line at 55% conversion, which means that enantiopure substrate can be obtained provided that a 5% yield can be sacrificed. The curves for $K = 3$ and 0.5 show that reversibility in particular, influences the e.e. of the substrate in an unfavorable manner.

7.1.1 Kinetic Resolution of Racemic Mixtures Catalyzed by Enzymes

A common way of producing enantiopure compounds is by resolution. Kinetic resolution of racemic mixtures is often performed by catalysis using hydrolytic enzymes. One characteristic feature of kinetic resolution is that the e.e. of both the product and the remaining substrate depend on the degree of conversion of the reaction [1]. This is in contrast to resolution via diastereomers or by asymmetric synthesis. Since the e.e. of the substrate fraction will sooner or later during a reaction reach 100%, this goal can always be reached, provided some loss of yield is acceptable. This is a characteristic and a clear advantage of kinetic resolution, when the yield is not the most important factor of the reaction, but the e.e. The crucial parameter for a kinetic resolution reaction is the enantiomeric ratio *E*, commonly called the enantioselectivity factor [2]. Enantioselectivity is, according to the IUPAC, 'the preferential formation in a chemical reaction of one stereoisomer over another' [3]. Enantioselectivity is most commonly used for asymmetric synthesis when a prochiral substrate is converted into a mixture of enantiomers in which one enantiomer predominates. However, according to the definition, enantioselectivity can also be used for kinetic resolution in which two enantiomers are converted to product at a different rate. In this case the reaction may also be called enantiomer selective or enantiospecific since the enzyme interacts differently with two different substrates, the two enantiomers. According to the IUPAC, 'a reaction is termed stereospecific if starting materials differing only in their configuration are converted into stereoisomeric products'. This definition should also cover kinetic resolution since the two enantiomers give different enantiomeric products. The *E*-value of kinetic resolution is considered to be constant throughout a resolution reaction [4a]. The *E*-value is basically the ratio of the specificity constants for reactions with the two enantiomers [5]. If, let us say, that the *E*-value is 19, then, for instance, the *R*-enantiomer reacts 19 times faster than the *S*-enantiomer. This means that, at the very start of the reaction, the product will contain a 19 : 1 ratio of the enantiomers or 95% of *R* and 5% of *S*, i.e. the e.e. is 90% [6].

7.1.2 Absolute Configurations in Resolution

The *E*-value does not in general provide information on which enantiomer is the faster reacting. However, information on the stereo preference of several lipases has been collected and it seems that most lipases follow the so-called Kazlauskas rule [7]. For secondary alcohols the structure of the faster reacting enantiomer of an enantiomeric pair is the one shown in Figure 7.2. It is important to realize that whether this has the *R* or *S* configuration depends on the CIP rules. (see Scheme 7.1 for relevant examples. When $R = Br$ or Cl the *R*/*S* notation changes, compared to 'small' or 'large' relative size. Provided that the *E*-value is high (e.g. $E \geq 20$) it can be used for predicting or even be taken as a proof of absolute configuration [8].

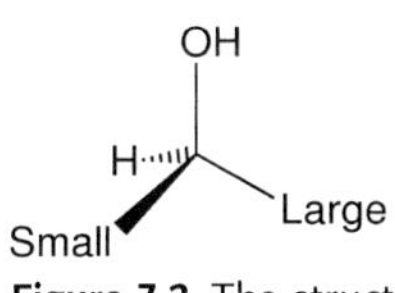

Figure 7.2 The structure of the faster reacting enantiomer in lipase-catalyzed esterification in kinetic resolution of racemic secondary alcohols or hydrolysis of the corresponding esters. 'Small' and 'large' refer to the relative size of the groups and not to the *R*/*S* notation.

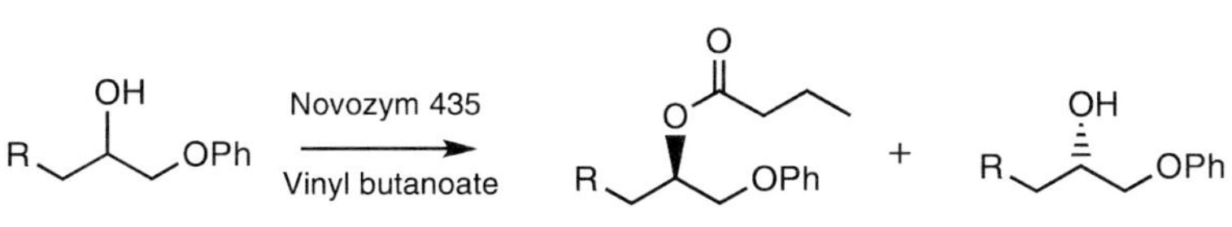

1 R = CH_3, **2** R = Et, **3** R = Br, **4** R = Cl, **a** alcohols, **b** butanoates

Scheme 7.1

7.2 Factors Affecting the Enantiomeric Ratio *E*

The enantioselectivity of a biocatalytic resolution or asymmetrization is primarily dependent on the enzyme and the structure of the substrate. Both of these can be changed in order to optimize the selectivity. The enzyme can be optimized by molecular genetic methods, while the substrate can be modified by organic chemical synthesis. These ways of optimizing selectivity will not be discussed in this chapter.

7.2.1 Is the *E*-value Really Constant?

The *E*-value should be calculated by measuring the e.e. of the remaining substrate and the product at different degrees of conversion [9]. However, it is common to measure the e.e. at one single degree of conversion, but it can also be done by using computer programs utilizing several measurements [10, 11]. The *E*-value is considered to be constant throughout a resolution and only dependent on the substrate and the enzyme. However, it has been discovered that the *E*-value may change during the reaction [12]. This feature can only be discovered if the *E*-value is measured at several degrees of conversion. If computer programs are used for calculating *E* based on several measurements, an average *E* will result. The *E*-value depends as mentioned primarily on the enzyme and the substrate, but also on the acyl donor used in acyl transfer reactions in organic solvents [13, 14]. Acyl donors like activated haloalkyl esters can give high *E*-values, but may be reversible to some extent. This is in particular harmful for the e.e. of the substrate fraction (see Figure 7.1). Enol esters, on the other hand, such as vinyl or isopropenyl esters, give irreversible reaction conditions, but may reduce the *E*-value due to acylation of amino groups of the enzyme [13]. The liberated acetaldehyde, when vinyl esters are used as acyl donors, may react with amino groups of the enzyme and, thus, reduce the selectivity [4b], but this is not always a problem [14]. The *E*-value also depends on the medium in which the reaction is carried out. In a hydrolysis reaction, this is usually water or a buffer, whereas for transesterification reactions this can be almost any organic solvent or mixture of solvents, including some water. During the reaction the medium will change, not because the actual solvent is changing, but because the enantiomers of the substrate are disappearing at a different rate and the enantiomers of product are formed in unequal quantities [15]. They also constitute the medium of the reaction and, consequently, this change can influence the selectivity of the reaction.

7.2.2 Influence of the Reaction Medium on the *E*-value

It is well recognized that the medium of the resolution reaction can influence the selectivity. The medium is primarily the solvent, which in hydrolysis mainly consists of water. In a transesterification reaction, however, the medium is organic solvents with varying amounts of water in it. The most relevant way to express the amount of water in the medium is by the water activity a_w. The a_w value can be set by adding a mixture of inorganic salts with different hydration numbers to the reaction [16]. It has been shown that enantiomer selectivity in resolutions can be increased when the water activity is increased [17]. The effect depends on which organic solvent is used, but when the water activity becomes too high, the acyl transfer reaction will stop.

7.2.3
Influence of Enzyme Immobilization on the *E*-value

Immobilization of enzyme catalysts is performed primarily in order to improve the technical properties of enzymes. It can be performed in various ways. Two distincly different methods are used: coupling to a carrier or cross-linking or by entrapment in a matrix or by membranes [18]. There are only a few reports on the dependence of selectivity. Reports claim that selectivity can be enhanced [19] or that it is not influenced. It seems reasonable to anticipate that each specific enzyme will behave differently and that more investigations are needed [20].

7.2.4
Enzyme Inhibition

Inhibition of lipases, both by the substrate or the product, has been observed. In alcoholysis of methyl propanoate with *n*-propanol catalyzed by *Candida antarctica* lipase B (CALB), the alcohol was found to inhibit the enzyme resulting in a dead-end complex [21]. Phosphate- and phosphonate-containing inhibitors are known to inhibit proteases. Studies of the inhibition of CALB have shown inhibition by diethyl *p*-nitrophenyl phosphate. The inactivation of the enzyme was caused by covalent binding of diethyl *p*-nitrophenyl phosphate in the active site [22].

7.2.5
Enantioselective Inhibition and Activation: Allosteric Effects

The addition of small molecules has been shown to change the enantioselectivity of certain enzyme-catalyzed reactions. It is believed that such molecules bind to a site in the protein different from the active site, which leads to a conformational change in the active site. Such enzymes are called allosteric enzymes, i.e. enzymes that comprise of multiple subunits and multiple active sites. Binding of a co-substrate or small molecule may cause an increase or decrease in the activity or selectivity of the enzyme.

Since enantiomer selectivity, the *E*-value, is the ratio of the reaction rate with the two enantiomers: either an allosteric increase in the reaction rate with one enantiomer or an allosteric decrease in the reaction rate with the other disfavored one will lead to an increase in *E*.

Enantioselective inhibition of *Candida rugosa* by dextromethorphan and levomethorphan resulting in an enhanced enantiomer selectivity has been reported in kinetic resolution 1989 [23]. Kinetic inhibition experiments have revealed that the molecular action of the base was a non-competitive inhibition (i.e. binding of the base to an allosteric site in the lipase) that caused the inhibition of the transformation of one enantiomer leading to the relative increase in the transformation rate of the other enantiomer [23]. L-Methioninol was found to increase the hydrolysis rate of the *(R)*-enantiomer in hydrolysis of 3-acetoxy nitriles catalyzed by lipase PS (*Pseudomonas* sp.) while it inhibited the hydrolysis of the *(S)*-enantiomer. It was suggested that the substrate and L-methioninol were combined to the enzyme at different sites and that conformational changes due to the binding of L-methioninol provided a change in the affinity towards the substrates [24]. Asymmetrization of 3-(3,4-dichlorophenyl)glutarate with immobilized CALB (Chirazyme L2, Roche) showed a loss of 30% activity within 18 h of the reaction, which was assumed to be due to product inhibition [25]. The decrease in enantioselectivity in CALB-catalyzed resolutions of secondary butanoates observed by Lundhaug *et al.* [26] was also suggested to be caused by inhibition of the enzyme by the enantiopure product alcohol. Akeboshi and co-workers [27] also reported that the enantioselectivity gradually declined, accompanying the progress of the hydrolysis of benzyl-protected primary alcohols and that the alcohol product inhibited the hydrolysis rate of the faster reacting enantiomer. This phenomenon was first reported in the 1930s when it was found that strychnine enhanced human liver esterase-catalyzed hydrolysis of methyl L-mandelate but not the hydrolysis of the D-isomer. These results also indicated an apparent allosteric binding of the enantiopure additive [28, 29].

Organic bases have also been reported to increase the enantioselectivity of lipase-catalyzed reactions in water-saturated organic solvents [30, 31]. It is known that both the activity and *E*-value of CALB (Novozym 435) increased dramatically upon addition of triethylamine (Et3N) to the water saturated organic reaction medium in a reaction between 2-phenyl-4-benzyloxazol-5(4*H*)-one and butan-1-ol. It was believed that the base was able to make an ion pair with the acidic by-product, thus avoiding the inhibitory effect of the acid by dissolving it and thereby removing it from the micro-environment of the enzyme [32]. The addition of Et_3N, crown ethers, and tris-(3,6 dioxaheptyl)amine to Novozym SP 435-catalyzed trans-esterification of 1-azido-3-phenoxy-2-propanol has also been reported to enhance enantioselectivity and reaction rates [33].

Kinetic experiments may be used for revealing the type of inhibition in enzymes. By inserting experimental data to the inverted Michaelis–Menten equation this gives straight-line plots (Lineweaver–Burk), which can be extrapolated to yield the characterizing constants of the enzyme. However, the Michaelis–Menten model cannot account properly for the kinetic properties of allosteric enzymes [34].

Kinetic studies of the inhibitory effect of 1-butanol in different solvents on CALB have been performed and the competitive inhibition constant K_i values obtained correlated with the calculated activity coefficients of the substrate, suggesting that desolvation of the alcohol was the changing condition [35].

7.2.6
The *E*-value of CALB is Influenced by R-Alcohols

When the butanoic ester of 1-phenoxy-4-methoxy-2-butanol was hydrolyzed by CALB catalysis it was discovered that the *E*-value changed during the conversion [12]. This effect seems to be quite general for a range of secondary esters. Moreover, when the corresponding alcohols (see Scheme 7.1) were esterified with vinyl butanoate catalyzed by CALB, the *E*-values also changed. During the conversion of the substrate the *E*-value increased when the reaction was hydrolysis and decreased during a transesterification reaction. The representative analyses from the reactions with 1-phenoxy-2-butanol (**1a**) and the corresponding butanoate (**1b**) catalyzed by Novozym 435 are shown in Figure 7.3.

A possible explanation was that the effect was connected to the changing concentration of the enantiomeric esters. However, no effect on *E* was observed when enantiopure (*R*)-**1b**, the faster reacting enantiomer of the butanoate of 1-phenoxy-2-butanol, was added to the transesterification reaction of 1-phenoxy-2-pentanol (**2a**). A similar result was obtained when enantiopure (*R*)-**4b**, the slower reacting enantiomer of the butanoate of 3-chloro-1-phenoxy-2-propanol, was added to the transesterification reaction of 1-phenoxy-2-butanol (**1a**). It was concluded that the effect of the changing *E*-value was not connected to the changing concentration of the ester. In order to verify this 3-chloro-1-phenoxy-2-propanol (**4a**) was esterified with vinyl butanoate with addition of an enantiopure (*R*)-alcohol such as (*R*)-**1a**, (*R*)-**2a**, (*R*)-**5a**, (*R*)-**6a**, and (*R*)-**7a** at 30% conversion (see Figure 7.4 and Scheme 7.2).

The reason for adding a different but structurally related (*R*)-alcohol was that the analysis and *E*-value calculations would be influenced by the added (*R*)-alcohol of the actual substrate. In the esterification of **4a** a decrease in *E* was observed

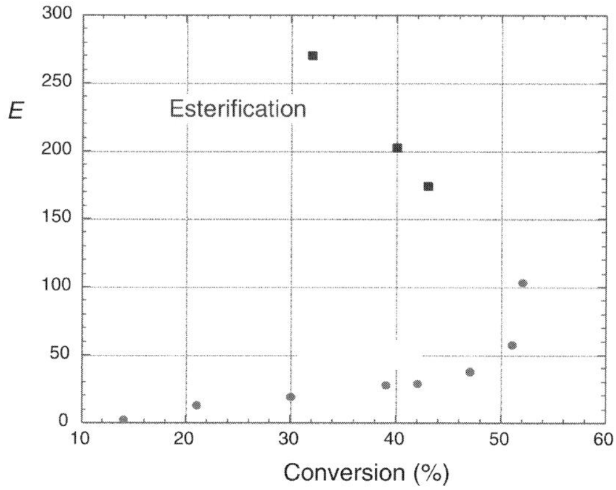

Figure 7.3 The *E*-values observed during esterification of 1-phenoxy-2-butanol (**1a**) catalyzed by Novozym 435 (squares) and the hydrolysis of the corresponding butanoate (**1b**) (circles).

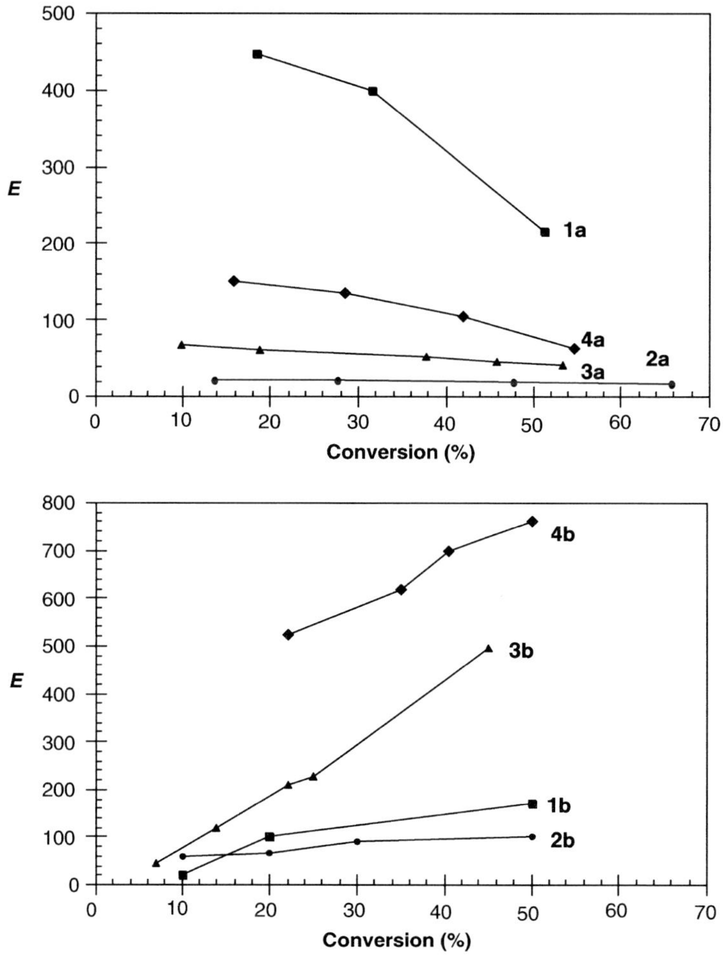

Figure 7.4 Upper panel: *E*-values at different degrees of conversion in transesterification reactions with vinyl butanoate of the alcohols **1a, 2a, 3a**, and **4a** catalyzed by Novozym 435 (this batch is different from the one used in Figure 7.3). Lower panel: *E*-values at different degrees of conversion in hydrolysis of the corresponding butanoates **1b, 2b, 3b**, and **4b** catalyzed by Novozym 435.

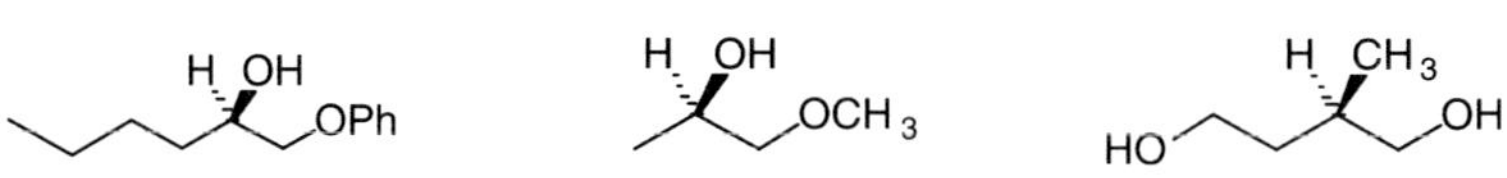

(*R*)-**5a** (*R*)-**6a** (*R*)-**7a**

Scheme 7.2

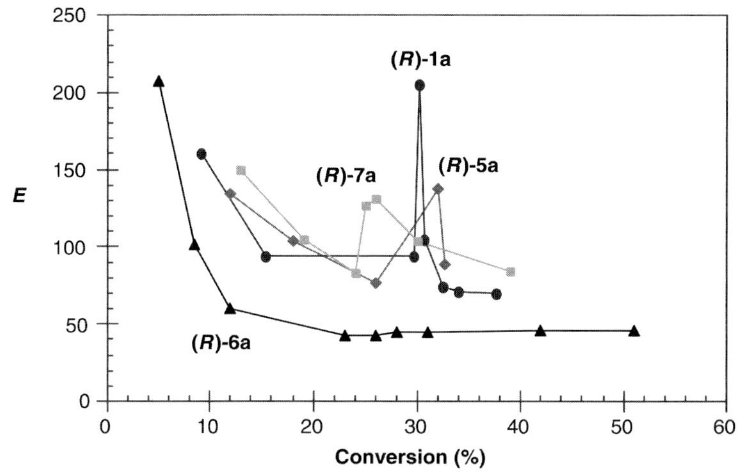

Figure 7.5 *E*-values at different degrees of conversion during transesterification of **4a** with Novozym 435. The enantiopure (*R*)-alcohol (*R*)-**1a**) was added at 30% conversion and the *E*-value increased from 94 to 205 and then decreased again. The effect was also shown for the addition of (*R*)-**5a** and (*R*)-**7a**: however, there was no effect for (*R*)-**6a**.

from 160 at 9% conversion to 94 at 30% conversion. (This decrease in *E* corresponded to a small difference in $\Delta\Delta G^{\#}$ of 1.55 kJ/mol (0.37 kcal/mol).) However, when (*R*)-**1a** was added to the transesterification reaction of **4a** at 30% conversion the *E*-value increased from 94 to 205 before it decreased again. Esterifications of **4a** with addition of (*R*)-1-phenoxy-2-hexanol ((*R*)-**5a**) and (*R*)-2-methyl-1,4-butanediol ((*R*)-**7a**) showed similar results. The addition of 1-phenoxy-2-pentanol ((*R*)-**2a**) and (*R*)-phenyl ethanol to the resolution of **4a** showed similar results, however with a smaller increase in *E*. The addition of (*R*)-1-methoxy-2-propanol ((*R*)-**6a**) surprisingly did not affect the selectivity. (Figure 7.5)

It was also surprising that the *E*-value in the case of (*R*)-1-phenoxy-2-hexanol ((*R*)-**5a**) increased and also dropped to the low level quickly after addition. It has been shown that **5a** is esterified by vinyl butanoate and CALB with almost no selectivity and with a low rate [14]. For that reason, it seems unlikely that the disappearance of the selectivity enhancement was only due to esterification of the alcohol. However, it is possible that the enantiopure alcohols after addition were bound to the enzyme and after a while brought into the bulk solution by the solvent. The observed quick decrease of effect after addition may therefore be a combination of the removal of the (*R*)-alcohol by esterification or by solvation.

Since this effect was observed with the immobilized preparation of CALB (Novozym 435) it was interesting to perform the same reactions with pure protein preparations of CALB to see if the immobilization of the enzyme influenced the changing *E*-value. The formulated CALB Novozym 525 F is a water solution of CALB containing 1–10% pure protein. After freeze drying at −80 °C this dry protein

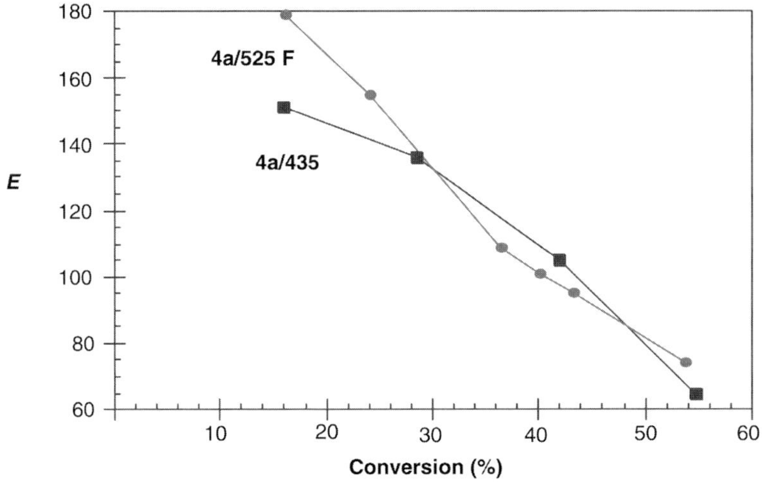

Figure 7.6 *E*-values from the transesterification reaction of **4a** with Novozym 435 (**4a**/435, squares) and with Novozym 525 F (**4a**/525 F, circles) in hexane with vinyl butanoate as the acyl donor. The *E*-value decreased during conversion in both reactions.

powder was used for catalyzing the transesterification reactions of **1a**, **2a**, **3a**, and **4a** with vinyl butanoate in hexane.

The *E*-values in all of the reactions decreased with increasing conversion. Figure 7.6 shows the *E*-values of the esterification of **4a** with Novozym 525 F (**4a**/525 F) plotted together with the resolution of **4a** with Novozym 435 (**4a**/435). It was inferred that immobilization did not significantly influence the reaction.

Since addition of the enantiopure (*R*)-alcohols increased the selectivity of CALB, it is likely that they were causing a conformational change in the enzyme, possibly due to an allosteric effect. However, an allosteric center is not mentioned in the structure details for CALB [36, 37].

Saturation transfer difference nuclear magnetic resonance experiments have shown that there is a strong interaction between the enzyme and the (*R*)-enantiomers of the alcohols, but not the (*S*)-enantiomers (H. W. Anthonsen, unpublished results). As described above, addition of a small amount of pure (*R*)-alcohols increased the *E*-value of kinetic resolution and, moreover, the effect disappeared quickly. The reason is probably that the alcohol moves into the active site where it is esterified. A future goal is to find an additive that can bind irreversibly to this 'unknown' allosteric site, thus causing a lasting effect.

7.2.7 Is a Changing *E* Caused by the Slow or the Fast Enantiomer?

When enantiomer selectivity in kinetic resolutions changes as the reaction conditions change, as discussed above, it may be due to an increased or decreased

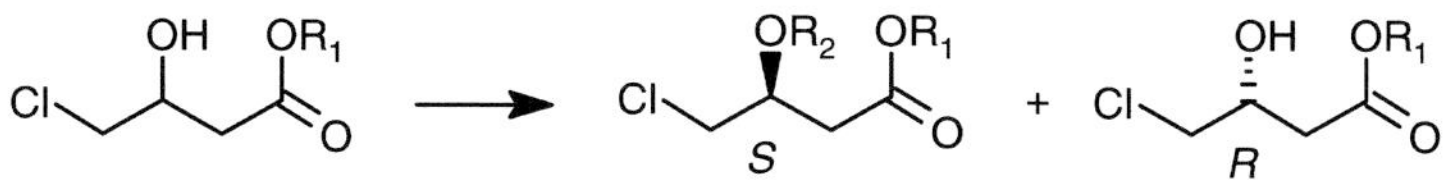

8a R_1 = Et
9a R_1 = *tert*-butyl

$R_2 = COC_3H_7$

Scheme 7.3

reaction rate of one or both of the enantiomers. The reaction rate of each of the enantiomers can be determined by calculating the concentration of the two enantiomers at different degrees of conversion from the equations below [38].

$$cv\text{F} = c \times (1 + ee_\text{p})$$

$$cv\text{S} = c \times (1 - ee_\text{p})$$

cvF and cvS are the conversions of the fast- and slow-reacting enantiomers (% converted of total concentration of substrate enantiomer), respectively, c is the conversion of the racemic substrate and ee_p is the enantiomeric excess of the product. The results can be plotted in Kaleidagraph 3.0 and the progress curves of different reactions are compared.

The effect of increasing the small substituent of **1a** by one CH_2 unit as in **2a** is shown in Figure 7.7. It is evident that the reason for the decrease in E is due to a slower reaction of the faster enantiomer [14]. This is the opposite of the effect observed when the ethyl ester of **8a** is changed into *tert*-butyl as in **9a** (Scheme 7.3). In this case the higher E-value of **9a** is due to a lowering of the rate of the slower enantiomer [39]. For **1a** and **2a** the slow enantiomers are almost equally slow, while for **8a** and **9a** the fast enantiomers are equally fast.

7.3 Asymmetrization of Prochiral Compounds

7.3.1 Asymmetrization of Prochiral Dicarboxylates: Single-Step Process

Asymmetrization of a prochiral dicarboxylic acid diester catalyzed by lipases, where the stereo center of the product is located on the acyl side, becomes a single-step process because the polar carboxylic acid and/or amide formed are not well accepted as substrates by the lipase. One example is the enantioselective hydrolysis or ammonolysis of diethyl 3-hydroxyglutarate, as shown in Scheme 7.4, a reaction which leads to the formation of a precursor for the important chiral side chain of atorvastatin, lipitor [40, 41]. The *S*-enantiomer was formed with high e.e. (98%), but unfortunately this is the undesired enantiomer for the production of the pharmaceutically important product. Only α-chymotrypsin gave a predominance of the

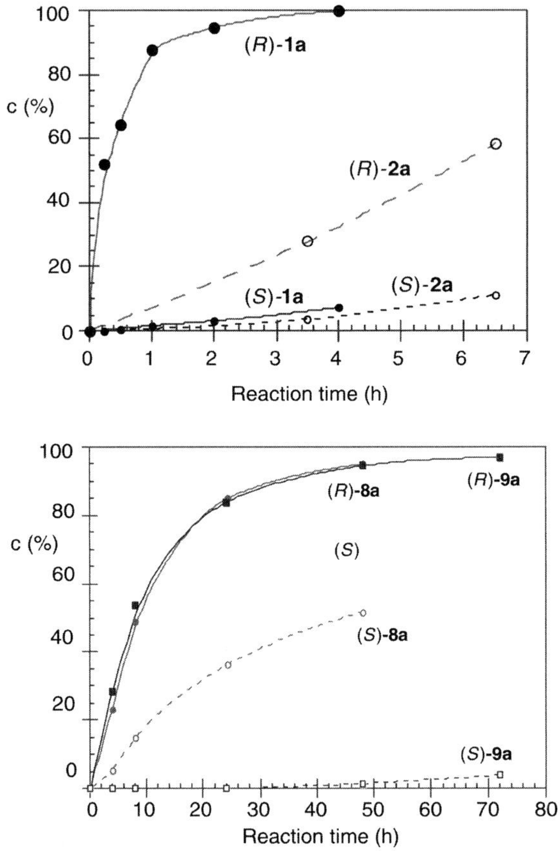

Figure 7.7 The upper panel shows the progress of esterification of the two enantiomers of 1-phenoxy-2-butanol (**1a**) catalyzed by Novozym 435. The graph labelled (*R*)-**1a** represents the product formed from (*R*)-**1a**. For comparison, the corresponding course of esterification of the two enantiomers of 1-phenoxy-2-pentanol (**2a**) is also shown. The lower panel shows a comparison between the esterifications of ethyl 4-chloro-3-hydroxybutanoate (**8a**) and the corresponding *tert*-butyl ester **9a**.

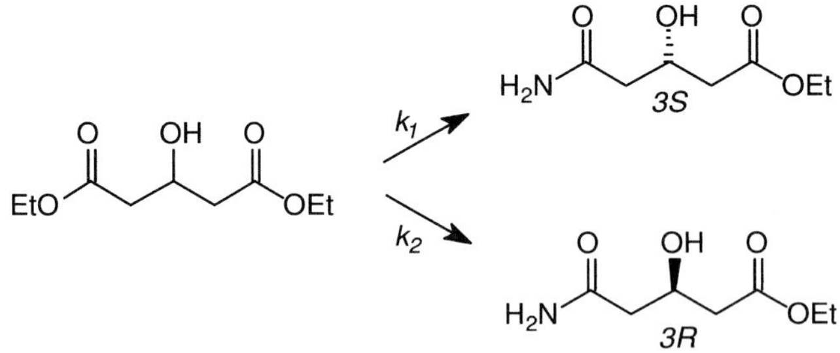

Scheme 7.4

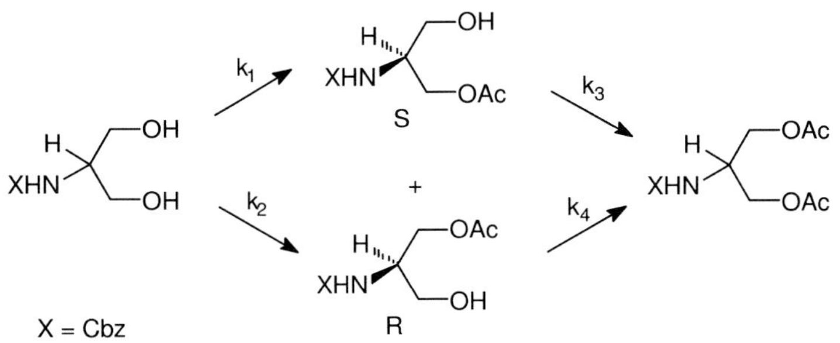

Scheme 7.5

R-enantiomer, but with a modest e.e. In such a single-step process the e.e. depends only on the ratio k_1/k_2 which is constant throughout the reaction since the starting material is achiral and since none of the products should give rise to allosteric effects.

7.3.2
Asymmetrization of Prochiral Diol: Double-Step Process

A prochiral diol can be converted into enantiomeric compounds as shown by the example in Scheme 7.5. Catalyzed by a hydrolytic enzyme like a lipase, it can be enantioselectively acetylated by vinyl acetate in organic solvent to yield a mixture of monoacetates. The chiral monoacetates will be formed at unequal rates and also react further at unequal rates. It is usually anticipated that if $k_1 > k_2$, then $k_4 > k_3$ and, moreover, that the ratio k_1/k_2 is constant throughout the reaction [4c].

Thus the e.e. of the monoacetate stage will increase due to both steps. When these anticipations apply, the e.e. of the monoacetates should increase throughout the reaction to reach 100% and, moreover, the yield of the two monoacetates ($R + S$) should increase before it eventually declines and reaches zero when the conversion has reached 100% and only the achiral diacetate will comprise the reaction mixture [4d] (Figure 7.8).

7.3.3
Is the e.e. Constant During Asymmetrization Reactions?

Recent studies using the primary diol shown in Scheme 7.5 and porcine pancreatic lipase (crude, Sigma Type II) as transesterification catalyst, indicate that the situation is more complex [42]. In one reaction, 60% of the diol was converted to enantiomer A of the monoacetate and the e.e. reached 90% after 30 min of conversion. Then enantiomer B was formed resulting in a drop of the e.e. to zero. During further reaction enantiomer B was in excess and the e.e. increased again, but then in favor of enantiomer B. Moreover, after 20 h of conversion, all of the diol was consumed and, surprisingly, the diacetate was virtually absent from the reaction mixture. These results indicate that it is not only important to monitor the course

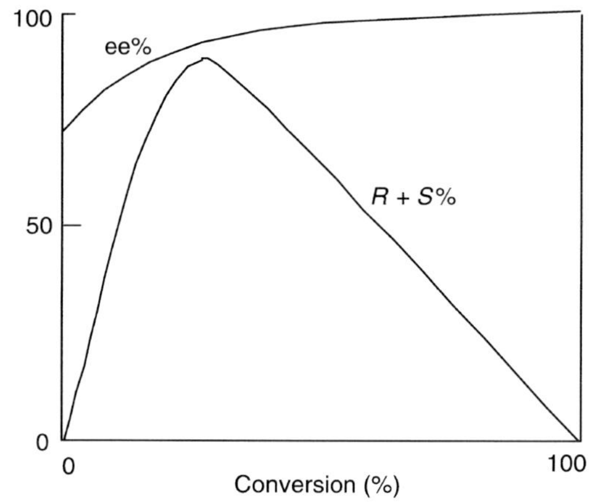

Figure 7.8 The graph shows how the e.e. of the mono-ester stage of a double-step process changes during conversion (upper trace) and the sum (%) of mono-esters present (lower trace) at various degrees of conversion.

of the reaction in resolutions, but also during asymmetrization. The unexpected results remain to be explained. They may be due to an allosteric effect as for resolution under 2.5. However, since the enzyme preparation used was not pure, the results may be due to the action of more than one enzyme. Moreover, the possibility of acyl migration should also be considered.

7.4 Conclusions

This chapter has dealt with the production of enantiopure compounds using hydrolases as catalysts. Two kinds of stereo discriminating reactions are discussed: enantiomer selective kinetic resolution and enantioselective asymmetrization of prochiral molecules.

The crucial parameter for kinetic resolution is the enantiomeric ratio, the *E*-value that is the ratio of the rate constants for reaction with the two enantiomers. A high *E*-value ensures high enantiomeric excess of the products. The success of asymmetrization depends on a high value for the ratio k_1/k_2, k_1 being the rate constant leading to the formation of one enantiomer and k_2 leading to the opposite. It is generally believed that both the *E*-value and the k_1/k_2 ratio does not change during the reaction: however, it has recently been shown that this is not always true. When the stereo center of a racemic ester is located in the alkyl part, liberation of the *R*-alcohol during hydrolysis leads to an increase in the *E*-value. On the other hand, removal of the *R*-alcohol during transesterification produces a decrease in *E*. This is due to an allosteric effect caused by the presence of *R*-alcohols.

Asymmetrization of prochiral diols by esterification also shows anomalous behavior. More experiments have to be carried out before final conclusions can be drawn.

Asymmetrization of prochiral molecules by hydrolysis when the stereo center is located on the acyl side is more straightforward, probably because the chiral monocarboxyl products are usually no substrates for a lipase.

The general rule seems to be that the *E*-value changes during resolution and that the e.e.-value depends on the degree of conversion during asymmetization. Principally this should not be restricted to enzymatic catalysis, but also be valid for reactions catalyzed by non-enzymatic catalysts.

References

1 Chen, C.-S., Fujimoto, Y., Girdaukas, G. and Sih, C.J. (1982) *Journal of the American Chemical Society*, **104**, 7294–7299.

2 Anthonsen, T. and Jongejan, J.A. (1997) *Methods in Enzymology*, **286**, 473–495.

3 IUPAC (1996) Basic Terminology of Stereochemistry, http://www.chem.qmul.ac.uk/iupac/stereo/ (accessed October 2nd, 2008).

4 (a) Faber, K. (2004) *Biotransformations in Organic Chemistry*, 5th edn, Springer-Verlag, Berlin, p. 39.
(b) Faber, K. (2004) *Biotransformations in Organic Chemistry*, 5th edn, Springer-Verlag, Berlin, p. 348.
(c) Faber, K. (2004) *Biotransformations in Organic Chemistry*, 5th edn, Springer-Verlag, Berlin, p. 35.
(d) Faber, K. (2004) *Biotransformations in Organic Chemistry*, 5th edn, Springer-Verlag, Berlin, p. 36.

5 Fersht, A. (1999) *Structure and Mechanism in Protein Science*, W.H. Freeman and Co., New York.

6 Anthonsen, T. (2001) *Basic Biotechnology*, 2nd edn (eds B. Kristiansen and C. Ratledge), Cambridge University Press, Cambridge, pp. 409–428.

7 Kazlauskas, R.J., Weissfloch, A.N.E., Rappaport, A.T. and Cuccia, L.A. (1991) *The Journal of Organic Chemistry*, **56**, 2656–2665.

8 Hoff, B.H. and Anthonsen, T. (1999) *Chirality*, **11**, 760–767.

9 Sih, C.J. and Wu, S.-H. (1990) *Topics in Stereochemistry*. **19**, 63–125.

10 Anthonsen, H.W., Hoff, B.H. and Anthonsen, T. (1996) *Tetrahedron: Asymmetry*, **7**, 2633–2638.

11 Anthonsen, H.W. (1996–1997) http://www.chem.ntnu.no (accessed October 2nd, 2008).

12 Waagen, V., Partali, V., Hansen, T.V. and Anthonsen, T. (1994) *Protein Engineering*, **7**, 589–591.

13 Hoff, B.H., Anthonsen, H.W., Anthonsen, T. and T. (1996) *Tetrahedron: Asymmetry*, **7**, 3187–3192.

14 Jacobsen, E.E., Hoff, B.H. and Anthonsen, T. (2000) *Chirality*, **12**, 654–659.

15 Jacobsen, E.E., van Hellemond, E.W., Moen, A.R., Prado, L.C.V. and Anthonsen, T. (2003) *Tetrahedron Letters*, **44**, 8453-8455.

16 Kvittingen, L., Sjursnes, B., Anthonsen, T. and Halling, P. (1992) *Tetrahedron*, **48**, 2793–2802.

17 Jacobsen, E.E., Anthonsen, T. and C. (2002) *Journal of Chemistry – Revue Canadienne de Chimie*, **80**, 577–581.

18 Hartmeier, W. (1986) *Immobilisierte Biokatalysatoren*, Springer-Verlag, Berlin.

19 Heinsman, N.W.J.T., Schröen, C.G.P.H., van der Padt, A., Franssen, M.C.R., Boom, R.M. and van't Riet, K. (2003) *Tetrahedron: Asymmetry*, **14**, 2699–2704.

20 Jacobsen, E.E., Andresen, L.S. and Anthonsen, T. (2005) *Tetrahedron: Asymmetry*, **16**, 847–850.

21 Bousquet-Dubouch, M.-P., Graber, M., Sousa, N., Lamare, S. and Legoy, M.-D. (2001) *Biochimica et Biophysica Acta*, **1550**, 90–99.

22 Patkar, S.A., Björkling, F., Zundel, M., Schulein, M., Svendson, A., Heldt-Hansen, H.P. and Gormsen, E. (1993) *Indian Journal of Chemistry*, **32B**, 76–80.

23 Guo, Z.-W. and Sih, C.J. (1989) *Journal of the American Chemical Society*, **111** (17), 6836–6841.

24 Itoh, T., Ohira, E., Takagi, Y., Nishiyama, S. and Nakamura, K. (1991) *Bulletin of the Chemical Society of Japan*, **64**, 624–627.

25 Homann, M.J., Vail, R., Morgan, B., Sabesan, V., Levy, C., Dodds, D.R. and Zaks, A. (2001) *Advanced Synthesis Catalysis*, **343**, 744–749.

26 Lundhaug, K., Overbeeke, P., Jongejan, J. and Anthonsen, T. (1998) *Tetrahedron: Asymmetry*, **9**, 2851–2856.

27 Akeboshi, T., Ohtsuka, Y., Ishihara, T. and Sugai, T. (2001) *Advanced Synthesis Catalysis*, **343**, 624–637.

28 Bamann, E. and Laeverenz, P. (1930) *Hoppe-Seyler's Zeitschrift fur Physiologische Chemie*, **193**, 201–214.

29 Ammon, R. and Fischgold, H. (1931) *Biochemische Zeitschrift*, **234**, 54.

30 Berger, B., Rabiller, C.G., Königsberger, K., Faber, K. and Griengl, H. (1990) *Tetrahedron: Asymmetry*, **1**, 541–546.

31 Maugard, T., Remaud-Simeon, M., Petre, D. and Monsan, P. (1997) *Tetrahedron*, **14**, 5185–5194.

32 Parker, M.-C., Brown, S.A., Robertson, L. and Turner, N.J. (1998) *Chemical Communications*, 2247–2248.

33 Pchelka, B.K., Loupy, A., Plenkiewicz, J. and Blanco, P.A.L. (2001) *Tetrahedron: Asymmetry*, **12**, 2109–2119.

34 Berg, J.M., Tymoczko, J.L. and Stryer, L. (2002) *Biochemistry*, W.H.Freeman, New York.

35 Garcia-Alles, L.F. and Gotor, V. (1998) *Biotechnology and Bioengineering*, **59**, 163-170.

36 Uppenberg, J., Hansen, M., Patkar, S. and Jones, T.A. (1994) *Structure*, **2**, 293–308.

37 Uppenberg, J., Öhrner, N., Norin, M., Hult, K., Kleywegt, G.J., Patkar, S., Waagen, V., Anthonsen, T. and Jones, T.A. (1995) *Biochemistr y*, **34**, 16838–16851.

38 Hoff, B.H. (1999) Ph.D. Thesis, NTNU, 78.

39 Hoff, B.H., Anthonsen, T. and (1999) *Tetrahedron: Asymmetry*, **10**, 1401–1412.

40 Jacobsen, E.E., Hoff, B.H., Riise Moen, A. and Anthonsen, T. (2003) *Journal of Molecular Catalysis. B: Enzymatic*, **21**, 55–58.

41 Moen, A. Riise, Hoff, B.H., Hansen, L.K., Anthonsen, T. and Jacobsen, E.E. (2004) *Tetrahedron: Asymmetry*, **15**, 1551–1554.

42 Lie, A., Ljones, T., Hoff, B.H., Anthonsen, T. and Jacobsen, E.E. (2008) to be published.

8
Kinetic Resolution of *Sec*-alcohol in Non-conventional Media

Maja Habulin, Mateja Primožič and Željko Knez

8.1
Introduction

The regioselectivity of enzymes, even on complex molecules, without any need for protecting groups is a fundamental strength of biocatalysis. Therefore with biocatalysis as a tool for chemical transformations performance the number of synthetic steps and thereby the occupation time of chemical reactors, which is an important factor of process economy, could be reduced. The potential utility of biocatalysis in the pharmaceutical industry has been widely recognized in the last decade. Considering the high catalytic power and exceptional stereo-, regio-, and chemoselectivity, the area of biotransformations in the pharmaceutical industry is now experiencing a significant growth due to the wide range of commercial applications [1]. Over the past decade the view of organic chemists has also changed with regard to the use of enzymes in organic synthesis, from downright ignorance to careful acceptance of this technology [2].

Green chemistry is searching for alternative, environmentally friendly reaction media as compared to traditional organic solvents. The benefits are an increase in reaction rates and lower reaction temperatures, as well as higher selectivities. Most chemical processes depend on solvents. The use of volatile organic solvents in chemical processes has a detrimental impact on the environment and human health, emphasizing the need for alternative clean manufacturing processes and environmentally benign technologies. Some of these solvents are miscible with water and therefore they must be stripped from water before it leaves the process, not only for ecological but also for economic reasons. Solvents must be recovered for recycling and reuse for an economically viable process. Ionic liquids (ILs) and supercritical fluids (SCFs) are alternative solvents that may also be used in green chemistry [3].

ILs are low melting point salts that represent an exciting new class of reaction solvents for catalysis. ILs are composed entirely from ions and possess negligible vapour pressures and, with proper selection of the wide range of possible cations and anions, the solvent properties of ILs could be controlled. ILs have been accepted as a new green chemical revolution, which has excited both academia

Modern Biocatalysis: Stereoselective and Environmentally Friendly Reactions.
Edited by Wolf-Dieter Fessner and Thorleif Anthonsen

ISBN: 978-3-527-32071-4

and chemical industries. This new chemical group can reduce the use of hazardous and polluting organic solvents due to their unique characteristics as well as taking part in various new syntheses [3]. One of the most exciting recent developments in the use of ILs for catalytic systems is the application of biocatalysis in these solvents [4] due to the good stability of enzymes in ILs.

SCFs, as solvents, show the most attractive properties, as they have gas-like low viscosities and high diffusivities and, simultaneously, they possess attractive liquid-like solubilizing power.

Supercritical carbon dioxide (SC-CO_2) is by far the most studied SCF, because of its economical, technical, environmental, and health advantages. Moreover, it is an excellent solvent for the transport of hydrophobic compounds. Furthermore, its critical pressure is relatively low (7.34 MPa) and it has an ambient critical temperature (304.45 K), which provides the mild conditions necessary for keeping its enzyme activity. It has GRAS (generally regarded as safe) status and can therefore be used as an alternative for organic solvents. One of the main advantages of using SC-CO_2 as a reaction medium lies in its fractionation potential. It exhibits a wide spectrum of solvent properties, due to the continuous variation of its density versus pressure and temperature.

In recent years, among non-conventional bioreaction media, the SCF/IL biphasic system has increasingly attracted attention as an emerging green and high-technology alternative to classical organic solvents for carrying out enzymatic reactions for the preparation of valuable and active materials [5]. Clean chemical processes, which provide directly pure products by using appropriate enzyme/IL systems in combination with SCFs, contain an exciting potential for developing the upcoming green chemical industry.

The volatile and non-polar SC-CO_2 forms different two-phase systems with non-volatile and polar ILs. The product recovery process with these systems is based on the principle that SC-CO_2 is soluble in ILs, but ILs are not soluble in SC-CO_2 [6]. Since most organic compounds are soluble in SC-CO_2, these products are transferred from the IL to the supercritical phase [7, 8].

Enzyme-catalyzed reactions based on such biphasic systems have been shown to be promising alternatives for developing green chemical processes because of their physical and chemical characteristics [9]. By combining these media with enzymes, the possibilities of carrying out integral green biocatalytic processes has been already demonstrated [10–12]. Such biphasic systems can be used for both the biotransformation and extraction of products simultaneously, even under extremely harsh conditions, because of the different miscibilities of ILs and SC-CO_2.

Chiral alcohols are useful starting materials for the synthesis of various biologically active compounds. The need for enantiomerically pure drugs and agrochemicals has increased in recent years [13]. Derivatives of enantiopure 1-phenylethanol are important chiral building blocks, which can be used as synthetic intermediates for the production of pharmaceuticals, fine-chemicals agrochemicals, and natural products. In particular (*R*)-1-phenylethanol is in widespread use as an ophthalmic preservative, an inhibitor of cholesterol intestinal adsorption, a solvatochromic dye, a fragrance, and so on.

Lipase B from *Candida antarctica* (CALB) has been shown to be an excellent enantioselective biocatalyst for the stereo-selective acylation of racemic alcohols [14, 15]. The most often used commercial preparation of CALB is Novozym 435, where the enzyme is immobilized on a macroporous acrylic resin and the matrix presents about 90% of the total mass.

In this research the kinetic resolution of 1-phenylethanol catalyzed by commercially available immobilized lipase from CALB was assayed in non-aqueous conditions in SC-CO_2 and IL/SC-CO_2 systems with the aim of studying the enantioselectivity of Novozym 435. The influence of different reaction parameters, such as pressure, the acyl donor/alcohol molar ratio and different ILs, on the enantiomerically pure compound (*R*)-1-phenylethyl acetate formation via kinetic resolution of 1-phenylethanol was investigated.

8.2 SCFs – Replacement for Organic Solvents in Biocatalysis

The breakthrough of biocatalysis in non-aqueous media started in the early 1980s [16–19] when enzyme-catalyzed reactions (e.g. esterification and transesterification) began to be performed in organic solvents (e.g. hexane and heptane). Changing the hydrophobicity of predominantly aqueous media through the addition of organic solvents causes the hydrophobic effects, which keep the hydrophobic residues buried in the core of the proteins and folded in an aqueous environment where enzymes are kinetically trapped in their native structure in organic solvents [20–23]. Enzymes are frequently more stable in organic solvents than in water.

The use of organic solvents for food and health products is nowadays being progressively restricted. Therefore, attempts have been made to use SCFs as alternative solvents [24, 25]. They could, like organic solvents, stabilize the enzymes and dissolve the hydrophobic compounds and are able to shift the thermodynamic equilibrium of reaction toward synthesis. Moreover, SCFs possess gas-like diffusivities and low viscosities, which leads to a reduction in the mass transfer resistance between the reaction mixture and the active sites of the enzyme, resulting in increased reaction rates. In addition, the dielectric constant and density of an SCF are very sensitive to pressure and temperature. Consecutively the reaction environment and, thus, the activity and specificity of enzymes could be controlled by changing the pressure and temperature. Another advantage of using SCFs as a solvent for enzyme-catalyzed reactions is simpler downstream processing in comparison with those in liquid solvents. Since SCFs are usually gases at ambient conditions, the solvent can be easily removed without leaving any residues in the product required for industrial manufacture, cosmetic and pharmaceutical applications. Among the solvents that can be used in supercritical conditions for conducting biocatalytic reactions, the most studied one has been CO_2, because of its low critical temperature. Apart from being non-toxic, easily available, and cheap, many enzymes have proven to be stable and active in it [26]. Enzymes as biocatalysts require narrow operation parameters. Elevated pressures and temperatures, as well as an extreme pH, may lead to deactivation of the protein. In SCFs there are direct effects of pressure on the enzyme activity, which may lead to denaturation and indirect effects of pressure on the enzymatic activity. In the case of SC-CO_2 only small direct effects of pressure with regard to enzyme inactivation are expected. The protein structure should retain on the whole and only local changes may occur. Those local changes may lead to another active state of a protein, which may posses an altered activity, specificity, and stability. Pressure is also likely to affect the reaction indirectly by changing either the rate constant or the reactants solubility. At higher pressures more solute–solvent interactions take place, resulting in a better solvent capacity.

Therefore, the pressure effect on the enantioselectivity of commercially available immobilized CALB (Novozym 435) for the kinetic resolution of 1-phenylethanol was studied in non-aqueous SC-CO_2.

8.3
Effect of Pressure

An enzyme stability test in SC-CO_2 was performed at different pressures. CALB was incubated overnight at certain conditions and afterwards its residual activity was determined. No direct influence of the pressure on its biocatalyst activity was found.

When performing the resolution of 1-phenylethanol, the conversion dependence on the pressure was noticed at the near-critical region when the reaction was performed in almost non-aqueous SC-CO_2. The pressure was varied from 6.7 to 19 MPa. At a temperature of 313.15 K the yield in the desired enantiopure compound ((*R*)-1-phenlyethyl acetate) rose when the pressure was increased up to 9 MPa due to the solubility change with pressure near the critical point. The phase behavior of the system also plays an important role in the reaction. The rate-limiting step of (*R*)-1-phenlyethyl acetate formation in SC-CO_2 was the diffusion of substrates from the reaction bulk through the boundary layer to the external surface of the enzyme support. Because of the enhanced mass transfer the reaction rate increased in the pressure range between 6.7 and 9 MPa, which can be observed from Figure 8.1.

Visual observations of the phase behavior at different reaction conditions are presented in Figure 8.2. The reaction was carried out in a liquid reaction bulk in the contact with the solid biocatalyst at 9 MPa. At a pressure close to 11 MPa at the chosen temperature and composition conditions the supercritical phase was attained for the whole reaction bulk.

On the basis of the biocatalyst stability results it was obvious that there were indirect effects that influenced the reaction, for example the density-dependent physical properties were affected by the change in pressure. An increase in pressure led to an enhanced solvent density with improvement in its solvating power in the reaction bulk. The solubility in the liquid reaction mixture increased with pressure as well. In the liquid subcritical region rich in CO_2, at pressures below

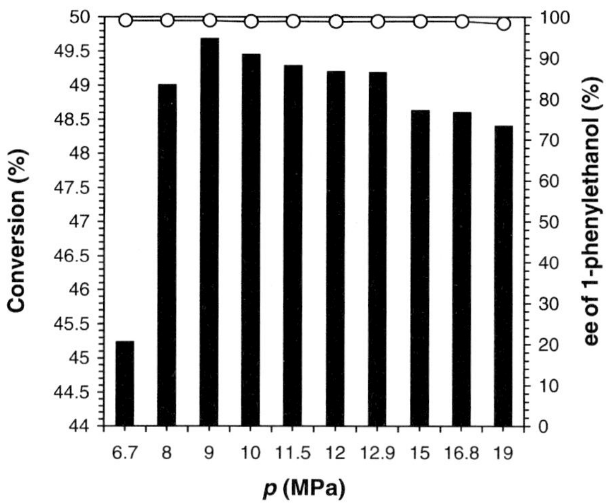

Figure 8.1 Pressure effect on the kinetic resolution of 1-phenylethanol, catalyzed by immobilized CALB in SC-CO_2 after five hours of reaction: bars – dependence of the conversion on the pressure; spots – pressure dependence of the enantiomeric excess for 1-phenylethanol.

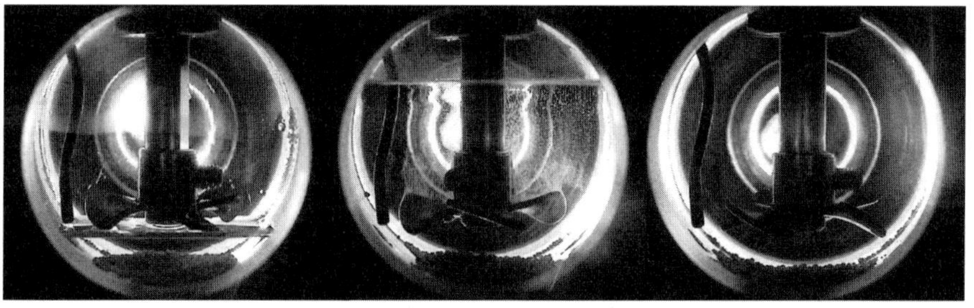

Figure 8.2 Visual vapour-liquid phase equilibrium observations of the reaction bulk: liquid reaction bulk in contact with the solid biocatalyst up to 6 MPa (left panel); liquid reaction bulk in contact with the solid biocatalyst up to 9 MPa; (middle panel); supercritical reaction bulk in contact with the solid catalyst upon 15 MPa (right panel).

9 MPa the diffusivity increased significantly and the surface tension and viscosity were reduced. Kinetic observations on the pressure effect showed that that liquid reaction mixture rich in SC-CO_2 led to higher performance than when diluted in a single supercritical phase. The highest conversion was obtained at 9 MPa. At pressures above 9 MPa the conversion slightly decreased: with a larger amount of CO_2 at higher pressures the dilution effect of the substrates appeared and the reaction was slowed down. At 15 MPa only one phase was present. In this case the higher density was counter-balanced by the substrates' mole fraction decrease in the enzyme phase.

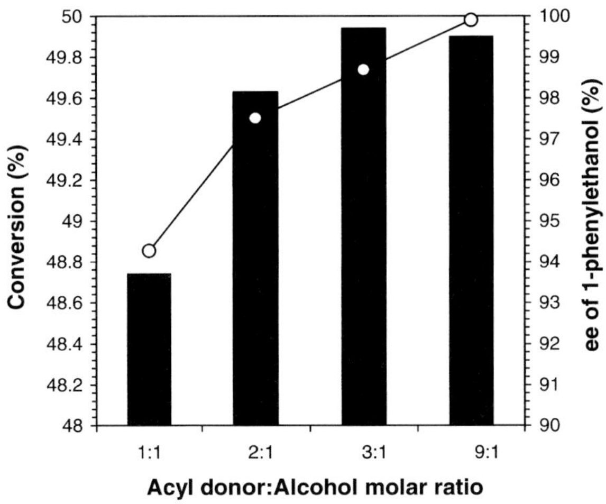

Figure 8.3 Influence of acyl donor/alcohol molar ratio on the kinetic resolution of 1-phenylethanol, catalyzed by immobilized CALB in SC-CO_2 after 5 hours of reaction: bars – dependence of the conversion on the pressure; spots – dependence of the enantiomeric excess on the pressure.

8.4
Effect of the Acyl Donor/Alcohol Molar Ratio

For a reversible reaction an increase in the acyl donor concentration results in higher product yields. In this case the chemical equilibrium is shifted towards synthesis. On the other hand, high concentrations of substrates may cause inhibition and the reaction is slowed down. For (*R*)-1-phenylethyl acetate formation the effect of the substrate vinyl acetate/1-phenylethanol molar ratio on the final conversion was studied. The results are presented on Figure 8.3. A higher yield of the enantiopure compound was achieved when raising the acyl donor molar concentration with respect to the alcohol concentration. A conversion of 49.9% was obtained at an acyl donor/alcohol molar ratio of 9/1. After 5 h of reaction at tested conditions a complete conversion of (*R*)-1-phenylethanol into the enantiopure (*R*)-1-phenylethyl acetate was attained. The enantiomeric excess for reactants (ee_R) was 99.9%.

8.5
ILs – Solvents for Sustainable Technology in Biocatalysis

ILs represent a new class of salts that are distinguished by a range of useful properties such as negligible vapor pressure, thermal stability, non-flammability, a high ionic conductivity, and remarkable solubility properties and have rapidly found a place of choice as valuable environmentally benign media substitutes for many volatile solvents [8, 27–30]. By modification of the cation and of the anion, their properties may be widely changed.

The commonly used cations for IL synthesis are based on imidazolium, ammonium, phosphonium, pyridinium, pyrrolidinium, and so on. ILs anions are either organic $[RCO_2]^-$ or inorganic and can be classified in two groups: polynuclear ($Al_2Cl_7^-$, $Al_3Cl_{10}^-$, $Au_2Cl_7^-$, $Fe_2Cl_7^-$ ...) and mononuclear anions which lead to neutral stoichiometric ILs (BF_4^-, PF_6^-, SbF_6^-, $N(CF_3SO_2)_2^-$ ...) [31–33].

The properties and behavior of the IL can be adjusted separately for each individual reaction and, therefore, they can be described as designer solvents. By choosing the correct IL, high product yields can be obtained and a reduced amount of waste can be produced in a given reaction. Often the IL can be recycled and this leads to a reduction in the costs of the processes. Reactions in ILs are not difficult to perform and usually require no special apparatus or methodologies. The reactions are often quicker and easier to carry out than in conventional organic solvents.

Several thermodynamic and kinetic behaviors of enzyme-catalyzed reactions performed in ILs, with respect to enzymatic reactions carried out in conventional solvents, could lead to an improvement in the process performance [34–37]. ILs showed an over-stabilization effect on biocatalysts [38] on the basis of the double role played by these neoteric solvents: ILs could provide an adequate microenvironment for the catalytic action of the enzyme (mass transfer phenomena and active catalytic conformation) and if they act as a solvent, ILs may be regarded as liquid immobilization supports, since multipoint enzyme-IL interactions (hydrogen, Van der Waals, ionic, etc.) may occur, resulting in a flexible supramolecular not able to maintain the active protein conformation [39]. Their polar and non-coordinating properties hold considerable potential for enantioselective reactions since profound effects on reactivities and selectivities are expected [40]. In recent years attention has been focused on the application of ILs as reaction media for enantioselective processes [41–43].

Extraction of the product from the IL in pure form can pose a problem. Water soluble compounds can be easily extracted with water. For separation of compounds with high vapor pressure the distillation could be used and higher temperatures would be required in order to extract products with low vapor pressures, which could result in decomposition of the product.

A great number of researchers have investigated the high-pressure phase behavior of IL/CO_2 systems and found that CO_2 is highly soluble in most ILs and that ILs are not measurably soluble in SC-CO_2. Volatile and non-polar SC-CO_2 together with non-volatile and polar ILs forms a new system with unique properties that have been utilized for extracting organic compounds from ILs using SC-CO_2 [3].

The influence of different ILs, based on N, N′dialkylimidazolium cations as reaction media, on the enantioselectivity of commercially available immobilized lipase from CALB for the kinetic resolution of 1-phenylethanol was studied. Further, the performance of the enzymatic kinetic 1-phenylethanol resolution by CALB in the system [bmim][PF_6]/SC-CO_2 was studied.

8.6 ILs, Based on the N, N′-Dialkylimidazolium Cations as Reaction Media

The use of ILs contributes many advantages in biocatalysis, including enhanced stability, activity, and stereoselectivity [41, 44, 45]. When combined with SC-CO_2, ILs showed a promising media for enhanced stereoselectivity and stability of biocatalysts [9, 46–48]. The use of SC-CO_2 and room temperature ILs as a combined bioreaction media has been widely demonstrated [5, 7, 8]. There has been much interest in research on an efficient enzymatic system in IL/CO_2 media for racemates kinetic resolution [49, 50]. Due to the wide spectrum of their physicochemical properties, several ILs, based on the N, N′-dialkylimidazolium cations were used for testing the enantioselectivity of CALB [34–36].

To screen different ILs for the kinetic resolution of 1-phenylethanol, ILs associated with mononuclear anions, such as tetrafluoroborate $[BF_4^-]$, hexafluorophosphate $[PF_6^-]$, and triflimide $[NTf_2^-]$, based on dialkylimidazolium cations, were used as reaction media at atmospheric pressure and 313.15 K. All the ILs examined were found to be proper media for the biocatalytical reaction system studied in terms of enantioselectivity. The inclusion of the CALB in the IL matrix may provide the protein with a conformation suited to its enantioselective action. The highest conversion was achieved in [bmim]$[PF_6]$ and in [emim]$[NTf_2]$ a slow enzyme deactivation was observed, as can be observed from Figure 8.4. Low hydrogen bond basicity minimizes interference with the internal hydrogen bonds of the enzyme [51]. In all three ILs, $[BF_4^-]$, $[PF_6^-]$, and $[NTf_2^-]$, the anions have low hydrogen bond basicity, as the $[BF_4^-]$ anion spreads its negative charge over four fluorine

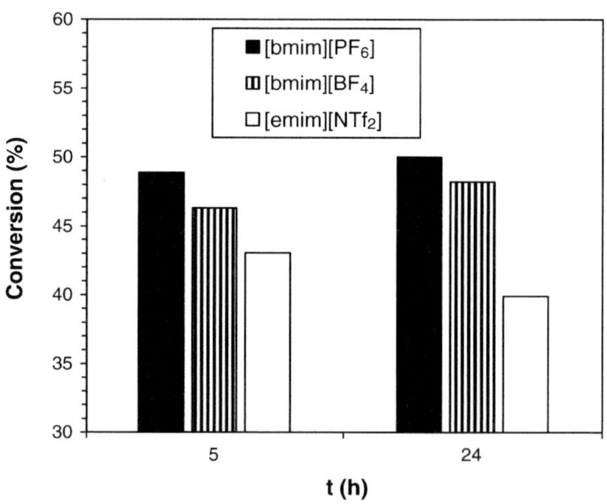

Figure 8.4 Three different ILs; [bmim]$[PF_6]$, [bmim]$[BF_4]$ and [emim]$[NTf_2]$ as reaction media for the enzymatic kinetic resolution of 1-phenylethanol by immobilized CALB after five and 24 hours of reaction.

atoms, the [NTf_2^-] anion over five atoms, and the [PF_6^-] anion over six fluorine atoms and great activity of CALB was detected.

The different properties of ILs, with regard to their polarity, hydrophobicity, and solvent miscibility behavior through combination with different anions, are the reason for the different biocatalyst activities. Good to excellent activity of CALB was observed with a decrease in polarity and hydrophobicity and a viscosity increase of the ILs. In [bmim][PF_6] a conversion of (*R*)-1-phenylethanol into the ester of 48.9% and an ee of 95.6% were achieved after 5 h and 100% of (*R*)-1-phenylethanol was converted into the enantiopure (*R*)-1-phenylethyl acetate after a 1-day reaction. Immobilized CALB exhibited excellent stability, activity, and selectivity towards the (*R*)-enantiomer of 1-phenylethanol in [bmim][PF_6]. In some research bis(trifluoromethylsulfonyl)imide-based ILs have been regarded as very suitable media for biocatalysis [39, 46, 50]. On the contrary, in the present work, lower suitability of the same IL was demonstrated. Since immobilized CALB catalyzed both hydrolytic and transesterification reactions, its enantioselectivity for long reaction times was lower.

8.7
ILs/SCFs Biphasic Systems as Promising Media for Biocatalysis

Enzyme transformations in ILS/SC-CO_2 biphasic systems, whereby enzyme molecules are 'immobilized' in the IL phase and substrates/products are transported by the SC-CO_2 phase, are described as a way for carrying out clean synthetic chemical processes to produce pure products [52].

8.8
The [bmim][PF_6]/SC-CO_2 System as a Reaction Medium

Kinetic resolution of 1-phenylethanol catalyzed by CALB was carried out in the IL/SC-CO_2 biphasic system. To prevent undesirable reactions and ensure better conversion of (*R*)-1-phenylethanol, [bmim][PF_6] was chosen for this kind of experiments. Because of the possible direct and indirect effects of the pressure on the activity of biocatalyst its influence was studied between 6 and 36.5 MPa. At all conditions examined a biphasic reaction medium was attained, which is illustrated in Figure 8.5. The enzyme was suspended in the IL phase, where the reaction took place. The substrates and products resided largely in the supercritical phase, which was also the extractive phase.

A strong influence of the pressure on the reaction was registered at pressures lower than 16 MPa, where an important increase in the CO_2 molar fraction in the liquid phase appeared. The phase behavior of CO_2/[bmim][PF_6] showed that the solubility of SC-CO_2 in this IL markedly increased up to 16 MPa, where up to 65% mol/mol of CO_2 was registered in the liquid phase. At 16 MPa the highest conversion of (*R*)-1-phenylethanol into the enantiopure (*R*)-1-phenylethyl acetate was the

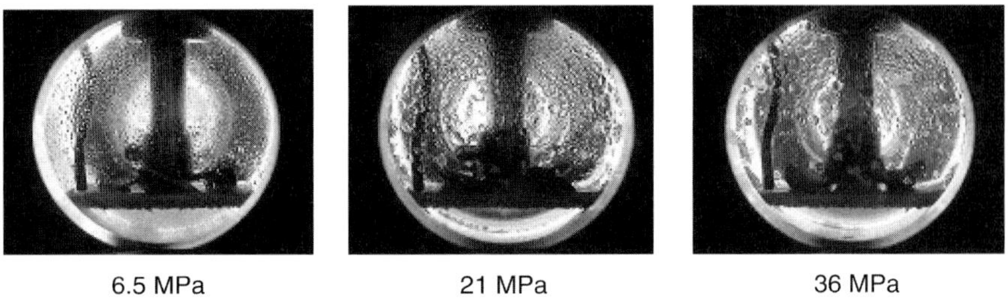

Figure 8.5 Visual vapour-liquid phase equilibrium observations of the reaction bulk, where liquid reaction bulk was in the contact with the solid biocatalyst and the upper phase was saturated with substrates and products, extracted by SC-CO_2.

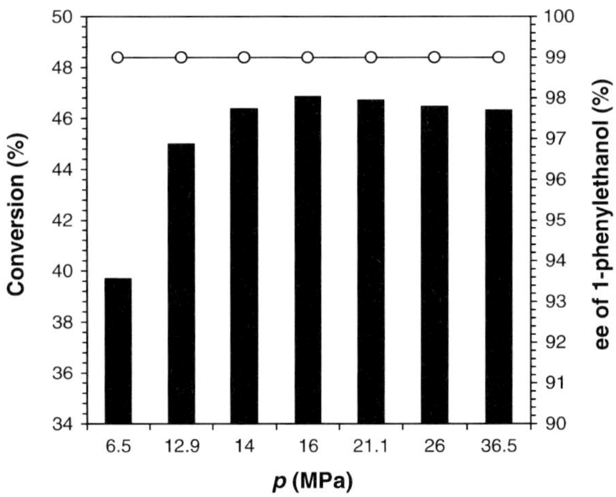

Figure 8.6 Performance of the enzymatic kinetic 1-phenylethanol resolution over immobilized CALB in [bmim][PF_6]/SC-CO_2 system at 313.15 K by varying the pressure: bars–dependence of the conversion on the pressure; spots–dependence of the enantiomeric excess on the pressure.

highest, approx. 46.8%, resulting in enantiomeric excess of 88.1%. From Figure 8.6 it may be observed that at pressures above 16 MPa no strong influence on the biocatalyst activity was detected.

Biocatalyst particle shape and size characterization before and after treatment in the two selected media was examined in order to ascertain the role of the solvent on enzyme structure. The crude biocatalyst preparation consisted of spherical micro-granules with a mean particle size of about 500 μm. No particle size modifications were observed by SEM when the biocatalyst was treated by SC-CO_2. After treatment in [bmim][PF_6] a lower mean particle size of about 430 μm was observed, probably due to partial carrier dissolution. Conversely, by exposing the biocatalyst particles to the combination of [bmim][PF_6]/SC-CO_2 no changes in the particle size were detected.

The IL/SC-CO_2 media for 1-phenylethanol resolution, catalyzed by CALB, offers a number of advantages, such as high resolution yields, simple downstream processing, guarantying an enzyme overstabilization effect and exceptional enzyme enantioselectivity with no transformation of the not desired enantiomer, albeit apparently lower conversions were registered in this medium compared to the ones, obtained in SC-CO_2.

8.9
Effect of Acyl Donor Concentration

The influence of acyl donor concentration in the IL/SC-CO_2 medium was exploited in order to achieve higher enantiopure yield (Figure 8.7) at 313.15 K and 16 MPa. The highest conversion was achieved at the same substrate composition as in SC-CO_2, namely vinyl acetate/1-phenylethanol molar ratio of 9/1. No (*S*)-1-phenylethanol conversion was detected at all tested conditions, which means an enantiomeric excess for products evaluated higher than 99.9%. Optimal conversion, achieved after five hours of reaction, was 47.2% and enantiomeric excess for reactants was 89.5%. The conversion could be maximized with higher amount of biocatalyst in the reaction mixture. As expected, after five hours of bioconversion, approximately complete conversion 49.9% of (*R*)-1-phenylethanol into the enantiopure (*R*)-1-phenylethyl acetate was achieved. Enantiomeric excess for reactants was 99.3%.

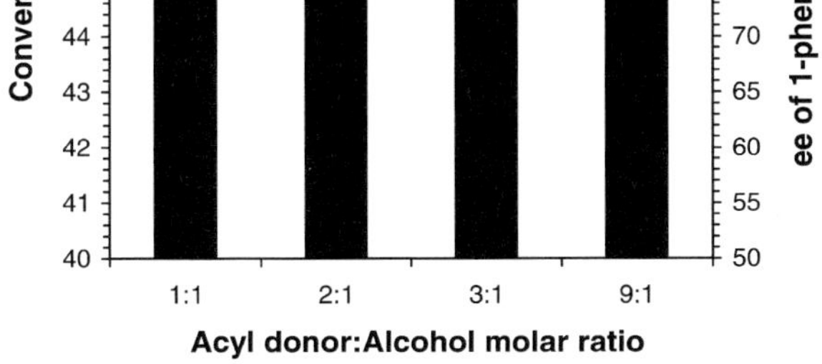

Figure 8.7 Acyl donor/alcohol molar ratio effect on the reaction of the enzymatic kinetic 1-phenylethanol resolution by CALB in [bmim][PF_6]/SC-CO_2.

8.10 Conclusion

Kinetic resolution of racemic 1-phenylethanol was successfully performed by CALB as the enantioselective biocatalyst using vinyl acetate as the acyl donor.

High yield, enantioselectivity and simple downstream processing make SC-CO_2 and IL/SC-CO_2 systems excellent media for the above mentioned kinetic resolution of *sec*-alcohol. Pressure effects the reaction to a certain extent, which was explained by exploiting thermodynamic arguments. Besides, the yield may be improved by increasing the acyl donor concentration.

The outlook of syntheses, based on using SC-CO_2 or appropriate IL/SC-CO_2 biphasic system for developing integral green chemical processes due to the physical and chemical characteristics of these neoteric solvents and the enhanced enzyme catalytic properties seems promising.

References

1 Zaks, A. and Dodds, D.R. (1997) *DDT*, **2**, 513–531.

2 Rasor, J.P. and Voss, E. (2001) *Applied Catalysis A: General*, **221**, 145–158.

3 Keskin, S., Kayrak-Talay, D., Akman, U. and Hortaçsu. Ö. (2007) *The Journal of Supercritical Fluids*, **43** (1), 150–180.

4 Gordon, C.M. (2001) *Applied Catalysis A: General*, **222** (1–2), 101–117.

5 Brennecke, J.F. and Maginn, E.J. (2001) *AIChE Journal*, **47** (11), 2384–2389.

6 Blanchard, L.A., Brennecke, Z. and Gu, J.F. (2001) *The Journal of Physical Chemistry B*, **105** (12), 2437–2444.

7 Blanchard, L.A. and Brennecke, J.F. (2001) *Industrial and Engineering Chemistry Research*, **40** (1), 287–292.

8 Blanchard, L.A., Hancu, D., Beckman, E.J. and Brennecke, J.F. (1999) *Nature*, **399** (6731), 28–29.

9 Lozano, P., De Diego, T., Carrié, D., Vaultier, M. and Iborra, J.L. (2002) *Chemical Communications*, **7**, 692–693.

10 Lozano, P., De Diego, T., Carrié, D., Vaultier, M. and Iborra, J.L. (2004) *Journal of Molecular Catalysis A – Chemical*, **214** (1), 113–119.

11 de los Ríos, A.P., Hernández-Fernández, F.J., Gómez, D., Rubio, M., Tomás-Alonso, F. and Víllora, G. (2007) *The Journal of Supercritical Fluids*, **43** (2), 303–309.

12 Reetz, M.T., Wiesenhofer, W., Francio, G. and Leitner, W. (2003) *Advanced Synthesis Catalysis*, **345** (11), 1221–1228.

13 Kataoka, M., Kita, K., Ada, M., Yasohara, Y., Hasegawa, J. and Shimizu, S. (2003) *Applied Microbiology and Biotechnology*, **62** (5–6), 437–445.

14 van Rantwijk, F., Lau, R.M. and Sheldon, R.A. (2003) *Trends in Biotechnology*, **21** (3), 131–138.

15 Ottosson, J. and Hult, K. (2001) *Journal of Molecular Catalysis B: Enzymatic*, **11** (4–6), 1025–1028.

16 Antonini, E., Carrea, G. and Cremonesi, P. (1981) *Enzyme and Microbial Technology*, **3** (4), 291–296.

17 Martinek, K., Semenov, A.N. and Berezin, I.V. (1981) *Biochimica et Biophysica Acta*, **658** (1), 76–89.

18 Martinek, K., Levashov, A.V., Khmelnitsky, Y.L., Klyachko, N.L. and Berezin, I.V. (1982) *Science*, **218** (4575), 889–891.

19 Zaks, A. and Klibanov, A.M. (1984) *Science*, **224** (4654), 1249–1251.

20 Klibanov, A.M. (2001) *Nature*, **409** (6817), 241–246.

21 Partridge, J., Moore, B.D. and Hailing, P.J. (1999) *Journal of Molecular Catalysis B: Enzymatic*, **6** (1–2), 11–20.

22 Griebenow, K., Klibanov, and A.M. (1996) *Journal of the American Chemical Society*, **118** (47), 11695–11700.

23 Wescott, C.R. and Klibanov, A.M. (1994) *Biochimica et Biophysica Acta*, **1206** (1), 1–9.

24 Chulalaksananukul, W., Condoret, J.S. and Combes, D. (1993) *Enzyme and Microbial Technology*, **15** (8), 691–698.

25 Knez, Ž. and Habulin, M.(2002) *The Journal of Supercritical Fluids*, **23** (1), 29–42.

26 Taniguchi, M., Masamichi, K. and Kobayashi, T. (1987) *Agricultural and Biological Chemistry*, **51** (2), 593–594.

27 Madeira Lau, R., van Rantwijk, F., Seddon, K.R. and Sheldon, R.A. (2000) *Organic Letters*, **2** (26), 4189–4191.

28 Wasserscheid, P. and Keim, W. (2000) *Angewandte Chemie – International Edition*, **39** (21), 3772–3789.

29 Welton, T. (1999) *Chemical Reviews*, **99** (8), 2071–2083.

30 Earle, M.J. and Seddon, K.R. (2000) *Pure and Applied Chemistry*, **72** (7), 1391–1398.

31 Olivier-Bourbigou, H. and Magna, L. (2002) *Journal of Molecular Catalysis A – Chemical*, **182** (1), 419–437.

32 Zhao, D.B., Wu, M. and Kou, Y. (2002) *Catalysis Today*, **74** (1–2), 157–189.

33 Liu, J.-F., Jönsson, J.A. and Jiang, G.-B. (2005) *Trends in Analytical Chemistry*, **24** (1), 20–26.

34 Eckstein, M., Sesing, M., Kragl, U. and Adlercreutz, P. (2002) *Biotechnology Letters*, **24** (11), 867–872.

35 Eckstein, M., Wasserscheid, P. and Kragl, U. (2002) *Biotechnology Letters*, **24** (10), 763–767.

36 Lozano, P., De Diego, T., Carrié, D., Vaultier, M. and Iborra, J.L. (2003) *Biotechnology Progress*, **19** (2), 380–382.

37 Persson, M. and Bornscheuer, U.T. (2003) *Journal of Molecular Catalysis B: Enzymatic*, **22** (1–2), 21–27.

38 Lozano, P., De Diego, T., Guegan, J.-P., Vaultier, M. and Iborra, J.L. (2001) *Biotechnology and Bioengineering*, **75** (5), 563–569.

39 De Diego, T., Lozano, P., Gmouh, S., Vaultier, M. and Iborra, J.L. (2005) *Biomacromolecules*, **6** (3), 1457–1464.

40 Baudequin, C., Baudoux, J., Levillain, J., Cahard, D., Gaumont, A.C. and Plaquevent, J.C. (2003) *Tetrahedron: Asymmetry*, **14** (20), 3081–3093.

41 Schöfer, S.H., Kaftzik, N., Wasserscheid, P. and Kragl, U. (2001) *Chemical Communications*, **5**, 425–426.

42 Sheldon, R. (2001) *Chemical Communications*, **23**, 2399–2407.

43 Kragl, U., Kaftzik, N., Schöfer, S.H., Eckstein, M., Wasserscheid, P., Hilgers, C. and Oggi, C. (2001) *Chemistry Today*, **19** (7–8), 22–24.

44 Howarth, J., James, P. and Dai, J.F. (2001) *Tetrahedron Letters*, **42** (42), 7517–7519.

45 Reetz, M.T., Wiesenhöfer, W., Franciò, G. and Leitner, W. (2002) *Chemical Communications*, **9**, 992–993.

46 Lozano, P., De Diego, T., Gmouh, S., Vaultier, M. and Iborra, J.L. (2004) *Biotechnology Progress*, **20** (3), 661–669.

47 Dzyuba, S.V. and Bartsch, R.A. (2003) *Angewandte Chemie – International Edition*, **42** (2), 148–150.

48 Garcia, S., Lourenco, N.M.T., Lousa, D., Sequerira, A.F., Mimoso, P., Cabral, J.M.S., Afonso, C.A.M. and Barreiros, S. (2004) *Green Chemistry*, **6** (9), 466–470.

49 Leitner W. (2004) *Pure and Applied Chemistry*, **76** (3), 635–644.

50 Lozano, P., De Diego, T., Larnicol, M., Vaultier, M. and Iborra, J.L. (2006) *Biotechnology Letters*, **28** (19), 1559–1565.

51 Kaar, J.L., Jesionowski, A.M., Berberich, J.A., Moulton, R. and Russell, A.J. (2003) *Journal of the American Chemical Society*, **125** (14), 4125–4131.

52 Lozano, P., De Diego, T., Gmouh, S., Vaultier, M. and Iborra, J.L. (2007) *International Journal of Chemical Reactor Engineering*, **5** (A53), 1–10.

9
Strategies for the Biocatalytic Lipophilization of Phenolic Antioxidants

Maria H. Katsoura, Eleni Theodosiou, Haralambos Stamatis and Fragiskos N. Kolisis

9.1
Introduction

Natural polyhydroxylated phenolic compounds, including flavonoid and phenolic glucosides as well as flavonolignans, are widely distributed among various plants known for their antioxidant activity and antimicrobial, anticarcinogenic, and anti-inflammatory properties [1, 2]. Because of their biological activities, these natural compounds can be used in pharmaceuticals and cosmetics as well as food additives [1]. However, their poor solubility and stability in different media, especially lipidic matrices and lipophilic media (fats, oils and emulsions), limit their application. Moreover, their biological activity seems to depend not only on their chemical structure [3, 4], but also on their degree of lipophilicity, which could enhance their uptake into cells or influence their interaction with proteins and enzymes [5, 6].

The modification of these natural polyhydroxylated compounds via acylation of the hydroxyl functions with aliphatic molecules not only increases their structural diversity, producing analogs that may be useful models for the study of structure–activity relationships, but also changes their physicochemical properties, increasing their solubility in lipophilic media. Moreover, the selective acylation of these natural compounds with various acyl donors could enhance their biological activities, such as their antioxidant and antimicrobial activity, as well as their pharmacological properties [5, 6].

The use of enzymes in non-aqueous organic media for the modification of various substrates, including natural compounds, has been introduced as an advantageous new approach in the last two decades [7, 8]. Several research groups have reported the feasibility of the enzymatic modification of various polyhydroxylated compounds, such as flavonoids, sugars, and glucosides, in both toxic and less toxic organic media using either lipases or proteases [9–12]. However, the use of organic solvents as media for the biocatalytic modification of such polyhydroxylated natural compounds has often shown several disadvantages, such as low yields and limited productivity that arise from the reduced solubility of such compounds in these media [12, 13].

Modern Biocatalysis: Stereoselective and Environmentally Friendly Reactions.
Edited by Wolf-Dieter Fessner and Thorleif Anthonsen

ISBN: 978-3-527-32071-4

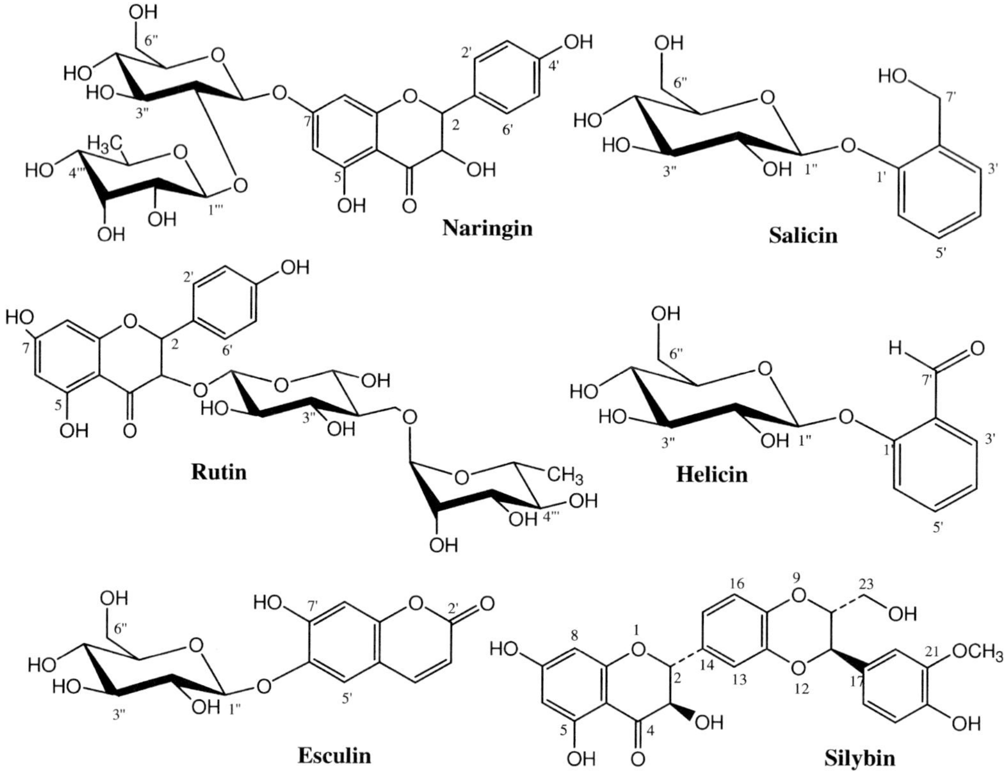

Figure 9.1 Structures of natural antioxidants.

On the other hand, several groups have reported the use of ionic liquids (organic salts consisting only of ions, liquid at or near room temperature) as media for the biocatalytic modification of polyhydroxylated compounds, including ascorbic acid [14], sugars, and glucosides [15, 16] as well as flavonoids [5, 17]. Because of their specific physicochemical characteristics, such as lack of vapor pressure, thermal stability, and their ability to dissolve many kinds of compounds, including polar or less polar organic compounds, ionic liquids can be used as alternative green media for various biotransformations [18–20]. It is interesting to note that, toxicological tests of a series of ionic liquids with imidazolium cations and various anions (including PF_6^- and BF_4^-) showed that most of the ionic liquids were less toxic than organic solvents by several orders of magnitude [21, 22]. The reduced toxicity of some ionic liquids is an important advantage for their use as media for the biocatalytic modification of compounds that can be used as food additives or in pharmaceutical and cosmetic formulations.

The use of various low-water reaction systems, including anhydrous organic solvents as well as imidazolium-based ionic liquids, is investigated for the biocatalytic preparation of lipophilic derivatives of various natural phenolic compounds (Figure 9.1) including phenolic and flavonoid glycosides (esculin, salicin, helicin, naringin, and rutin) as well as flavonolignan (silybin). The effects of various reaction parameters on the conversion yield and the regioselectivity of the biocatalytic process in different reaction media are discussed.

9.2 Materials and Methods

9.2.1 Materials

Immobilized lipase B from *Candida antarctica* (CALB) (Novozym 435) was obtained from Novo Nordisk Novozymes. All chemicals and ionic liquids used were of the highest available purity.

9.2.2 Enzymatic Acylation Procedure

The enzymatic acylation of natural polyhydroxylated compounds in ionic liquids ([bmim]BF_4 and [bmim]PF_6), and organic solvents was carried out in stirred flasks according to the procedure reported elsewhere [5, 6, 17].

9.2.3 Analytical Methods

Quantitative analysis of reaction mixtures was performed by high-performance liquid chromatography (HPLC) and a diode array ultraviolet detector, as reported elsewhere [5, 6, 17].

9.2.4 Purification and Chemical Structure Determination of Esters

Highly purified esters were obtained by preparative HPLC [5, 17]. The chemical structure of the purified monoesters was determined by ^{13}C and ^{1}H nuclear magnetic resonance (NMR) and their molecular masses were determined by a mass spectroscopy (MS) technique as reported elsewhere [5, 17].

9.3 Results and Discussion

Herein we discuss the enzymatic modification of natural polyhydroxylated compounds in both organic and ionic liquid media catalyzed by commercially available immobilized CALB.

9.3.1
Modification of Natural Antioxidants in Organic Solvents

9.3.1.1 Enzymatic Acylation of Rutin and Silybin with Dicarboxylic Acids

Silybin, the most biologically active compound of the flavonolignan mixture silymarin, has three phenolic, one primary aliphatic, and one secondary aliphatic hydroxyl group, while rutin has a sugar moiety with only secondary hydroxyl groups. A plethora of studies support that CALB exhibits selectivity towards not only primary OH groups, but towards secondary OH groups as well [5, 23–26]. The lipase preferentially acylates the primary OH group with less steric hindrance and the secondary OH group is acylated next where there is no primary OH available [27]. The regioselectivity of the biocatalyst is affected by the type of the sugar moiety attached to the aglycon flavonoid, in respect to the presence of primary OH groups as well as steric obstructions [5, 12, 28]. Apart from the nature of the polyhydroxylated natural compound, the nature of the acyl donor could influence the performance of the biocatalytic reaction [9, 17]. The acylation of silybin and rutin with three dicarboxylic acids (hexanedioic acid, dodecanedioic acid, and hexadecanedioic acid) is investigated in *tert*-amyl alcohol. Table 9.1 shows the effect of the nature of the antioxidant and acyl donor on the performance of the reaction.

Higher conversion yields were observed for all dicarboxylic acids used in the case of silybin, as compared to rutin, due to the presence of primary hydroxyls in this flavonolignan for which the lipase shows higher specificity. Analysis of the chemical structures of purified products by NMR revealed that the acylation of rutin occurred at the 4‴′-OH of the rhamnose moiety, while the aylation of silybin occurred at its unique primary hydroxyl at position C23. It must be noted that, in the case of silybin, both monoester and diester formation was observed and identi-

Table 9.1 Effect of the reaction medium on the conversion yield after 96 h of incubation for the acylation of rutin (7.5 mM) and silybin (7.5 mM for tert-amyl alcohol and *tert*-butyl alcohol and 30 mM for acetone) with dicarboxylic acids catalyzed by immobilized CALB at 50 °C.

Reaction medium	Conversion %					
	Rutin			Silybin		
	C_6	C_{12}	C_{16}	C_6	C_{12}	C_{16}
tert-Amyl alcohol	13.1	12.3	10.5	32.0	30.7	38.0
tert-Butyl alcohol	6.2	8.0	6.4	17.7	13.8	22.4
Acetone	n.d.[a]	n.d.[a]	n.d.[a]	31.3	23.3	45.5

a Not determined.
The substrates were used in equimolar amounts.

fied by MS spectroscopy. Irrespectively of the system used, the oxidized form of silybin (2,3-dehydrosilybin) was also present. When silybin was used as the substrate a higher conversion yield was observed when hexadecanedioic acid was used as the acyl donor, while in the case of rutin no significant effect of the carbon chain length on the conversion yield was observed.

9.3.1.2 Effect of Organic Solvent

In order to investigate the effect of organic solvent, *tert*-amyl alcohol, *tert*-butyl alcohol, and acetone, were chosen as reaction media, mainly because of their compatibility with the reaction studied, as well as their low toxicity [29].

As can been seen in Table 9.1 the use of *tert*-butyl alcohol led to lower conversion yields both for rutin and silybin as compared to *tert*-amyl alcohol. In the case of acetone, higher conversion yields were obtained for C_{12} and C_{16} dicarboxylic acids. The solubility capacity of acetone for silybin is about 10 times higher (~40 g/l) than *tert*-amyl alcohol and *tert*-butyl alcohol. The higher solubility of silybin in acetone is advantageous, as the amount of esters obtained was high, reaching values of 4.5, 3.4, and 6.6 g/l for hexanedioic, dodecanedioic, and hexadecanedioic acid, respectively, after 96 h of incubation.

9.3.1.3 Effect of Substrate Concentration

In order to investigate the effect of the substrates' molar ratio on the reaction performance, the concentrations of the enzyme and antioxidant were kept constant, whereas that of the acyl donor was increased. In the case of rutin, independently of the molar ratio, used the acylation proceeded with low yields, while only monoesters were formed in all reaction systems used. The same behavior was observed in the case of silybin using *tert*-amyl alcohol and *tert*-butyl alcohol as reaction media (data not shown). In contrast, when acetone was used as medium the conversion yield for the acylation of silybin increased as the molar ratio was increased (Table 9.2).

Table 9.2 Effect of the molar ratio of the acyl donor to flavonolignan on the conversion yield and reaction rate after 96 h of incubation for the acylation of silybin (30 mM) with dicarboxylic acids catalyzed by immobilized CALB in acetone at 50 °C.

Molar ratio	Conversion % (initial reaction rates in mmol/h/g of biocatalyst)		
	C_6	C_{12}	C_{16}
1	31.3 (0.02)	23.3 (0.02)	45.5 (0.04)
2	44.5 (0.02)	33.2 (0.04)	56.5 (0.04)
5	59.0 (0.04)	42.5 (0.04)	66.8 (0.04)

9.3.2 Modification of Natural Antioxidants in Ionic Liquid Media

9.3.2.1 Enzymatic Acylation of Natural Polyhydroxylated Compounds

In order to investigate the efficiency and selectivity of the enzymatic acylation of various polyhydroxylated natural compounds in ionic liquid media the transesterification of salicin, helicin, esculin, and naringin, as well as silybin catalyzed by immobilized CALB was investigated using vinyl butyrate as the acyl donor. Two ionic liquids were used, the water-miscible 1-butyl-3-methylimidazolium tetrafluoroborate ([bmim]BF_4) and 1-butyl-3-methylimidazolium hexafluorophosphate ([bmim]PF_6), in which the solubility of water is limited.

The percentage conversion yield of the enzymatic acylation of the forenamed compounds in ionic liquids as compared to acetone is depicted in Figure 9.2.

As can be seen, immobilized CALB efficiently catalyzes the acylation of all polyhydroxylated compounds in the ionic liquid media used, leading to high conversion yields. The reaction rates for the enzymatic acylations are summarized in Table 9.3. Higher reaction rates were obtained for all polyhydroxylated compounds tested when [bmim]BF_4 was used as the reaction medium. It is interesting to note that the solubilities of esculin, salicin, helicin, naringin, and silybin at 60 °C were approximately 68, 23, 40, 100, and 82 mM, respectively, in [bmim]BF_4. On the other hand, in [bmim]PF_6 and acetone, in which the reactions rates were lower, the solubilities of the aforenamed compounds were 20, 5.5, 19.5, 1.5, and 7.2 mM and 6.5,

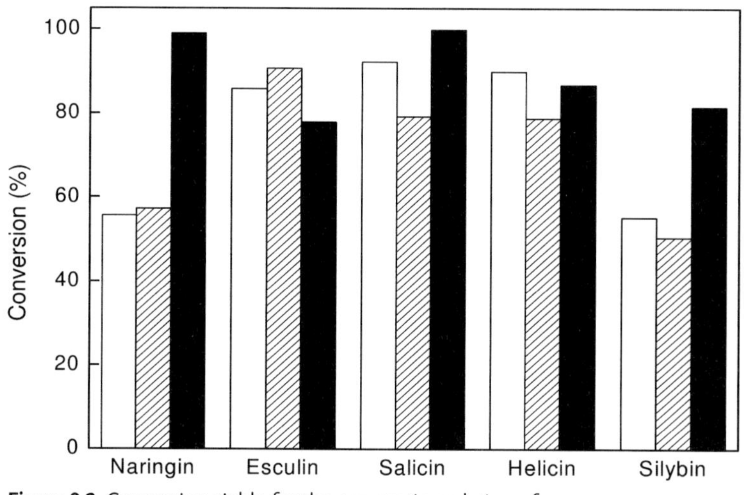

Figure 9.2 Conversion yields for the enzymatic acylation of different polyhydroxylated natural compounds (30 mM) with vinyl butyrate (300 mM) catalyzed by immobilized CALB in (a) [bmim]BF_4 (white bars) and (b) [bmim]PF_6 (patterned bars) at 60 °C and acetone (black bars) at 50 °C, after 72 h of incubation.

Table 9.3 Initial reaction rates and the fraction of ester present under the form of monoester for the acylation of various polyhydroxylated natural compounds (30 mM) with vinyl butyrate catalyzed by immobilized CALB in [bmim]BF_4 and [bmim]PF_6 at 60 °C and in acetone at 50 °C after 72 h of reaction.

Substrates	Initial reaction rates (mmol/h/g of biocatalyst) (mol% of monoester produced[a])		
	[bmim]BF_4	[bmim]PF_6	Acetone
Naringin	0.06 (86.8)	0.05 (63.0)	0.04 (45.5)
Esculin	0.04 (98.1)	0.03 (80.6)	0.03 (78.1)
Helicin	0.15 (68.3)	0.10 (64.6)	0.03 (52.8)
Salicin	0.08 (23.6)	0.05 (41.9)	0.03 (28.4)
Silybin	0.27 (100.0)	0.07 (100.0)	1.58 (100.0)

a Refers to percentage of the total ester formed.
The molar ratio of the substrates was 10.

2, 19.5, 12, and 8.3 mM, respectively. For the enzymatic acylation of all polyhydroxylated compounds studied in the present work, higher reaction rates were observed in ionic liquid media where the solubility of the substrates was higher, which is in accordance with the results reported by Park and Kazlauskas [16] for the enzymatic acetylation of glucose.

The monoacylated derivative (6″-*O*-butyrate) was detected as the major product by HPLC analysis for all enzymatic reactions, while a diacylated derivative was also detected and confirmed by MS analysis [17]. As it can be seen in Table 9.3, higher regioselectivity of the process was observed in [bmim]BF_4 than in [bmim]PF_6, while the selectivity was even lower in acetone. The lower regioselectivity of the enzymatic acylation of polyhydroxylated compounds in [bmim]PF_6 as well as in the organic solvent could be related to the lower solubility of unmodified phenolic substrates in these media, compared to that of their monoacylated derivatives [5, 16, 17]. On the other hand, the enhanced solubility of phenolic substrates in [bmim]BF_4 could explain the increased regioselectivity observed in this ionic liquid.

The enhanced solubility of all natural polyhydroxylated compounds tested in ionic liquids as compared to traditional organic solvents used for the enzymatic modification of these compounds can be highly useful for the production of great amounts of their acylated derivatives in a single-step biocatalytic process. It is noteworthy that, when the concentration of the above phenolic substrates was near their solubility limit, the amount of their monoacylated derivatives formed in [bmim]BF_4 reached values up to 30.0 g/l (in the case of naringin), which were considerably higher than those reported for organic media [10, 30].

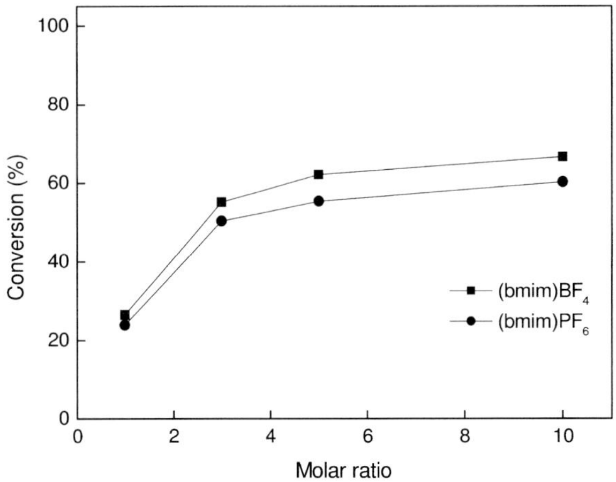

Figure 9.3 Effect of the molar ratio of vinyl butyrate to silybin (30 mM) on the conversion yield for its enzymatic acylation catalyzed by immobilized CALB in the ionic liquids [bmim]BF_4 and [bmim]PF_6 at 60 °C after 72 h of incubation.

9.3.2.2 Effect of Substrate Concentration

The effect of the molar ratio of vinyl butyrate to silybin on the conversion yield in the ionic liquids [bmim]BF_4 and [bmim]PF_6 is depicted in Figure 9.3.

As can be seen in both ionic liquids used, the increase in the molar ratio of the substrates from 3 to 15 had a positive effect on the conversion yield. The same positive effect was also observed in the case of naringin (data not shown) and can be attributed to a thermodynamic shift of the equilibrium towards the synthesis of products due to acyl donor excess [6, 29].

In addition to the molar ratio increase, the increase in the antioxidant concentration was observed to have a positive effect on the conversion yield using both silybin and naringin (data not shown).

9.3.2.3 Effect of Acyl Donor Nature: Synthesis of Hybrid Antioxidants

In order to investigate the effect of the nature of the acyl donor on the enzymatic acylation of naringin or silybin in ionic liquids several free fatty acids as well as their corresponding vinyl or methyl esters were used as acyl donors.

The results show that, both in the case of naringin and silybin, the acyl donor chain length greatly affects the conversion yield of the enzymatic process. Higher conversion yields were achieved in both ionic liquid media using short-chain acyl donors (up to four carbon atoms), while the use of long-chain acyl donors resulted in substantially lower conversion yields (Table 9.4).

However, in the case of acetone (results shown in Table 9.4) the enzyme shows a preference towards intermediate or high chain length fatty acids, which is in accordance with other reports for the enzymatic acylation of polyhydroxylated compounds catalyzed by immobilized CALB in conventional organic solvents

Table 9.4 Effect of the acyl donor chain length on the enzymatic acylation of silybin (30 mM) catalyzed by immobilized CALB in the ionic liquids [bmim]BF_4 and [bmim]PF_6 at 60 °C and in acetone at 50 °C after 72 h of incubation.

Acyl donors	Conversion (%)		
	[bmim]BF_4	[bmim]PF_6	Acetone
Caprylic acid	8.9	9.8	12.4
Lauric acid	5.8	6.4	21.5
Palmitic acid	1.1	1.5	28.4
Oleic acid	1.6	13.6	14.2
Vinyl butyrate	26.5	23.9	53.6
Vinyl laurate	3.9	3.8	75.4
Methyl oleate	2.3	23.7	24.6

The substrates were used in equimolar amounts.

[10, 23]. The different enzyme behavior observed in the case of ionic liquids can be attributed to the lower solubility of long chain acyl donors in these media, compared to the less polar organic solvents used for the enzymatic modification of natural polyhydroxylated compounds. Due to the low solubility of long chain acyl substrates in ionic liquids, a two-phase system was formed, which is expected to decrease the availability of substrates to the enzyme and therefore the biocatalytic acylation of phenolic compounds [5].

In addition to free fatty acids and their esters as previously described, compounds with antioxidant potency were also used as acyl donors. In order to investigate the synthesis of hybrid antioxidants, the enzymatic acylation of esculin catalyzed by immobilized CALB in the ionic liquids [bmim]BF_4 and [bmim]PF_6 was investigated using several phenolic acids as well as their corresponding vinyl esters.

As can be seen in Figure 9.4, the enzymatic synthesis of hybrid antioxidants of esculin is feasible in ionic liquids, resulting in good conversion yields. The conversion yields obtained were higher in ionic liquid [bmim]PF_6 as compared to [bmim]BF_4 for all phenolic acids tested, while the use of phenolic acid esters in transesterification reactions led to higher conversion yields in both ionic liquid media used.

9.4 Conclusions

The biocatalytic acylation of polyhydroxylated natural compounds, including flavonoid and phenolic glucosides as well as flavonolignans, is feasible in both

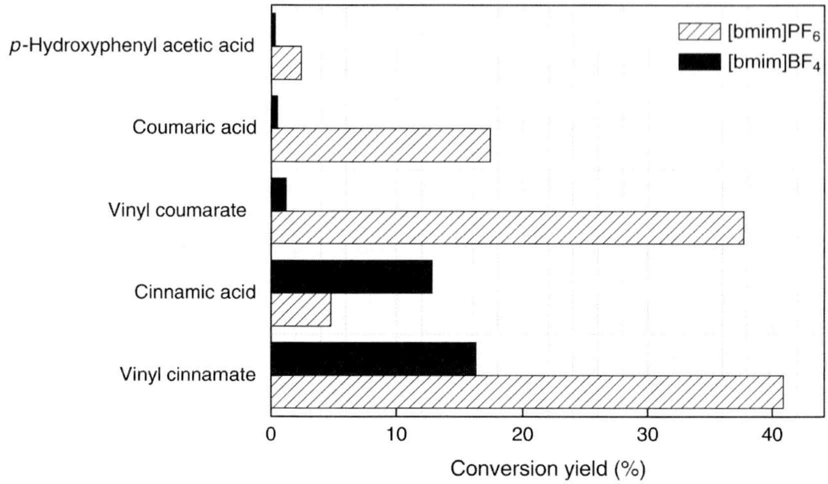

Figure 9.4 Conversion yields for the enzymatic acylation of esculin (25 mM) with different phenolic acids (50 mM) catalyzed by immobilized CALB in (a) [bmim]BF_4 (black bars) and (b) [bmim]PF_6 (patterned bars) after 72 h of incubation at 60 °C.

organic solvents and ionic liquid media. A variety of acyl donors such as dicarboxylic acids, free fatty acids, or their vinyl esters, as well as phenolic acids, which are compounds with antioxidant potency, can be used for the regioselective acylation of polyhydroxylated natural compounds and the preparation of lipophilic derivatives with biological activity. The effectiveness and the regioselectivity of the biocatalytic modification depend strongly on various reaction parameters, including the nature of the reaction medium, the nature of the acyl donor, the substrate concentrations, and the solubility of the substrates. The use of ionic liquids, such as [bmim]BF_4, in which the solubility of natural polyhydroxylated compounds is greatly enhanced, can be advantageous for the production of larger amounts of lipophilic derivatives in a single-step biocatalytic process.

References

1 Parke, D.V., Basu, T.L., Temple, A.J. and Garg, M.L. (eds) (1999) *Antioxidants in Human Health and Disease*, CABI Publishers, New York.

2 Pietta, P.-G (2000) *Journal of Natural Products*, **63**, 1035–1042.

3 Heim, K.E., Tagliaferro, A.R. and Bibilya, D.J. (2002) *The Journal of Nutritional Biochemistry*, **13**, 572–584.

4 Rice-Evans, C., Miller, N.J. and Paganga, G. (1996) *Free Radical Biology & Medicine*, **20**, 933–956.

5 Katsoura, M.H., Polydera, A.C., Tsironis, L., Tselepis, A.D. and Stamatis, H. (2006) *Journal of Biotechnology*, **123**, 491–503.

6 Mellou, F., Lazari, D., Skaltsa, H., Tselepis, A.D., Kolisis, F.N. and Stamatis,

H. (2005) *Journal of Biotechnology*, **116**, 295–303.
7 Klibanov, A.M. (2001) *Nature*, **409**, 241–246.
8 Schmid, A., Dordick, J.S., Hauer, B., Kiener, A., Wubbolts, M. and Witholt, B. (2001) *Nature*, **409**, 258–268.
9 Ardhaoui, M., Falcimaigne, A., Engasser, J.M., Moussou, P., Pauly, G. and Ghoul, M. (2004) *Journal of Molecular Catalysis B: Enzymatic*, **29**, 63–67.
10 Ardhaoui, M., Falcimaigne, A., Ognier, S., Engasser, J.M., Moussou, P., Pauly, G. and Ghoul, M. (2004) *Journal of Biotechnology*, **110**, 265–271.
11 Ardhaoui, M., Falcimaigne, A., Ognier, S., Engasser, J.M., Moussou, P., Pauly, G. and Ghoul, M. (2004) *Biocatalysis and Biotransformation*, **22**, 253–259.
12 Gao, C., Mayon, P., MacManus, D.A. and Vulfson, E.V. (2001) *Biotechnology and Bioengineering*, **71**, 235–243.
13 Danieli, B., Luisetti, M., Sampognaro, G., Carrea, G. and Riva, S. (1997) *Journal of Molecular Catalysis B: Enzymatic*, **3**, 193–201.
14 Park, S. and Kazlauskas, R.J. (2003) *Current Opinion in Biotechnology*, **14**, 432–437.
15 Kim, M.J., Choi, M.Y., Lee, J.K. and Ahn, Y. (2003) *Journal of Molecular Catalysis B: Enzymatic*, **26**, 115–118.
16 Park, S. and Kazlauskas, R.J. (2001) *Journal of Organic Chemistry*, **66**, 8395–8401.
17 Katsoura, M.H., Polydera, A.C., Katapodis, P., Kolisis, F.N. and Stamatis, H. (2007) *Process Biochemistry*, **42**(9), 1326–1334.
18 Gordon, C.M. (2001) *Applied Catalysis A: General*, **222**, 101–117.
19 Itoh, T., Nishimura, Y., Ouchi, N. and Hayase, S. (2003) *Journal of Molecular Catalysis B: Enzymatic*, **26**, 41–45.
20 Wilkes, J.S. (2004) *Journal of Molecular Catalysis A – Chemical*, **214**, 11–17.
21 Jarstoff, B., Störmann, R., Ranke, J., Mölter, K., Stock, F., Oberheitmann, B., Hoffmann, J., Nüchter, M., Ondruschka, B. and Filser, J. (2003) *Green Chemistry*, **5**, 136–142.
22 Ranke, J., Mölter, K., Stock, F., Bottin-Weber, U., Poczobutt, J., Hoffmann, J., Ondruschka, B., Filser, J. and Jastorff, B. (2003) *Ecotoxicology and Environmental Safety*, **58**, 396–404.
23 Kontogianni, A., Skouridou, V., Sereti, V., Stamatis, H. and Kolisis, F.N. (2003) *Journal of Molecular Catalysis B: Enzymatic*, **21**, 59–62.
24 McCabe, R.W. and Taylor, A. (2004) *Enzyme and Microbial Technology*, **35**, 393–398.
25 Uppenberg, J., Ohmer, N., Norin, M., Hult, K., Kleywegt, G.J., Patkar, S., Waagen, V., Anthomen, T. and Jones, T.A. (1995) *Biochemistry*, **34** (51), 16838–16851.
26 Teng, R.W., Bui, T.K.H., McManus, D., Armstrong, D., Mau, S.L. and Bacic, A. (2005) *Biocatalysis and Biotransformation*, **23**, 109–116.
27 Nakajima, N., Ishihara, K., Itoh, T., Furuya, T. and Hamada, H. (1998) *Journal of Bioscience and Bioengineering*, **87** (1), 105–107.
28 Stamatis, H., Sereti, V. and Kolisis, F.N. (2001) *Journal of Molecular Catalysis B: Enzymatic*, **11**, 323–328.
29 Kontogianni, A., Skouridou, V., Sereti, V., Stamatis, H. and Kolisis, F.N. (2001) *European Journal of Lipid Science and Technology*, **103**, 655–660.
30 Otto, R.T., Scheib, H., Bornscheuer, U.T., Pleiss, J., Syldatk, C. and Schmid, R.D. (2000) *Journal of Molecular Catalysis B: Enzymatic*, **8**, 201–211.

10
Biocatalysis Applied to the Synthesis of Nucleoside Analogs

Vicente Gotor

10.1
Introduction

Nowadays, many research groups employ biotransformations for preparing different kinds of organic compounds that are difficult to prepare by conventional chemical procedures. It is well recognized that biocatalysis is of special relevance in fine-chemical industry for manufacturing enantiopure compounds. Among the enzymes used for organic synthesis, lipases have demonstrated a great versatility in enzymatic hydrolysis, transesterification, or aminolysis reactions [1]. Other biocatalysts, such as oxidoreductases and some lyases, are also of interest. Of special relevance is the use of these biocatalysts for the preparation of pharmaceuticals [2].

Selective biotransformations of polyhydroxylated compounds, such as carbohydrates, have been used since some time ago for the activation or protection of some of their hydroxyl groups, allowing different chemical transformations to be carried out without the need for using tedious protection and deprotection steps. Monoesters of some monosaccharides and disaccharides have been obtained using lipases or proteases in polar organic solvents [3]. In addition, the utility of biotransformations in other natural products such as nucleosides [4] or steroids [5] has also been reviewed.

Modified nucleosides have attracted much attention as antiviral and anticancer agents. Consequently, extensive changes have been made to both the heterocyclic base and the sugar moiety in order to reduce the toxicity and viral drug resistance associated with certain nucleoside analogs. Until recently, only nucleosides possessing the natural β-D configuration have been studied as chemotherapeutic agents due to their easy access. However, the discovery of lamivudine (an α-L deoxynucloside), the first compound with the unnatural β-L configuration approved by the Food and Drug Administration for use in combined therapy against human immunodeficiency virus type 1 (HIV-1) and hepatitis B virus (HBV), has sparked tremendous interest in the synthesis of β-L-nucleosides. As a result, several L-nucleosides are currently undergoing clinical trials as potential antiviral or antitumor agents [6]. Figures 10.1 and 10.2 show some nucleosides of clinical interest.

Modern Biocatalysis: Stereoselective and Environmentally Friendly Reactions.
Edited by Wolf-Dieter Fessner and Thorleif Anthonsen

ISBN: 978-3-527-32071-4

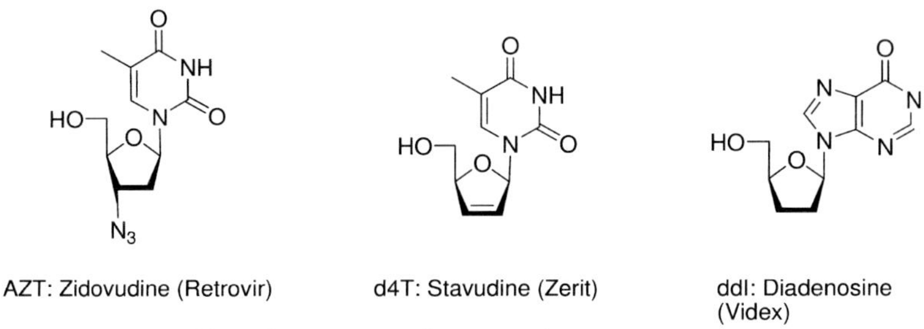

Figure 10.1 Partial list of FDA-approved anti-HIV drugs.

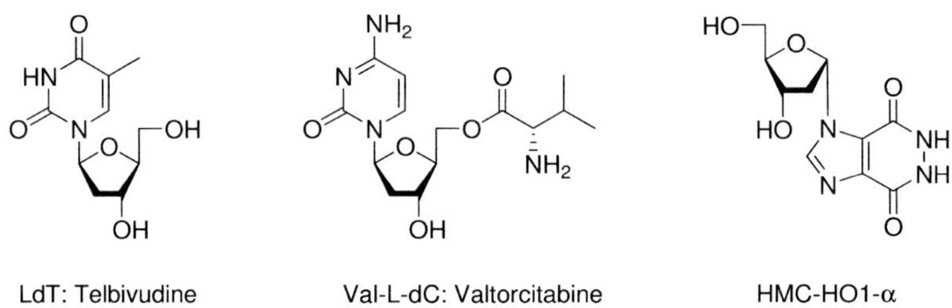

Figure 10.2 Some L-nucleosides in clinical trials as anti-HBV.

In recent years biotransformations have also shown their potential when applied to nucleoside chemistry [7]. This chapter will give several examples that cover the different possibilities using biocatalysts, especially lipases, in order to synthesize new nucleoside analogs. The chapter will demonstrate some applications of enzymatic acylations and alkoxycarbonylations for the synthesis of new analogs. The utility of these biocatalytic reactions for selective transformations in nucleosides is noteworthy. In addition, some of these biocatalytic processes can be used not only for protection or activation of hydroxyl groups, but also for enzymatic resolution of racemic mixtures of nucleosides. Moreover, some possibilities with other biocatalysts that can modify bases, such as deaminases [8] or enzymes that catalyze the synthesis of new nucleoside analogs via transglycosylation [9] are also discussed.

10.2
Chemoenzymatic Modification of the Sugar

As modified oligonucleotides have become a major field of investigation for chemists, methods for the suitable protection/deprotection applied to the synthesis of nucleoside monomers have become equally important. Selective protection of a multifunctional compound is a challenging problem in organic synthesis. In nucleoside chemistry, selective manipulation of the hydroxyl groups on the carbohydrate moiety over amino groups of the bases is synthetically challenging and requires a multistep protocol. In this sense, enzymes are a good alternative since biocatalyzed processes are environmentally acceptable working under mild conditions demonstrating high chemo- and regioselectivity and are usually compatible with organic solvents showing recycling possibilities.

This first section will study different regioselective processes of several types of nucleosides depending on the lipase used. Application of biotransformations over these compounds has acquired great importance in order to prepare new derivatives with interesting pharmacological activities. Two lipases, namely, *Candida antarctica* type B (CALB) and *Pseudomonas cepacea,* free (PSL) or *Pseudomonas cepacea,* immobilized (PSLC), are selective towards one of the two hydroxyl groups of different 2′-deoxynucleosides. Thus, it is possible to prepare the acylated compounds in 5′-position with CALB [10], whereas PSL is selective towards the secondary hydroxyl group [11]. Vinyl or oxime esters can be used as acyl donors. The first example of the application of this process with different oxime esters is shown in Scheme 10.1. The solvent plays an important role in this process and dioxane is usually used in these reactions. It is remarkable that, in the case of ribonucleosides (X=OH), it is not possible to achieve high regioselectivities in the 2′ or 3′-position with PSL, but CALB again catalyzes the process towards the primary hydroxyl group in the 5′-position. In addition, a theoretic study to explain the abnormal regioselectivity of PSL to the less reactive hydroxyl group has recently been published [12].

Another application is that the introduction of functional groups of physiological interest, such as carbamates, can be carried out from the corresponding carbonates (Scheme 10.2). For the synthesis of the starting carbonates a similar strategy to the regioselective enzymatic acylation of 2′-deoxynucleosides is used: in this case oxime carbonates are employed as alkoxycarbonylating agents [13]. The synthesis of carbamates takes place using the corresponding vinyl carbonate derivatives, because these activated carbonates can react in the absence of catalyst with amines or ammonia to obtain the nucleoside–carbamate analogs in quantitative yields [14]. This enzymatic alkoxycarbonylation is of great utility in activating hydroxyl groups

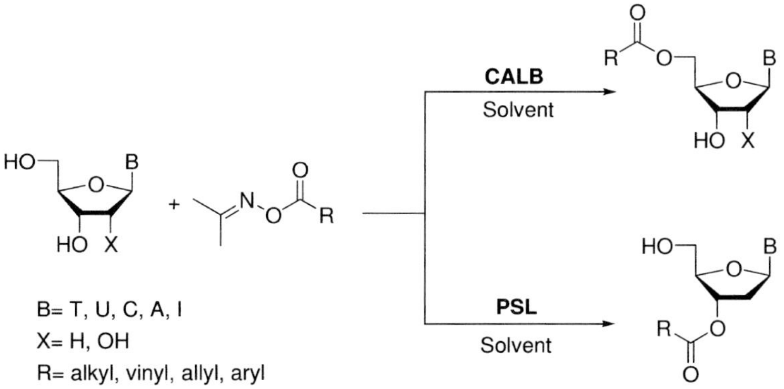

Scheme 10.1 Regioselective acylation of 2′-deoxynucleosides.

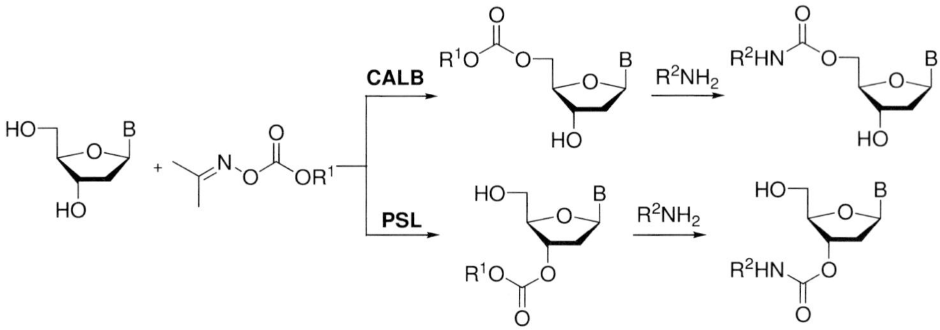

Scheme 10.2 Regioselective preparation of carbamate derivatives of 2′-deoxynucleosides.

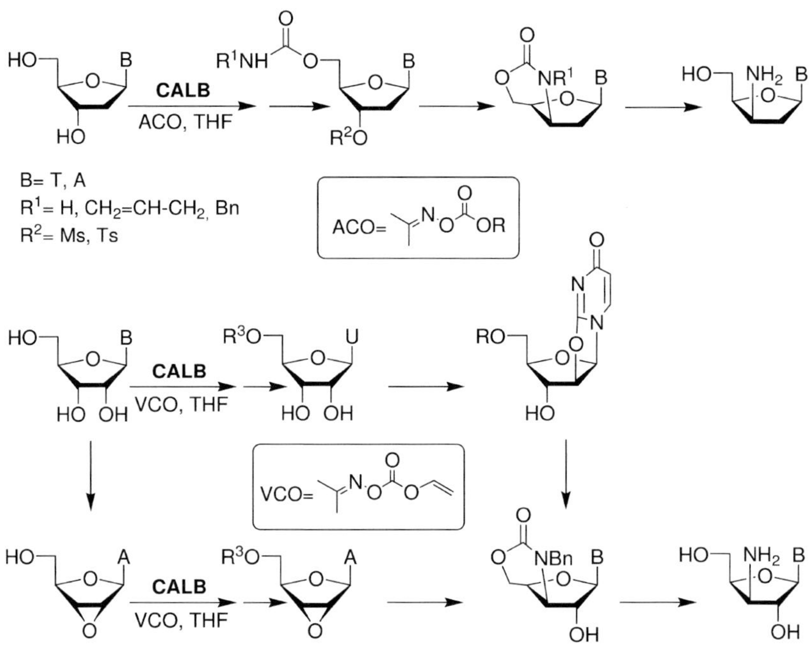

Scheme 10.3 Chemoenzymatic preparation of amino-xylonucleoside derivatives.

or also protecting them when benzyloxime carbonate (R^1=CH_2Ph) is used as the alkoxycarbonylating agent.

Moreover, this chemoenzymatic process yields new aminonucleosides that are very difficult to prepare by chemical procedures alone [15]. Scheme 10.3 shows a general chemoenzymatic procedure for obtaining 3′-amino-*xylo*-nucleosides. The key step is based on the 5′-directed intramolecular nucleophilic substitution at the 3′-activated position of the nucleoside. This methodology is applicable to ribonucleosides and 2′-deoxyribonucleosides and shows the utility of a combination of biocatalysis and chemical catalysis for the preparation of products of high added value.

Among the limited protecting groups available, the levulinyl group is frequently chosen to protect the 3′- and/or 5′-hydroxyl of the nucleosides for solution-phase oligonucleotide synthesis. The strategy for the synthesis of 3′-*O*-levulinyl nucleosides (2′-deoxy or 2′-*O*-protected) first involves protection of the 5′-hydroxy group as a 5′-*O*-dimethoxytrityl (5′-*O*-DMTr) derivative. Subsequent treatment with levulinic acid or levulinic anhydride and dicyclohexylcarbodiimide (DCC) followed by removal of the 5′-*O*-DMTr group under acidic media affords the 3′-protected nucleosides [16]. On the other hand, direct reaction of the corresponding unprotected nucleosides with levulinic acid, followed by separation of the 3′,5′-diacylated products by column chromatography and treatment of the residue with DMTrCl to remove the 3′-acylated compound and purification again by flash chromatography, furnished the 5′-*O*-levulinyl derivatives in low yields. A short and convenient synthesis of 3′- and 5′-*O*-levulinyl-2′-deoxynucleosides [17] has been developed in Scheme 10.4 from the corresponding 3′,5′-di-*O*-levulinyl derivatives by regioselective enzymatic hydrolysis, avoiding several tedious chemical protection/deprotection steps. *Candida antarctica* lipase A (CALA) has been identified as an ideal biocatalyst for the resolution of sterically hindered compounds [18] and in some cases has shown an opposite enantioselectivity to CALB [19]. In addition, the regioselective acylation of 2′-deoxynucleosides showed opposite regioselectivity using either CALA or CALB.

Moreover, these compounds can also be prepared through enzymatic acylation from the natural nucleosides using the corresponding oxime ester derivative as the acyl donor (Scheme 10.5). This process takes place with high regioselectivity and excellent yields [20]. In summary, the enzymatic procedure is an efficient and regioselective method for the synthesis of 3′- or 5′-*O*-levulinyl-protected nucleosides. As a result, these key building blocks needed for solution-phase synthesis of oligonucleotides are easily accessible in high yield and purity.

In some cases vinyl esters have been successfully used for these regioselective processes (Scheme 10.6). Thus, CALB catalyzes the selective monobenzoylation at the 5′-hydroxyl group of 2′-deoxynucleosides using vinyl benzoate as the acyl

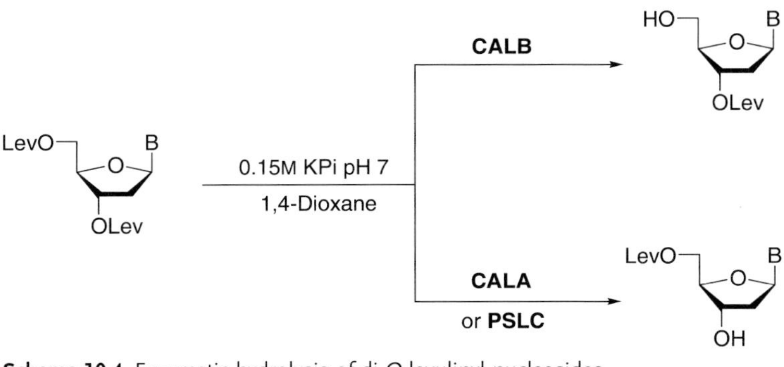

Scheme 10.4 Enzymatic hydrolysis of di-*O*-levulinyl nucleosides.

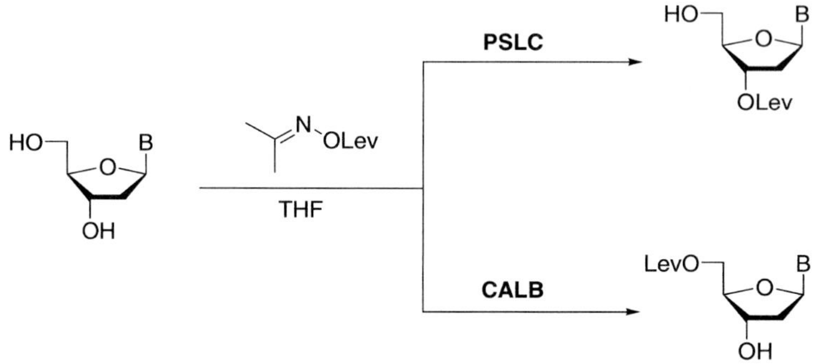

Scheme 10.5 Preparation of monolevulinyl 2′-deoxynucleosides through enzymatic acylation.

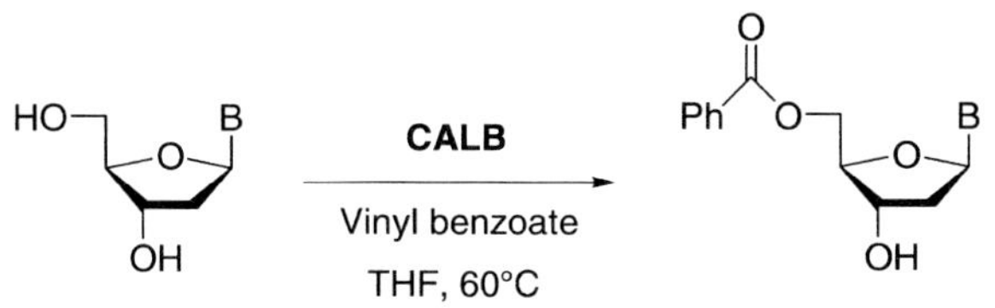

B= T, C^{Bz}, A^{Bz}, G^{Ibu}

Scheme 10.6 Synthesis of 5′-benzoyl-2′-deoxynucleosides.

transfer reagent in quantitative yields [21]. In this case the application of acetonoxime benzoate as an acylating agent for other nucleosides furnished moderate yields with slow reaction rates. The easy scalability of the process and the fact that both the acylating agent and enzyme can be recovered and reused after each reaction makes the new process atom efficient [22] and very attractive for industrial applications.

The overall approach for the synthesis of 3′-*O*-DMTr derivatives has been carried using these benzoyl derivatives as starting material. Introduction of the DMTr group at the 3′-hydroxyl position is accomplished by treatment of 5′-*O*-benzoyl-2′-deoxynucleosides with DMTrCl in pyridine at 70 °C. The choice of this high reaction temperature furnished an improved reaction rate and better overall yield. Next, treatment with sodium methoxide in MeOH at 0 °C removed the 5′-*O*-benzoyl group [23]. Under these conditions, 3′-*O*-DMTr-thymidine is isolated in a 90% yield after chromatography (Scheme 10.7).

However, under similar reaction conditions the *N*-benzoyl-cytidine derivative furnished a mixture of *N*-benzoyl-3′-*O*-DMTr-deoxycytidine and the corresponding *N*-unprotected nucleoside. However, this limitation was avoided using the corresponding levulinic derivative because the protecting group was easily removed using hydrazine (Scheme 10.8).

It has been described that, in some cases, the acylation of one hydroxyl group of the sugar moiety in a nucleoside derivative can increase its biological activity

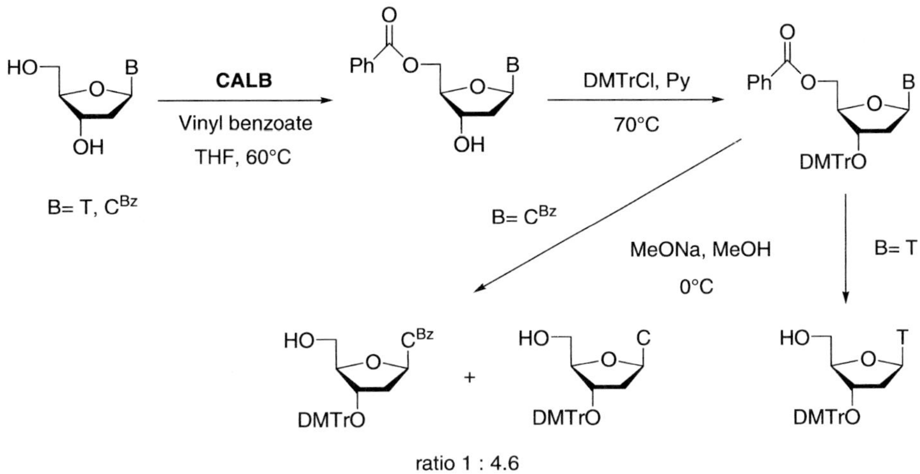

Scheme 10.7 Syntheses of 3′-*O*-DMTr 2′-deoxynucleosides.

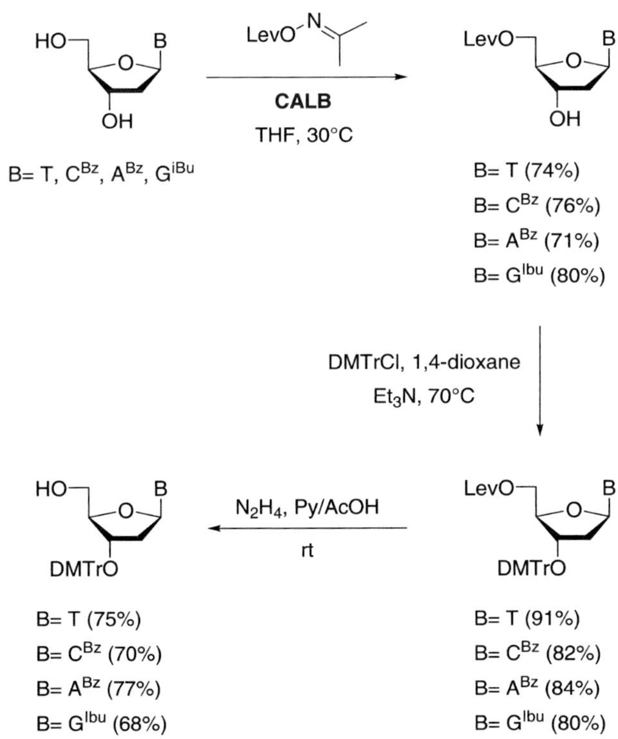

Scheme 10.8 Syntheses of 3′-*O*-DMTr derivatives from 5′-*O*-levulinyl derivatives.

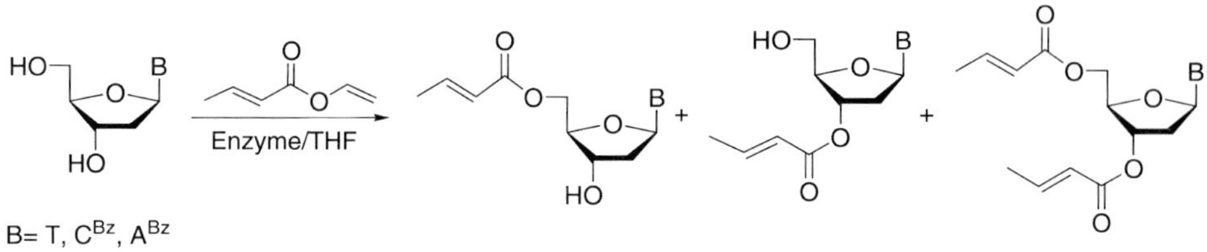

B= T, C^{Bz}, A^{Bz}

Scheme 10.9 Synthesis of *O*-crotonyl 2′-deoxynucleosides.

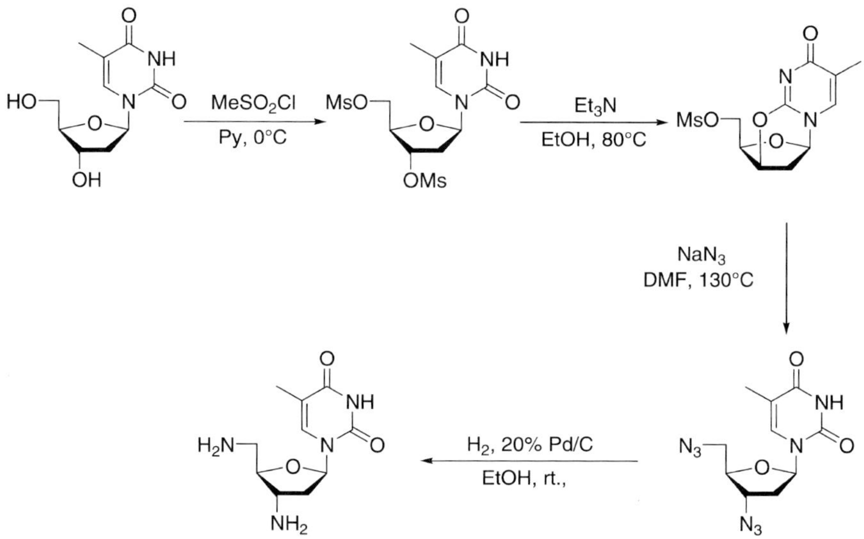

Scheme 10.10 Chemical preparation of 3′,5′-diamino-3′,5′-trideoxythymidine.

compared with the unmodified analog [24]. Regioselective syntheses of several *O*-crotonyl 2′-deoxynucleoside derivatives have been efficiently achieved using a biocatalytic methodology. While CALB affords the 5′-*O*-acylated compounds, PSLC provides the 3′-*O*-crotonylated analogs [25]. Since classical chemical approaches did not work appropriately due to side isomerization reactions, a mixture of both lipases was used for achieving a useful synthetic route toward diacylated nucleosides (Scheme 10.9).

Aminonucleosides can offer interesting physiological properties. Some routes for the synthesis of new analogs of this kind of compounds have been described [26].A modified route for the preparation of 3′,5′-diamino-3″,5′-trideoxythymidine is outlined in Scheme 10.10. The reaction, which takes place in four steps with an overall yield of 63%, is an improvement on the earlier described process [27].

In addition, regioselective enzymatic acylation of pyrimidine 3′,5′-diaminonucleoside derivatives is possible depending on the biocatalyst employed in the process [28]. *N*-5′-Acylated products were obtained using CALB as a biocatalyst, whereas PSLC was selective towards the *N*-3′ position. Molecular modeling studies

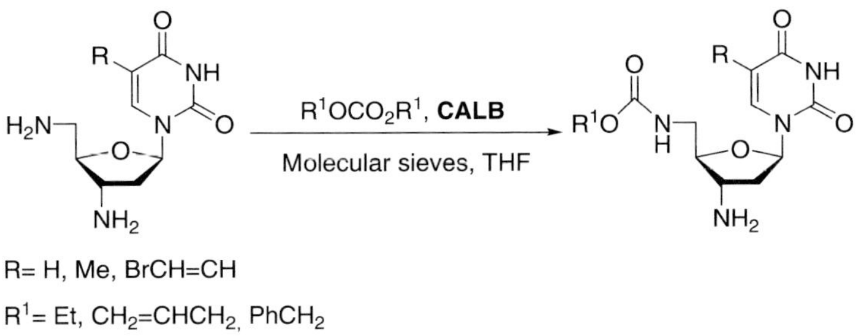

Scheme 10.11 Enzymatic alkoxycarbonylation of 3′,5′-diaminonucleosides.

were developed in order to explain the behavior of CALB in these enzymatic processes [29]. On the other hand, alkoxycarbonylation of amines yields carbamates that are valuable intermediates in medicinal chemistry, as this functionality is part of various therapeutic agents since it allows, for example, a better permeability across the cellular membranes. Moreover, these processes also allow the protection of the amino group in mild conditions. The first regioselective enzymatic alkoxycarbonylation of primary amino groups was achieved with pyrimidine 3′,5′-diaminonucleosides using CALB and non-activated homocarbonates allowing the synthesis of several *N*-5′-carbamates with moderate to high yields (Scheme 10.11) [30].

10.3 Resolution and Anomeric Separation

Parallel kinetic resolution (PKR), a concept that has been introduced for reactions where starting from a racemic mixture can allow the preparation of two different compounds at the same reaction rate [31], has been applied for the separation of a mixture of β-D/L-deoxynucleosides. A practical synthesis of β-L-3′- and β-L-5′-*O*-levulinyl-2′-deoxynucleosides has been described for the first time [32] through enzymatic acylation and/or hydrolysis processes. It is remarkable that the different behavior exhibited by PSL in the acylation of D- and L-nucleosides allows the parallel kinetic resolution of D/L nucleoside racemic mixtures. Scheme 10.12 shows a PKR of a 1 : 1 mixture of D and L nucleosides via an acylation reaction for furnishing easily separable compounds. This methodology would have tremendous potential for both research and industrial applications in the nucleic acid field.

In addition, a similar strategy allows the separation of α- and β-anomers of a mixture. Thus, an efficient and high yield protocol for the synthesis and separation of 3′- and/or 5′-protected α-2′-deoxynucleosides has been developed through regioselective acylation/deacylation processes catalyzed by enzymes [33]. PSLC was found to be highly chemo- and regioselective toward the 3′-position of the β-2′-deoxynucleoside derivatives, whereas the same lipase displayed opposite regioselectivity towards the 5′-position for the corresponding α-anomer (Scheme 10.13). The different R_f of both acylated anomers allows an easy separation.

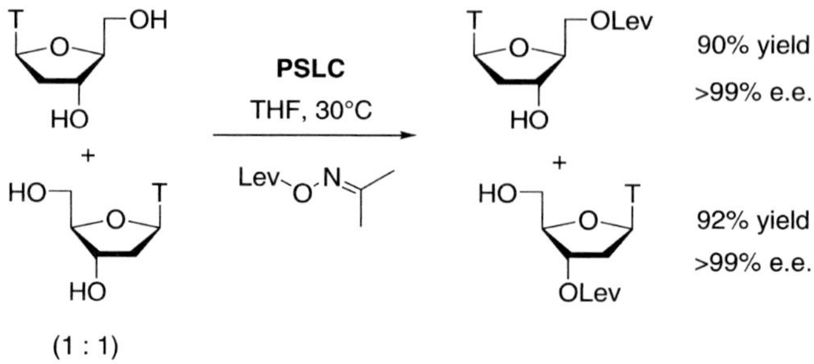

Scheme 10.12 Enzymatic separation of D/L-thymidine racemic mixture.

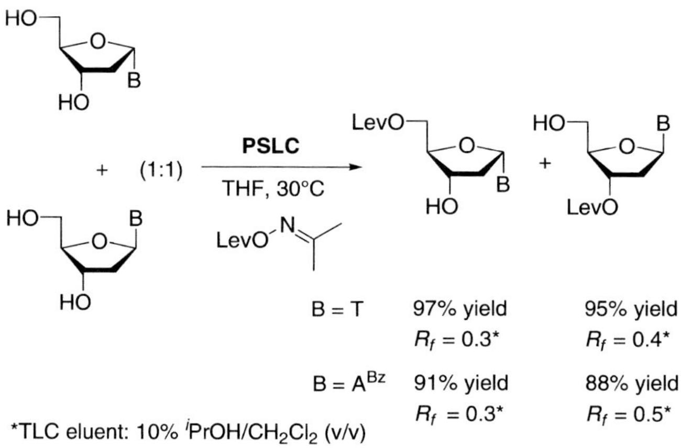

Scheme 10.13 Enzymatic separation of a α/β-anomeric mixture.

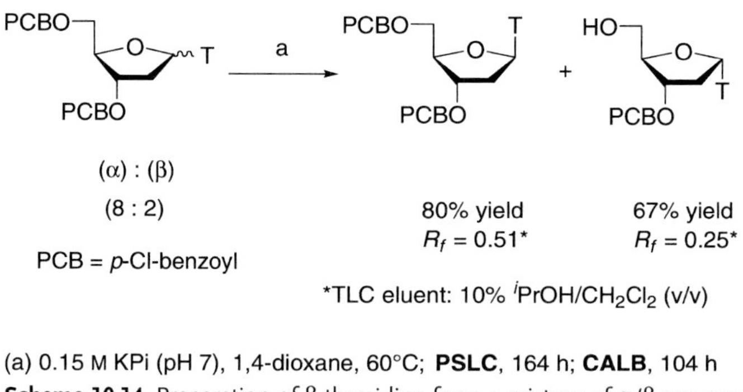

Scheme 10.14 Preparation of β-thymidine from a mixture of α/β-anomers.

A practical example of separation of a mixture of α/β-anomers of the mother liquor collected from the commercial-scale synthesis of β-thymidine is shown in Scheme 10.14. PSLC is used as a biocatalyst because this lipase has shown the desired opposite regioselectivity toward the hydrolysis of α- and β-nucleosides. The hydrolysis reaction takes place with excellent selectivity toward the 5′-*p*-

chlorobenzoate group of the α anomer furnishing 3′-*O*-(*p*-chlorobenzoyl)-α-D-thymidine as an exclusive product without a detectable amount of completely hydrolyzed α-thymidine. Despite the long reaction time, the β anomer is recovered unmodified. The pure α-nucleoside was isolated in 67% yield after chromatography. Next, the hydrolysis reaction with CALB was tried for two reasons: firstly, to show that CALB could be recycled several times to reduce the overall cost of the process and, secondly, CALB is commercially available at a reasonable price [33]. The hydrolysis of a mixture of the products with CALB under similar conditions was found to be faster than the PSLC-mediated reaction and afforded the same products. A complete conversion of the starting material to the products with CALB in 104 h instead of the 164 h required for the reaction with PSLC was observed. The faster reaction rate with CALB was expected to reduce the overall cycle time in the plant for the hydrolysis process and assist in meeting thelower cost objectives. After chromatographic separation, the conventional base-catalyzed hydrolysis of the products furnished β-thymidine and α-thymidine in excellent yields.

10.4
Biotransformations that Modify the Base

Due to the importance of modified nucleosides that are also in the base, the development of new biocatalytic processes applied to the synthesis of derivatives modified in the base is of great interest. Adenosine deaminase (ADA) and adenylate deaminase (AMPDA) are biocatalysts that catalyze the hydrolytic deamination of purine nucleosides and nucleotides. Some applications of these deaminases for the preparation and transformation of compounds structurally related to nucleosides with potential antitumor and antiviral activities have been described in the last few years [8].

ADA and AMPDA are enzymes involved in the purine cycle and they catalyze the hydrolytic deamination of purine nucleosides and nucleotides. In the last few years these biocatalysts have shown a great utility in biocatalysis, because they display broad substrate specificity and their use can also be extended to carbocyclonucleosides and acyclonucleosides [34]. The general process that takes place using these enzymes is outlined in Scheme 10.15.

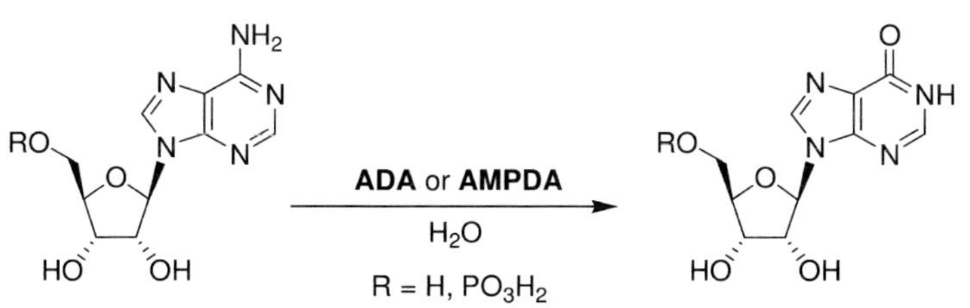

Scheme 10.15 Hydrolytic deamination of ADA and AMPDA of adenosine and adenylic acid.

ADA is able to transform other 6-substituted purine ribosides containing halogen, methoxy or hydroxylamino groups in inosine, although at lower rate [35]. In addition, the influence of the 5′-hydroxyl group of the sugar moiety is remarkable. Thus, it was demonstrated some time ago that 5′-*O*-protected or 5′-deoxy-5′-substituted nucleosides are not adequate substrates for this enzyme [36]. However, protected nucleosides on the 2′- and 3′-position, such as the corresponding 2′,3′-isopropylidene adenosines [37], are adequate substrates when 5′-hydroxy or 5′-amino groups are present in the nucleoside, although in the latter case the process takes place with more difficulty (Scheme 10.16).

Deaminases can be very practical for the transformation of nucleosides in other derivatives in combination with chemical catalysis. An interesting example is the synthesis of 2′,3′-dideoxyinosine starting from 2′-deoxyadenosine. The process takes place through a combination between two biocatalytic reactions followed by conventional chemical reactions in a multistep process. First, two enzymatic processes take place (Scheme 10.17). Again, CALB was the most effective catalyst in achieving the regioselective acylation in the 5′-position in comparison with other lipases [38]. Deoxygenation of the unprotected 3′-hydroxyl group was achieved

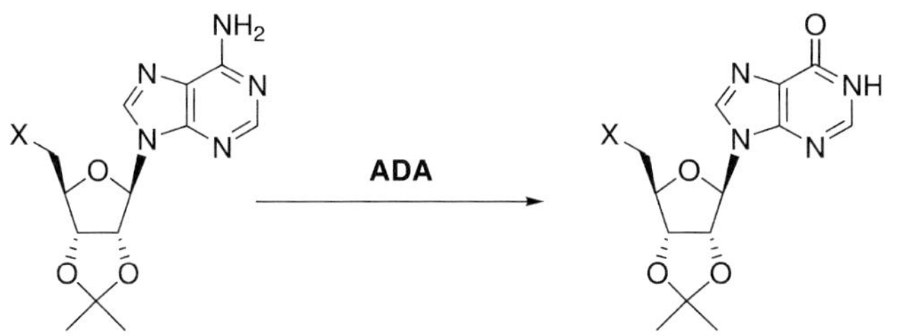

X =OH, NH_2

Scheme 10.16 Deamination of 2′,3′-isopropylidene derivatives.

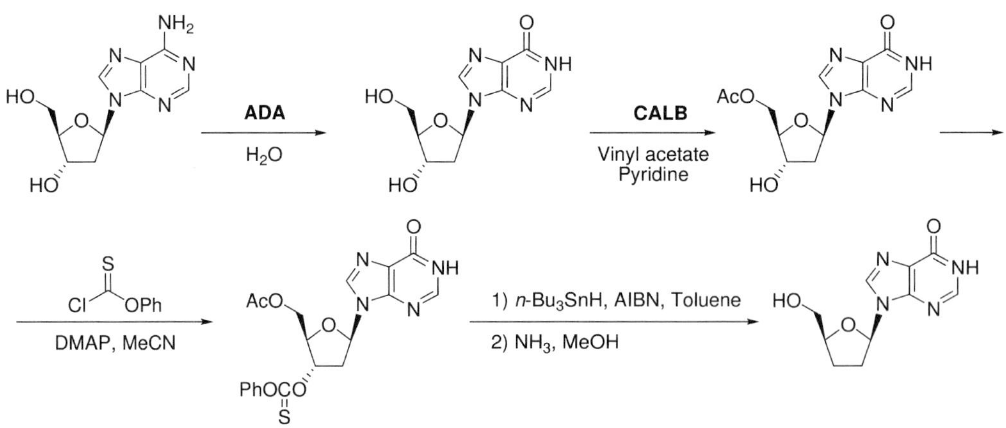

Scheme 10.17 Synthesis of 2′,3′-dideoxyinosine in a multistep process.

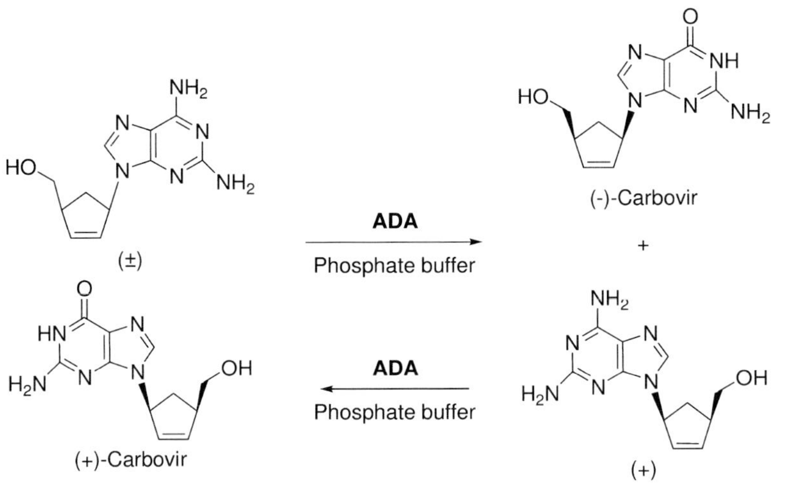

Scheme 10.18 Enzymatic resolution of (±)-carbovir.

chemically by the formation of the corresponding phenylthiocarbonate and further reaction with tributyltin hydride. This chemical process is frequently used for the deoxygenation of different types of nucleosides [39].

In addition, ADA has also been used in the resolution of some nucleoside analogs derived from 6-aminopurines [40]. In this context, an interesting example is shown in Scheme 10.18. This strategy has been successfully developed for carrying out the resolution of (±)-carbovir. Transformation of the unmodified (+)-amino compound under different reaction conditions allows the preparation of (+)-carbovir, which is less active as an HSV-1 inhibitor than the *levo* enantiomer (Scheme 10.18) [41].

10.5 Transglycosylation for the Synthesis of Nucleosides

In this section some examples are discussed in order to prepare nucleoside analogs by base interchange through a transglycosylation reaction. In these processes two different types of intramolecular enzymes are usually employed: nucleoside phosphorylases and *N*-2′-deoxyribosyl transferases.

Nucleoside phosphorylases catalyze the reversible phosphorolysis in nucleosides and the transferase reaction involving purine or pyrimidine bases [42]. Scheme 10.19 shows the general synthetic strategy of these processes. Although, nucleoside phosphorylases show a broad substrate specificity, *N*-2′-deoxyribosyl transferases specifically catalyze the exchange of the base from a 2′-deoxyribosylnucleoside with another purine or pyrimidine [43]. Two different mechanisms involved in the synthesis of 2′-deoxyribonucleosides by means of base interchange have been proposed. One of them involves a covalent catalysis analog to the one known for

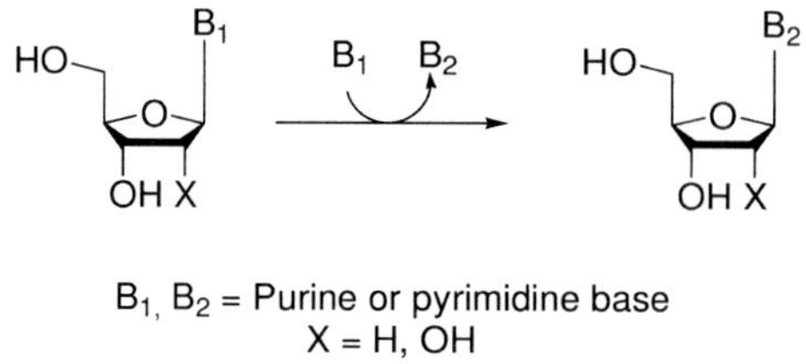

Scheme 10.19 Synthesis of nucleoside analogs by using nucleoside phosphorylases.

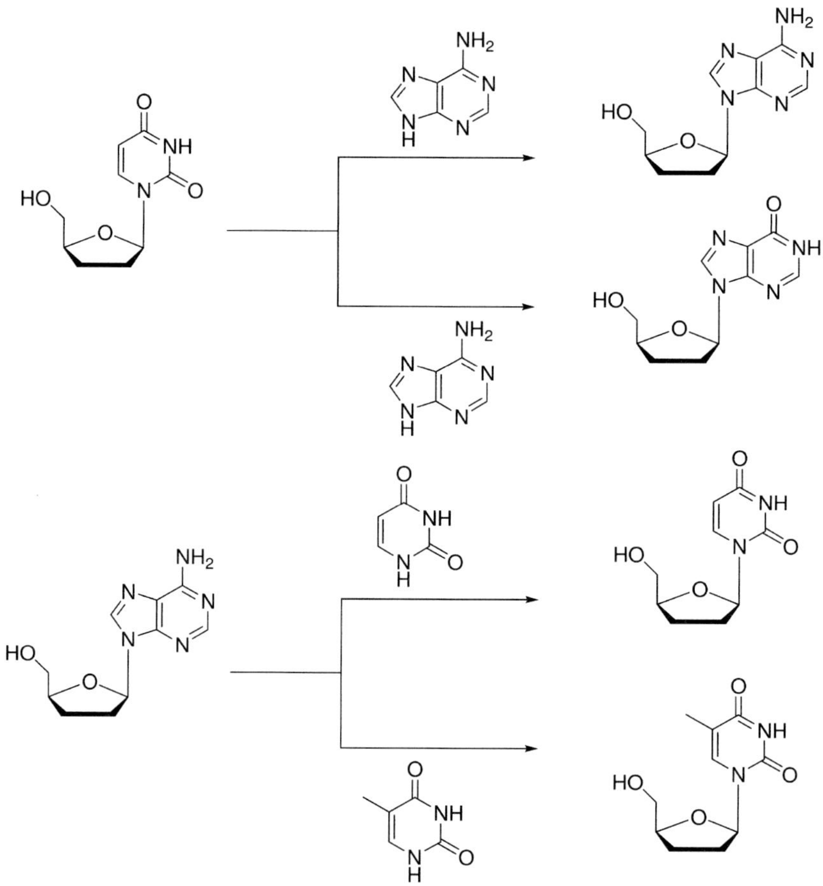

Scheme 10.20 Preparation of nucleoside analogs using *E. coli* strains.

glycosidases, where either a glutaryl or an aspartyl residue is involved in the catalysis [44]. The other possibility involves two different enzymes: thymidine nucleoside phosphorylase and purine nucleoside phosphorylase [45].

The preparation of different nucleoside analogs by interchange purine–pyrimidine or pyrimidine–purine bases has been well documented [46]. Scheme 10.20 depicts the versatility of this reaction using whole cells as biocatalysts. In

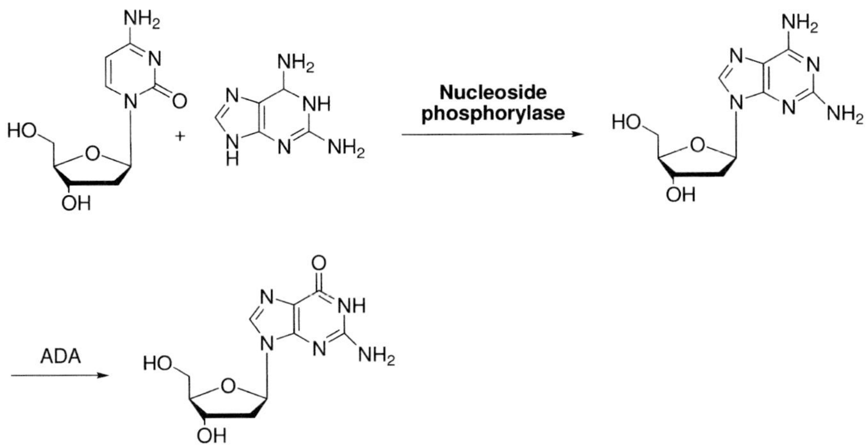

Scheme 10.21 Preparation of nucleoside analogs through a multistep process.

this case, the *Escherichia coli* strain presents great versatility for the preparation of different nucleoside analogs [47].

In some cases the combination of two biocatalysts in a multistep process allows, in an indirect manner, the preparation of some compounds or increases in their yields. Thus, Scheme 10.21 shows that the synthesis of a nucleoside analog containing a hypoxanthine base takes place in a two-step fashion using a combination of both nucleoside phosphorylase and adenine deaminase, obtaining the final analog in high yield. However, the direct preparation of this compound with hypoxanthine takes place with a very low yield due to the low solubility of this base [48].

10.6 Summary

The possibility of carrying out enzymatic processes under mild conditions and their high selectivity make biocatalytic reactions very attractive for performing some transformations that are difficult to achieve by chemical procedures alone. This chapter has shown the utility of three different types of enzymes for selective transformations on nucleosides applied to the preparation of new analogs. It is well established that lipases are the biocatalysts mostly employed by organic chemists, especially in the resolution of racemic mixtures. This chapter has reviewed some of the processes concerning the versatility of lipases for the regioselective protection or activation of hydroxyl groups in deoxynucleosides. In addition, in recent years some applications for the resolution of D/L mixtures and separation of anomers have emerged and these processes will probably have great applicability in industry, since lipases are able to carry out this separation in mild conditions. Moreover, in the last few years two types of enzymes, deaminases, which convert

6-substituted purine nucleosides into the corresponding 6-oxo analogs and glycosyltransferases, which carry out the preparation of nucleoside analogs by base interchange through a transglycosylation reaction, have emerged as useful biocatalysts for the synthesis of new derivatives of physiological interest.

References

1 Gotor-Fernández, V. and Gotor, V. (2007) Use of lipases in organic synthesis, in *Industrial Enzymes: Structure, Function and Application* (eds J. Polaina and A.P. MacCabe), Springer, Dordrech, The Netherland, Chapter 18.

2 Gotor, V. (2002) *Organic Process Research & Development*, **6**, 420–426.

3 Wong, C.-H. (1995) *Pure and Applied Chemistry*, **67**, 1609–1616.

4 (a) Ferrero, M. and Gotor, V. (2000) *Monatshefte für Chemie*, **131**, 585–618. (b) Ferrero, M. and Gotor, V. (2000) *Chemical Reviews*, **100**, 4319–4348.

5 Ferrero, M. and Gotor, V. (2000) Biocatalytic synthesis of steroids, in *Stereoselective Biocatalysis* (ed. R.N. Patel), Marcel Dekker, New York, USA, Chapter 11.

6 (a) Gumina, G., Chong, Y., Choo, H., Song, G.-Y. and Chu, C.K. (2002) *Current Topics in Medicinal Chemistry*, **2**, 1065–1086. (b) Sommadossi, J.-P. (2002) *Recent Advances in Nucleosides* (ed. C.K. Chu), Elsevier, Amsterdam, pp. 417–432. (c) Lee, K. and Chu, C.K. (2001) *Antimicrobial Agents and Chemotherapy*, **45**, 138–144. (d) Standing, D.N., Bridges, E.G., Placidi, L., Faraj, A., Loi, A.G., Pierra, C., Dukhan, D., Gosselin, G., Imbach, J.-L., Hernández, B., Juodawlkis, A., Tennant, B., Korba, B., Cote, P., Cretton-Scott, E., Schinazi, R.F., Myers, M., Bryant, M.L. and Sommadossi, J.-P. (2001) *Antiviral Chemistry & Chemotherapy*, **12**, 119–129. (e) Gumina, G., Song, G.-Y. and Chu, C.K. (2001) *FEMS Microbiology Letters*, **202**, 9–15.

7 Lavandera, I., García, J., Fernández, S., Ferrero, M., Gotor, V. and Sanghvi, Y. (2005) Current protocols in nucleic acid chemistry, in *Protection of Nucleosides for Oligonucleotide Synthesis*, John Wiley & Sons, Ltd, 2.11.1–2.11.36.

8 Santaniello, E., Ciuffeda, P. and Alessandrini, L. (2007) Deaminating enzymes of the purine cycle as biocatalysts for chemoenzymatic synthesis and transformations of antiviral agents structurally related to purine nucleosides, in *Biocatalysis in the Pharmaceutical and Biotechnology Industries* (ed. R. Patel), CRC Press, Chapter 17.

9 Cordezo, L.A., Fernández-Lucas, J., García Burgos, C.A., Alcantara, A.R. and Sinisterra, J.V. (2007) Enzymatic synthesis of modified nucleosides in biocatalysis, in *The Pharmaceutical and Biotechnology Industries* (ed. R. Patel), CRC Press, Chapter 14.

10 Morís, F. and Gotor, V. (1993) *Journal of Organic Chemistry*, **58**, 653–660.

11 Morís, F. and Gotor, V. (1992) *Synthesis*, **7**, 626–628.

12 Lavandera, I., Fernández, S., Magdalena, J., Ferrero, M., Grewal, H., Savile, C.K., Kazlauskas, R.J. and Gotor, V. (2006) *ChemBioChem*, **7**, 693–698.

13 (a) Morís, F. and Gotor, V. (1992) *Journal of Organic Chemistry*, **57**, 2490–2492. (b) Morís, F. and Gotor, V. (1992) *Tetrahedron*, **48**, 9869–9876.

14 (a) García-Allés, L., Morís, F. and Gotor, V. (1993) *Tetrahedron Letters*, **34**, 6337–6338. (b) García-Allés, L. and Gotor, V. (1995) *Tetrahedron*, **51**, 307–316.

15 García-Allés, L., Magdalena, J. and Gotor, V. (1996) *Journal of Organic Chemistry*, **61**, 6980–6986.

16 (a) Reese, C.B. and Song, Q. (1999) *Nucleic Acids Research*, **27**, 963–971. (b) Reese, C.B. and Song, Q. (1999) *Journal of the Chemical Society – Perkin Transactions*, **1**, 1477–1486.

17 García, J., Fernández, S., Ferrero, M., Sanghvi, Y.S. and Gotor, V. (2002)

Journal of Organic Chemistry, **67**, 4513–4519.

18 Domínguez de María, P., Carboni-Oerlemans, C., Tuin, B., Bargeman, G., Van de Meer, A. and Van Gemert, R. (2005) *Journal of Molecular Catalysis B: Enzymatic*, **37**, 36–46.

19 de Gonzalo, G., Brieva, R., Sánchez, V.M., Bayod, M. and Gotor, V. (2001) *Journal of Organic Chemistry*, **66**, 8947–8953.

20 García, J., Fernández, S., Ferrero, M., Sanghvi, Y.S. and Gotor, V. (2003) *Tetrahedron: Asymmetry*, **14**, 3533–3540.

21 García, J., Fernández, S., Ferrero, M., Sanghvi, Y.S. and Gotor, V. (2004) *Tetrahedron Letters*, **45**, 1709–1712.

22 Trost, B.M. (2002) *Accounts of Chemical Research*, **35**, 695–705.

23 Díaz-Rodríguez, A., Fernández, S., Sanghvi, Y.S., Ferrero, M. and Gotor, V. (2006) *Organic Process Research & Development*, **10**, 581–587.

24 (a) Hamamura, E.K., Prystasz, M., Verheyden, J.P.H., Moffat, J.G., Yamaguchi, K., Uchida, N., Sato, K., Nomura, A., Shiratori, O., Takese, S. and Katagiri, K. (1976) *Journal of Medicinal Chemistry*, **19**, 654–662.
(b) Hamamura, E.K., Prystasz, M., Verheyden, J.P.H., Moffat, J.G., Yamaguchi, K., Uchida, N., Sato, K., Nomura, A., Shiratori, O., Takese, S. and Katagiri, K. (1976) *Journal of Medicinal Chemistry*, **19**, 667–674.

25 Díaz-Rodríguez, A., Lavandera, I., Fernández, S., Sanghvi, Y.S., Ferrero, M. and Gotor, V. (2005) *Tetrahedron Letters*, **46**, 5835–5838.

26 Herdewijn, P., Balzarini, J., Pauwels, R., Janssen, G., Van Aerschot, A. and De Clercq, E. (1989) *Nucleosides Nucleotides*, **8**, 1231–1257.

27 Lavandera, I., Fernández, S., Ferrero, M. and Gotor, V. (2003) *Tetrahedron*, **59**, 5449–5456.

28 Lavandera, I., Fernández, S., Ferrero, M. and Gotor, V. (2001) *Journal of Organic Chemistry*, **66**, 4079–4082.

29 Lavandera, I., Fernández, S., Magdalena, J., Ferrero, M., Kazlauskas, R.J. and Gotor, V. (2005) *ChemBioChem*, **6**, 1381–1390.

30 Lavandera, I., Fernández, S., Ferrero, M. and Gotor, V. (2004) *Journal of Organic Chemistry*, **69**, 1748–1751.

31 Dehli, J.R. and Gotor, V. (2002) *Chemical Society Reviews*, **31**, 365–370.

32 García, J., Fernández, S., Ferrero, M., Sanghvi, Y.S. and Gotor, V. (2004) *Organic Letters*, **6**, 3759–3762.

33 García, J., Fernández, S., Ferrero, M., Sanghvi, Y.S. and Gotor, V. (2006) *Journal of Organic Chemistry*, **71**, 9765–9771.

34 Santaniello, E., Ciuffreda, P. and Alessandrini, L. (2005) *Synthesis*, **4**, 509–526.

35 Magire, M.-H. and Sim, M.K. (1971) *European Journal of Biochemistry*, **23**, 22–29.

36 Cory, J.G. and Suhadolnik, R.J. (1965) *Biochemistry*, **4**, 1729–1732.

37 Ciuffreda, P., Loseto, A. and Santaniello, E. (2002) *Tetrahedron*, **58**, 5767–5771.

38 Ciuffreda, P., Casati, S. and Santaniello, E. (1999) *Bioorganic and Medicinal Chemistry Letters*, **9**, 1577–1582.

39 Robins, M.J., Wilson, J.S. and Hansske, F. (1983) *Journal of the American Chemical Society*, **105**, 4059–4065.

40 Moon, H.R., Ford, H. and Marquez, V.E. (2000) *Organic Letters*, **2**, 3793–3796.

41 Vince, R. and Brownell, J. (1990) *Biochemical and Biophysical Research Communications*, **168**, 912–916.

42 Pal, S. and Nair, V. (1997) *Biocatalysis and Biotransformation*, **15**, 147–158.

43 Short, S.A., Amnstrong, S.R., Ealick, S.E. and Porter, D.J. (1996) *Journal of Biological Chemistry*, **271**, 4978–4987.

44 Danzin, C. and Cardinaud, R. (1976) *European Journal of Biochemistry*, **62**, 365-372.

45 Walter, M.R., Cook, W.J., Cole, L.B., Short, S.A., Koszalka, G.W., Krenitsky, T.A. and Ealick, S.E. (1990) *Journal of Biological Chemistry*, **265**, 14016–14022.

46 Hene, W.J. and Wong, C.H. (1989) *Journal of Organic Chemistry*, **54**, 4692–4695.

47 Shirae, H., Kobayashi, K., Shiragami, H., Irie, Y., Yasuda, N. and Yokozeki, K. (1989) *Applied and Environmental Microbiology*, **55**, 419–424.

48 Yokozeki, K. and Tsuji, T. (2000) *Journal of Molecular Catalysis B: Enzymatic*, **10**, 207–213.

11
Efficient Fructooligosaccharide Synthesis with a Fructosyltransferase from *Aspergillus aculeatus*

Francisco J. Plou, Miguel Alcalde, Iraj Ghazi, Lucía Fernández-Arrojo and Antonio Ballesteros

11.1
Introduction

Glycosyltransferases (EC 2.4) catalyze the transfer of a glycosyl donor to an acceptor molecule thereby forming a new glycosidic bond with great regio- and/or stereoselectivity [1, 2]. Glycosyltransferases are classified into three main mechanistic groups depending on the nature of the donor molecule: Leloir-type glycosyltransferases, which require sugar nucleotides (for example UDP-glucosyltransferases), non-Leloir glycosyltransferases, which use sugar-1-phosphates (for example phosphorylases) and transglycosidases, which employ non-activated sugars such as sucrose, lactose, or starch [3]. Transglycosidases present the same mechanism as retaining glycosidases, resulting in the net retention of the stereochemical anomeric configuration [4]. According to the Henrissat classification [5], which is based on amino acid sequence comparisons and comprises more than 2500 enzymes, both transglycosidases and glycosidases (EC 3.2) are grouped in the 'glycoside hydrolase family' (GH family). Although the normal function of transglycosidases is the transfer of a glycosyl residue to another sugar acceptor (transfer reaction), they can also use water as the acceptor of the glycosyl–enzyme intermediate (hydrolysis reaction). Only an enzyme that possesses a significant transglycosylation to hydrolysis ratio should be considered a transglycosidase [6].

Fructosyltransferases are transglycosidases that transfer the fructosyl moiety of sucrose yielding fructose oligomers (fructooligosaccharides, FOS) and/or polymers (fructans, for example inulin or levan) [7]. Enzymes with transfructosylating activity are present in many higher plants (asparagus, Jerusalem artichoke, chicory, onion, and so on) and microorganisms, (especially fungi [8–10]) and belong to families 32 and 68 of the GH family in the Henrissat classification [5]. The assignation of a particular fructooligosaccharide-producing enzyme to the fructosyltransferases (EC 2.4.1.9) or β-fructofuranosidases (invertases, EC 3.2.1.26) groups still remains controversial and should be based on its transferase to hydrolysis ratio, especially at low substrate concentrations [11].

Modern Biocatalysis: Stereoselective and Environmentally Friendly Reactions.
Edited by Wolf-Dieter Fessner and Thorleif Anthonsen

ISBN: 978-3-527-32071-4

Fructooligosaccharides are fructose oligomers with a terminal glucose group in which two to four fructofuranosyl moieties are linked via β(2 → 1)-glycosidic bonds [7, 12]. Commercial fructooligosaccharides are mainly composed of 1-kestose, nystose, and 1^F-fructofuranosylnystose (Scheme 11.1) [13]. They are produced on a multiton scale from sucrose using transfructosylating enzymes from *Aspergillus* and related fungi (*Aspergillus niger, Aspergillus oryzae, Aureobasidium pullulans,* and so on) [7–10, 14]. There is also interest in the development of novel fructooligosaccharides with improved properties. In this context, ^6G-type fructooligosaccharides (also called neo-fructooligosaccharides) consist mainly of neo-kestose (neo-GF2) and neo-nystose (neo-GF3), in which a fructosyl unit is β(2 → 6) bound to the glucose moiety of sucrose or 1-kestose, respectively [15, 16]. ^6F-type fructooligosaccharides (^6F-fructooligosaccharides) occur naturally in various food products: they are linear or branched β-(2,6)-linked fructooligosaccharides (the first is 6-kestose) [17, 18]. Recently, the regioselective enzymatic synthesis of ^6F-type fructooligosaccharides with the β-fructofuranosidase from *Schwanniomyces occidentalis* has been

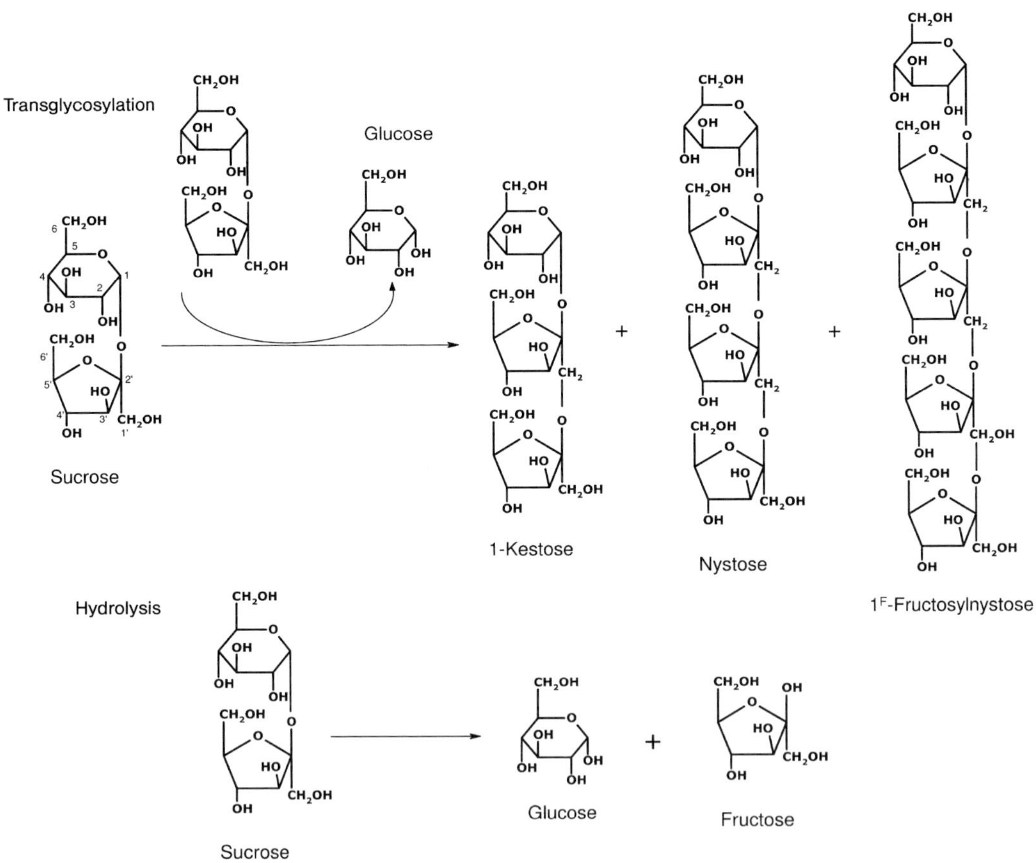

Scheme 11.1 Transfructosylation and hydrolytic reactions catalyzed by fructosyltransferase from *A. aculeatus*.

reported [19]. Fructooligosaccharides are non-cariogenic, have a sweetness of about 40–60% compared with sucrose, and, more importantly, exhibit prebiotic properties.

In the context of functional foods, a prebiotic is a non-digestible food ingredient that beneficially affects the host by selectively stimulating the growth and/or the activity of certain types of bacteria in the colon (basically genus *Bifidobacterium* and *Lactobacillus*) [20, 21]. Prebiotics produce positive effects on human health as the metabolism of these bacteria releases short-chain fatty acids (acetate, propionate, and butyrate) and L-lactate [22, 23] with different implications for health. Between them, they exert protective effects against colorectal cancer and bowel infectious diseases by inhibiting putrefactive and pathogen bacteria, improve the bioavailability of essential minerals, or enhance the glucid and lipid metabolism [24, 25]. Besides the fructooligosaccharides [26] and galactooligosaccharides (GOS) [27], which are commercialized in the USA and Europe, Japan produces other 'emerging' prebiotics (isomaltooligosaccharides, soybean oligosaccharides, lactosucrose, gentiooligosaccharides, and xylooligosaccharides, to cite some) the market of which will be soon extended [28].

Pectinex Ultra SP-L is a commercial enzyme preparation from *Aspergillus aculeatus* that is used in the food industry for reducing viscosity in fruit juice processing. It contains different pectinolytic and cellulolytic enzymes [29]. In addition, the existence of fructosyltransferase activity in Pectinex Ultra SP-L has been reported by several authors [30–32]. In recent years, we have investigated the purification, characterization, and application of the fructosyltransferase from *A. aculeatus* contained in this commercial preparation [33].

11.2
Purification of Fructosyltransferase in Pectinex Ultra SP-L

The fructosyltransferase present in Pectinex Ultra SP-L was purified using a combination of DEAE-Sepharose, Mono-Q HR 5/5 and Sephacryl S-100 columns [33]. The fructosyltransferase was found to be a minor protein, as it only represented 0.4% (w/w) of the total protein in the extract. The enzyme was purified 107-fold from the crude Pectinex and the recovered activity was about 37%. The purified protein was blotted from an SDS–polyacrylamide gel electrophoresis (PAGE) gel to a polyvinylidene fluoride PVDF membrane and analyzed using an automatic sequencer for N-terminal amino acids. The peptide sequence for the first 20 amino acids was determined and compared with the sequence of known proteins using BLASTP (http://expasy.org/tools/blast). The results of the alignment are shown in Table 11.1. This amino acid sequence showed a significant homology with the N-terminal of β-fructofuranosidase from *A. niger* ATCC (American Type Culture Collection) 20611 [34] and with an invertase from *A. niger* B60 [35].

Its relative electrophoretic mobility in denaturing and non-denaturing polyacrylamide gels indicated that it had a dimeric structure with an estimated molecular mass of 65 kDa per monomer (Figure 11.1). This was confirmed by

Table 11.1 Alignment of N-terminal amino acid sequences from *A. aculeatus* and *A. niger* fructosyltransferases and invertases.

Enzyme		N-terminal	
Fructosyltransferase from *Aspergillus aculeatus*	1	LDTTAPPXFXLSTLPXXXLF	20
Fructosyltransferase from *Aspergillus niger* ATCC 20611	3	LDTTAPPPTNLSTLPNNTLF	22
Invertase from *Aspergillus niger* B60	6	DYNVAPP–NLSTLPNGSLF	23

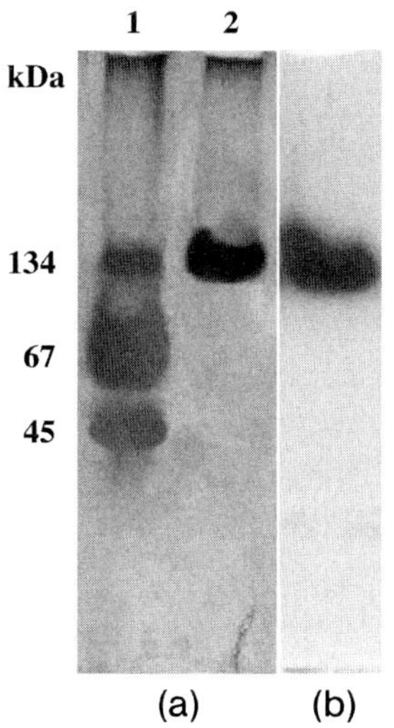

Figure 11.1 Native polyacrylamide gel (5%) electrophoresis of purified fructosyltransferase. (a) Lane 1, molecular weight markers (45 kDa ovalbumin, 67 kDa bovine serum albumin (BSA), and 134 kDa BSA dimer); lane 2, pure fructosyltransferase. (b) Pure fructosyltransferase (one lane from the same gel was cut and stained for fructosyltransferase activity).

gel-filtration chromatography on Sephacryl S-100 using standard proteins. Fructosyltransferase purified from Pectinex Ultra SP-L was found to be a glycoprotein. The glycosylation degree was deduced from the difference between the molecular mass of the monomer and that of the deglycosylated form obtained after treatment with peptide-*N*-glycosidase F (PNGase F) (52 kDa, assessed by SDS-PAGE). This suggests that the enzyme contains about 20% (w/w) carbohydrate. The isoelectric point of the enzyme was estimated to be in the range 3.8–4.2 on the basis of relative mobility on a 4% polyacrylamide gel under denaturing conditions.

Fructosyltransferases and β-fructofuranosidases from various sources show substantial differences in their molecular properties (subunit molecular weight,

Table 11.2 Molecular properties of fructosyltransferases and β-fructofuranosidases from different sources.

Source	Subunit MW (kDa)[a]	Native MW (kDa)[b]	pI	Reference
Arthrobacter sp. K1	52	52	4.3	[37]
Asparagus officinalis L.	64	64	–	[38]
A. aculeatus	65	134	4	[33]
Aspergillus foetidus	90	180	–	[39]
Aspergillus nidulans, S-form	78	185	~5	[40]
A. niger ATCC 20611	100	340	–	[41]
A. niger B60	115	230	–	[35]
A. niger IMI303386	125	125	~5	[42]
A. niger AS0023	125	250–750	–	[43]
A. niger from mouldy lemons	47	95	–	[44]
A. oryzae ATCC 76080	87	87	–	[45]
Azotobacter chroococum ATCC 4412	57	59	–	[46]
Bacillus macerans	66	66	–	[47]
Bifidobacterium infantis ATCC 15697	70	70	4.3	[48]
Bifidibacterium infantis JCM 7007	75	232	–	[49]
C. utilis	150	300	–	[36]
Clostridium perfringens	37	37	–	[50]
Lactobacillus reuteri CRL 1100	58	58	–	[51]
S. cerevisae	60	120	–	[52]
Schizosaccharomyces pombe	205	205	–	[53]
Schwanniomyces occidentalis	77	150	–	[54]
	85	85	–	[19]

a Approximate molecular weight obtained from SDS-PAGE.
b Approximate molecular weight obtained from native-PAGE or gel filtration.

isoelectric point, degree of deglycosylation, and so on). Table 11.2 summarizes some characteristics of the purified fructosyltransferase from *A. aculeatus* compared with related enzymes. Many of them are dimeric or multimeric enzymes: the monomer only shows activity in a few cases, for example in *S. occidentalis* [19] or in *Candida utilis* [36].

11.3
Properties of Fructosyltransferase from *A. aculeatus*

11.3.1
Substrate Specificity

The substrate specificity of the fructosyltransferase from *A. aculeatus* was estimated using a variety of di-, tri-, and tetrasaccharides at 100 g/l. The assay was based on the detection of reducing sugars by the dinitrosalicylic acid method [55] and was adapted to 96-well microplates [56]. As both the transglycosylation and hydrolytic reactions liberate reducing sugars (Scheme 11.1), the measured activity was a combination of both processes.

The pure enzyme displayed the preference sucrose > raffinose > 1-kestose > nystose (in a ratio 100 : 24 : 10 : 6). Other carbohydrates such as turanose, cellobiose, melibiose, leucrose, methyl-α-D-glucopyranoside, and stachyose were also tested but their activity was negligible. Thus, the enzyme only recognized sugars containing a sucrose moiety in their chemical structure. The trisaccharide raffinose is equivalent to a sucrose molecule substituted at the C-6 hydroxyl group, whereas 1-kestose and nystose can be considered as different sucrose derivatives, fructosylated at the C-1′-position. The specific activity diminished with longer oligosaccharides, being negligible with the sucrose-containing tetrasaccharide stachyose, which suggests limitations for the binding of large molecules in the active site of this fructosyltransferase.

11.3.2
Effect of pH and Temperature

The optimum pH measured with sucrose as substrate ranged from 5.0 to 7.0, which is similar to that reported with other fructosyltransferases [7, 43, 57]. No significant activity was found at pH values below 3.5 and above 9.5. Regarding the pH stability, it was observed that the enzyme was more stable in the presence of carbohydrates, which mimics the operational conditions of fructooligosaccharide synthesis. Thus, the enzyme retained more than 90% of its initial activity after 24 h incubation at moderately acid and neutral pH values (from 4.5 to 7.5), whereas in the absence of the carbohydrate it retained less than 50% of the initial activity in the same pH range [33]. The optimum temperature was in the range 50–70 °C, which is slightly higher than that reported for related enzymes [58]. The enzyme was significantly stable up to 60 °C ($t_{1/2} > 24$ h), but temperatures greater than 65 °C inactivated the enzyme.

11.3.3
Influence of Chemicals

The fructosyltransferase from *A. aculeatus* displayed activity without the addition of any metal ions. However, some effects were observed in its susceptibility to mono- and divalent cations [33]. For example, Mn^{2+}, K^{+}, and Co^{2+} caused a 1.4–1.9-fold increase in the activity, whereas low concentrations of Hg^{2+} and Zn^{2+} produced 35–60% inhibition. It was also found that the enzyme was slightly activated by several non-ionic and anionic surfactants such as sodium dodecylsulphate (1.5-fold at 10 mM), sodium deoxycholate (1.4-fold at 1 mM), and Triton X-100 (1.4-fold at 5% w/v). Moreover, it was resistant to low concentrations (1–10 mM) of reducing agents such as dithiothreitol and β-mercaptoethanol.

11.3.4
Kinetic Behavior

Both fructooligosaccharide synthesis and sucrose hydrolysis are catalyzed by most of the fructosyltransferases and β-fructofuranosidases (invertases) in the presence of sucrose. The transferase : hydrolase ratio, which determines the maximum yield of fructooligosaccharide, depends basically on two parameters: the concentration of sucrose and the intrinsic enzyme properties, that is its ability to bind the nucleophile (to which a fructose is transferred) and to exclude H_2O from the acceptor binding site [11].

In order to investigate the behavior of fructosyltransferase from *A. aculeatus* in more detail, the kinetic constants (K_m and k_{cat}) for both transfer and hydrolysis were determined. The reaction rates (expressed in U/mg protein) versus sucrose concentration up to 1.75 M were measured (Figure 11.2). The reactions were analyzed by high-performance liquid chromatography (HPLC) to discriminate between hydrolytic and transferase activities. The hydrolytic and transfructosylating initial rates were calculated from the amount of free and transferred fructose, respectively. The latter was calculated by subtracting the amount of free fructose from the amount of glucose. As shown in Figure 11.2, the transglycosylation and hydrolysis profiles did not fit well with Michaelis–Menten kinetics. However, the adjustment improved substantially when applying the Hill transformation of the Michaelis–Menten equation $\left[V = V_{max}[S]^h\left(K_m^h + [S]^h\right)^{-1}\right]$, where h is the Hill coefficient. This has been also observed with other fructosyl-transferring enzymes [59].

The K_m and k_{cat} values for the transferase and hydrolytic activities are given in Figure 11.2. With regard to the hydrolytic reaction, the K_m value (27 mM) is in the range of those reported for β-fructofuranosidases from *Saccharomyces cerevisae* (26 mM) [60], *Pichia anomala* (16 mM) [61], or *Arxula adeninivorans* (41 mM) [62] and higher than those observed for invertases from *S. occidentalis* (4.9 mM) [19] or *C. utilis* (1–2 mM) [63]. However, the k_{cat} for hydrolysis (775 s^{-1}) was lower than those previously described for well-known invertases (for example 9430 s^{-1} for *S. cerevisae* invertase [60]). Although the K_m for the transfer reaction (535 mM) was higher than those reported for enzymes exhibiting notable transfructosylating activity (for example 290 mM for *A. niger* fructosyltransferase [41]), the main feature of *A. aculeatus* fructosyltransferase was its significant k_{cat} value for the transfer reaction (16 200 s^{-1}). This implies that, at high sucrose concentrations (above 1 M, 342 g/l), the rate of transfer is nearly 20-fold that of hydrolysis, which explains the low concentration of free fructose that was observed with this enzyme in the HPLC analyses.

11.3.5
Fructooligosaccharide Production

Maximal fructooligosaccharide production for any particular enzyme depends basically on the relative rates of the transfructosylation and hydrolysis reactions

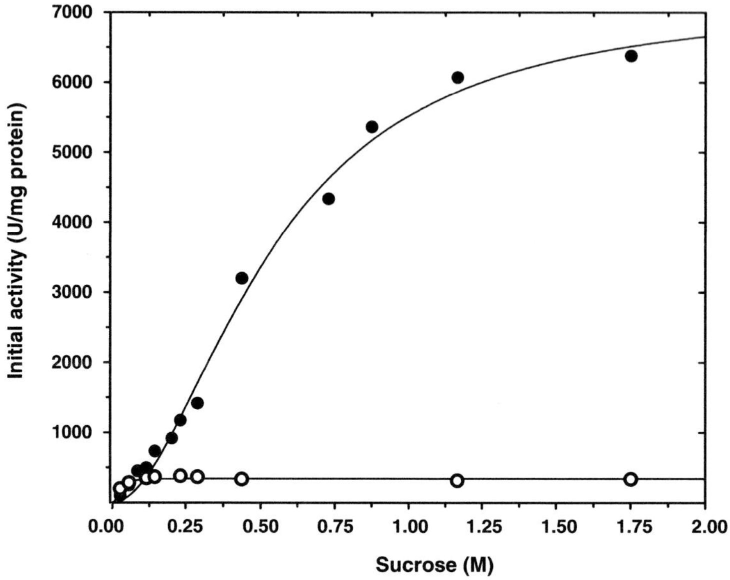

Reaction	K_m (mM)	k_{cat} (s^{-1})	h	k_{cat}/K_m (mM^{-1}/s^{-1})
Transglycosylation	535 ± 45	16 200 ± 900	1.9 ± 0.1	30 ± 4
Hydrolysis	27 ± 3	775 ± 25	2.3 ± 0.8	29 ± 4

Figure 11.2 Kinetic plots (initial rate versus sucrose concentration) of fructosyltransferase from *A aculeatus*. Transfer activity (•) and hydrolytic activity (o). Reactions were carried out in 0.2 M sodium acetate buffer (pH 5.5) at 60 °C. Kinetic parameters were calculated estimating a molecular mass of 135 kDa for the active enzyme. Adapted from Ref. [33].

[42]. The pure fructosyltransferase from *A. aculeatus* was assayed for the production of fructooligosaccharides. A concentrated sucrose solution (600 g/l) containing pure enzyme (5 U/ml) was incubated at 60 °C and the transglycosylation products were investigated in detail by HPLC. A series of fructooligosaccharides (1-kestose, nystose, and 1^F-fructosylnystose) was formed [56]. The analysis presented a maximum fructooligosaccharide production of 60.7% (w/w), referred to total carbohydrates in the reaction mixture. The rest of the carbohydrates in the mixture were fructose (marginal), glucose, and the remaining sucrose. These values compared well with those reported using the crude Pectinex Ultra SP-L under similar experimental conditions [32, 56] and with fructosyltransferases from other *Aspergillus* strains [8, 10, 64, 65].

11.4
Immobilization of Fructosyltransferase from *A. aculeatus*

11.4.1
Sepabeads EC-EP as Immobilization Carriers

For industrial application of carbohydrate-active enzymes, an effective immobilization method is required for facilitating the continuous processing and reuse of the biocatalyst [66, 67]. In this context, the use of available carriers for covalent immobilization is of great interest for minimizing enzyme leakage from the support, considering that these reactions take place in aqueous media. Sepabeads EC-EP are polymethacrylate-based carriers with a high density of reactive epoxide groups for protein covalent coupling. Compared with other acrylic epoxy-activated polymers, Sepabeads EC-EP possess a high mechanical and osmotic stability, low compressibility, and do not swell in water. Furthermore, the raw materials applied for the production of these supports are included in the European Union list of resins allowed for the processing of foodstuffs.

Two epoxy-activated related supports, Sepabeads EC-EP3 and EC-EP5 (Figure 11.3), were assayed for the immobilization of *A. aculeatus* fructosyltransferase. By combination of nitrogen isotherms and mercury porosimetry analyses, the textural properties of both carriers were determined (Table 11.3). As shown, both samples

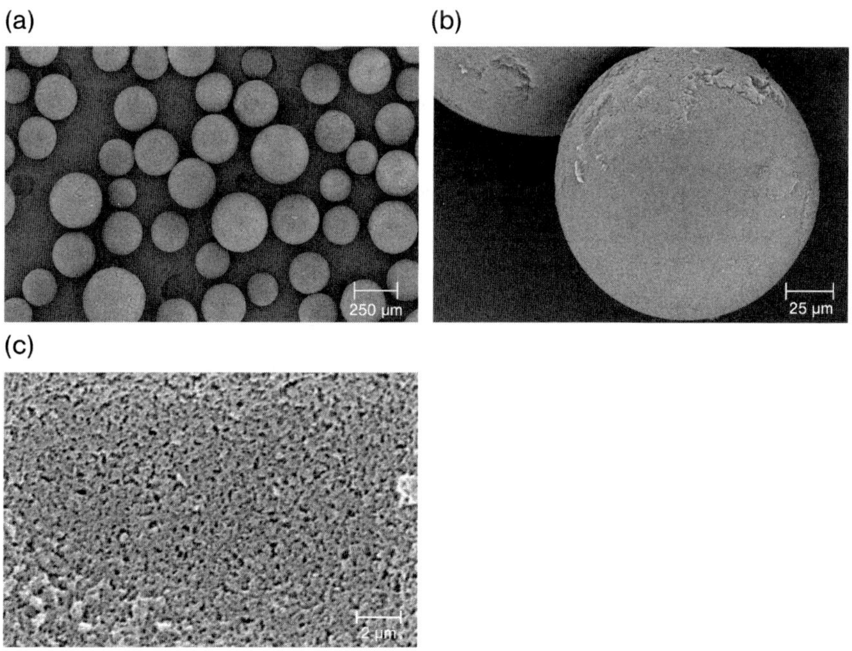

Figure 11.3 Scanning electron micrographs of Sepabeads EC-EP05: (a) 60×, (b) 500×, and (c) 8000×.

Table 11.3 Main properties of the Sepabeads EC-EP carriers.

Properties	EC-EP3	EC-EP5
Content of epoxide groups (μmol/g)[a]	106	23
Average particle size (μm)[b]	77	139
S_{BET} (m^2/g)[c]	43	9
Density of epoxide groups (μmol/m^2)	2.5	2.4
Pore volume (cm^3/g)[d]	1.19	1.67
Average pore size (nm)	130	800
Water content (%)[e]	60	65

a Provided by the supplier.
b Determined by mercury porosimetry, considering a symmetric distribution of particle size.
c Specific area measured by N_2 adsorption.
d By combination of N_2 isotherms and mercury porosimetry.
e Determined by Karl–Fisher titration.

were essentially macroporous, although the total pore volume of Sepabeads EC-EP5 (1.67 cm^3/g) was significantly higher than that of Sepabeads EC-EP3 (1.19 cm^3/g). The average pore size of both supports was also different: the maximum in the pore size distribution curve was 800 nm for Sepabeads EC-EP5 and 130 nm for Sepabeads EC-EP3.

11.4.2 Effect of pH and Ionic Strength on Immobilization

The coupling of enzymes to epoxy-activated carriers is commonly carried out at high ionic strength, because a salt-induced association between the macromolecule and the support surface increases the effective concentration of nucleophilic groups on the protein close to the epoxide reactive sites, thus favoring the immobilization process [68, 69]. However, the salt concentration needed for immobilizing an enzyme is highly dependent on the nature of the biocatalyst [69].

For the above reasons, the effect of pH and ionic strength on the immobilization of the fructosyltransferase from *A. aculeatus* on Sepabeads EC-EP was studied. Although this enzyme constitutes only a minor percentage (0.4% w/w) of the total proteins present in Pectinex Ultra SP-L, the commercial preparation was used directly (without enzyme purification) in order to develop a simple process that could be easily scaleable in the industry. It was considered that, although the other proteins present in Pectinex Ultra SP-L could also be immobilized in Sepabeads EC-EP, they would not interfere in fructooligosaccharide synthesis experiments.

In order to bind the enzyme to the support using different functional groups, the immobilization was performed at two pH values (5.5 and 9.0), adjusting the pH of commercial Pectinex Ultra SP-L with potassium phosphate or sodium carbonate, respectively. At pH 5.5, the reactive groups of the protein are the carboxylic heads of the aspartic and glutamic side chains, as well as the C-terminal α-carboxylic group [70]. At pH 9.0, the amino and thiol groups in the protein are able to bind to the support. The buffer concentration was varied over the range 0.2–1.0 M. A total of 30 mg of protein were mixed with 1 g of support. Figure 11.4 shows the protein recovery at 0.3 and 0.5 M buffer concentration after 24 h incubation. Surprisingly, the amount of protein bound to the support was higher at lower ionic strength. Maximums of 34 and 58% of the 'offered' protein were bound to Sepabeads EC-EP3 and EC-EP5, respectively. In general, Sepabeads EC-EP5 retained more protein than EC-EP3 under the same experimental conditions, probably as a consequence of its higher porosity. The effect of enzyme immobilization on the porosity of the supports was also analyzed: a slight decrease (approximately 4%) in the pore volume of Sepabeads EC-EP3 was produced as a consequence of enzyme immobilization. The total bound protein was larger at pH 9.0 than at pH 5.5.

It is worth noting that a low buffer concentration also resulted in higher activity with both supports [56]. The highest activity of the biocatalyst (15.2 U/g) was achieved with Sepabeads EC-EP5, using 0.3 M sodium carbonate (pH 9.0). With Sepabeads EC-EP3, a slightly acidic pH (5.5) and low buffer concentration (0.3 M) yielded the highest biocatalyst activity (4.2 U/g). The incubation time was also analyzed: a further increase from 24 to 72 h did not result in a significant improvement in biocatalyst activity.

Based on the above results at low ionic strength, the direct immobilization of Pectinex Ultra SP-L, the pH of which is 4.8, on Sepabeads EC-EP was assayed

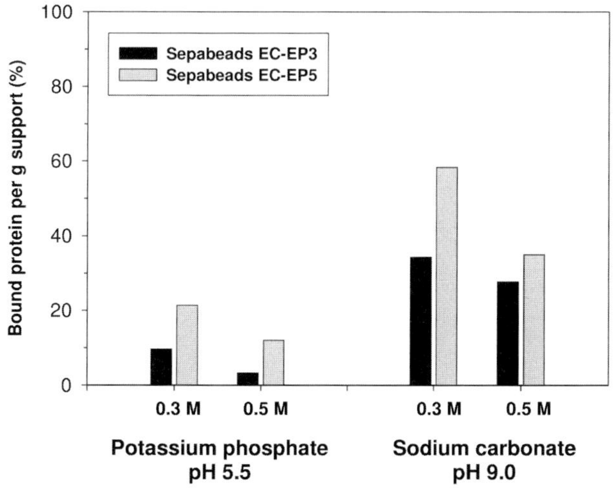

Figure 11.4 Effect of buffer concentration on the immobilization on Sepabeads EC-EP carriers of transfructosylating activity in Pectinex Ultra SP-L. Immobilization conditions: 30 mg of protein added per gram of support, 24 h, room temperature, and roller shaking. Adapted from Ref. [56].

without the addition of buffer or salt to control pH and ionic strength. The protein loading (mg) per gram of support was varied over the range 60/1–200/1. Interestingly, the biocatalyst activity reached 25.9 and 18.8 U/g with Sepabeads EC-EP5 and EC-EP3, respectively [56].

Although fructosyl-transfer enzymes have been immobilized by different techniques such as adsorption [71], entrapment [72], or covalent attachment [73], the biocatalyst activity per mass unit is not commonly reported. However, Chiang *et al.* [74] described a maximum activity of 77 U/g for *A. niger* β-fructofuranosidase covalently attached to methacrylamide-based polymeric beads. More recently, fructosyltransferase from *A. aculeatus* present in Pectinex Ultra SP-L was covalently bound to Eupergit C [75]: although the activity recovery was high, the activity per gram of biocatalyst was not reported.

11.4.3
Application of Immobilized Biocatalysts to Fructooligosaccharide Synthesis

Two immobilized biocatalysts, obtained using the direct method on Sepabeads EC-EP3 and EC-EP5, in the transfuctosylation reaction on sucrose were tested. A high substrate concentration (630 g/l) was assayed in order to favor transglycosylation activity. The progress of the reaction with Sepabeads EC-EP3 is depicted in Figure 11.5. It was observed that the fructooligosaccharide concentration reached a maximum value of 387 g/l after 36 h, with a weight ratio of 1-kestose/nystose 1^F-fructofuranosylnystose of 62/37/1. At this reaction time, the percentage of fructooligosaccharide in the solids, when referred to the total carbohydrates in the mixture, was 61.5%. After 150 h the mass distribution of the carbohydrates was 57–58% fructooligosaccharide, 29–31% glucose, 9.5–10.5% sucrose, and 2–3% fructose. Similar fructooligosaccharide yields have been reported with other immobilized fructosyltransferases [74, 75].

11.5
Fructooligosaccharide Production Using Sugar Beet Syrup and Molasses

11.5.1
Sugar Beet Syrup and Molasses as Low-cost Feedstock for Fructooligosaccharide Synthesis

Sugar beet syrup is the juice obtained after sugar beet juice evaporation, from which sucrose is crystallized. Molasses is the main by-product of the sugar industry. Both sources are notably less expensive than pure sucrose for production of fructooligosaccharide. The use of molasses for fructooligosaccharide synthesis has only been explored by Shin *et al.* in cultures of *A. pullulans* cells [9]. Sugar syrup and molasses are viscous solutions containing sucrose and other carbohydrates (Figure 11.6). Molasses has also more unidentified components and some particu-

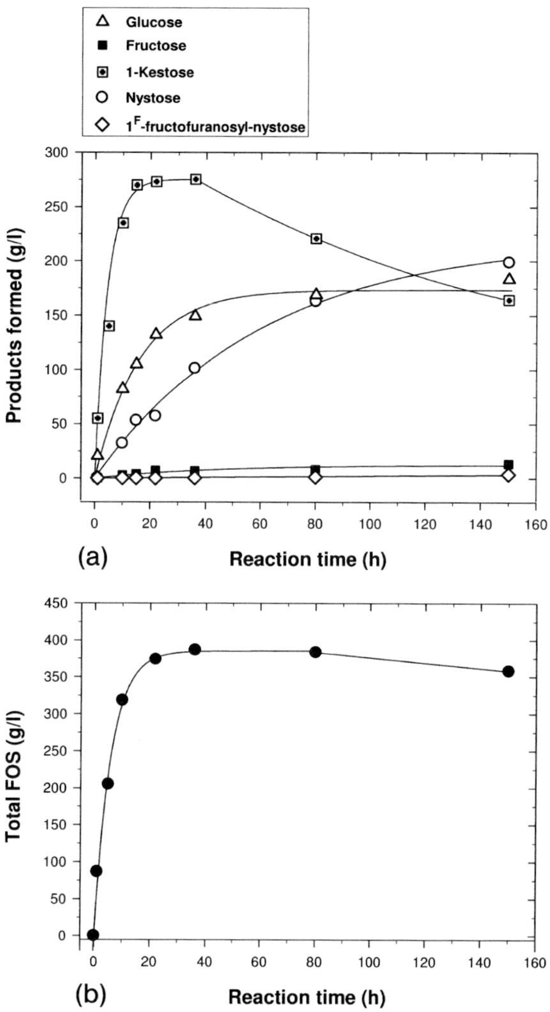

Figure 11.5 Batch synthesis of fructooligosaccharides catalyzed by immobilized Pectinex Ultra SP-L in Sepabeads EC-EP3. (a) Product distribution and (b) total fructooligosaccharide. Experimental conditions: 630 g/l of sucrose, 0.3 U/ml (standard DNS assay), 50 mM sodium acetate buffer (pH 5.4), and 60 °C. Adapted from Ref. [56].

late materials (especially carbonates). In this work, the sugar syrup and molasses contained 620 and 570 g/l sucrose, respectively.

The initial pHs of the sugar syrup and molasses were 7.5 and 8.9, respectively. It was observed that fructooligosaccharide production from syrup and molasses increased substantially when moving from the starting pH to pH 5.5 (by the addition of glacial acetic acid).

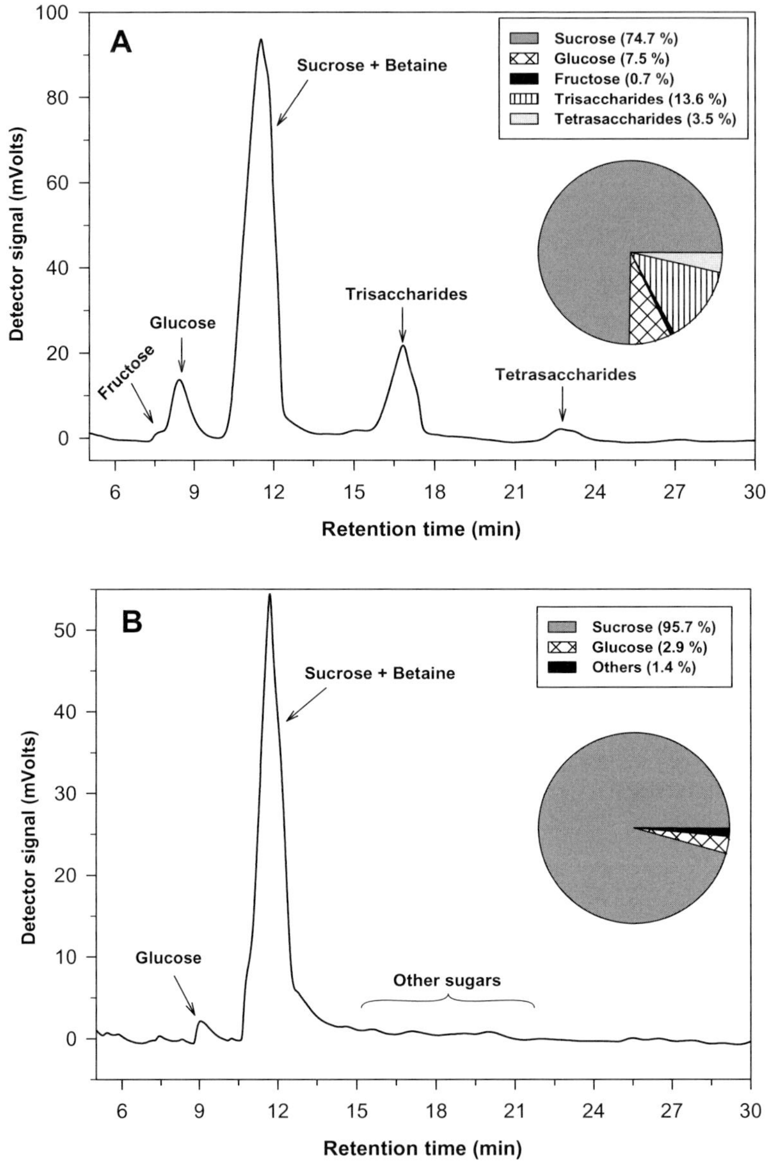

Figure 11.6 HPLC chromatograms of (a) the beet sugar syrup and (b) the molasses used in this work. Inset: carbohydrate distribution (in weight).

11.5.2
Batch Production of Fructooligosaccharide

In order to reduce the viscosity of the feedstock and to have efficient mixing, several dilutions from syrup and molasses were prepared by adding distilled water to the substrate and adjusting the pH to 5.5. The kinetics of batch fructooligosaccharide production adding 0.2 ml water per milliliter of substrate were analyzed. Table 11.4 summarizes the concentration of total fructooligosaccharide. In the case of the syrup, the fructooligosaccharide concentration reached a maximum value of 387 g/l after 30 h (229 g/l of 1-kestose, 149 g/l of nystose, and 9 g/l of 1^F-fructofuranosylnystose). At this reaction time, the percentage of fructooligosaccharide was 56.0%, when referred to the total carbohydrates in the mixture. With regard to the molasses, the maximum fructooligosaccharide concentration was 235 g/l (62 g/l of 1-kestose, 143 g/l of nystose, and 30 g/l of 1^F-fructofuranosylnystose) after 65 h, which corresponded to 49.2% fructooligosaccharide. After 140 h the reactions were close to equilibrium, with fructooligosaccharide percentages of 49 and 42% for syrup and molasses, respectively. The results are comparable to those using pure sucrose as the substrate with soluble and immobilized Pectinex Ultra SP-L [56].

As syrup and molasses are colored materials and the fructooligosaccharide specifications for human nutrition are very restrictive in this regard, the fructooligosaccharide obtained by the methods described above could be employed for animal feed. It is well known that the administration of fructooligosaccharides to the diets of some animals results in an improvement in feed efficiency and a reduction in diarrhea and smell in feces [76]. In order to extend the use of

Table 11.4 Fructooligosaccharide production using beet sugar syrup and molasses.

Reaction time (h)	Fructooligosaccharide	
	Syrup (g/l) (%)	Molasses (g/l) (%)
7	253.1 (36.6%)	96.2 (20.1%)
24	382.9 (55.3%)	190.1 (39.8%)
30	387.6 (56.0%)	223.7 (46.8%)
65	357.9 (51.7%)	235.5 (49.3%)
115	347.8 (50.2%)	223.2 (46.7%)
140	341.7 (49.3%)	202.2 (42.3%)

Experimental conditions: pH 5.5, 5.0 U/ml (standard DNS assay), 60 °C. The substrate was diluted by adding 0.2 ml of water per milliliter of substrate.
Data adapted from Ref. [32]
The fructooligosaccharide percentages refer to the total amount of carbohydrates in the mixture.

fructooligosaccharides as animal feed additives, it is necessary to minimize production costs. Sugar beet syrup and molasses are cheap and available sources of sucrose and are adequate feedstock for fructooligosaccharide production of animal-feed grade.

11.6 Conclusions

The fructosyltransferase mined from the commercial preparation Pectinex Ultra SP-L is quite stable towards pH, temperature, and the presence of chemicals. The *A aculeatus* fructosyltransferase showed a high transferase:hydrolase ratio that provides it with great potential for oligosaccharide synthesis. The enzyme can be easily immobilized on epoxy-activated supports for better performance. Further studies are required to draw mechanistic conclusions on the nature of the kinetics observed with this fructosyltransferase.

Acknowledgments

The authors thank Drs Moreno Daminati and Paolo Caimi (Resindion S.R.L., Milan, Italy) for providing us with the Sepabeads EC-EP polymers and for technical help. The authors are grateful to Ramiro Martínez (Novozymes A/S, Spain) for providing us with the Pectinex Ultra SP-L. Project BIO2004-03773-C04-01 from the Spanish Ministry of Education and Science and a grant from Genoma España, the National Foundation for Promoting Genomics and Proteomics, supported this research. The authors are also grateful to the Spanish Agency for International Cooperation for a research fellowship.

References

1 Seibel, J., Jordening, H.J. and Buchholz, K. (2006) *Biocatalysis and Biotransformation*, **24**, 311–342.

2 Planas, A. and Faijes, M. (2002) *Afinidad*, **59**, 295–313.

3 Plou, F.J., Martin, M.T., Gomez de Segura, A. Alcalde, M. and Ballesteros, A. (2002) *Canadian Journal of Chemistry*, **80**, 743–752.

4 MacGregor, E.A. (2005) *Biologia*, **60**, 5–12.

5 Coutinho, P.M. and Henrissat, B. (1999) Carbohydrate-active enzymes: an integrated database approach, in *Recent Advances in Carbohydrate Bioengineering* (eds H.J. Gilbert, G. Davies, B. Henrissat, B. Svensson), The Royal Society of Chemistry, Cambridge, pp. 3–12.

6 Feng, H.Y., Drone, J., Hoffmann, L., Tran, V., Tellier, C., Rabiller, C. and Dion, M. (2005) *The Journal of Biological Chemistry*, **280**, 37088–37097.

7 Antosova, M. and Polakovic, M. (2001) *Chemical Papers-Chemicke Zvesti*, **55**, 350–358.

8 Sangeetha, P.T., Ramesh, M.N. and Prapulla, S.G. (2005) *Process Biochemistry*, **40**, 1085–1088.

9 Shin, H.T., Baig, S.Y., Lee, S.W., Suh, D.S., Kwon, S.T., Lim, Y.B. and Lee, J.H.

(2004) *Bioresource Technology*, **93**, 59–62.

10 Fernandez, R.C., Maresma, B.G., Juarez, A. and Martinez, J. (2004) *Journal of Chemical Technology and Biotechnology*, **79**, 268–272.

11 Ballesteros, A., Plou, F.J., Alcalde, M., Ferrer, M., Garcia-Arellano, H., Reyes-Duarte, D. and Ghazi, I. (2006) Enzymatic synthesis of sugar esters and oligosaccharides from renewable resources, in *Biocatalysis in the Pharmaceutical and Biotechnological Industries* (ed. R. Patel), CRC Press, pp. 465–490.

12 Yun, J.W. (1996) *Enzyme and Microbial Technology*, **19**, 107–117.

13 Crittenden, R.G. and Playne, M.J. (1996) *Trends in Food Science and Technology*, **7**, 353–361.

14 Velasco, J. and Adrio, J.L. (2002) Microbial enzymes for food-grade oligosaccharide biosynthesis, in *Microorganisms for Health Care, Food and Enzyme Production* (ed. J.L. Barredo), Research Signpost, Kerala, India.

15 Park, M.C., Lim, J.S., Kim, J.C., Park, S.W. and Kim, S.W. (2005) *Biotechnology Letters*, **27**, 127–130.

16 Grizard, D. and Barthomeuf, C. (1999) *Food Biotechnology*, **13**, 93–105.

17 Marx, S.P., Winkler, S. and Hartmeier, W. (2000) *FEMS Microbiology Letters*, **182**, 163–169.

18 Katapodis, P. and Christakopoulos, P. (2004) *World Journal of Microbiology and Biotechnology*, **20**, 667–672.

19 Alvaro-Benito, M., de Abreu, M., Fernandez-Arrojo, L., Plou, F.J., Jimenez-Barbero, J., Ballesteros, A., Polaina, J. and Fernandez-Lobato, M. (2007) *Journal of Biotechnology*, **132**, 75–81.

20 Roberfroid, M.B. (2001) *The American Journal of Clinical Nutrition*, **73**, 406S–409S.

21 Gibson, G.R. and Ottaway, R.A. (2000) *Prebiotics: New Developments in Functional Foods*, Chandos Publishing, Oxford.

22 Probert, H.M. and Gibson, G.R. (2002) *Letters in Applied Microbiology*, **35**, 473–480.

23 Roberfroid, M. (2002) *Digestive and Liver Disease*, **34**, S105–110.

24 Tuohy, K.M., Rouzaud, G.C.M., Bruck, W.M. and Gibson, G.R. (2005) *Current Pharmaceutical Design*, **11**, 75–90.

25 Grizard, D. and Barthomeuf, C. (1999) *Reproduction, Nutrition, Development*, **39**, 563–588.

26 Tungland, B.C. (2003) Fructooligosaccharides and other fructans: structures and occurrence, production, regulatory aspects, food applications, and nutritional health significance, in *Oligosaccharides in Food and Agriculture* (eds G. Eggleston and G.L. Coté), The American Chemical Society, Washington, pp. 135–152.

27 Cruz, R., Cruz, V.D., Belote, J.G., Khenayfes, M.D., Dorta, C., Oliveira, L.H.D., Ardiles, E. and Galli, A. (1999) *Bioresource Technology*, **70**, 165–171.

28 Rastall, R.A. and Hotchkiss, A.T. Jr (2003) Potential for the development of prebiotic oligosaccharides from biomass, in *Oligosaccharides, in Food and Agriculture* (eds G. Eggleston and G.L. Coté), The American Chemical Society, Washington, pp. 44–53.

29 Okai, A.A.E. and Gierschner, K. (1991) *Zeitschrift Fur Lebensmittel-Untersuchung Und -Forschung*, **192**, 244–248.

30 Hang, Y.D. and Woodams, E.E. (1995) *Biotechnology Letters*, **17**, 741–745.

31 Tanriseven, A. and Gokmen, F. (1999) *Biotechnology Techniques*, **13**, 207–210.

32 Ghazi, I., Fernandez-Arrojo, L., Gomez de Segura, A., Alcalde, M., Plou, F.J. and Ballesteros, A. (2006) *Journal of Agricultural and Food Chemistry*, **54**, 2964–2968.

33 Ghazi, I., Fernandez-Arrojo, L., Garcia-Arellano, H., Ferrer, M., Ballesteros, A. and Plou, F.J. (2007) *Journal of Biotechnology*, **128**, 204–211.

34 Yanai, K., Nakane, A., Kawate, A. and Hirayama, M. (2001) *Bioscience Biotechnology and Biochemistry*, **65**, 766–773.

35 Boddy, L.M., Berges, T., Barreau, C., Vainstein, M.H., Dobson, M.J., Ballance, D.J. and Peberdy, J.F. (1993) *Current Genetics*, **24**, 60–66.

36 Chavez, F.P., Pons, T., Delgado, J.M. and Rodriguez, L. (1998) *Yeast*, **14**, 1223–1232.

37 Fujita, K., Hara, K., Hashimoto, H. and Kitahata, S. (1990) *Agricultural and Biological Chemistry*, **54**, 913–919.

38 Shiomi, N. (1982) *Carbohydrate Research*, **99**, 157–169.

39 Rehm, J., Willmitzer, L. and Heyer, A.G. (1998) *Journal of Bacteriology*, **180**, 1305–1310.

40 Chen, J.S., Saxton, J., Hemming, F.W. and Peberdy, J.F. (1996) *Biochimica et Biophysica Acta – Protein Structure and Molecular Enzymology*. **1296**, 207–218.

41 Hirayama, M., Sumi, N. and Hidaka, H. (1989) *Agricultural and Biological Chemistry*, **53**, 667–673.

42 Nguyen, Q.D., Rezessy-Szabo, J.M., Bhat, M.K. and Hoschke, A. (2005) *Process Biochemistry*, **40**, 2461–2466.

43 Hocine, L.L., Wang, Z., Jiang, B. and Xu, S.Y. (2000) *Journal of Biotechnology*, **81**, 73–84.

44 Rubio, M.C. and Maldonado, M.C. (1995) *Current Microbiology*, **31**, 80–83.

45 Chang, C.T., Lin, Y.Y., Tang, M.S. and Lin, C.F. (1994) *Biochemistry and Molecular Biology International*, **32**, 269–277.

46 Cejudo, M.G., de la Vega, F.J. and Paneque, A. (1991) *Enzyme and Microbial Technology*, **13**, 267–271.

47 Park, J.P., Oh, T.K. and Yun, J.W. (2001) *Process Biochemistry*, **37**, 471–476.

48 Warchol, M., Perrin, S., Grill, J.P. and Schneider, F. (2002) *Letters in Applied Microbiology*, **35**, 462–467.

49 Imamura, L., Hisamitsu, K. and Kobashi, K. (1994) *Biological and Pharmaceutical Bulletin*, **17**, 596–602.

50 Ishimoto, M. and Nakamura, A. (1997) *Bioscience Biotechnology and Biochemistry*, **61**, 599–603.

51 de Gines, S.C., Maldonaldo, M.C. and de Valdez, G.F. (2000) *Current Microbiology*, **40**, 181–184.

52 Taussig, R. and Carlson, M. (1983) *Nucleic Acids Research*, **11**, 1943–1954.

53 Moreno, S., Sanchez, Y. and Rodriguez, L. (1990) *The Biochemical Journal*, **267**, 697–702.

54 Klein, R.D., Deibel, M.R., Sarcich, J.L., Zurcherneely, H.A., Reardon, I.M. and Heinrikson, R.L. (1989) *Preparative Biochemistry*, **19**, 293–319.

55 Sumner, J.B. and Howell, S.F. (1935) *The Journal of Biological Chemistry*, **108**, 51–54.

56 Ghazi, I., Gómez de Segura, A., Fernández-Arrojo, L., Alcalde, M., Yates, M., Rojas-Cervantes, M.L., Plou, F.J. and Ballesteros, A. (2005) *Journal of Molecular Catalysis B: Enzymatic*, **35**, 19–27.

57 Fujishima, M., Sakai, H., Ueno, K., Takahashi, N., Onodera, S., Benkeblia, N. and Shiomi, N. (2005) *The New Phytologist*, **165**, 513–524.

58 Fernandez, R.C., Ottoni, C.A., Matsubara, E.S., da Silva, R.M.S., Carter, J.M., Magossi, L.R., Wada, M.A.A., Rodrigues, M.F.D., Maresma, B.G. and Maiorano, A.E. (2007) *Applied Microbiology and Biotechnology*, **75**, 87–93.

59 van Hijum, S.A.F.T., Szalowska, E., van der Maarel, M.J.E.C. and Dijkhuizen, L. (2004) *Microbiology-SGM*, **150**, 621–630.

60 Reddy, A. and Maley, F. (1996) *The Journal of Biological Chemistry*, **271**, 13953–13958.

61 Rodriguez, J., Perez, J.A., Ruiz, T. and Rodriguez, L. (1995) *The Biochemical Journal*, **306**, 235–239.

62 Boer, E., Wartmann, T., Luther, B., Manteuffel, R., Bode, R., Gellissen, G. and Kunze, G. (2004) *Antonie van Leeuwenhoek International Journal of General and Molecular Microbiology*, **86**, 121–134.

63 Belcarz, A., Ginalska, G., Lobarzewski, J. and Penel, C. (2002) *Biochimica et Biophysica Acta – Protein Structure and Molecular Enzymology*, **1594**, 40–53.

64 Vannieuwenburgh, C., Guibert, A. and Combes, D. (2002) *Bioprocess and Biosystems Engineering*, **25**, 13–20.

65 Ouarne, F. and Guibert, A. (1995) *Zuckerindustrie*. **120**, 793–798.

66 Gomez de Segura, A., Alcalde, M., Plou, F.J., Remaud-Simeon, M., Monsan, P. and Ballesteros, A. (2003) *Biocatalysis and Biotransformation*, **21**, 325–331.

67 Martin, M.T., Plou, F.J., Alcalde, M. and Ballesteros, A. (2003) *Journal of Molecular Catalysis B: Enzymatic*, **21**, 299–308.

68 Mateo, C., Abian, O., Fernandez-Lorente, G., Pedroche, J., Fernandez-Lafuente, R. and Guisan, J.M. (2002) *Biotechnology Progress*, **18**, 629–634.

69 Wheatley, J.B. and Schmidt, D.E. (1999) *Journal of Chromatography*, **849**, 1–12.A

70 Gómez de Segura, A., Alcalde, M., Yates, M., Rojas-Cervantes, M.L., Lopez-Cortes, N., Ballesteros, A. and Plou, F.J. (2004) *Biotechnology Progress*, **20**, 1414–1420.

71 Platkova, Z., Polakovic, M., Stefuca, V., Vandakova, M. and Antosova, M. (2006) *Chemical Papers-Chemicke Zvesti* **60**, 469–472.

72 Hayashi, S., Tubouchi, M., Takasaki, Y. and Imada, K. (1994) *Biotechnology Letters*, **16**, 227–228.

73 Hayashi, S., Hayashi, T., Kinoshita, J., Takasaki, Y. and Imada, K. (1992) *Journal of Industrial Microbiology*, **9**, 247–250.

74 Chiang, C.J., Lee, W.C., Sheu, D.C. and Duan, K.J. (1997) *Biotechnology Progress*, **13**, 577–582.

75 Tanriseven, A. and Aslan, Y. (2005) *Enzyme and Microbial Technology*, **36**, 550–554.

76 Spiegel, J.E., Rose, R., Karabell, P., Frankos, V.H. and Schmitt, D.F. (1994) *Food Technology*, 85–89.

12
Hydantoin Racemase: the Key Enzyme for the Production of Optically Pure α-Amino Acids

Francisco Javier Las Heras-Vázquez, Josefa María Clemente-Jiménez, Sergio Martínez-Rodríguez and Felipe Rodríguez-Vico

12.1
Introduction

Optically pure D-amino acids are valuable intermediates for the preparation of semi-synthetic antibiotics, pesticides, and other products of interest for the pharmaceutical, food, and agrochemical industries [1, 2]. These compounds are named non-natural amino acids and are becoming increasingly important. Typical examples of D-amino acids are D-(phenyl or substituted phenyl) glycines as intermediates of antibiotics such as semi-synthetic penicillins and cephalosporins, D-valine for the synthesis of the insecticide fluvanilate, D-alanine for the production of the dietetic sweetening agent alitame, or D-citrulline and D-homocitrulline used in the potent luterizing hormone releasing hormone (LH-RH) antagonists.

There are two main routes for the production of D-amino acids: chemical synthesis and enzymatic catalysis. As regards conventional chemical synthesis, unless asymmetrical starting compounds or catalysts are used, a mixture of the D- and L-stereoisomers is obtained in equal proportions. The racemic mixture is therefore optically inactive and the stereoisomers must be separated. The separation of the enantiomers by classical crystallization of diastereomeric salts is the most costly production step and in any case this method can only yield 50% of the desired enantiomer [3]. Enzymatic synthesis can solve the above problems, providing optical purity of the D-amino acid and a 100% yield, as well as mild, environment-friendly reaction conditions.

One of the most widely used enzymatic methods for D-amino acid production is the 'hydantoinase process' [4]. The great advantage of this process is that, potentially, any optically pure D-amino acid can be obtained using the corresponding substrate from a wide spectrum of D,L-5-monosubstituted hydantoins, which are readily accessible by chemical synthesis [5]. In this cascade of reactions the chemically synthesized D,L-5-monosubstituted hydantoin ring is first hydrolyzed by a stereoselective hydantoinase enzyme (D-hydantoinase). Further hydrolysis of the resulting *N*-carbamoyl D-amino acid to the free D-amino acid is catalyzed

Modern Biocatalysis: Stereoselective and Environmentally Friendly Reactions.
Edited by Wolf-Dieter Fessner and Thorleif Anthonsen

ISBN: 978-3-527-32071-4

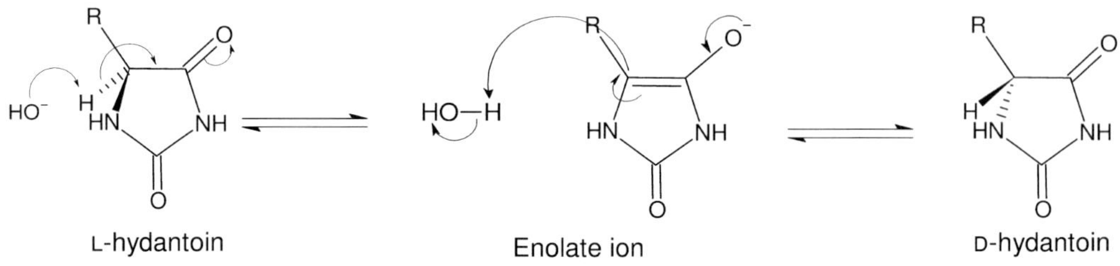

Figure 12.1 Reaction mechanism of the 5-monosubstituted hydantoins keto-enol tautomerism under alkaline conditions.

Table 12.1 Racemization rate constants (k_{rac}) and corresponding half-life times ($t_{1/2,rac}$) for various hydantoins [7].

5-Substituted hydantoins	Corresponding amino acid	k_{rac} (h^{-1})	$t_{1/2,rac}$ (h)
p-Hydroxyphenylhydantoin	*p*-Hydroxy-phenylglycine	2.26	0.12
Phenylhydantoin	Phenylglycine	2.59	0.27
Hydroxymethylhydantoin	Serine	0.43	1.60
Phenylethylhydantoin	Homo-phenylalanine	0.144	4.8
Benzylhydantoin	Phenylalanine	0.14	5.00
Methylthioethylhydantoin	Methionine	0.12	5.82
Neopentylhydantoin	γγ-Methyl leucine	0.11	6.4
1-Hydroxyethylhydantoin	*allo*-Threonine	0.067	6.41
Trimethylsilylmethylhydantoin	L-3-Trimethylsilyl-alanine	0.049	10.4
3′-Ureidopropylhydantoin	Citrulline	0.044	14.26
1′-Methylethylhydantoin	*allo*-Isoleucine	0.043	15.84
Imidazolylmethylhydantoin	Histidine	0.032	16.09
Isobutylhydantoin	Leucine	0.020	21.42
Methylhydantoin	Alanine	0.012	33.98
Ethylhydantoin	2-Aminobutyric acid	0.018	38.51
Isopropylhydantoin	Valine	0.0108	55.90

by a highly enantiospecific *N*-carbamoyl D-amino acid amidohydrolase (D-carbamoylase). Concurrent to hydantoinase hydrolysis of the D-configurated 5-monosubstituted hydantoin, the L-enantiomer is depleted by chemical and/or enzymatic racemization.

Chemical racemization of the 5-monosubstituted hydantoins proceeds via keto-enol tautomerism under alkaline conditions, as shown in Figure 12.1 [6]. The racemization velocity is highly dependent on the bulkiness and electronic factors of the substituent in 5-position (see Table 12.1) and is usually a very slow process [7]. High velocities of chemical racemization have only been observed for D,L-5-phenyl- and D,L-5-*p*-hydroxy-phenylhydantoin, because of the resonance stabilization by the 5-substituent, while all other hydantoins take many hours to racemize [8]. Consequently, although the hydantoinase process has been successfully used for producing 100% D-phenyl- and D-*p*-hydroxy-phenylglycine [9], only 50% of the

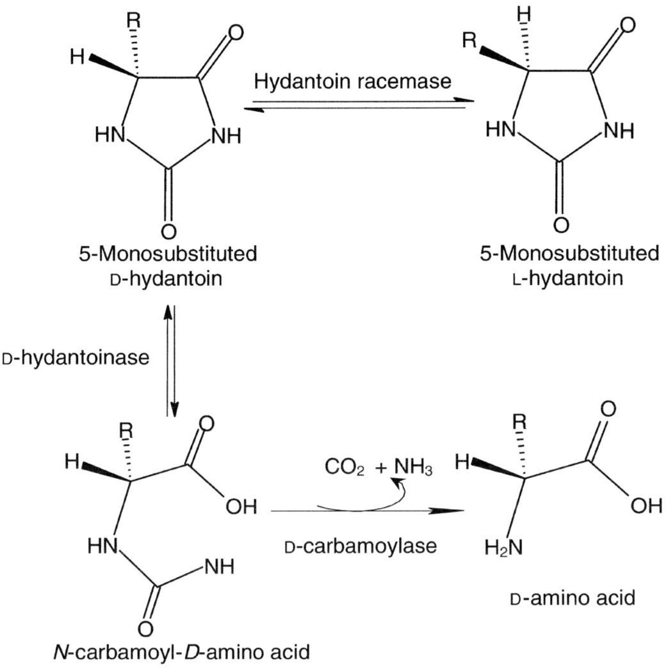

Figure 12.2 Reaction scheme for the hydrolysis of D,L-5-monosubstituted hydantoin derivatives to the corresponding D-amino acids.

remaining D,L-5-monosubstituted hydantoins are converted to the corresponding D-amino acid, while the other 50% correspond to L-hydantoin, which is not hydrolyzed by D-hydantoinase. For these hydantoins a faster racemization is possible via an enzymatic equilibration reaction incorporating a third enzyme named hydantoin racemase, in the presence of D-hydantoinase and D-carbamoylase (Figure 12.2). This enzymatic racemization of those D,L-5-monosubstituted hydantoins having a very slow chemical racemization velocity has been shown in several microorganisms by total conversion and 100% production of optically pure D-amino acids [10–12], suggesting that a hydantoin racemase might be responsible for the fast racemization of the substrates. In this way, complete and fast racemization is obtained, the velocity of hydrolysis is increased, and accumulation of one isomer (L-hydantoin) is avoided (Figure 12.3).

12.2 Search for New Hydantoin Racemases and Molecular Characterization

Given the important role that hydantoin racemase plays in the production of optically pure D-amino acids, allowing the racemization of 5-monosubstituted

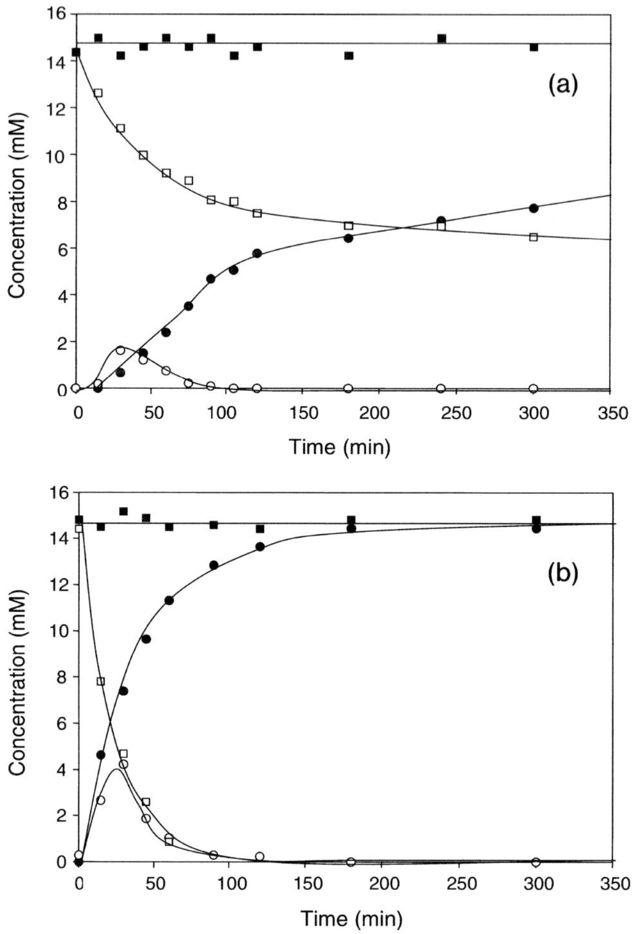

Figure 12.3 Reaction profile of D-methionine production from D,L-methylthioethylhydantoin (D,L-MTEH) using (a) a double system with D-hydantoinase and D-carbamoylase enzymes and (b) a triple system with hydantoin racemase enzyme as third enzyme. Symbols: ●, D-methionine; ○, *N*-carbamoyl D-methionine; □, D,L-methylthioethylhydantoin; ■, sum of all three [9].

hydantoins under unfavorable physiological conditions for chemical racemization, our laboratory started the search for such enzymes in different sources. Knowledge of the biochemical, physical, and thermodynamic characteristics of hydantoin racemase enzymes will allow a better design of biocatalysts, such as immobilized enzymes or whole cell biocatalysts. Previous works have studied two hydantoin racemases from both the molecular and biochemical angles, but these were involved in the production of L-amino acids [13–16]. As for the presence of hydantoin racemase involved in optically pure D-amino acid production, a putative hydantoin racemase was described together with D-hydantoinase and D-carbamoylase in a DNA fragment from *Agrobacterium* sp. IP-I 671 [17]. However, from the biochemical and enzymatic points of view there was no information on hydantoin racemases involved in the production of D-amino acids.

In 2001 two different genome sequences *of Agrobacterium tumefaciens* C58 and *Sinorhizobium meliloti* were reported [18–20]. In the linear chromosome of each genome, together with other protein-coding genes, our laboratory found two putative hydantoin racemases with high sequence homology to previously described hydantoin racemases. Four fragments were amplified by a polymerase chain reaction and cloned in expression plasmids. While both sequences from *A. tumefaciens* showed hydantoin racemase activity in the recombinant cells, only one was active in *S. meliloti*. Recently a hydantoin racemase from *Microbacterium liquefaciens* AJ 3912 has also been purified [21].

These six hydantoin racemases are the only ones isolated and studied to date and they show a high homology in their amino acid sequences (Figure 12.4), with identity percentages of over 40%. This increases to 80% when comparing hydantoin racemase 1 from *A. tumefaciens* C58 (AtHyuA1) [9, 22] and the one from *S. meliloti* CECT 4114 (SmeHyuA) [23] or those from *Arthrobacter aurescens* DSM 3747 (AauHyuA) [15] and from *M. liquefaciens* AJ 3912 (MliHyuA) [24]. All six hydantoin racemases have shown an amino acid sequence length of between 236 and 247 residues. The phylogenetic tree generated by comparing the complete amino acid sequences of hydantoin racemases revealed the existence of three branches (Figure 12.5): one comprises AtHyuA1 and SmeHyuA, a second branch comprise AauHyuA and MliHyuA, together with hydantoin racemase from *Pseudomonas* sp. NS671 (PspHyuE) [13], and the third branch comprises only hydantoin racemase 2 from *A. tumefaciens* C58 (AtHyuA2) [25].

The genetic organization of the enzymes involved in the hydantoinase process (*hyu*), including hydantoin racemase (hyuA), hydantoinase (hyuH) and *N*-carbamoylase (hyuC) genes has been reported. There are two clearly different distributions of the genes, depending on whether the *hyu* gene cluster is involved in L- or D-amino acid production (Figure 12.6). Thus, two of the three L-specific *hyu* gene clusters described corresponding to *A. aurescens* DSM 3747 [26] and *M. liquefacines* AJ 3912 [24] are very similar. These two *hyu* gene clusters and that of *Pseudomonas* sp. NS671 [13] have the same orientation and it was suggested that they constitute an operon, harboring a putative hydantoin transporter gene in addition to hydantoin racemase, L-*N*-carbamoylase and L-hydantoinase genes. On the other hand, the non-selective hydantoinase from *Pseudomonas* sp. NS671 consists of two subunits, hyuA and hyuB. It shows similarity to *N*-methylhydantoinases, which belong to the superfamily of ATP-dependent cyclic amidases [27]. However, in the D-specific *hyu* gene cluster D-*N*-carbamoylase and D-hydantoinase genes are transcribed from a common intercistronic promoter region in opposite orientation [17].

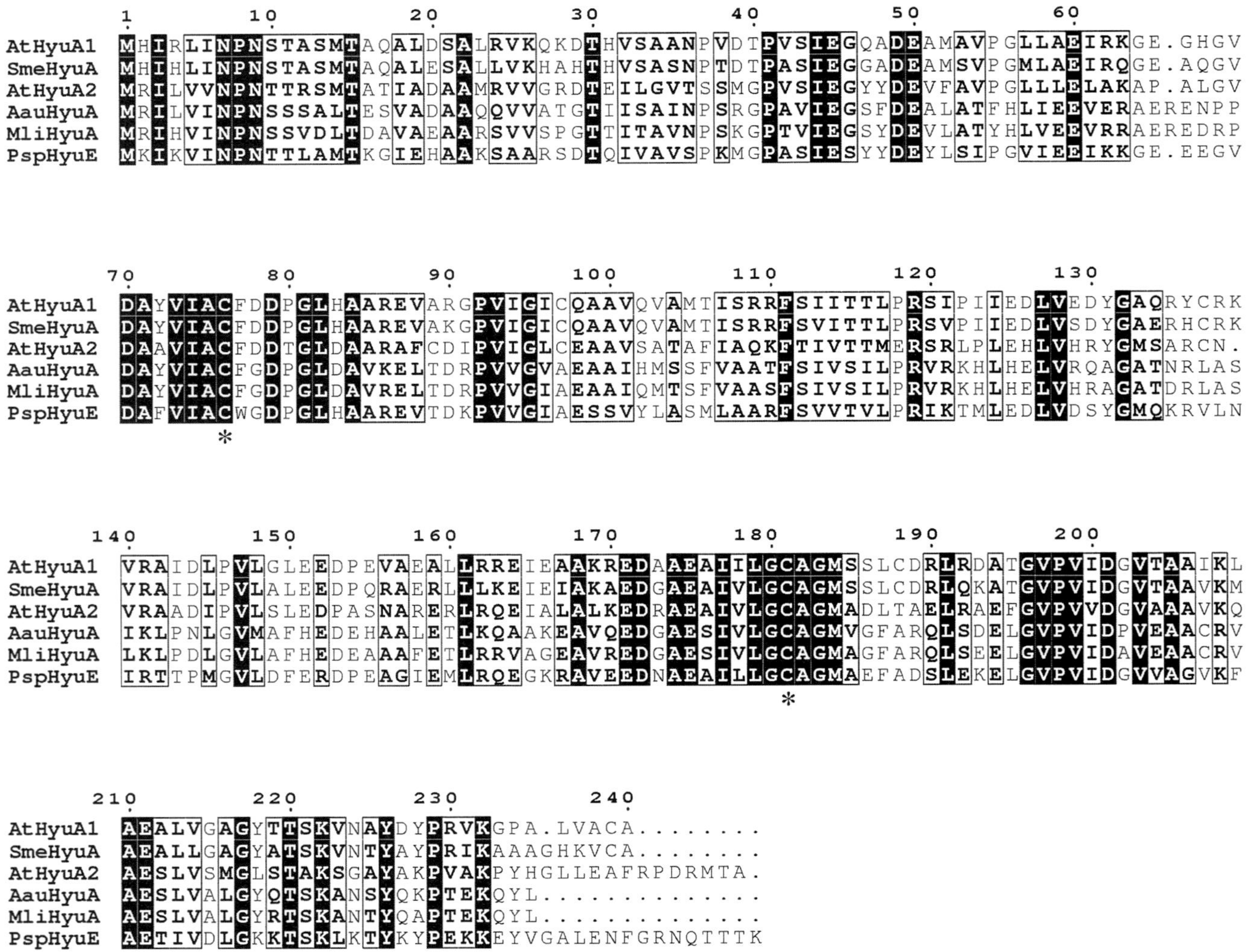

Figure 12.4 Multiple alignment of the amino acid sequences of hydantoin racemases. Hydantoin racemase 1 from *A. tumefaciens* C58 (AtHyuA1), GenBank accession no. AY436503; hydantoin racemase from *S. meliloti* CECT 4114 (SmeHyuA), GenBank accession no. AY393697; hydantoin racemase 2 from *A. tumefaciens* C58 (AtHyuA2), GenBank accession no. AY436504; hydantoin racemase from *A. aurescens* DSM 3747 (AauHyuA), GenBank accession no. AF146701; hydantoin racemase from *M. liquefaciens* AJ 3912 (MliHyuA), GenBank accession no. BD181023; hydantoin racemase from *Pseudomonas* sp. NS671 (PspHyuE), GenBank accession no. M84731.

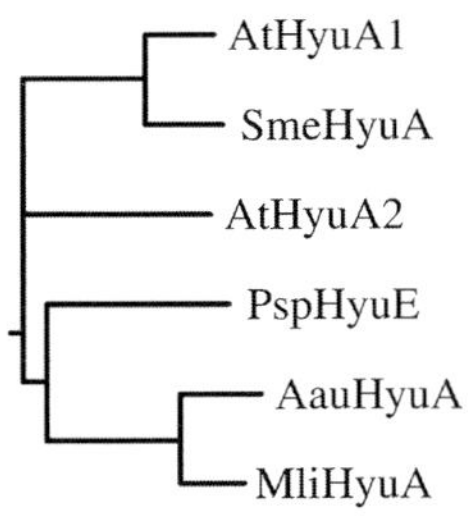

Figure 12.5 Phylogenetic analysis of the amino acid sequences of hydantoin racemases characterized to date. For abbreviations of the enzymes, see the caption to Figure 12.4.

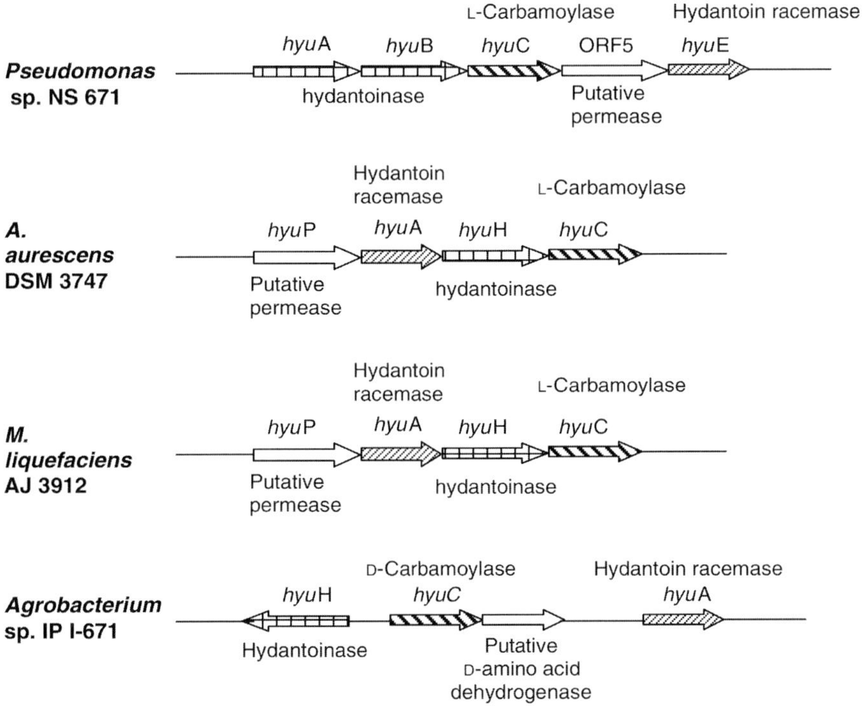

Figure 12.6 Genetic organization of the *hyu* gene clusters from *Pseudomonas* sp. NS671, *A. aurescens* DSM 3747, *M. liquefaciens* AJ 3912 and *A. tumefaciens* IPI-671. Genes and their transcriptional orientation are symbolized by arrows. Genes encoding related proteins are represented by arrows with identical patterns.

12.3 Biochemical Characterization of Hydantoin Racemase Enzymes

The apparent molecular mass of the subunit of the different hydantoin racemases is very similar (Table 12.2) and larger than that calculated from the amino acid sequence, with values of 25 000–27 000 Da. These differences can be explained by a 'molten globule' state, where the enzymes in sodium dodecyl sulfate–polyacrylamide electrophoresis gel (SDS–PAGE) conditions are compact denatured structures but with near native secondary structure, significant loss of tertiary structure, and increased exposure of hydrophobic surface area. In SDS–PAGE these partially denatured hydantoin racemases moved slower than if they were totally denatured. The relative molecular mass of the native enzymes measured by size exclusion chromatography (high-performance liquid chromatography or HPLC) has been estimated at 100 000 Da, with the exception of that reported for *Pseudomonas* sp. (190 000 Da) and for *A. aurescens* DSM 3747 (175 000 Da). Thus, a tetrameric structure is deduced for hydantoin racemase enzymes from *Agrobacterium, Sinorhizobium,* and *Microbacterium,* whereas the *Pseudomonas* sp. hydantoin racemase enzyme has been described as hexameric and the *A. aurescens* DSM 3747 hydantoin racemase enzyme as hexameric, heptameric, or octameric (Table 12.2).

Although on the whole the hydantoin racemases have shown high thermal stability, with optimal activity at 55 °C (Table 12.2), the optimal temperatures for the ones from *Pseudomonas* and *Sinorhizobium* decrease to 45 and 40 °C, respectively. However, the optimal pH is higher than 8, except for both hydantoin racemases from *Agrobacterium*. This low alkaline pH avoids chemical racemization. Consequently, the racemization of the D- or L-5-monosubstituted hydantoins in an industrial process will only occur enzymatically.

The activity of the purified hydantoin racemase enzymes has been assayed in the presence of different metal ions, the chelating agent EDTA, and the reducing compound dithiothreitol, in order to study whether they are metalloenzymes. Most

Table 12.2 Biochemical properties of hydantoin racemases described to date.

Strain	System chirality	Molecular weight of Monomers (kDa)	Number of subunits	Optimal pH	Optimal temperature (°C)
Pseudomonas sp. NS671	L	32	6	9.5	45
A. aurescens DSM 3747	L	32.1	6, 7, and 8	8.5	55
A. tumefaciens C58 1	D	32	4	7.5	55
A. tumefaciens C58 2	D	27	4	7.5	55
M. liquefaciens AJ 3912	L	27	4	8.2	55
S. meliloti CECT 4114	D	31	4	8.5	40

of the metal ions had no significant effect on the enzymes and an inhibitory effect was only detected in the presence of Cu^{2+}, Hg^{2+}, Pb^{2+}, and Zn^{2+}. No effect was detected after enzyme incubation with dithiothreitol, with the exception of the *Arthrobacter* enzyme, which was stimulated. However, the insignificant effect of the metal chelating agent EDTA on the activity of the hydantoin racemases would indicate that they are not metalloenzymes.

12.4
Substrate Enantioselectivity and Kinetic Analysis of Hydantoin Racemases

The natural substrates for hydantoin racemases are as yet unknown, but for industrial purposes their ability for racemizing different 5-monosubstituted hydantoins with very low chemical racemization has been studied (Figure 12.7). Of the six hydantoin racemases, only the *Arthrobacter* one has shown better racemization for hydantoins with aromatic substituents, the enzyme from *Microbacterium* was only studied with one substrate and the other enzymes racemized the aliphatic substrates faster (see the example of AtHyuA1 in Figure 12.8). For *Agrobacterium* hydantoin racemases 1 and 2 a reciprocal dependence between kinetic activity and the substituent chain length was detected for hydantoins with aliphatic substituents. The highest racemization velocity corresponded to the shortest length in the substituent chain. However, racemization of D- and L-hydantoins took longer with AtHyuA2 than with AtHyuA1. This faster racemization by AtHyuA1 is corrobo-

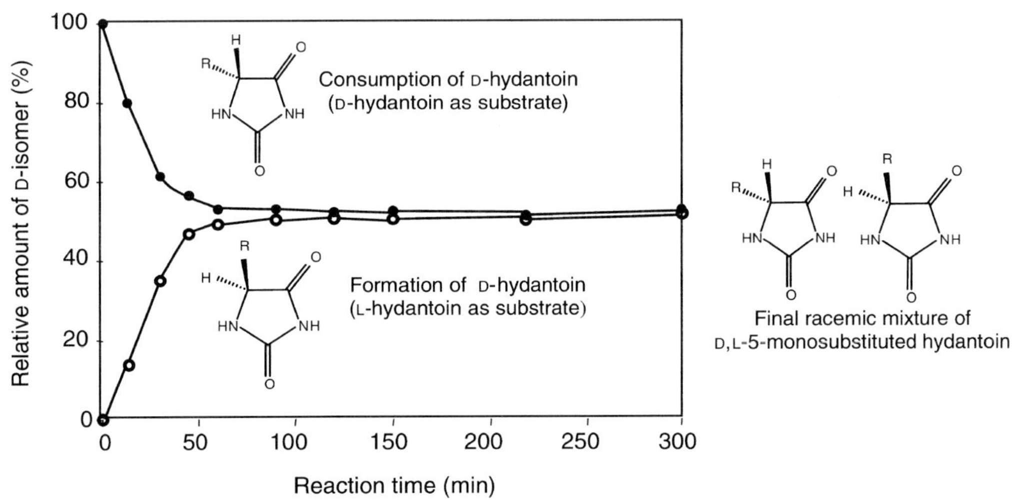

Figure 12.7 Reaction profile of the enzymatic racemization of D- and L-5-monosubstituted hydantoin to the racemic mixture by hydantoin racemase enzyme monitored by chiral HPLC (see the methodology in [28]).

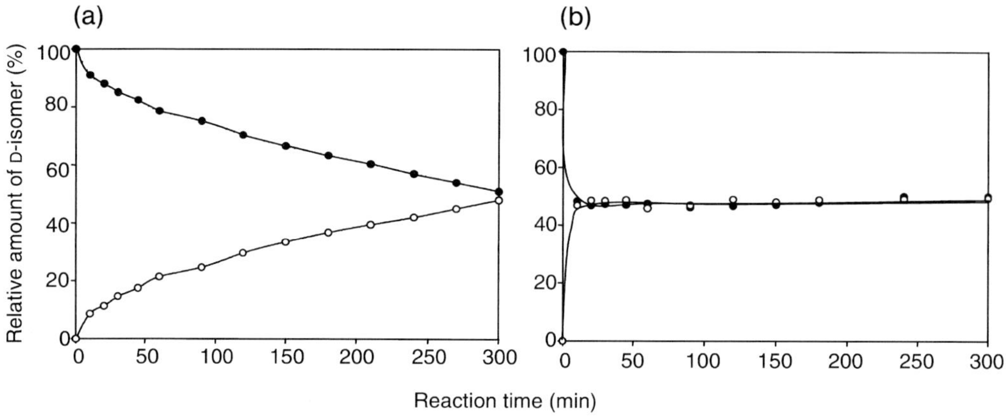

Figure 12.8 Enzymatic racemization of the D-isomer (●) and L-isomer (○) of (a) 5-benzylhydantoin and (b) 5-ethylhydantoin by hydantoin racemase 1 from *A. tumefaciens* C58 (AtHyuA1).

rated by more favorable kinetic parameters, with lower K_m and higher k_{cat} values than for AtHyuA2 (Table 12.3).

A singular effect of inhibition by 5-monosubstituted hydantoins has been detected for hydantoin racemase from *Sinorhizobium*. The D-isomer of the potential substrate 5-ethylhydantoin had no detectable racemization under standard conditions, while its L-isomer (L-5-ethylhydantoin) showed the fastest racemization of all hydantoins analyzed. On the other hand, neither the D- nor the L-isomer of 5-methylthioethylhydantoin showed any racemization in the presence of the *Sinorhizobium* hydantoin racemase enzyme (Figure 12.9). These substrates towards which the enzyme had no detectable activity were studied as possible inhibitors of the hydantoin racemase. No inhibition effect was detected by D-5-ethylhydantoin for the substrate L-5-ethylhydantoin. However, inhibition by D- and L-5-methylthioethylhydantoin was studied at varying concentrations of L-5-ethylhydantoin. Double-reciprocal plots of the data indicated that both D- and L-5-methylthioethylhydantoin are competitive inhibitors of L-5-ethylhydantoin as the variable substrate (Figure 12.10). Substrate inhibition processes have been described for *Pseudomonas* sp. NS671 and *A. aurescens* DSM 3747 hydantoin racemases, even at low substrate concentrations. However, this phenomenon has not been detected either for *Agrobacterium* hydantoin racemase. Hydantoin racemase from *Sinorhizobium* has shown no activity with D- and L-5-methylthioethylhydantoin substrate as was previously reported for *Pseudomonas* sp. NS671 hydantoin racemase towards either isomer of 5-isopropylhydantoin. However, this is the only case in which substrates that the enzyme is not capable of racemizing have been studied as inhibitors of hydantoin racemase activity.

Table 12.3 Kinetic parameters of *A. tumefaciens* C58 hydantoin racemase 1 and 2 (ATHyuA1 and AtHyuA2).

Substrate[a]		Enzyme	K_m (mM)	k_{cat} (s^{-1})	k_{cat}/K_m ($s^{-1}mM^{-1}$)
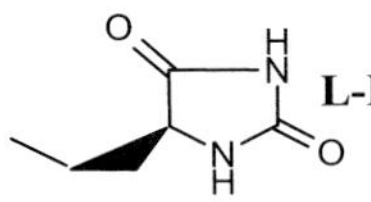	L-EH	AtHyuA1 AtHyuA2	4.45 ± 0.060 19.42 ± 2.34	13.64 ± 1.53 1.81 ± 0.01	3.06 ± 0.63 0.09 ± 0.01
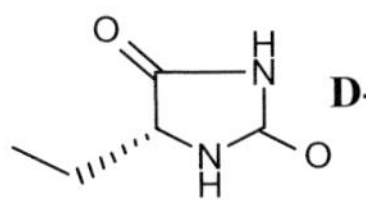	D-EH	AtHyuA1 AtHyuA2	4.26 ± 0.70 12.54 ± 1.81	16.47 ± 1.47 1.80 ± 0.10	3.87 ± 0.59 0.14 ± 0.03
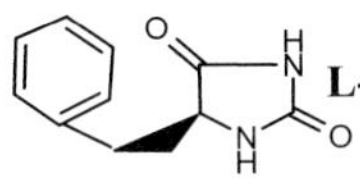	L-BH	AtHyuA1 AtHyuA2	5.56 ± 1.45 18.42 ± 4.80	1.07 ± 0.23 0.18 ± 0.02	0.19 ± 0.04 0.01 ± 0.00
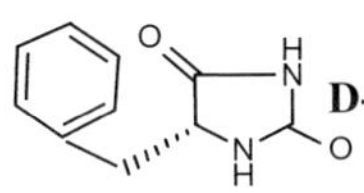	D-BH	AtHyuA1 AtHyuA2	4.71 ± 0.62 20.77 ± 5.47	2.36 ± 0.43 0.46 ± 0.07	0.50 ± 0.10 0.02 ± 0.00
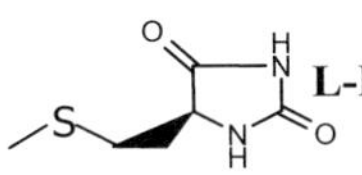	L-MTEH	AtHyuA1 AtHyuA2	5.41 ± 1.06 1.90 ± 1.48	1.57 ± 0.37 0.78 ± 0.07	0.35 ± 0.08 0.07 ± 0.01
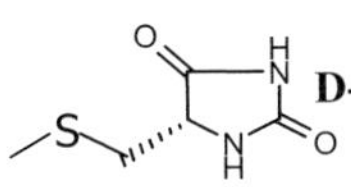	D-MTEH	AtHyuA1 AtHyuA2	4.47 ± 0.96 6.31 ± 0.31	2.03 ± 0.45 0.5 ± 0.01	0.45 ± 0.12 0.08 ± 0.01
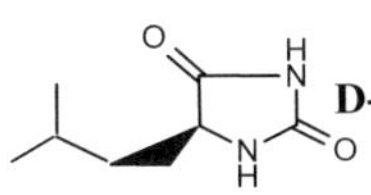	D-IBH	AtHyuA1 AtHyuA2	1.23 ± 0.20 3.02 ± 0.78	2.05 ± 0.53 0.48 ± 0.02	1.66 ± 0.13 0.16 ± 0.05
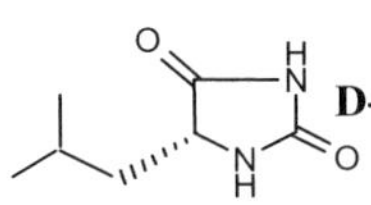	D-IBH	AtHyuA1 AtHyuA2	4.58 ± 0.51 6.79 ± 0.60	4.16 ± 0.67 0.83 ± 0.03	0.90 ± 0.08 0.12 ± 0.01

a D- and L-EH: D- and L-5-ethylhydantoin; D- and L-BH: D- and L-5-benzylhydantoin; D- and L-MTEH: D- and L-5-methylthioethylhydantoin; D- and L-IBH: D- and L-5-isobutylhydantoin. These parameters were determined at 40 °C for 15 min at pH 7.5. The k_{cat} was defined as the D- or L-5-monosubstituted hydantoin (mmol) racemized per second and of enzyme (mmol) at 40 °C.

12.5
Proposal for a Reaction Mechanism of Hydantoin Racemase Enzymes

The amino acid sequences of known hydantoin racemases present two highly conserved cysteines around positions 75 and 180 (see the asterisks in Figure 12.4). The enzymes involved in the racemization/epimerization of different substrates such as glutamate racemase and diaminopimelate epimerase present two cyste-

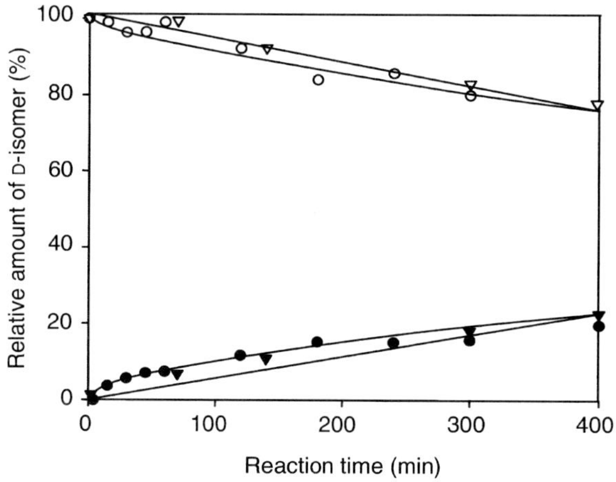

Figure 12.9 Enzymatic racemization by hydantoin racemase from *S. meliloti* CECT 4114 of the D-isomer (○) and L-isomer (●) of 5-methylthioethylhydantoin. Chemical racemization of the D-isomer (▽) and L-isomer (▼) of each substrate was also measured at the same intervals.

ines involved in the catalytic center. To demonstrate the role of these two cysteines in the enzyme activity, our group has carried out structural, kinetic, and binding studies showing the importance of these residues in the recognition and racemization of the substrates by hydantoin racemase enzyme [29, 30]. For these studies hydantoin racemase from *S. meliloti* was selected.

Comparison of the amino acid sequence of SmeHyuA with those of aspartate or glutamate racemases of well-known activity revealed low similarity percentages. However, two completely conserved cysteines have been shown to be involved in the racemization process for glutamate and aspartate racemases and, thus, belong to the active site. In order to study the role of the two cysteines of hydantoin racemase in racemization, our laboratory developed mutants of these cysteines in *Sinorizhobium*. Thus, cysteine 76 and cysteine 181 were mutated to serine and alanine, thereby obtaining four mutants (C76S, C76A, C181S, and C181A). Activity assays with mutants show a drastic decrease in activity of the enzyme, indicating that both cysteine 76 and 181 are essential for catalysis. Substitution by a residue able to transfer a proton (serine) retained a small fraction of the activity, whereas substitution by an alanine resulted in no detectable activity, thus indicating that the presence of a proton donor group in that position is critical for the catalysis of this enzyme. Similar behavior has been detected in other enzymes acting by a two-base mechanism.

The binding affinities of SmeHyuA C76A and C181A active site mutants (alanine mutants) were evaluated for further insight into the role of each residue. Binding of the above substrates to both mutants was also measured by isothermal titration calorimetry which showed that the presence of a proton at the 5-position of the substrate and not in the lateral chain is the critical factor for proper binding.

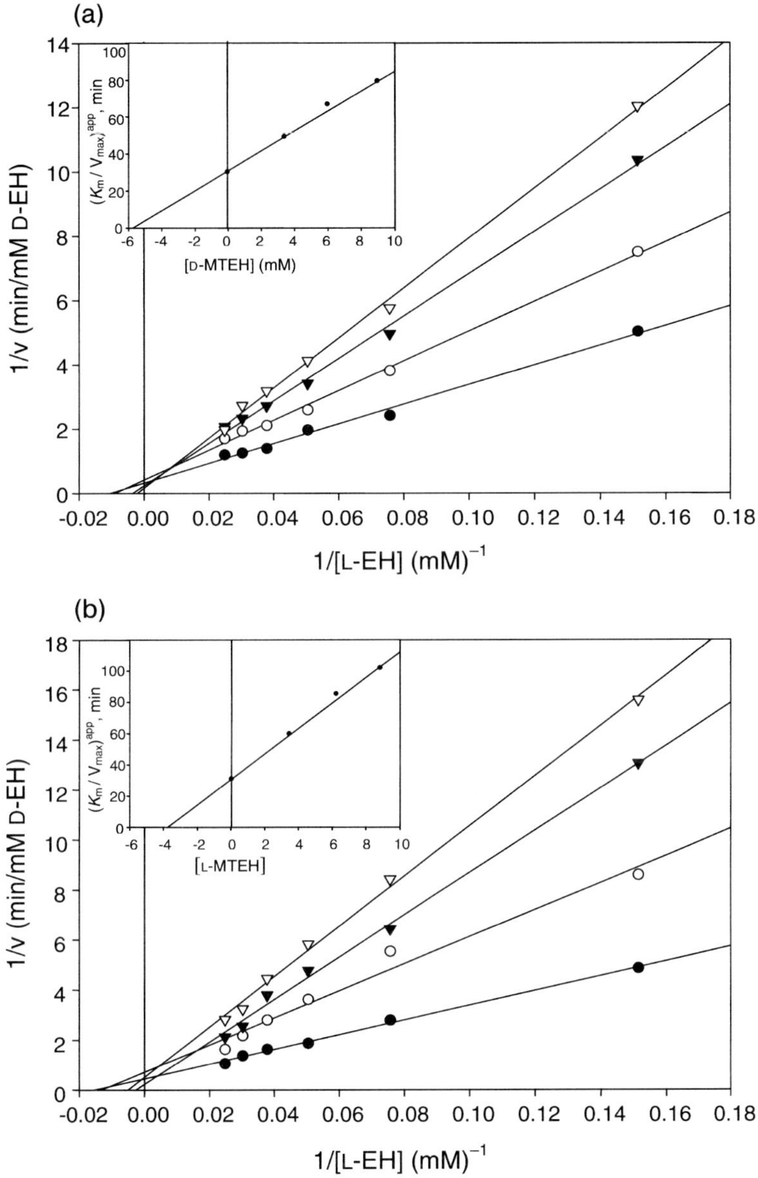

Figure 12.10 (a) D-5-Methylthioethylhydantoin (D-MTEH) and (b) L-5-methylthioethylhydantoin (L-MTEH) inhibition patterns with L-5-ethylhydantoin (L-EH) as the variable substrate. The D- and L-MTEH concentrations were varied in the presence of a constant concentration of L-EH: (●) no MTEH added, (○) 3 mM MTEH, (▼) 6 mM MTEH and (▽) 9 mM MTEH. *Sinorhizobium meliloti* hydantoin racemase activity was assayed at 40 °C for 15 min at pH 8.5. The insets show the plot of the apparent K_m/V_{max} ($(K_m/V_{max})^{app}$) versus the inhibitor concentration (D- or L-MTEH), giving K_i values of 5.75 and 3.76 mM for (a) and (b), respectively. Reactions were performed in triplicate and yielded kinetic constants with less than 10% standard error from the mean.

Additional binding experiments conducted by fluorescence measurements with C76A mutant and D- and L-isopropylhydantoin and L-ethylhydantoin (D-ethylhydantoin is an inhibitor) showed that this mutant is unable to bind the D-isomers of the substrates. The same experiments carried out with C181A mutant proved that this mutant was not able to bind the L-isomers. These results indicate that cysteine 76 is responsible for the recognition of D-isomers of the 5-monosubstituted hydantoins, whereas cysteine 181 recognizes the L-isomers.

To confirm these results we also analyzed the stability of the mutant by unfolding studies with guanidinium hydrochloride as denaturant. The addition of D-isopropylhydantoin to C76A or L-isopropylhydantoin to C181A increased the stability of the active site mutants when incubated in the presence of guanidinium hydrochloride. These experiments again suggest that cysteine 76 is a key residue for the recognition of D-isomers of the 5-monosubstituted hydantoins, whereas cysteine 181 is indispensable for the recognition of L-isomers. In order to ascertain how these two cysteines are arranged in the catalytic center, a three-dimensional homology model for SmeHuyA was developed using other previously resolved racemases, such as aspartate racemase from *Pyrococcus* and glutamate racemase from *Aquifex* (Figure 12.11) [30]. This model supported the hypothesis that cysteines 76 and 181 are located opposite one another and further confirmed the 'two-base mechanism' proposed for SmeHyuA. Moreover, this catalytic behavior was inferred from the observations in a previous work with hydantoin racemase from *A. aurescens* [15], where reactions followed by nuclear magnetic resonance in D_2O showed that the solvent isotope was efficiently incorporated into the product enantiomer, but not into the substrate enantiomer, regardless of which one served as substrate.

From these results, the deduced model is that of a two-base mechanism (Figure 12.12). In this model, when a D-isomer of a 5-monosubstituted hydantoin is

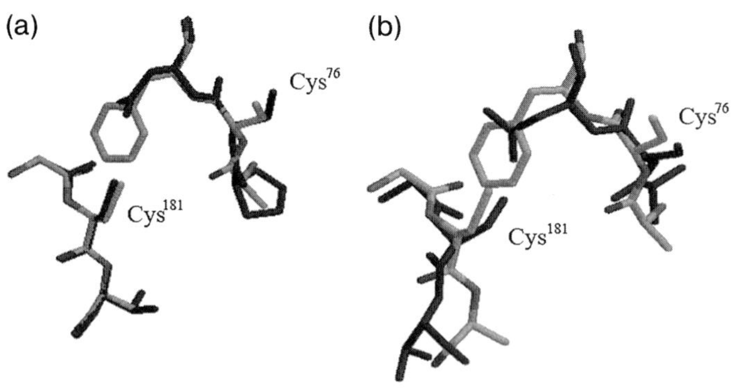

Figure 12.11 Superposition of active sites from the different proteins. The active site of SmeHyuA model (gray) is compared with that of (a) aspartate racemase from *Pyrococcus horikoshii* (1FJLA, black) and that of (b) glutamate racemase from *Aquifex pyrophilus* (1B73A, black).

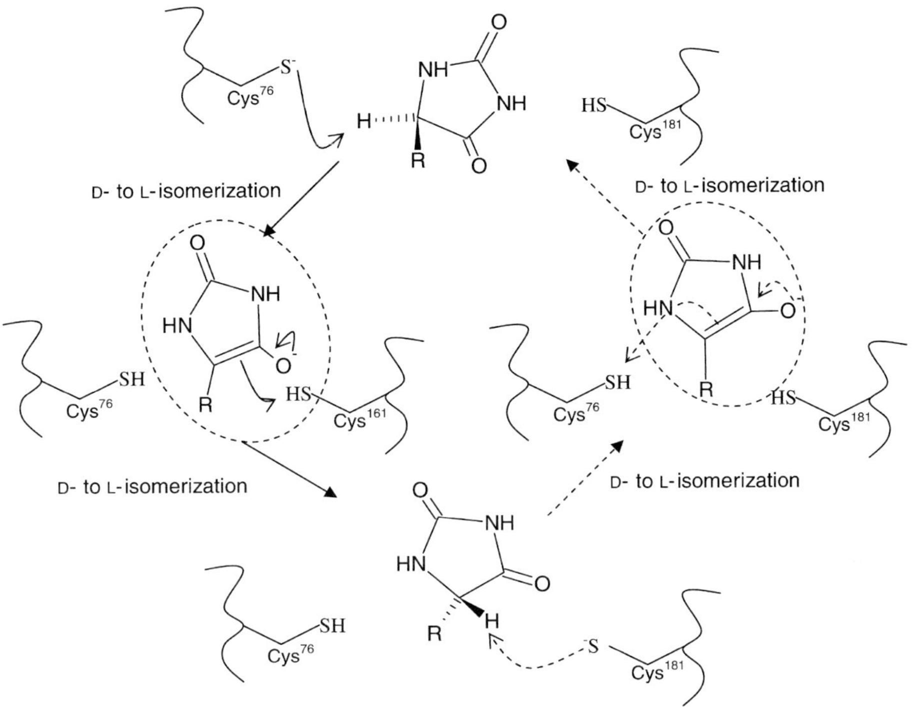

Figure 12.12 Proposed racemization mechanism of hydantoin racemase using a two-base mechanism. The hypothetical character of the intermediate structure is indicated by the dashed lines surrounding the figure.

available, cysteine 76 (in a thiolate form) would act as a base and retrieve a proton. A plane intermediate of the substrate would be formed, as has been proposed for the chemical racemization of the hydantoins [8] and for the racemization of D- and L-glutamate by glutamate racemase. Cysteine 181 would act as an acid adding a proton to the opposite side of the substrate, thus producing the L-5-monosubstituted hydantoin. On the other hand, the binding to cysteine 181 and the retrieval of a proton would carry out the isomerization of the L-isomer of the substrate. In this case, this residue would act as the base and cysteine 76 would donate a proton to the putative intermediate formed.

12.6 Design of a Tailormade Recombinant Biocatalyst Including Hydantoin Racemase Enzymes for Optically Pure D-Amino Acid Production

Once several hydantoin racemases had been studied and characterized, their behavior, together with the other enzymes involved in the hydantoinase process, was analyzed with a view to producing optically pure D-amino acids. Our group developed a three-step enzymatic reaction for D-amino acid production using a recombinant whole cell biocatalyst after separate expression in three different *Escherichia coli* strains of D-hydantoinase, D-carbamoylase, and hydantoin racemase from *Agrobacterium* [9]. However, enzyme production requires three individual cultivations and transport of the reaction intermediate might be a limiting factor. Alternatively, the three enzymes can be co-expressed in the same host cell and directly used as biocatalysts. The first recombinant system co-expressing these three genes in *E. coli* was employed to produce L-amino acids cloning hydantoin racemase together with L-hydantoinase and L-carbamoylase, all from *A. aurescens* [16]. More recently, D-hydantoinase and D-carbamoylase from *Flavobacterium* sp. and hydantoin racemase from *M. liquefaciens* have been co-expressed in *E. coli* [31]. These two systems co-express the three genes after cloning in plasmids with different antibiotic resistance genes. This strategy involves adding several antibiotics to the medium, resulting in high selective pressure in the growth of the recombinant cells. In addition, both of these recombinant systems have shown intermediate accumulation as a result of weak D-carbamoylase activity, which has been described as a limiting factor in the overall process [32].

In order to overcome these two problems our group designed a biocatalyst for the production of optically pure D-amino acids from D,L-5-monosubstituted hydantoins by co-expressing the three genes in one plasmid in a polycistronic structure [33]. Two whole cell recombinant systems were designed, both of which contain two common genes, D-hydantoinase and D-carbamoylase from *A. tumefaciens* BQL9. The third gene was hydantoin racemase 1 (AthyuA1) [22] for the first system and hydantoin racemase 2 (AthyuA2) [25] for the second one, both from *A. tumefaciens* C58. The induction conditions of the polycistronic structures were optimized and several *E. coli* strains were analyzed as hosts for the recombinant plasmids. Both constructions showed the highest activity after 8 h induction in the culture at 34 °C. However, the optimal inductor concentration was different, 0.1 mM of isopropyl-beta-D-thiogalactopyranoside (IPTG) for system 1 and 0.2 mM for system 2. Four *E. coli* strains (DH5α, JM109, TOP10F, and BL21) were studied as hosts for the recombinant systems and *E. coli* strain BL21 proved to be the best for both systems. Stability studies of both recombinant plasmids in this strain have demonstrated that the recombinant cells have grown for at least 90 generations without ampicillin pressure after the initial inoculation.

The optimum temperature for both whole cell systems was around 55 °C, reaching 100% conversion from D,L-methylthioethylhydantoin to D-methionine at pH 8. System 1 was able to hydrolyze all the substrate D,L-methylthioethylhydantoin and achieve total conversion to obtain optically pure D-methionine in about 100 min (Figure 12.13a). System 2 was slower, taking about 200 min to attain 100% conversion of the substrate to the product (Figure 12.13b). Other hydantoins, with both aliphatic and aromatic substituents in carbon 5, were hydrolyzed to the corresponding optically pure D-amino acid (Figure 12.14). All the substrates studied, whether of slow or fast chemical racemization, were converted by both systems to the corresponding D-amino acids without accumulation of the intermediate

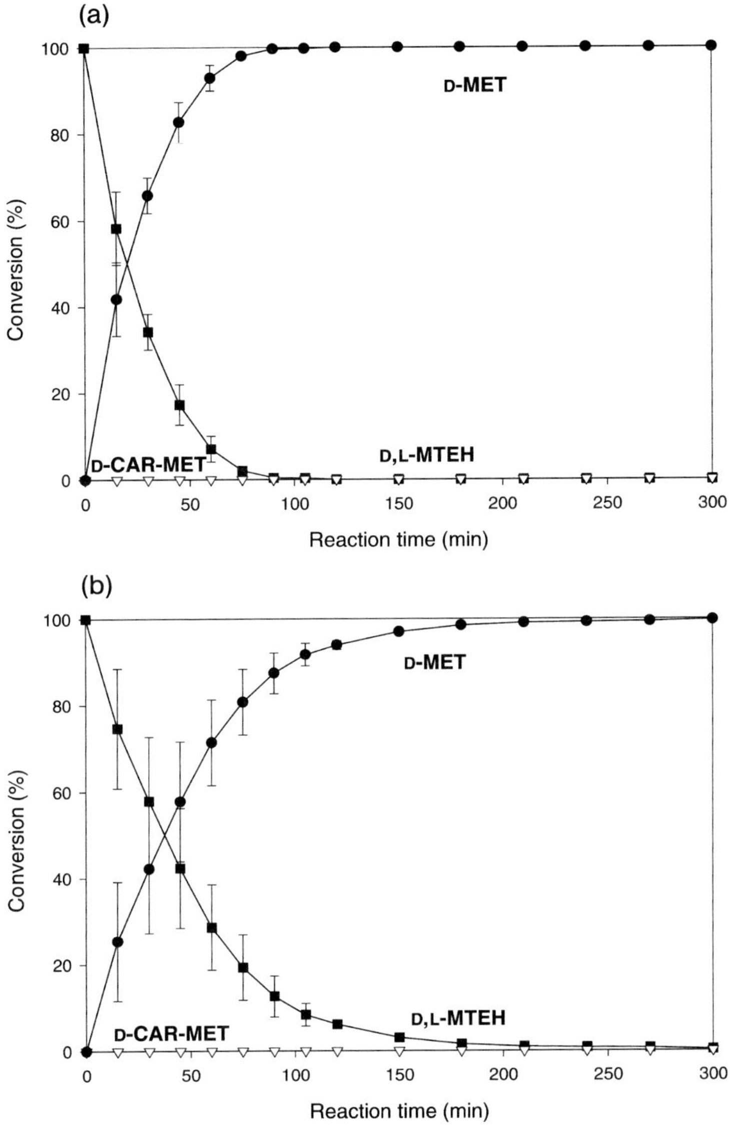

Figure 12.13 Comparison of the reaction profile of D-methionine (D-met) production from 15 mM D,L-methylthioethylhydantoin (D,L-MTEH) without D-carbamoyl-methionine (D-car-met) accumulation using (a) system 1 and (b) system 2.

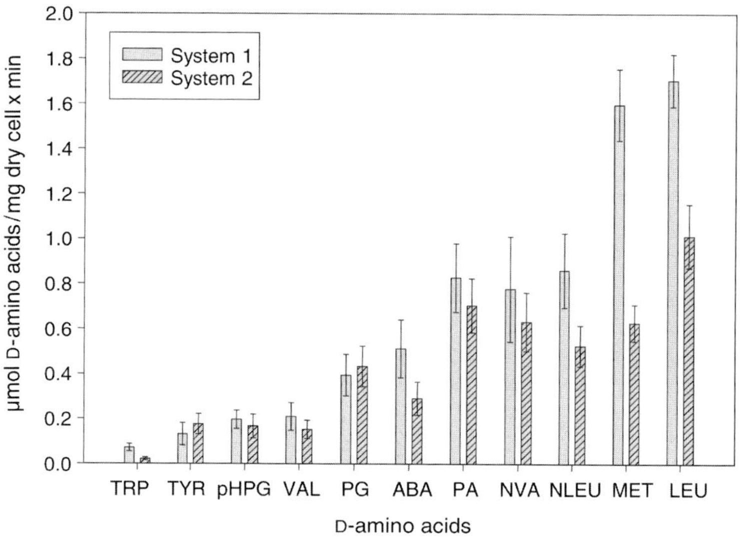

Figure 12.14 Initial reaction rate for the production of different optically pure D-amino acids from 5-monosubstituted hydantoins using systems 1 and 2. TRP, D-tryptophan; TYR, D-tyrosine; pHPG, D-*p*-hydroxyphenyl glycine; VAL, D-valine; PG, D-phenylglycine; ABA, D-aminobutyric acid; PA, D-phenylalanine; NVA, D-norvaline; NLEU, D-norleucine; MET, D-methionine; LEU, D-leucine. Reactions were performed in triplicate and error bars represent the standard error from the mean.

N-carbamoyl-D-amino acid. This is the first time that this bottleneck has been overcome. There was no substrate inhibition due to the higher overexpression of D-carbamoylase compared to the other two enzymes involved in this biotransformation. Both systems showed higher reaction rates for the conversion of aliphatic amino acids than for aromatic ones.

System 1 was able to hydrolyze the 5-monosubstituted hydantoins faster than system 2 for the production of almost all the D-amino acids studied. System 1 was slightly slower than system 2 only for the production of the aromatic amino acids D-tyrosine and D-phenylglycine. This agrees with previously described results, finding that AtHyuA1 enzyme (included in system 1) was more viable for industrial application than AtHyuA2 (included in system 2) due to its higher substrate affinity and racemization velocity [25].

The behavior of the recombinant biocatalyst on a larger scale is of considerable economic interest. For this reason D-methionine production was analyzed from 300 mM D,L-methylthioethylhydantoin (52.3 g/l), in a reaction volume 300 times that used at the laboratory scale (300 ml). When system 1 was induced in optimal conditions the product yield of D-methionine reached 100% in 6 h and there was no D-carbamoyl-methionine accumulation (Figure 12.15a). However, when the

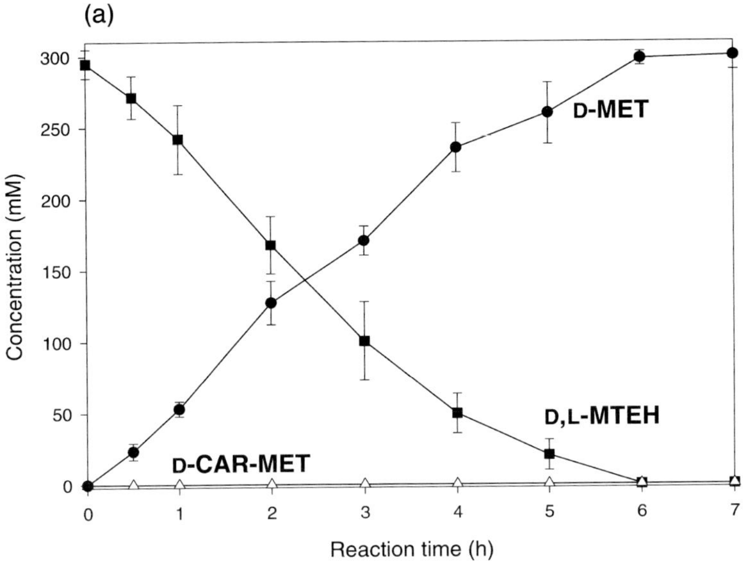

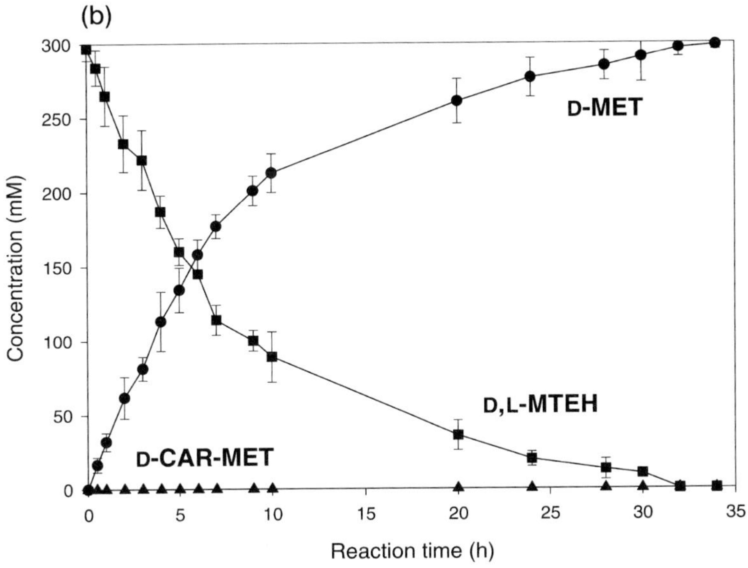

Figure 12.15 Profile of D-methionine (D-met) production from 300 mM D,L-methylthioethylhydantoin (D,L-MTEH) without D-carbamoyl-methionine (D-car-met) accumulation using system 1 (a) induced with 0.1 mM IPTG and (b) non-induced in a large-scale reaction with a final volume of 300 ml. Reactions and measurements were carried out in triplicate and error bars represent the standard error from the mean.

inducer IPTG was not added, this system showed less activity, reaching total conversion at approximately 32 h (Figure 12.15b).

The development of this multienzymatic system for the production of D-amino acids from any D,L-5-monosubstituted hydantoin allows the hydantoinase process to produce not only two amino acids, such as D-phenylglycine and D-*p*-hydroxy-phenylglycine (as explained at the beginning of the chapter), but also many non-natural D-amino acids that could be components of potential pharmaceuticals.

Acknowledgments

The authors gratefully acknowledge funding of this research by grants from the Ministerio de Educacion y Ciencia, Spain (BIO 2007-67009) and from Consejeria de Innovacion, Ciencia y Tecnologia, Andalusia region, Spain (P07-CVI-2651).

References

1 Syldatk, C., Läufer, A., Müller, R. and Höke, H. (1990) *Advances in Biochemical Engineering/Biotechnology*, **41**, 29–75.

2 Bommarius, A.S., Schwarm, M. and Drauz, K. (1998) *Journal of Molecular Catalysis B: Enzymatic*, **5**, 1–11.

3 May, O., Verseck, S., Bommarius, A. and Drauz, K. (2002) *Organic Process Research & Development*, **6**, 452–457.

4 Altenbuchner, J., Siemann-Herzberg, M. and Syldatk, C. (2001) *Current Opinion in Biotechnology*, **12**, 559–563.

5 Breuer, M., Ditrich, K., Habicher, T., Hauer, B., Kebeler, M., Stürmer, R. and Zelinski, T. (2004) *Angewandte Chemie (International Edition in English)*, **43**, 788–824.

6 Ware, E. (1950) *Chemical Reviews*, **46**, 403–470.

7 Pietzsch, M. and Syldatk, C. (2002) In *Enzyme Catalysis in Organic Synthesis* (eds K. Drauz and H. Waldmann), Wiley-VCH Verlag GmbH, Weinheim, pp. 761–799.

8 Lazarus, R.A. (1990) *The Journal of Organic Chemistry*, **55**, 4755–4757.

9 Martinez-Rodriguez, S., Las Heras-Vazquez, F.J., Clemente-Jimenez, J.M., Mingorance-Cazorla, L. and Rodriguez-Vico, F. (2002) *Biotechnology Progress*, **18**, 1201–1206.

10 Möller, A., Syldatk, C., Schulze, M. and Wagner, F. (1988) *Enzyme and Microbial Technology*, **10**, 618–625.

11 Battilotti, M. and Barberini, U. (1988) *Journal of Molecular Catalysis*, **43**, 343–352.

12 Hils, M., Müch, P., Altenbuchner, J., Syldatk, C. and Mattes, R. (2001) *Applied Microbiology and Biotechnology*, **57**, 680–688.

13 Watabe, K., Ishikawa, T., Mukohara, Y. and Nakamura, H. (1992) *Journal of Bacteriology*, **74**, 3461–3466.

14 Watabe, K., Ishikawa, T., Mukohara, Y. and Nakamura, H. (1992) *Journal of Bacteriology*, **74**, 7989–7995.

15 Wiese, A., Pietzsch, M., Syldatk, C., Mattes, R. and Altenbuchner, J. (2000) *Journal of Biotechnology*, **80**, 217–230.

16 Wilms, B., Wiese, A., Syldatk, C., Mattes, R. and Altenbuchner, J. (2001) *Journal of Biotechnology*, **86**, 19–30.

17 Hils, M., Müch, P., Altenbuchner, J., Syldatk, C. and Mattes, R. (2001) *Applied Microbiology and Biotechnology*, **57**, 680–688.

18 Goodner, B., Hinkle, G., Gattung, S., Miller, N., Blanchard, M., Qurollo, B., Goldman, B.S., Cao, Y., Askenazi, M., Halling, C., Mullin, L., Houmiel, K., Gordon, J., Vaudin, M., Iartchouk, O., Epp, A., Liu, F., Wollam, C., Allinger, M.,

Doughty, D., Scott, C., Lappas, C., Markelz, B., Flanagan, C., Crowell, C., Gurson, J., Lomo, C., Sear, C., Strub, G., Cielo, C. and Slater, S. (2001) *Science*, **294**, 2323–2328.

19 Wood, D.W., Setubal, J.C., Kaul, R., Monks, D., Chen, L., Wood, G.E., Chen, Y., Woo, L., Kitajima, J.P., Okura, V.K., Almeida, N.F., Jr, Zhou, Y., Bovee, D., Sr, Chapman, P., Clendenning, J., Deatherage, G., Gillet, W., Grant, C., Guenthner, D., Kutyavin, T., Levy, R., Li, M., McClelland, E., Palmieri, A., Raymond, C., Rouse, G., Saenphimmachak, C., Wu, Z., Gordon, D., Eisen, J.A., Paulsen, I., Karp, P., Romero, P., Zhang, S., Yoo, H., Tao, Y., Biddle, P., Jung, M., Krespan, W., Perry, M., Gordon-Kamm, B., Liao, L., Kim, S., Hendrick, C., Zhao, Z., Dolan, M., Tingey, S.V., Tomb, J., Gordon, M.P., Olson, M.V. and Nester, E.W. (2001) *Science*, **294**, 2317–2323.

20 Galibert, F., Finan, T.M., Long, S.R., Puhler, A., Abola, P., Ampe, F., Barloy-Hubler, F., Barnett, M.J., Becker, A., Boistard, P., Bothe, G., Boutry, M., Bowser, L., Buhrmester, J., Cadieu, E., Capela, D., Chain, P., Cowie, A., Davis, R.W., Dreano, S., Federspiel, N.A., Fisher, R.F., Gloux, S., Godrie, T., Goffeau, A., Golding, B., Gouzy, J., Gurjal, M., Hernandez-Lucas, I., Hong, A., Huizar, L., Hyman, R.W., Jones, T., Kahn, D., Kahn, M.L., Kalman, S., Keating, D.H., Kiss, E., Komp, C., Lelaure, V., Masuy, D., Palm, C., Peck, M.C., Pohl, T.M., Portetelle, D., Purnelle, B., Ramsperger, U., Surzycki, R., Thebault, P., Vandenbol, M., Vorholter, F.J., Weidner, S., Wells, D.H., Wong, K., Yeh, K.C. and Batut, J. (2001) *Science*, **293**, 668–672.

21 Suzuki, S., Onishi, N. and Yokozeki, K. (2005) *Bioscience, Biotechnology, and Biochemistry*, **69**, 530–836.

22 Las Heras-Vazquez, F.J., Martinez-Rodriguez, S., Mingorance-Cazorla, L., Clemente-Jimenez, J.M. and Rodriguez-Vico, F. (2003) *Biochemical and Biophysical Research Communications*, **303**, 541–547.

23 Martinez-Rodriguez, S., Las Heras-Vazquez, F.J., Mingorance-Cazorla, L., Clemente-Jimenez, J.M. and Rodriguez-Vico, F. (2004) *Applied and Environmental Microbiology*, **70**, 625–630.

24 Suzuki, S., Takenaka, Y., Onishi, N. and Yokozeki, K. (2005) *Bioscience, Biotechnology, and Biochemistry*, **69**, 1473–1482.

25 Martinez-Rodriguez, S., Las Heras-Vazquez, F.J., Clemente-Jimenez, J.M. and Rodriguez-Vico, F. (2004) *Biochimie*, **86**, 77–81.

26 Wiese, A., Syldatk, C., Mattes, R. and Altenbuchner, J. (2001) *Archives of Microbiology*, **176**, 187–196.

27 May, O., Habenicht, A., Mattes, R., Syldatk, C. and Siemann, M. (1998) *Biological Chemistry*, **379**, 7743–7747.

28 Martinez-Rodriguez, S., Clemente-Jimenez, J.M., Rodriguez-Vico, F. and Las Heras-Vazquez, F.J. (2007) *D-Amino Acids: A New Frontier in Amino Acid and Protein Research* (eds R. Konno, H. Brückner, A. D'Aniello, G. Fisher, N. Fuji and H. Homma), Nova Science Inc, New York, pp. 573–577.

29 Andujar-Sanchez, M., Martinez-Rodriguez, S., Las Heras-Vazquez, F.J., Clemente-Jimenez, J.M., Rodriguez-Vico, F. and Jara-Perez, V. (2006) *Biochimica et Biophysica Acta*, **1764**, 292–298.

30 Martinez-Rodriguez, S., Andujar-Sanchez, M., Neira, J.L., Clemente-Jimenez, J.M., Jara-Perez, V., Rodriguez-Vico, F. and Las Heras-Vazquez, F.J. (2006) *Protein Science*, **15**, 2729–2738.

31 Nozaki, H., Takenaka, Y., Kira, I., Watanabe, K. and Yokozeki, K. (2005) *Journal of Molecular Catalysis B: Enzymatic*, **32**, 213–218.

32 Park, J.H., Kim, G.J. and Kim, H.S. (2000) *Biotechnology Progress*, **16**, 564–570.

33 Martinez-Gomez, A.I., Martinez-Rodriguez, S., Clemente-Jimenez, J.M., Pozo-Dengra, J., Rodriguez-Vico, F. and Las Heras-Vazquez, F.J. (2007) *Applied and Environmental Microbiology*, **73**, 1525–1531.

13
Chemo-enzymatic Deracemization Methods

Davide Tessaro, Gianluca Molla, Loredano Pollegioni and Stefano Servi

13.1
Introduction

In recent times, the term deracemization has been used for describing techniques becoming increasingly important for the preparation of chiral compounds as single enantiomers. The term refers to a process in which one single enantiomer is obtained when starting from a racemate [1].

The two more common strategies for achieving such an objective are deracemization by stereoinversion or deracemization by dynamic kinetic resolution DKR (Scheme 13.1).

Deracemization by stereoinversion is a process in which one form (S_f) of the racemic starting material (R_f + S_f) is enantioselectively transformed into an intermediate (S_i) which can in turn react to give the form of opposite configuration (R_f). An example of this method could be the selective oxidation of one enantiomer of a racemic secondary alcohol and the subsequent reduction with a catalyst of opposite stereopreference [2].

Deracemization by DKR is in principle a kinetic resolution process in which the non-transformed enantiomer is racemized *in situ*. The conditions are that a chiral catalyst promotes the transformation of one enantiomer (R_s) into the product (R_p) while the other enantiomer is racemized at a comparable rate and the racemic mixture (R_s + S_s) is restored. The product (R_p) is not racemized under the same conditions. While a simple kinetic resolution yields a maximum of 50% of the product, with this technique a 100% conversion can be reached. Although the majority of chiral molecules of industrial interest are still prepared by kinetic resolution, the continuous development of industrial enzymes and racemizing processes fosters new chemo- or biocatalytic systems for DKR to appear. A great impulse for deracemization methods based on the one-pot/two-steps resolution racemization process was brought about by Bäckvall *et al.* over a 10-year period [3].

The method is of general applicability in the deracemization of secondary alcohols and amines and consists of a lipase-catalyzed irreversible acylation and *in situ* racemization of the non-reacted enantiomer catalyzed by a ruthenium catalyst,

Modern Biocatalysis: Stereoselective and Environmentally Friendly Reactions.
Edited by Wolf-Dieter Fessner and Thorleif Anthonsen

ISBN: 978-3-527-32071-4

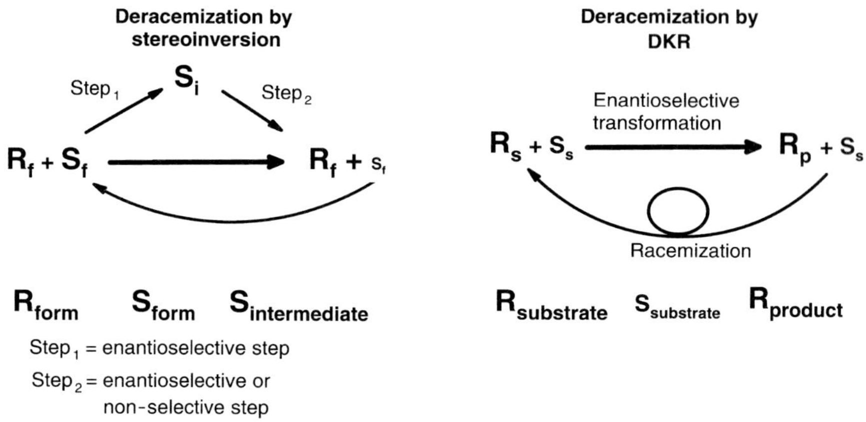

Scheme 13.1 Deracemization by stereoinversion and by DKR.

usually in an amount of 2 mol %. The two steps occur in the same medium, an organic solvent.

Many other deracemization methods based on a two-enzyme system or in the successive enzymatic resolution/base-catalyzed racemization have been reported. They are often regarded as having a reduced environmental impact, not requiring a transition metal catalyst in the racemization step [4].

Unlike a kinetic resolution process, where in principle both enantiomers of the same compound can be obtained, a deracemization process gives access to one single enantiomer. The possibility of securing the enantiomer of opposite configuration depends on the availability of the enzymes involved: frequently enzymes with complementary stereopreferences are not available.

In this chapter deracemization methods for the preparation of multifunctional compounds (α- and β-hydroxy acids, α-hydroxynitriles, and α-amino acids) are discussed (Scheme 13.1).

13.2 Deracemization Methods for α- and β-Hydroxy Acids

Many α- and β-hydroxy acids are found as natural products or incorporated into bioactive compounds and are required in both absolute configurations [5]. The biocatalytic asymmetric synthesis of these compounds has been reported in detail [6]. Since the racemic forms are easily available through conventional synthetic methods, the application of deracemization methods appears a convenient approach for obtaining single enantiomers.

13.2.1
Deracemization of Hydroxy Acids by DKR (Hydrolytic Enzymes + Ruthenium-based Racemization Catalysts)

The method of combined enzyme- and transition metal-catalyzed reactions widely applied to the DKR of secondary alcohols has also been applied to the DKR of α-hydroxy acid esters **rac-1**. The principle is based on the enantioselective acylation catalyzed by *Pseudomonas* species lipase (PS-C from Amano Ltd) using *p*-Cl-phenyl acetate as an acyl donor in cyclohexane combined with *in situ* racemization of the non-acylated enantiomer catalyzed by ruthenium compounds [7]. Under these conditions, various α-hydroxy esters of type **1** were deracemized in moderate to good yields and high enantioselectivity (Scheme 13.2).

The use of the ruthenium catalysts is advantageous since it does not need the presence of a base as a co-catalyst. The presence of a base could induce racemization of the formed acetates by enolization.

The same methodology was also applied to the DKR of β-hydroxyesters. In the latter case, the reaction was carried out in tandem with an aldol reaction and the β-hydroxyester formed, after neutralization, underwent DKR using the immobilized lipase from *Candida antarctica* and a ruthenium catalyst [8].

The β-acetoxy esters **2** were obtained in yields over the range 69–75%, based on the aldehyde and with a high enantiomer excess (ee). In addition, in this case the absence of a base as a co-catalyst is essential in order to avoid retro-aldol reaction of the product. The scope of this methodology is not restricted to aromatic hydroxy acids, as the non-aromatic cyclohexyl analog was also successfully deracemized (Scheme 13.3).

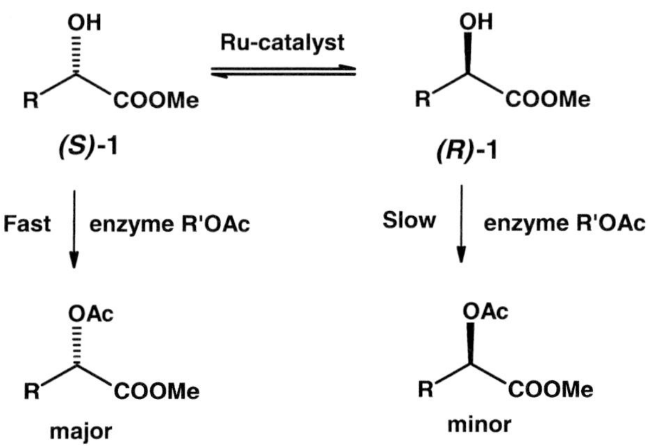

R = Ar, Cyclohexyl lipase acylation in cyclohexane;
2% racemization catalyst Ru-based 70–80% yields; 99% ee

Scheme 13.2 Deracemization of α–hydroxy acids by DKR (hydrolytic enzymes + ruthenium-based racemization catalyst).

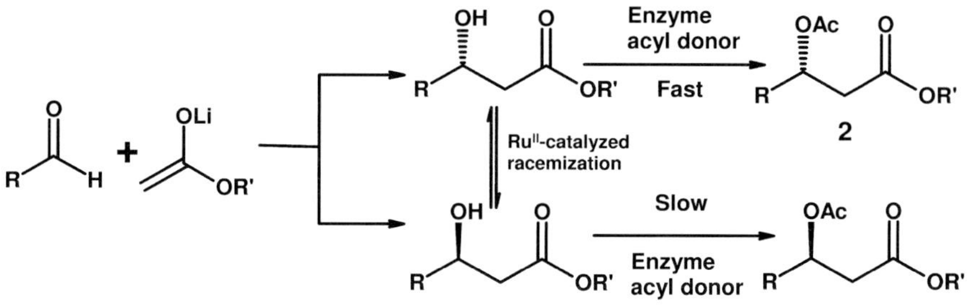

Scheme 13.3 DKR of *in situ*-formed β-hydroxyesters coupling a lipase enantioselective irreversible acylation with ruthenium-catalyzed racemization (R = Ph, *p*-OMe-Ph, $PhCH_2$-).

13.2.2
Deracemization of Hydroxy Acids by DKR with a Two-enzyme System

Deracemization of mandelic acid with the combined action of two enzymes has been reported. *rac*-Mandelic acid is acylated by a *Pseudomonas* sp. lipase in diisopropyl ether. After solvent removal the mixture of mandelic acid enriched in the R-form and the *O*-acetyl derivative of the S-configuration are subjected to the mandelate racemase-catalyzed racemization in aqueous buffer. In these conditions only the non-acetylated hydroxy acid is racemized. In order to obtain (*S*)-*O*-acetylmandelic acid in an 80% isolated yield and a >98% e.e. the process must be repeated four times [9].

This example of a classical cyclic resolution–racemization methodology indicates the limitations of a multistep enzyme-catalyzed reaction when the reaction medium, the reagents, or the reaction conditions are not mutually compatible.

Mandelate racemase, which is probably the best known of all racemizing enzymes [10], shows an activity limited to compounds structurally related to mandelates [11]. The presence of an aromatic group or an unsaturation in the α-position is essential for activity.

In order to extend the two-enzyme system to other 2-hydroxy acids, a racemase with a broader activity was found in *Lactobacillus paracasei*. This was exploited for deracemization of 2-hydroxy-4-phenylbutanoic acid and 3-phenyllactic acid, which are important synthetic intermediates. In addition, in this case the procedure requires a kinetic resolution step and a successive racemization step. *O*-Acetyl derivatives of the absolute (*S*)-configuration can be obtained in two successive repeating cycles. Yields are around 60%. Of course the *O*-acetyl derivatives of opposite configuration can be obtained when the lipase-catalyzed reaction is applied in the hydrolysis direction. Obtaining the *O*-acetyl derivatives of the absolute (*R*)-configuration requires an additional acetylation step of the initially resolved and racemized (*S*)-hydroxy acid [12].

13.2.3
Deracemization of Hydroxy Acids by Stereoinversion

α-Hydroxy acids of the (*R*)-configuration can be obtained through deracemization by stereoinversion of the racemic lactate analogs **3** (Scheme 13.4). The deracemization takes place through a selective oxidation catalyzed by L-specific glycolate oxidase followed by a reductive step catalyzed by D-lactate dehydrogenase [13]. In this procedure the NADH co-factor is then regenerated *in situ* by the formate dehydrogenase/ammonium formate method (Scheme 13.4).

The method is similar to the deracemization of α-amino acids (see Section 13.4.1) with a combination of amino acid oxidase and amino transferase or amino acid dehydrogenase. Although the co-factor regeneration is considered a solved problem, the application of the method to obtaining α-hydroxy acids of wide structural variety and of both configurations is also in this case dependent on enzyme availability [13].

Alternatively, a 'one-pot, single-step' deracemization of *sec*-alcohols has been achieved by employing two different microorganisms in a single reaction vessel. However, the number of examples of this type is limited and the oxidation and reduction steps are usually performed sequentially in a one-pot, two-step procedure. For instance, racemic mandelic acid was deracemized in the presence of whole cells of *Pseudomonas polycolor* and *Micrococcus freudenreichii* [14]. Separate experiments showed that *P. polycolor* was responsible for the oxidation, while *M. freudenreichii* was needed for reduction of the corresponding α-keto acid. After 24 h, (*R*)-mandelic acid **4** was isolated in a 60% yield and 99% e.e. [14].

The two-enzyme system was also used to convert L-lactate into D-lactate with a yield better than 97%. L-Lactate is oxidized by L-lactate dehydrogenase to give pyruvate. The keto acid is reduced in an electrochemical system at the cathode to racemic lactate and NADH is oxidized to NAD^+ at the anode. The continuous

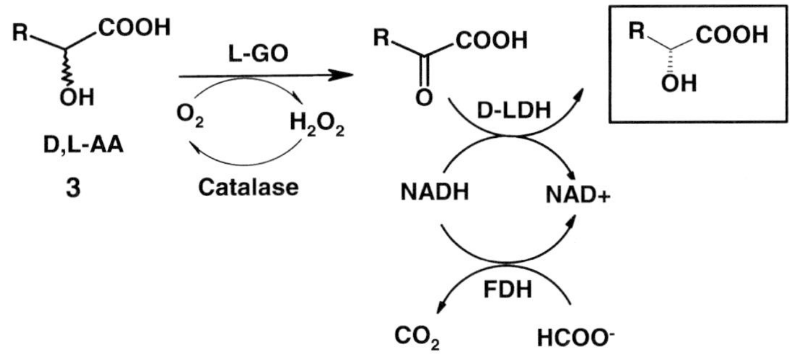

D-LDH D-Lactate dehydrogenase L-GO L-Glycolate oxidase
FDH Formate dehydrogenase

Scheme 13.4 Deracemization of hydroxy acids by stereoinversion.

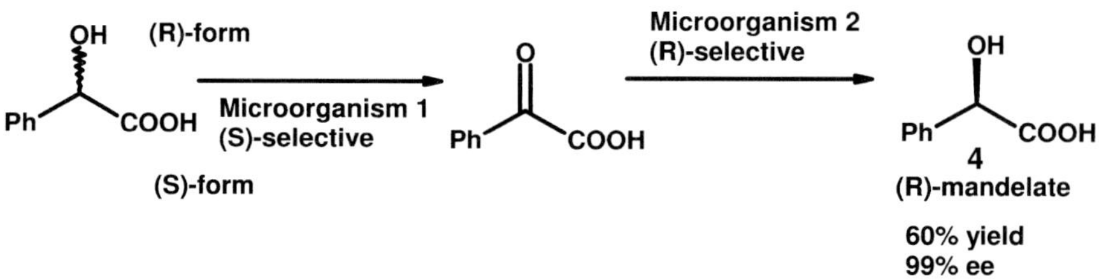

Scheme 13.5 Whole cell deracemization by stereoinversion.

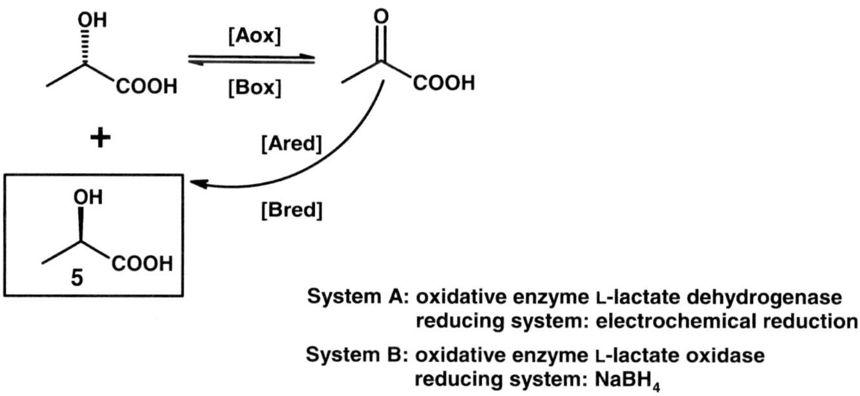

Scheme 13.6 Lactate chemoenzymatic deracemization.

oxidation of L-lactate allows the accumulation of the D-enantiomer. The application of the method to deracemization or complete inversion of chiral 2-hydroxy acids requires specific dehydrogenases [15] (Scheme 13.5).

In a conceptually related method, the oxidation of the L-lactate is catalyzed by the commercially available L-specific lactate oxidase from *Aerococcus viridans* and the reduction of pyruvate to *rac*-lactate is performed with $NaBH_4$ in the same solution. In repeating cycles D-lactate **5** is obtained as the sole product and in an excellent e.e. [16] (Scheme 13.6). This concept has recently been applied to the deracemization of α-amino acids [17].

13.2.4
Deracemization of Hydroxy Acids by Microbial Stereoinversion

Microbial stereoinversion consists of a deracemization process in which a single whole cell system is applied for the two-step inversion of the configuration of one enantiomer, usually a compound containing a secondary alcohol group. Examples of a two-enzyme system is known for deracemization of mandelates where a single microorganism is used [18].

Recently the microorganism *Candida parapsilosis* was used for the deracemization of α- and β-hydroxy esters. A series of racemic aryl and aryl substituted

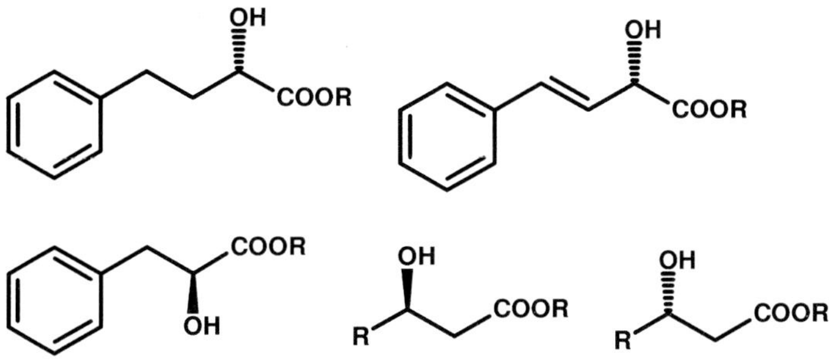

Scheme 13.7 Hydroxy acids produced by microbial stereoinversion.

α-hydroxy esters were deracemized by the whole cells of *C. parapsilosis* to give the corresponding optically pure (*S*)-hydroxy esters in good yields (65–85%) and e.e. (90–99%). The mechanism of deracemization involves two different enzymatic reactions: oxidation of one of the antipodes by an (*R*)-oxidase to the keto intermediate followed by its subsequent reduction by a (*S*)-specific reductase enzyme [19] (Scheme 13.7).

In order to clarify the proposed mechanism, the reduction of ethyl 3-oxo-3-phenyl propionate was carried out, resulting in the exclusive formation of the (*S*)-enantiomer confirming the presence of (*S*)-specific reductase. When the pure (*S*)-enantiomer was used as the substrate for deracemization, the product recovered retained the configuration as well as the optical purity without any loss, clearly supporting the presence of an (*S*)-specific reductase, which does not oxidize the (*S*)-hydroxy ester to the corresponding keto ester.

The generality of the reaction was established by using different substituents in the standard substrate ethyl 3-hydroxy 3-phenyl propionate. The presence of electron donating and electron withdrawing groups in the *para* position does not affect the deracemization reaction, while substitution in the *ortho* position hinders the same. On a preparative scale, up to 500 mg of racemic substrate incubated with the cells were deracemized with isolated yields close to 80%.

13.3 Deracemization of α-Hydroxy Nitriles

Cyanohydrins are versatile and important building blocks in organic synthesis [20]. Consequently, their enantioselective synthesis has attracted considerable attention [21]. In particular, the use of hydroxynitrile lyase has proved to be a general and reliable method for obtaining α-hydroxy nitriles of both configurations [22]. An interesting approach is the lipase- and base-catalyzed synthesis of acyl-cyanohydrin for a synthetic DKR [23]. This is a combination of two reaction systems: the dynamic, base-catalyzed equilibrium between acetone cyanohydrin, acetone, HCN, aldehyde and a racemic cyanohydrin and the lipase-catalyzed enantioselective and irreversible acylation of the hydroxyl group. The combination yields the

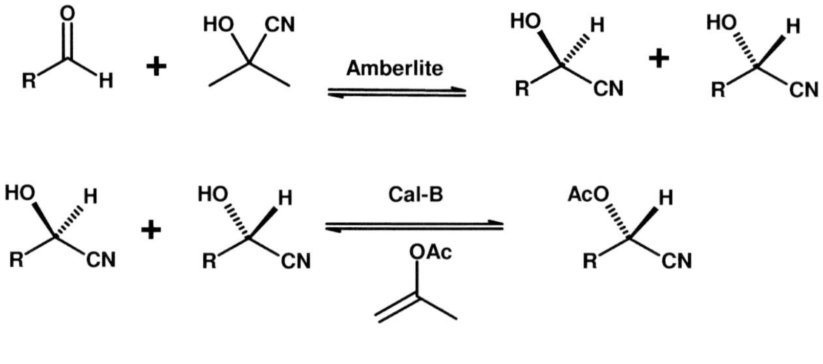

R = Ar, naphtyl, furyl, cyclohexyl

Scheme 13.8 Deracemization of α-hydroxy nitriles.

corresponding stable cyanohydrin ester. This methodology was first published in 1991 and has been quoted frequently [23].

However, the reaction has long reaction times and it suffers from several experimental details which lower the reaction yields. Recently improvements in the original procedures allowed (*S*)-mandelonitrile acetate to be obtained from the corresponding *rac*-alcohol in a 97% yield and 98% e.e. while also shortening the reaction time. Best results were obtained with (lipase from *C. antarctica* type B) (CALB) adsorbed over celite [24] (Scheme 13.8).

A similar reaction procedure has also been successfully applied to the deracemization of cyanohydrines containing heterocyclic structures [25].

13.4 Deracemization of α-Amino Acids

For several reasons α-amino acids are ideal substrates for deracemization methods. They racemize easily by base catalysis under a number of conditions and they are racemized in Nature by the intervention of specific amino acid racemases. They are also recognized as substrates by oxidative enzymes to give the corresponding oxo-acids, in turn substrates for amino transferases and amino acid dehydrogenases. Several industrial preparations of L- and D-amino acids are based on processes of deracemization [26] or of separate two-steps resolution–racemization [27].

13.4.1 Deracemization of α-Amino Acids by Stereoinversion

13.4.1.1 Deracemization by Stereoinversion via the Two-enzyme System D-Amino Acid Oxidase and L-Amino Transferase

D-Amino acid oxidase is a highly stereoselective flavoenzyme the properties of which are described in detail in Section 13.5.1. The enzymatic production of α-imino acids or α-keto acids from the corresponding amino acids is a convenient way of producing these compounds. The α-keto acids can be transformed into amino acids in a reaction catalyzed by an amino transferase in the presence of a stoichiometric amount of a second amino acid acting as an amino donor. The α-imino acids instead can give an amino acid by means of an amino acid dehydrogenase. Both reactions can be considered overall as equivalent to reductive aminations.

A combination of D-amino acid oxidase and L-amino transferase is an example of a deracemization by stereoinversion. The product is an L-amino acid. The reaction catalyzed by amino transferase has an equilibrium constant close to unity, a very unpractical situation leading to uncomplete transformation and to the production of almost inseparable mixtures of amino acids (at least two, the amino acid product and the amino acid used as an amino donor). For preparative purposes it is therefore mandatory to shift the equilibrium to the product side. A recent example of a deracemization procedure based on this coupled enzymatic system is the preparation of L-2-naphthyl-alanine **6** as illustrated in Scheme 13.9 [28]. The reaction occurs in one pot with initial oxidation of the D-amino acid catalyzed by D-amino acid oxidase from *Rhodotorula gracilis*. The hydrogen peroxide that is formed in stoichiometric amounts is decomposed by catalase. The α-keto acid is the substrate for L-aspartate amino transferase (L-Asp amino transferase), which is able to use L-cysteine sulfinic acid **7** as an amino donor.

The reaction is driven towards the products by spontaneous decomposition of the sulfinyl-pyruvate to pyruvate and sulfur dioxide. The deracemization is complete yielding L-2-naphthyl-alanine in a 95% yield and an e.e. of 99.5%. The preparative significance of the transformation is reduced by the low space–time yield due to the extremely low water solubility of the naphthyl amino acid. However, a solid-to-solid bioconversion is possible in principle [28]. Previous examples of a similar combination of enzymes have been published by Fotheringham *et al.* [29].

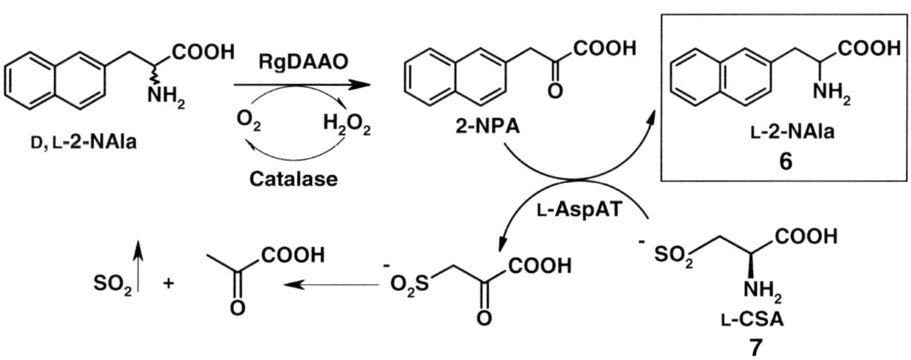

Scheme 13.9 Deracemization by stereoinversion using three different enzymes in an equilibrium-shifted transamination.

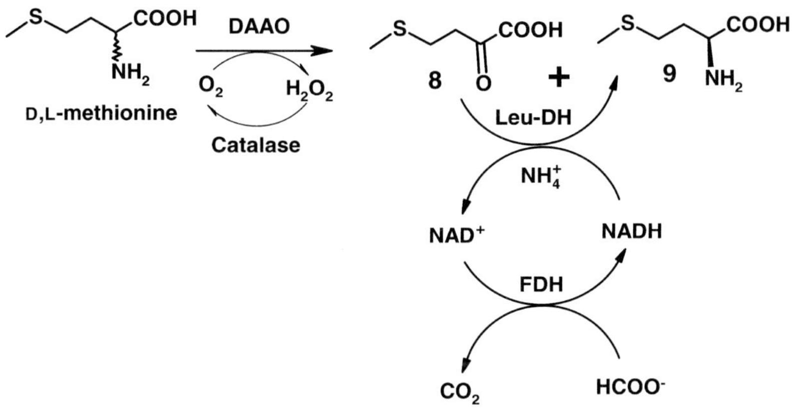

Scheme 13.10 Deracemization of D,L-methionine by stereoinversion.

13.4.1.2 **Deracemization by Stereoinversion via the Two-enzyme System D-Amino Acid Oxidase and L-Leucine Dehydrogenase**

In a related approach, D,L-methionine can be efficiently deracemized to obtain the L-enantiomer using a multienzyme system consisting of D-amino acid oxidase, catalase, leucine dehydrogenase, and formate dehydrogenase. The α-keto acid **8** produced from the oxidation of the D-form is transformed into L-methionine **9** in the presence of ammonia, leucine dehydrogenase, and a stoichiometric amount of NADH. The NAD^+ thus formed is recycled to NADH with ammonium formate and formate dehydrogenase [30] (Scheme 13.10).

The coupling of these two enzymatic systems could find many more applications due to the availability of amino acid dehydrogenases of broader specificity [31]. A series of amino acid dehydrogenases with D-specificity for the preparation of D-amino acids has been applied to the reductive amination of α-keto acids [32]. However, the deracemization of *rac*-amino acids exploiting this type of enzyme requires an amino acid oxidase with L-specificity, which is a rare enzymatic activity. As an alternative the α-oxo acid, usually available through difficult synthetic procedures, can be used directly.

13.4.1.3 **Deracemization by Stereoinversion via the Three-enzyme System L-Amino Acid Oxidase, D-Amino Transferase and Amino Acid Racemase**

The preparation of D-amino acids with the above three-enzyme system requires enzymes with opposite stereochemical selectivity and a suitable amino acid as a donor. While D-amino acid oxidase is an enzyme, the function of which in Nature is mainly related to the elimination of D-amino acids, L-amino acid oxidases are usually found in aggressive animals (snakes). Bacterial L-amino acid oxidases often show a specific activity that is too low for preparative purposes [33]. Moreover, D-amino transferases are less common than the L-specific ones and require more expensive D-amino acids as amino donors.

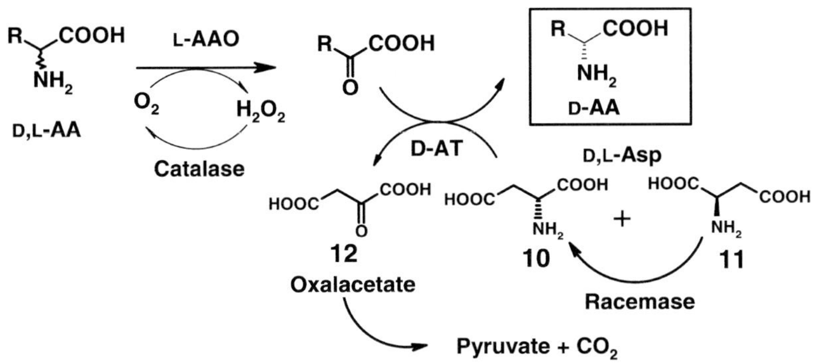

Scheme 13.11 Deracemization by stereoinversion finalized to the preparation of D-amino acids.

Several inventive procedures have been used for overcoming these difficulties, offering interesting multistep enzyme-catalyzed reactions for deracemization of amino acids. The requirement of an amino donor of D-configuration can be solved by its *in situ* generation from the L-form by using an amino acid racemase.

The procedure reported in Scheme 13.11 describes deracemization of an amino acid involving oxidation with an L-specific enzyme and transamination with a D-amino transferase using D-aspartate **10**, which is generated from L-aspartate **11** by aspartate racemase, as the amino donor. The oxidative enzyme is defined as an L-amino acid deaminase, a flavoprotein from *Proteus myxofaciens* [34]. The transamination reaction is shifted towards the product since the oxalacetate **12** formed decarboxylates spontaneously to give pyruvate and carbon dioxide.

13.4.2
Deracemization of α-Amino Acids via DKR

The requisites for successful DKR are that the starting material is racemizable under the reaction conditions while the product is configurationally stable under the same conditions. Due to the industrial importance of amino acids, a number of studies have been devoted to the problem of effective racemization of the unused enantiomer that remained after a standard kinetic resolution [35]. However, racemization methods based on high temperature or extreme pH values are unsuitable for *in situ* racemization. Thus, if enzymes are involved, the racemization reaction must be carried out under mild conditions, using the catalytic action of a second enzyme (racemase) or a base, exploiting the difference in the pK_a value of the α-methine carbon due to difference in the structure between the substrate and product.

13.4.2.1 Deracemization of α-Amino Acids via Enzyme-catalyzed DKR Coupled with *In Situ* Racemization

Several effects can contribute to lowering the pK_a of the proton on the α-carbon of an amino acid derivative, as the requisite for racemization. Both the transformation of the amine into a NH–CO bond either in a cyclic structure or in an open chain and the presence of the carboxylate in the form of an ester, amide, or thio ester have the effect of an accelerated racemization of the amino acid derivative. The presence of a C=C bond in the β–γ-position also has a strong effect. In order to obtain an effective racemization under the conditions of pH and temperature at which the enzyme is most active, a combination of more than one structural modification is often required. Substrates for which a general method of an enzyme-catalyzed kinetic resolution coupled with *in situ* racemization is successful are 5-monosubstituted hydantoins **13** (containing both activating NH–CO and CO–NH increments), 4-substituted-2-phenyloxazolin-5-one **14** (in which the amino acid nitrogen is in the form of an imino group and the carboxylate is present as a lactone bond) and N-Boc-amino acid thioesters **15** (in which the amino group is in the NH–CO form, the carboxylate is in the form of a thioester and the R-group is aromatic). The enzymatic kinetic resolution of amino acid esters coupled with an aldehyde-mediated *in situ* racemization and the enzymatic kinetic resolution of *N*-acyl amino acids **16** coupled with *N*-acyl-amino acid racemase-assisted racemization complete the group of potentially effective deracemizations by DKR for amino acids (Scheme 13.12).

The Hydantoinase–Carbamoylase System for D-Amino Acid Synthesis 5-Monosubstituted hydantoins are α-amino acids cyclically protected at both the carboxyl- and the α-amino group. They can be easily prepared from an aldehyde and isocyanate or by the Bucherer–Bergs synthesis and similar methods. Indeed, the hydantoin synthesis is also a practical method for the preparation of the racemic amino acid. Enzymes belonging to the dihydro-pyrimidinase family hydrolyze hydantoins to the carbamoyl amino acid. The latter can be hydrolyzed in turn to the amino acid by a second enzyme, a carbamoylase. Both enzymes can discriminate between enantiomers and, if their action is cooperative, either the L- or the D-amino acid can be obtained (Scheme 13.10) [36]. What makes the system of special interest is that the proton in the 5-position of the hydantoin ring (it will become the α-hydrogen in the α-amino acid) is considerably more acidic than conventional protons in amino acid esters or amides and much more acidic than the amino acid itself. Thus, the hydantoin can be often racemized *in situ* at slightly basic pH where the enzymes are still stable and active. If these conditions are met,

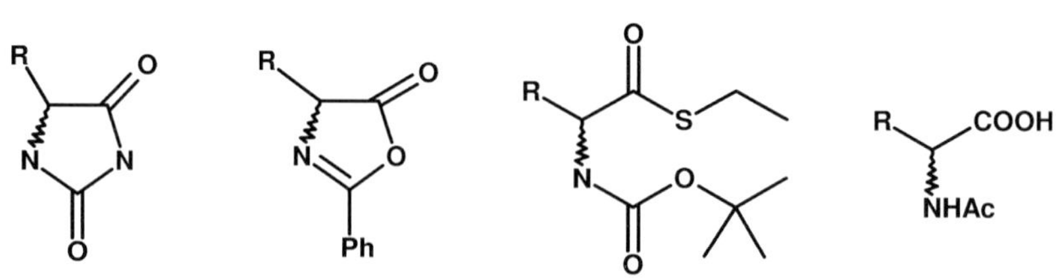

13 14 15 16

Scheme 13.12 Substrates for enzymatic DKR described in Section 13.4.2.

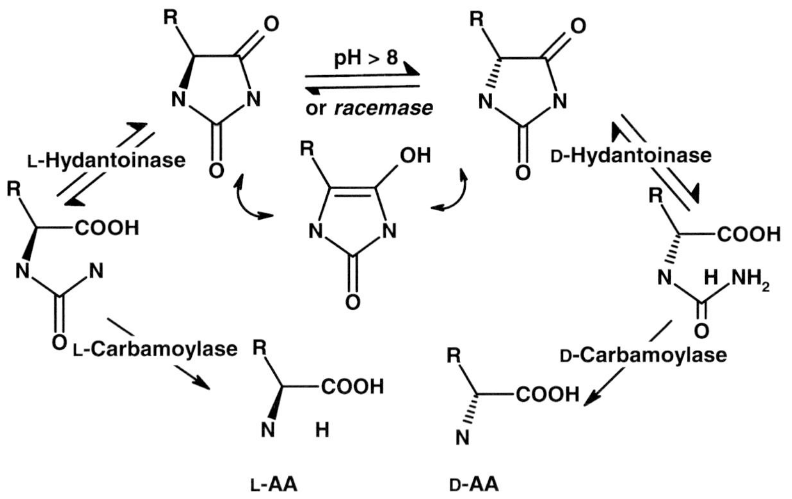

Scheme 13.13 Hydantoinase–Carbamoylase system for the deracemization of α-amino acids via DKR with *in situ* racemization.

the amino acid can be obtained in one single enantiomeric form in a 100% yield. A bacterial strain was developed at the industrial level and is now applied for the large-scale production of D-*p*-hydroxy-phenylglycine and other D-amino acids.

The use of the whole cell system is necessary due to the limited stability of the carbamoyl-hydrolyzing enzyme and to the different requirements for optimal reactivity. Best carbamoylase activity is obtained by lowering the reaction pH slightly, which is regulated from the development of carbon dioxide from the first reaction step. The D-hydantoinase–D-carbamoylase process has proved successful for the preparation of a large number of D-amino acids [37] (Scheme 13.13).

Hydantoinase–Carbamoylase System for L-Amino Acid Synthesis Despite a number of reports of strains with L-selective hydantoin-hydrolyzing enzymes [38] the commercial application of the hydantoinase process is still restricted to the production of D-amino acids. Processes for the production of L-amino acids are limited by low space–time yields and high biocatalyst costs. Recently, a new generation of an L-hydantoinase process was developed based on a tailor-made recombinant whole cell biocatalyst. Further reduction of biocatalyst cost by use of recombinant *Escherichia coli* cells overexpressing hydantoinase, carbamoylase, and hydantoin racemase from *Arthrobacter* sp. DSM 9771 were achieved. To improve the hydantoin-converting pathway, the level of expression of the different genes was balanced on the basis of their specific activities. The system has been applied to the preparation of L-methionine: the space–time yield is however still limited [39]. Improvements in the deracemization process from *rac*-5-substituted hydantoins to L-amino acids still requires a more selective L-hydantoinase.

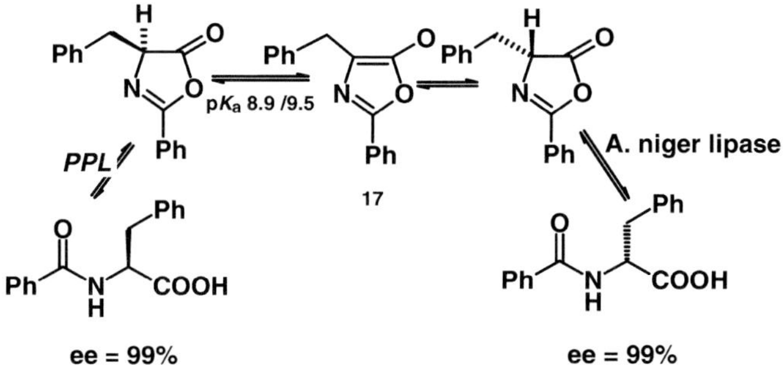

Scheme 13.14 4-Substituted-2-phenyloxazolin-5-one as substrates for lipases or esterases, with opposite selectivity allowing DKR with *in situ* racemization.

Enzymatic Kinetic Resolution of Phenyl Oxazolinones Coupled with *In Situ* Base-catalyzed Racemization Phenyl oxazolinones are a protected form of amino acids that can be obtained by cyclization of *N*-acyl-amino acids. Their structure is similar to hydantoins with the relevant difference that the carboxylate is in a lactone-type (–CO–O–) form while in hydantoins it is in a lactame type form (–CO–N–). Lipases are known to be not well suited for amide bond transformation. Therefore oxazolinones were designed as alternative structures for a DKR similar to the one described for hydantoins but involving lipases as resolving agents. Rapid enolization under slightly basic conditions should ensure the possibility for DKR. The oxazolinone **17** derived from phenylalanine is an excellent substrate for porcine pancreatic lipase and lipase from *Aspergillus niger*, although the transformation required rather long reaction times (20 h) (Scheme 13.14). The interest in this specific method is that the two enzymes show opposite enantiopreference with this substrate.

Altogether the hydrolysis has useful characteristics only for the preparation of phenylalanine [40]. Other oxazolinones are transformed with a much lower e.e. and longer reaction times.

An alternative approach to the DKR of phenyl oxazolinones is represented by alcoholysis in organic solvents. The lipase from *Pseudomonas cepacia* was used for this purpose. In organic solvent with low water activity non-enzymatic hydrolysis proceeds very slowly but the rate of enolization of the C-4 proton is sufficiently rapid so that 100% of the substrate is converted into product. By using methanol as the nucleophile, methanolysis of oxazolones proceeded at a useful rate to furnish methyl esters of *N*-benzoyl-L-α-amino acids. The optical purity of the products ranged from 66 to 98% e.e. [41].

A method based on the DKR of 4-substituted-2-phenyloxazolin-5-one has been applied to the large-scale synthesis of *N*-benzoyl-L-*t*-leucine-butyl ester **19**, an important non-natural amino acid intermediate of several biomimetic peptides. The procedure is based on the alcoholysis of 4-*t*-butyl-2-phenyloxazolin-5-one **18** with *n*-butanol catalyzed by the lipase from *Mucor miehei* (Scheme 13.15) [42]. The unreactive enantiomer is racemized *in situ* by a base-catalyzed reaction. The

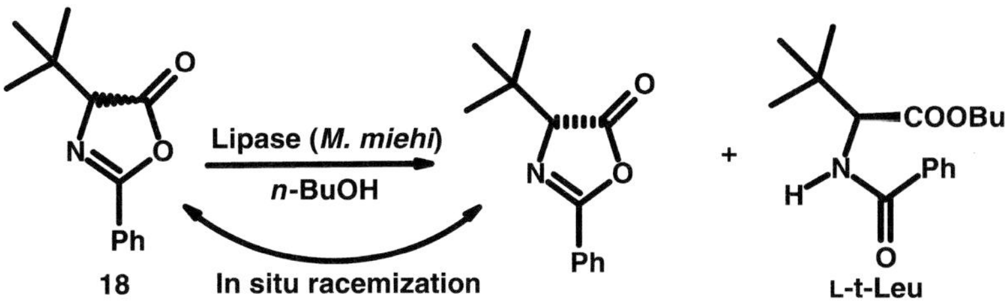

Scheme 13.15 4-*t*-Butyl-2-phenyloxazolin-5-one as substrates of lipase from *M. miehei* for the DKR with *in situ* racemization.

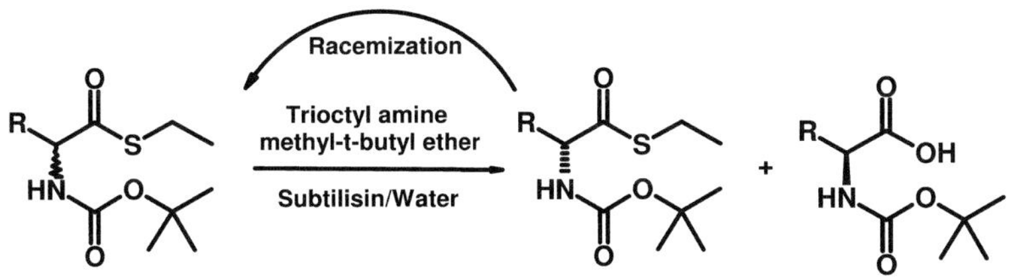

Scheme 13.16 Enzymatic kinetic resolution of *N*-Boc-amino acid thioesters coupled with base-catalyzed racemization.

reaction can be performed at a substrate concentration of 200 g/l in 24 h. Yields are 90% and the e.e. is >95%.

Enzymatic Kinetic Resolution of *N*-Boc-Amino Acid-Thioesters Coupled with Base-catalyzed Racemization Recently, a new method leading to the preparation of a number of aryl-glycines of the L-configuration has been published. The method is based on the hydrolysis of *N*-Boc-amino acid thioesters **15** catalyzed by an industrial preparation of the protease subtilisin (Scheme 13.16) [43].

In thioesters, the acidity of the α-hydrogen is higher as compared to to the corresponding oxo-esters, amides, or acids. The enzymatic transformation of a thioester into the corresponding carboxylate with a higher pK_a of the α-proton makes up the basis for DKR, provided that the enzymatic systems withstand the basic conditions required for substrate racemization. This concept has been applied in the DKR of α-alkyl thioesters. For application in amino acid chemistry racemic compounds were designed for this novel application. Moreover, subtilisin Carlsberg showed a good activity on these *N*-Boc amino acid derivatives with reaction rates comparable to the one observed with the regular esters. The reaction was performed in a water–methyl tertiary-butyl ether (MTBE) two-phase system in the presence of subtilisin. The reaction reached 50% conversion after 24 h: at this point the mixture was depleted of the L-enantiomer, which is the substrate of the enzyme. When trioctyl-amine was added as a base to the reaction mixture, such racemizing conditions generated the L-enantiomer and the enzymatic hydrolysis started again reaching complete conversion after 24 h. The *N*-Boc-L-phenylglycine was isolated in a 95% yield and >99% e.e. The structural requirements are an aryl or vinyl group on the α-carbon, a variety of alkyl groups on the sulfur atom, and a carbonyl group on the nitrogen. In addition, in this case an application to the preparation of the D-enantiomers would require catalysis by an enzyme with opposite enantioselectivity.

The Enzymatic Kinetic Resolution of Amino Acid Esters Coupled with Aldehyde-Mediated *In Situ* Racemization The reaction of the amino group of amino acid derivatives with aromatic aldehydes can lead to racemization through the reversible formation of imines. DKR can be obtained by combining the racemization with an enzyme-catalyzed preferential hydrolysis of one enantiomer. This principle has been applied to the DKR of a series of amino acid esters using pyridoxal phosphate or salicylic aldehyde. As an example, the esters were treated with an industrial preparation of subtilisin in the presence of 20 mol% of pyridoxal in a 19 : 1 mixture of 2-methyl-2-propanol and water. During the reaction course the hydrolyzed amino acid is insoluble and precipitates in the reaction medium. The remaining ester is racemized by imine formation with pyridoxal phosphate. The rate of racemization of the amino acid is only marginal due to the higher pK_a of the α-proton and the low solubility. Yields exceeding 90% of the L-enantiomer are obtained with an e.e. ranging from 90 to 98%. With this method Phe, Tyr, Leu, NorVal, and NorLeu of the L-configuration have been obtained. The drawback of the method is the requirement of a sensible amount of the expensive racemization catalyst (Scheme 13.17) [44].

A similar concept was applied to the DKR of phenyl glycine esters catalyzed by lipases in an aminolysis reaction [45].

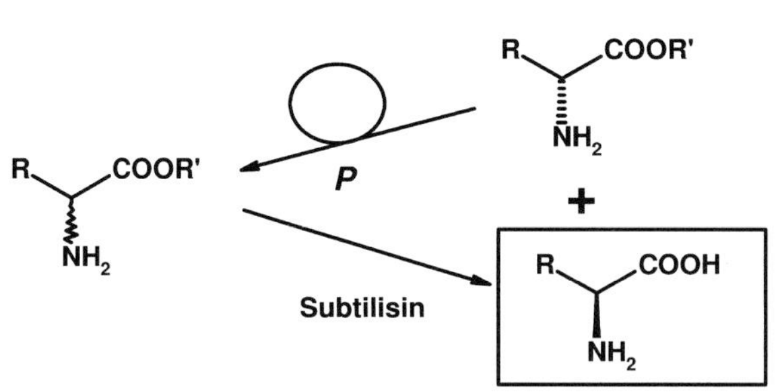

P **= pyridoxal phosphate** **R' = benzyl, n-butyl, propyl**

RCHNH$_2$COOH = Phe, Tyr, Leu, NorLeu, NorVal

Schem 13.17 Kinetic resolution of amino acid esters coupled with pyridoxalphosphate mediated *in situ* racemization.

Enzymatic Kinetic Resolution of *N*-Acyl Amino Acids Coupled with Racemization by *N*-Acyl Amino Acid Racemase Acylases are enzymes hydrolysing the *N*-acetyl derivatives of amino acids. They require the free carboxylate for activity and have long been used for the kinetic resolution of amino acids. The unreacted enantiomer is usually racemized in a separate step by treatment with acetic anhydride. While acylases from hog kidney have an L-specificity, bacterial acylases with L- and D-specificity of various origins have been isolated and used for the kinetic resolution of *N*-acetyl amino acids. An industrial process for the production of L-Met and other proteinogenic and non-proteinogenic L-amino acids such as L-Val, L-Phe, L-Norval, or L-aminobutyric acid has been established. Currently, several hundred tons per year of L-methionine are produced by this enzymatic conversion using an enzyme membrane reactor [46].

The starting material for the acylase process is a racemic mixture of *N*-acetyl-amino acids **20** which are chemically synthesized by acetylation of D,L-amino acids with acetyl chloride or acetic anhydride in alkali via the Schotten–Baumann reaction. The kinetic resolution of *N*-acetyl-D,L-amino acids is achieved by a specific L-acylase from *Aspergillus oryzae*, which only hydrolyzes the L-enantiomer and produces a mixture of the corresponding L-amino acid, acetate, and *N*-acetyl-D-amino acid. After separation of the L-amino acid by a crystallization step, the remaining *N*-acetyl-D-amino acid is recycled by thermal racemization under drastic conditions (Scheme 13.18) [47]. In a similar process racemic amino acid amides are resolved with an L-specific amidase and the remaining enantiomer is racemized separately. Although the final yields of the L-form are beyond 50% of the starting material in these multistep processes, the efficiency of the whole transformation is much lower than a DKR process with *in situ* racemization. On the other hand, the structural requirements for the free carboxylate do not allow the identification of derivatives racemizable *in situ*: therefore, the racemization requires

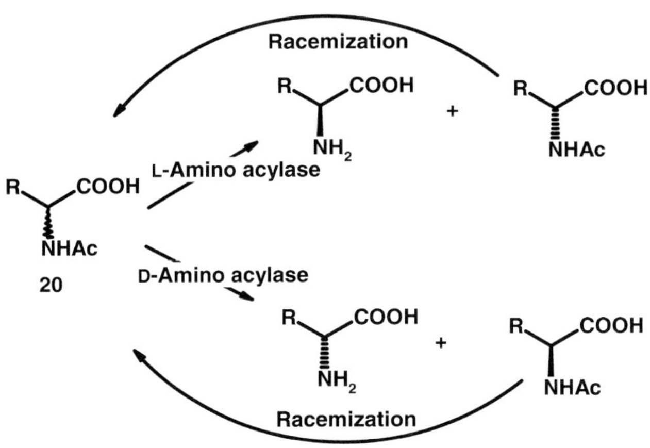

Scheme 13.18 The *N*-acylase-catalyzed kinetic resolution of *N*-acyl amino acids coupled with racemase induced racemization.

the application of specific enzymes. A whole cell system has been developed expressing the genes encoding for *N*-acyl amino acid racemase and L-aminoacylase from *Deinococcus radiodurans* in *E. coli*. This system was used for the deracemization of D,L-*N*-Ac-homophenylalanine. The L-amino acid was obtained in quantitative yield when the expression of the two enzymes was regulated properly. The cells could be used several times in the biotranformation.

Amino Acid Oxidase and Chemical *In Situ* Reduction of the Initially Formed Imino Compound A simple and interesting procedure for the deracemization of α-amino acids was introduced by Soda [48], who combined the oxidation of the D-enantiomer of D,L-proline to dehydro-proline with a chemical reduction of the imine **21** in on -pot, thereby restoring the racemic mixture. If the reaction in the first step is completely enantioselective, the e.e. of the amino acid after one cycle is 50%. Repeating the reaction in successive cycles raises the e.e. close to 100% (Scheme 13.19).

In a similar way, D,L-pipecolic acid was converted to L-pipecolic acid [49]. However, the use of sodium borohydride, which slowly decomposes in aqueous solution, causes a significant rise in the pH value, which has to be controlled and which interferes with the necessary repeated cycles. Turner and Fotheringham expanded the applicability of the method, using milder and water stable reducing agents [50] and applying amino acid oxidases with either D- or L-specificity.

Thus, using L-amino acid oxidase from *P. myxofaciens* and various amine–borane complexes or D-amino acid oxidase from porcine kidney and sodium cyanoborohydride, the preparation of several natural and non-natural enantiopure D- and L-amino acids was achieved, respectively [51]. In a more recent report, several β- and γ-substituted α-amino acids were deracemized using D-amino acid oxidase from *Trigonopsis variabilis* and sodium cyanoborohydride or sodium borohydride [52] (Scheme 13.20).

Aiming at the development of more efficient whole cell biocatalysts, deracemization of 4-chlorophenylalanine was reported using *E. coli* cells expressing an

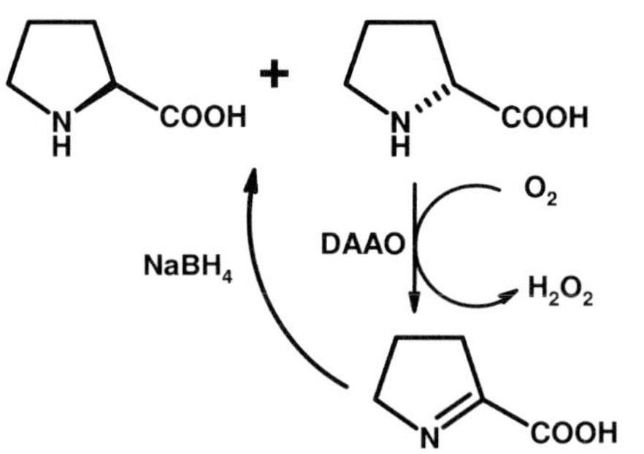

21

Scheme 13.19 Chemoenzymatic deracemization with D-amino acid oxidase and chemical reduction.

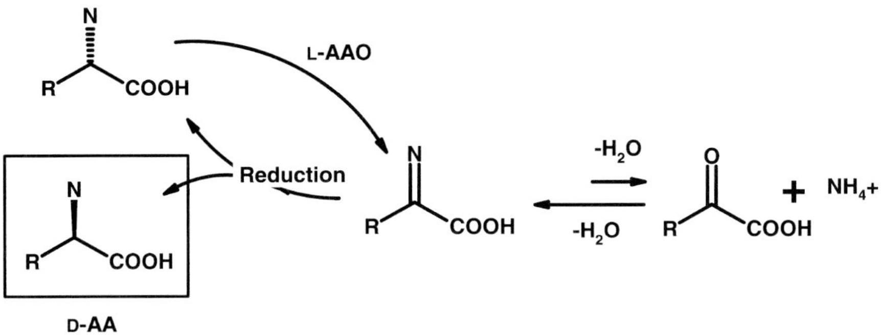

Scheme 13.20 Chemoenzymatic deracemization with L-amino acid oxidase and chemical reduction.

L-amino acid amino transferase from *Sinorhizobium meliloti* ATCC 51124. The enantiopure L-α-amino acid was obtained in high optical yield via the tandem action of D-amino acid dehydrogenase from the *E. coli* host cell (induced by L-Ala in the medium) and the cloned L-amino acid amino transferase from *S. meliloti* [53].

13.5 Useful Enzymes for Deracemization Methods

13.5.1 Amino Acid Oxidases

The enzymes discussed in this paragraph are D- and L-amino acid oxidases, the presence of which has been detected in numerous organisms and tissues. They were only recently identified in bacteria and the corresponding enzymes were isolated (for a review see [54]). These enzymes are flavin adenine dinucleotide (FAD)-containing flavoproteins that catalyze the oxidative deamination of amino acids (Scheme 13.21).

The flavin co-factor is reduced with concomitant oxidation of the amino acid into the corresponding imino acid, which spontaneously hydrolyzes to the α-keto acid and ammonia: the flavin co-factor is then reoxidized by molecular oxygen with the production of hydrogen peroxide. Both D- and L-amino acid oxidases deaminate a variety of amino acids, particularly those having a hydrophobic side chain, but are practically inactive on acidic amino acids (which are deaminated by specific glutamate and aspartate oxidases) and are strictly stereospecific.

13.5.1.1 D-Amino Acid Oxidase (EC 1.4.3.3)

The most investigated D-amino acid oxidases are those from the yeasts *R. gracilis* and *T. variabilis* and the one from pig kidney. In solution D-amino acid oxidases are homodimers, with each monomer constituted by ~350–370 residues and containing one molecule of non-covalently bound FAD co-factor. Mammalian and

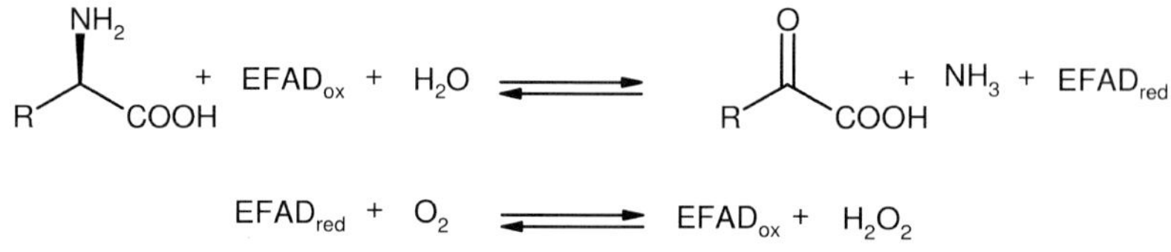

Scheme 13.21 Reaction catalyzed by D-amino acid oxidase and L-amino acid oxidase.

fungal D-amino acid oxidases differ significantly in their affinity for the co-enzyme FAD, which is most tightly bound to those from microorganisms and in kinetic efficiency (the V_{max} values on D-alanine as a substrate at air saturation and 25 °C are 110, 60 and 9.5 units/mg protein for *R. gracilis, T. variabilis* and pig kidney D-amino acid oxidases, respectively). Yeast D-amino acid oxidases show a high stability in the 6.0–8.2 pH range and a decrease in stability at higher pH values, while pig kidney D-amino acid oxidase shows a maximum stability at pH 9.5 and a decrease at acidic pH values. *Rhodotorula gracilis* D-amino acid oxidase is stable up to 35 °C, *T. variabilis* D-amino acid oxidase is stable up to 40 °C (it is fully inactivated at 65 °C), and the stability of pig kidney D-amino acid oxidase is greater than that of *R. gracilis* D-amino acid oxidase. D-Amino acid oxidases from different sources show different substrate specificity. Known microbial D-amino acid oxidases were recently classified into two groups [55]: D-amino acid oxidases from microorganisms such as *Fusarium oxysporum, C. parapsilosis,* and *Candida boidinii* are highly specific for D-Ala, while *R. gracilis* and *T. variabilis* D-amino acid oxidases show maximum activity with D-Met, D-Trp, D-Val, and D-Phe. The kinetic mechanism of the reaction catalyzed by D-amino acid oxidase on D-Ala is consistent with a ternary complex (sequential) mechanism: the rate-limiting step is the rate constant for product release from the reoxidized enzyme for pig kidney D-amino acid oxidase and the rate constant for flavin reduction for *R. gracilis* D-amino acid oxidase.

According to structural and sequence similarities of the FAD-binding region, D-amino acid oxidase is classified as a member of the glutathione reductase family (subgroup GR_2). Each D-amino acid oxidase monomer can be divided into a FAD-binding (basal) domain and a dimerization (apical) domain [56]. The basal domain contains the typical nucleotide-binding domain, while the apical domain is formed by a pseudo-β-barrel structure consisting of two twisted β-sheets: residues belonging to this region form the major part of the active site and the whole dimer interface. Structural and mechanistic studies on D-amino acid oxidases from various sources suggest that the enzymatic reaction involves a direct hydride transfer from the substrate D-amino acid to the oxidized flavin. The most striking feature of D-amino acid oxidases is that no active site residues are directly involved in the chemical steps of catalysis (Figure 13.1a). The D-amino acid side chain makes contacts with the wide cavity representing the upper part of the active site cleft, which is mostly hydrophobic: this provides a rationale as to why *R. gracilis* D-amino acid oxidase prefer D-amino acids carrying an aromatic and/or hydrophobic side chain to polar ones and for its capacity to accommodate large compounds,

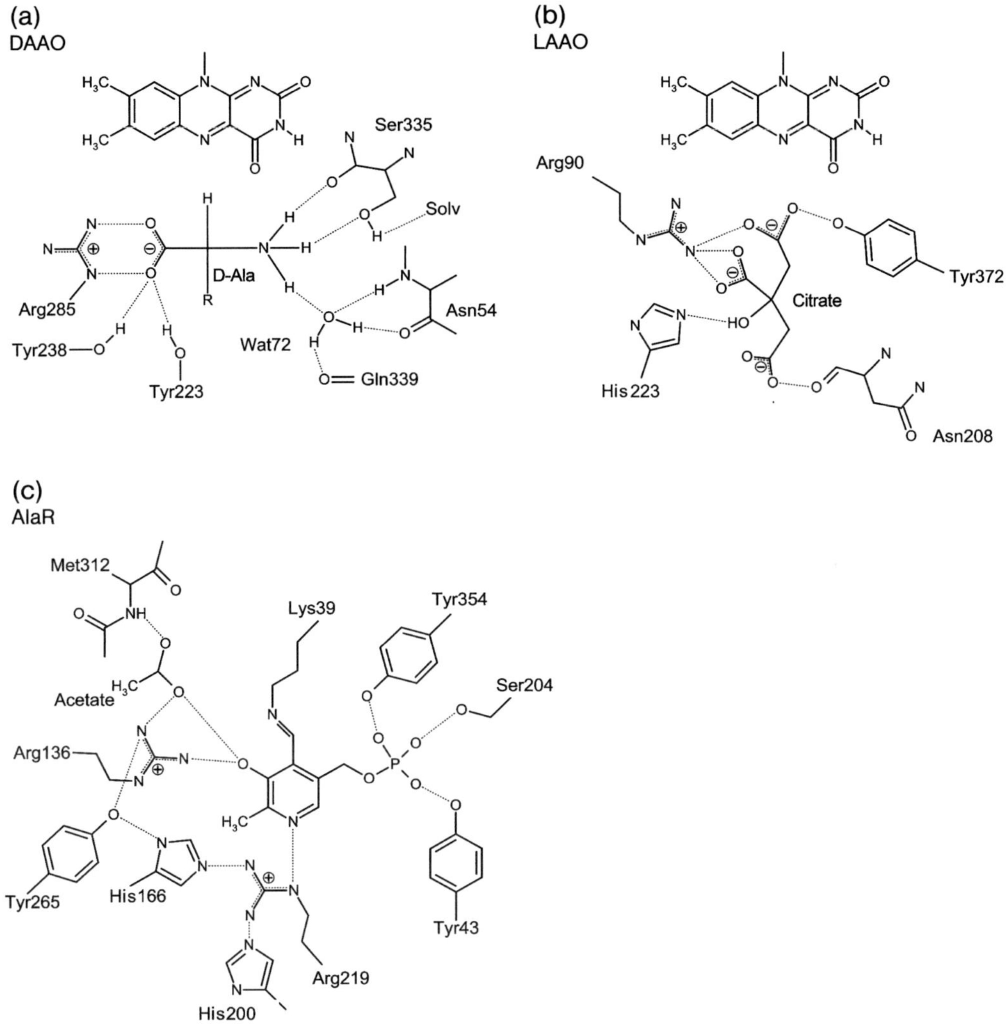

Figure 13.1 Schematic representation of interactions at the active site of (a) *R. gracilis* D-amino acid oxidase in complex with CF_3-D-alanine (pdb 1COL), (b) *C. rhodostoma* L-amino acid oxidase in complex with citrate (pdb IF8R), and (c) *B. stearothermophilus* alanine racemase (pdb 1SFT).

as in the case of the antibiotic cephalosporin C. It also explains the reluctance of *R. gracilis* D-amino acid oxidase to act on amino acids carrying a polar and, in particular, a charged side chain.

Protein engineering studies have allowed modulation of D-amino acid oxidases' oligomerization state, stability (which is significantly increased in the immobilized form), FAD binding, and substrate specificity (for a review see [57]). The residue

Met213 of *R. gracilis* D-amino acid oxidase has been proposed as the most important residue for determining the substrate specificity of the enzyme. In fact, the M213R mutant evolved the ability to oxidize acidic D-amino acids such as D-Asp. Recently, *R. gracilis* D-amino acid oxidase was employed in the resolution of racemic mixtures of D,L-naphthyl amino acids and the M213G mutant showed a higher catalytic efficiency on D-naphthyl alanine and D-naphthyl glycine [28]. Subsequently, five mutants displaying an altered substrate specificity were selected by a directed evolution approach of *R. gracilis* D-amino acid oxidase: these D-amino acid oxidase variants have a better catalytic efficiency for all the substrates used [57].

13.5.1.2 **L-Amino Acid Oxidase (EC 1.4.3.2)**

L-Amino acid oxidase is a flavoenzyme that catalyzes the oxidative deamination of L-amino acids. L-Amino acid oxidase activities have been detected in mammals, birds, reptiles, invertebrates, molds, and bacteria [54]. L-Amino acid oxidases show the typical absorption spectrum due to the presence of a molecule of non-covalently bound FAD per subunit (with maxima at 465 and 380 nm): they behave like flavoprotein oxidases, as in the case of D-amino acid oxidase. L-Amino acid oxidase isolated from rat liver was reported to utilize flavin mononucleotide (FMN) as a co-enzyme, but since it is more active on L-hydroxy acids than on amino acids, it was thus considered as an L-hydroxy acid oxidase. Even a partially purified L-amino acid oxidase from turkey liver appeared to have FMN as a co-factor.

Snake venom L-amino acid oxidases are the best-studied members of this protein family but their function is still debated: they play a role in inducing apoptosis and affecting platelets and are postulated to be toxins. L-Amino acid oxidases from bacteria, fungi, and plants are involved in the utilization of nitrogen sources. Snake venom L-amino acid oxidases are usually homodimeric flavoglycoproteins with a molecular mass of ~110–150 kDa: the scattering in molecular mass is partially due to the different sugar content. Snake venom L-amino acid oxidases also show different isoelectric points (from 4.4 to 8.1) and frequently more than one type of the enzyme is produced. Snake venom L-amino acid oxidases share sequence similarity around the FAD-binding site with human monoamine oxidase and microbial L-amino acid oxidases and with mouse interleukin 4-induced *Fig1*-protein, but no similarity is observed with D-amino acid oxidases.

The best substrates for ophidian L-amino acid oxidases are aromatic or, most generally, hydrophobic amino acids, polar and basic amino acids being deaminated at much lower rates: Glu, Asp, and Pro are not oxidized by L-amino acid oxidase. L-Amino acid oxidase is also active on ring-substituted aromatic amino acids, as well as on seleno cysteinyl derivatives. The substrate specificity depends on the source of the enzyme (e.g. *Ophiophagus hannah* L-amino acid oxidase also oxidizes Lys and Orn) and on the pH. The V_{max} of the reaction of *Crotalus adamanteus* L-amino acid oxidase at pH 8.5 is ~12 and ~6 units/mg protein on L-Arg and L-Leu, respectively, while a higher activity is observed for the L-amino acid oxidase from *O. hannah* on L-His (~60 units/mg protein). The kinetic mechanism of L-amino acid oxidase from *C. adamanteus* follows that in Scheme 13.13 and is substantially similar to that reported for D-amino acid oxidase [54]. At high pH and low oxygen concentration (<10^{-4} M) a ping-pong mechanism is operative, whereas at low pH and higher oxygen concentration (>5×10^{-3} M) a sequential mechanism is effective. The catalytic mechanism of flavin reduction in the L-amino acid oxidase reaction was proposed to be similar to that of D-amino acid oxidase.

L-Amino acid oxidase undergoes two types of reversible inactivation: one obtained by raising the pH from 5.5 to 7.5 and the temperature from 25 to 38 °C, while the second one is caused by storage at −5 to −60 °C and depends on the pH (it is favored by acidic pH values) and on the ionic composition of the storage buffer. In both cases reactivation is achieved by incubating the enzyme at pH 5 and 38 °C for 1 h. Snake venom L-amino acid oxidases should be maintained in the dark at 4 °C and near neutral pH to avoid inactivation.

The X-ray crystallographic structure of *Calloselasma rhodostoma* L-amino acid oxidase indicates that it is functionally a dimer in which each subunit is constituted by three domains: an FAD-binding domain, a substrate-binding domain, and a helical domain [58]. A funnel is formed at the interface between the latter domains and provides substrate access to the active site. Thus, the mode of substrate binding (Figure 13.1b) and access significantly differs between L-amino acid oxidase and D-amino acid oxidase.

Recently, L-amino acid deaminase (EC 1.4.3.x) activities have been identified, particularly from the *Proteus* genus [59]. This enzyme, constituted by ~370 residues, is an FAD-containing L-amino acid oxidase flavoprotein that uses molecular oxygen to convert L-amino acids into the corresponding α-keto acids and ammonia but does not produce hydrogen peroxide. L-amino acid deaminase prefers amino acids with aliphatic, aromatic, and sulfur-containing side chains (the best substrates are L-Leu, L-Phe, L-Met, and L-Trp) and its kinetic efficiency is quite low because of the low V_{max} value (≤2 units/mg protein).

13.5.2 Amino Acid Racemases

A variety of amino acid racemases have been identified in bacteria, archaea, and eukaryotes. They are classified into two groups: pyridoxal 5′-phosphate (PLP) -dependent and -independent enzymes. Therefore, racemization can be achieved via two mechanisms: through a chirally unstable Schiff base intermediate with an aromatic aldehyde serving as co-factor PLP (Scheme 13.22a) and by a two-base mechanism without co-factor (Scheme 13.22b).

13.5.2.1 PLP-dependent Racemases

Alanine Racemase (EC 5.1.1.1) Alanine racemase is a PLP-dependent bacterial enzyme that catalyzes the racemization of L- and D-alanine exclusively (for a review see [60]). The enzyme from *Salmonella typhimurium* is an exception since it also accepts L-Ser, L-*homo*-Ser, and L-Cys as substrates. Alanine racemase, the most studied member of the PLP-dependent racemases, plays a major role in the bacterial growth by providing D-Ala for the peptidoglycan assembly and cross-linking.

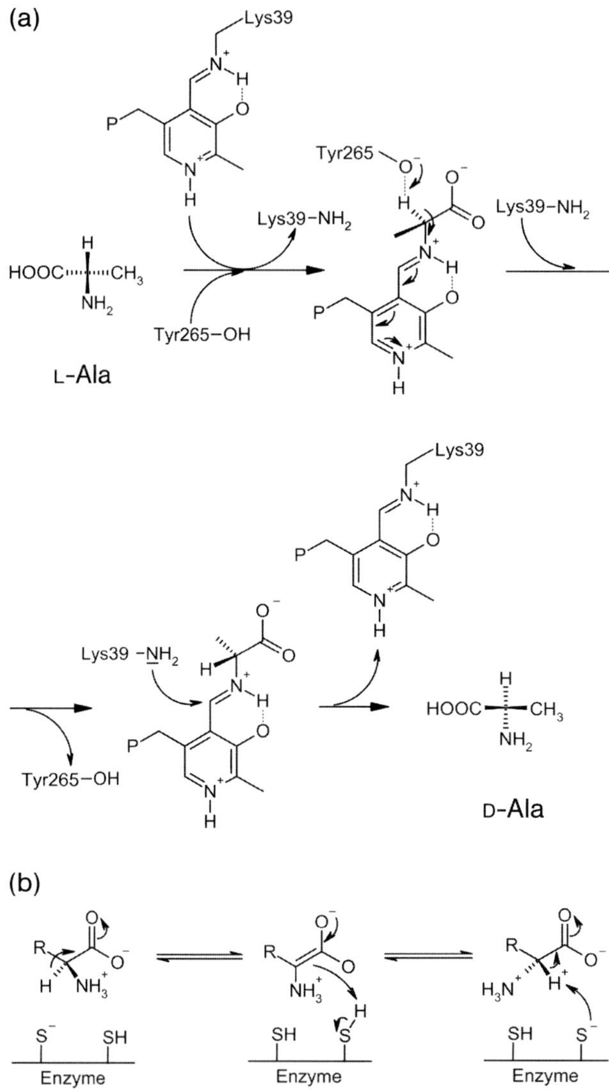

Scheme 13.22 Amino acid racemization by (a) PLP-dependent alanine racemase and (b) PLP-independent α-amino acid racemase.

Alanine racemase was purified and cloned form various sources. Interestingly, two distinct genes were identified in a number of genome sequences, for example, *E. coli*, *Bacillus subtilis*, *Pseudomonas aeruginosa*, and so on.

Site-directed mutagenesis experiments and X-ray crystallography of alanine racemase from *Bacillus stearothermophilus* (Figure 13.1c) [61] suggested that the enzyme uses two catalytic bases, namely Tyr265 and PLP-binding Lys39. Lys39 abstracts the α-H from D-Ala and adds the α-H to the intermediate to form D-Ala: Tyr265 is instead the corresponding catalytic residue acting on L-Ala. This mechanism requires that the α-H abstracted from the substrate is transferred between the side chain of Lys39 and that of Tyr265, otherwise the enzyme reaction would stop after a single turnover. A second mechanism was also proposed, in which the carboxyl group of the substrate participates in catalysis mediating the α-H transfer between the two catalytic residues. Furthermore, alanine racemase is a bifunctional enzyme, that is, besides racemization, it also catalyzes transamination at low pH (~6).

Alanine racemases are homodimeric enzymes of an apparent molecular mass of ~76 kDa, containing two molecules of PLP as co-enzyme [61]: the known enzymes from different sources are highly homologous. With regard to the kinetic properties, the K_m values for D- and L-Ala determined at 30 °C for the *Pseudomonas fluorescens* alanine racemase are ~12 and 19 mM, respectively and the V_{max} values for racemization are ~1200 and 2200 units/mg protein, respectively. The thermophilic alanine racemase from *B. stearothermophilus* is quite stable to heat treatment (up to 75 °C for 1 h) while the mesophilic one from *B. subtilis* is stable up to 55 °C under the same conditions.

Serine Racemase (EC 5.1.1.16) Serine racemases have been discovered in both bacteria and eukaryotes (for a review see [60, 62]. In the latter organisms, serine racemase catalyzing the conversion of L-Ser to D-Ser was at first discovered in the silkworm *Bombyx mori:* it is a PLP-dependent racemase which is also active on L-Ala (~6% of the activity on L-Ser). A serine racemase was also purified from rat brain (and a serine racemase cDNA was cloned from mouse brain). Mammalian serine racemase shows sequence similarity with L-threonine dehydratase from various sources: all the active site residues of the latter enzyme are also conserved in mouse serine racemase. Mammalian serine racemase is a member of the fold-type II group of PLP enzymes (similarly to L-threonine dehydratase, D-serine dehydratase, and so on) and distinct from alanine racemase, which belongs to the fold-type III group. Mouse serine racemase shows a low kinetic efficiency: the K_m values for L- and D-Ser are ~10 and 60 mM, respectively and the V_{max} values with L- and D-Ser are 0.08 and 0.37 units/mg protein (less than 0.1% of those of alanine racemase on L- and D-Ala, see above).

On the other hand, bacterial serine racemase from *Enterococcus gallinarum* (VanT) is instead homologous to alanine racemase. It is constituted by 698 amino acids with an N-terminal domain containing 10 predicted transmembrane segments and a C-terminal domain showing 31% of sequence identity with alanine racemase from *B. stearothermophylus*. The active site residues of alanine racemase are also conserved in VanT: this observation indicates that the reaction catalyzed by VanT probably proceeds via a two-base mechanism analogously to that proposed for alanine racemase (see above). Interestingly, its physiological role is related to vancomycin resistance.

13.5.2.2 PLP-independent Racemases

The members of this class follow a two-base mechanism involving two cysteine residues as the conjugated catalytic acid and base for the abstraction of the α-H from both amino acids.

Glutamate Racemase (EC 5.1.1.14) Glutamate racemase genes have been cloned from a number of bacterial sources [60]. Glutamate racemases are monomeric enzymes of 28–30 kDa molecular mass. Glutamate racemases from *Pediococcus pentosaceus* is absolutely specific for glutamate and the one from *E. coli* is instead activated (~100-fold) by UDP-*N*-acetyl-muramoyl-L-alanine, a precursor of the peptidoglycan layer of the cell walls. A region of 21 amino acids at the N-terminus of the *E. coli* enzyme is responsible for the activation.

Glutamate racemase contains two active site cysteines per enzyme molecule and one of these is essential for catalysis: a thiolate from one cysteine abstracts the α-H of the substrate and the other cysteine thiol delivers a proton to the opposite face of the resulting carbanionic intermediate. X-ray diffraction studies and site-directed mutagenesis experiments of *Aquifex pyrophilus* glutamate racemase indicated that Cys73 is responsible for the deprotonation of D-Glu and Cys184 is responsible for the deprotonation of L-Glu. Because of the large separation between the pK_a values of the α-H of the amino acid (~21) and that of a thiol (~10), an unidentified system is required for facilitating proton abstraction. Asp10 and His186 of *Lactobacillus fermenti* glutamate racemase have been proposed for assisting the function of Cys73 and Cys184, respectively, although crystallographic data on *A. pyrophilus* glutamate racemase indicate that the histidine is too far away from the second cysteine for a direct interaction.

With regard to the kinetic properties of *L. fermenti* glutamate racemase, the V_{max} value is ~70 s^{-1} (~130 units/mg protein) and the K_m is 0.3 mM for both L- and D-Glu [63]. The highest kinetic efficiency is observed at pH 8–8.5.

Aspartate Racemase (EC 5.1.1.13) Aspartate racemase produces D-Asp from the corresponding L-isomer for the biosynthesis of the peptidoglycan layer of bacterial cell walls. This enzyme has been identified in various *Lactobacillus* and *Streptococcus* strains, as well as in archaea such as *Desulfurococcus* and *Thermococcus* strains [60]. Analogously to glutamate racemase, *Streptococcus thermophylus* aspartate racemase does not require co-factors and contains an essential cysteine, thus suggesting that a two-base mechanism is also active in aspartate racemase. The X-ray diffraction analysis of *Pyrococcus horikosii* aspartate racemase shows that cysteines 82 and 194 are located on both sides of a cleft between two protein domains where they probably act as the catalytic acid and base. Interestingly, PLP-dependent aspartate racemases have been recently reported, specifically from *Thermoplasma acidophilum* and *Scarpharca broughtonii*.

Additional Racemase Activities Further useful amino acid racemases are [62]: phenylalanine racemase (EC 5.1.1.11), the only racemase that requires ATP to activate the substrate, proline racemase (EC 5.1.1.4) from *Clostridium stricklandii*, and amino acid racemase (EC 5.1.1.10) from *Pseudomonas putida*. It shows a wide substrate specificity comprising D-Ala, D-Phe, D-Arg, D-Asp, L-*nor*-Leu, L-Ser, L-Met, and L-Orn, arginine racemase (EC 5.1.1.9) active on L-Lys, L-Arg, L-Orn, and so on, and racemases active on α-amino acid derivatives, such as the one active on α-amino lactams (EC 5.1.1.15) and that active on α-acetyl amino acids (NAAR).

13.5.2.3 Mandelate Racemase (EC 5.1.2.2)

Mandelate racemase is a PLP-independent well-characterized racemase (for a review see [62, 64]. Because of its remarkable stability and broad substrate specificity, it has been considered an ideal candidate for deracemization reactions. The gene encoding *P. putida* mandelate racemase has been cloned and its three-dimensional structure has been solved: it is an octameric enzyme constituted by subunits of 39 kDa, with each subunit consisting of a circular arrangement of eight parallel β-sheets. In the active site a catalytically active Mg^{2+} ion is positioned through the interaction with an array of aspartate, glutamate, and lysine residues. Besides its natural substrate mandelate, it also racemizes the corresponding amide derivatives. Furthermore, the aromatic system of mandelate may be expanded to a naphthyl system and heteroaromatic analogs of mandelate, as well as vinyl glycolate, can also be accepted by the enzyme. The broad substrate tolerance is mainly due to a remarkable plasticity of the hydrophobic binding pocket within the active site. A predictive substrate model for mandelate racemase has been proposed [64].

As stated above for glutamate racemase, the main obstacle during the interconversion of mandelate enantiomers is the abstraction of the α-H (pK_a ~29) for generating the corresponding achiral enolate intermediate. This step is facilitated by the binding of both enantiomers within the active site through a tight network of hydrogen bonds and salt bridges involving the residues Glu317 and Lys164 and Mg^{2+}.

13.5.3 Transaminases

Amino transferases (transaminases) catalyze the PLP-dependent reversible transfer of an amino and keto group between an amino acid donor and a keto acid acceptor yielding new amino acid and keto acid product, according to the equation in Scheme 13.23.

When the amino donor has the L-configuration and an L-amino acid transaminase is used, the product amino acid will have an L-configuration and when the amino donor has the D-configuration and a D-amino transaminase is used, the product amino acid will have a D-configuration. The equilibrium constant for the reaction is typically close to unity.

L-Amino transaminase activities are ubiquitous in Nature since they are involved in the biosynthesis of most natural amino acids. On the other hand, D-amino transaminase activities have been identified in bacteria, mostly in the *Bacillus* strains and are involved in the production of D-amino acids for the peptidoglycan layer of the cell wall. The mechanism is well established, with PLP shuttling

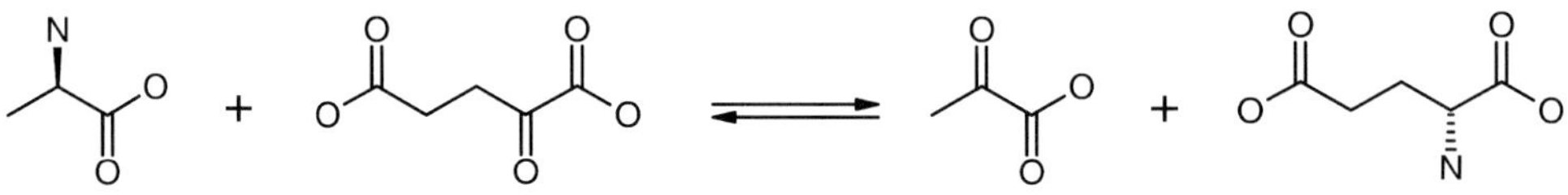

Scheme 13.23 Reaction catalyzed by amino transferase.

between pyridoxal and pyridoxamine forms in the reversible transfer of amino and keto groups between an amino acid and a keto acid substrate pair. The reaction proceeds by ping-pong kinetics.

Transaminases possess many features appropriate for efficient biocatalysts, such as high turnover numbers and no requirement for external recycling of the co-factor. Because of the wide substrate tolerance of many amino transferases such as tyrosine amino transferase and branched-chain amino transferases from *E. coli*, these enzymes have been largely employed in the enantiospecific preparation of non-proteinogenic amino acids. These include straight-chain alkyl, diacid, branched-chain, aromatic, and bifunctional amino acids [65].

13.5.3.1 L-Amino Transferases (EC 2.6.1.x)

The most commonly used L-amino transferase activities for L-amino acid synthesis are the ones from *E. coli*, which can be used in whole cell or immobilized systems. They include the following. First, aspartate amino transferase (EC 2.6.1.1), the *E. coli* gene, named *aspC*, which encodes for a 88 kDa protein that is active as homodimer. Branched-chain amino transferase (EC 2.6.1.42), the enzyme from *E. coli*, which is a hexamer composed of identical subunits of ~34 kDa and has a three-dimensional structure that has been solved, Tyrosine amino transferase (EC 2.6.1.5): in *E. coli* the *tyrB* gene encodes for the protein involved in the final step of Tyr biosynthesis, that is the transamination of *p*-OH-phenylpyruvate to tyrosine using L-Glu as the amino donor. L-Ornithine δ-amino transferase (EC 2.6.1.13): it catalyzes the reversible transamination of the δ-amino group from L-ornithine to α-ketoglutarate producing L-glutamate-γ-semialdehyde and L-Glu. This activity has been identified in many bacteria, mainly belonging to the *Bacillus* strains. The *B. subtilis rocD* gene encodes for a 44 kDa protein. A number of other amino transferases have been identified: 4-amino-butyrate-2-ketoglutarate transaminase (EC 2.6.1.19), which has been used to prepare the herbicide L-phosphinothricin and D-phenylglycine amino transferase from *Pseudomonas stutzeri*, which shows a very narrow substrate tolerance and causes the stereoinversion of the transaminated amino acid (it catalyzes the transamination of D-Phe using α-ketoglutarate as amino donor to form L-Glu) [66].

13.5.3.2 D-Amino Transferases (EC 2.6.1.21)

Among the D-amino transferase activities, the one from *Bacillus* sp. YM1 has been widely investigated: the gene has been cloned, the protein has been purified and its three-dimensional structure has been solved [67]. This enzyme is a dimer of overall molecular mass 65 kDa and each monomer requires one PLP molecule as a co-enzyme. It does not show sequence identity with L-amino transferase. The fold of D-amino transferase is completely different from that of any of the other known enzymes that utilize PLP. However, there are some striking similarities between the active sites of D-amino transferase and the corresponding enzymes that transaminate L-amino acids, such as L-aspartate transaminase. Most recently the D-amino transferases from *Staphylococcus haemolyticus, Bacillus licheniformis* and *Bacillus sphaericus* have been studied. Due to their broad substrate specificity, D-amino transferases are useful in the production of D-amino acids, especially when a thermostable enzyme is employed.

13.6 Summary and Outlook

Deracemization methods constitute an efficient access to chiral compounds as single enantiomers starting with a racemic mixture. Entirely enzymatic methods consist of multistep enzyme-catalyzed reactions as in the deracemization of amino acids catalyzed by an amino acid oxidase and an amino transferase (Section 13.4.1). The limitation of this method comes from the availability of the oxidative enzymes with both stereochemical preferences, in the thermodynamic equilibrium conditions in the reaction catalyzed by amino transferases, and in the stereochemical preference of this enzyme. When an enzymatic resolution process is coupled with a racemization step, a specific racemizing enzyme can be used. The advantage is that the process can be performed in one single reaction act, in mild conditions (see for instance Sections 13.2.1 and 13.4.1.3). However, racemases are in general delicate enzymes, with very narrow substrate specificities (Section 13.5.2). Excellent results can be obtained when a hydrolytic enzyme is coupled with a base in racemizing conditions. Besides the industrial system hydantoinase/carbamoylase for the DKR of 5-substituted hydantoins leading to D-amino acids, other methods relying on a base-catalyzed racemization like oxazolone and thioester hydrolysis are promising for large-scale applications. The constraint again is the availability of hydrolytic enzymes with opposite enantiospecificity. These methods compare favorably with well-known procedures based on transition metal catalysis.

References

1 (a) Pamies, O. and Bäckvall, J.-E. (2003) *Chemical Reviews*, **103**, 3247–3262.
(b) Faber, K. (2001) *Chemistry – European Journal*, **7**, 5004–5010.
(c) Huerta, F.F., Minidis, A.B.E. and Backvall, J.E. (2001) *Chemical Society Reviews*, **30**, 321–331.
(d) Strauss, U.T, Felfer, U. and Faber, K. (1999) *Tetrahedron, Asymmetry*, **10**, 107–117.
(e) Azerad, R. and Buisson, D. (2000) *Current Opinion in Biotechnology*, **11**, 565–571.
(f) El Gihani, M.T. and Williams, J.M.J. (1999) *Current Opinion in Chemical Biology*, **3**, 11–15.

(g) Stecher, H. and Faber, K. (1997) *Synthesis*, **1**, 1–16.
(h) Caddick, S. and Jenkins, K. (1996) *Chemical Society Reviews*, **25**, 447–456.
(i) Ward, R.S. (1995) *Tetrahedron Asymmetry*, **6**, 1475–1490.
(j) Noyori, R., Tokunaga, M. and Kitamura, M. (1995) *Bulletin of the Chemical Society of Japan*, **68**, 36–55.

2 Gruber, C., Lavandera, I., Faber, K. and Kroutil, W. (2006) *Advanced Synthesis & Catalysis*, **348**, 1789–1805.

3 (a) Larsson, A.L.E. and Persson, B.A. (1997) *Angewandte Chemie (International Edition in English)*, **36**, 1211–1212.
(b) Norinder, J., Bogar, K., Kanupp, L. and Backvall, J.E. (2007) *Organic Letters*, **9**, 5095–5098.

4 Arosio, D., Caligiuri, A., D'Arrigo, P., Pedrocchi-Fantoni, G., Rossi, C., Saraceno, C., Servi, S. and Tessaro, D. (2007) *Advanced Synthesis & Catalysis*, **349**, 1345–1348.

5 (a) Ström, K., Sjögren, J., Broberg, A. and Schnürer, J. (2002) *Applied and Environmental Microbiology*, **68**, 4322–4327.
(b) Dieuleveux, V., van der Pyl, D., Chataud, J. and Gueguen, M. (1998) *Applied and Environmental Microbiology*, **64**, 800–803.
(c) Valls, N., Lopez-Canet, M., Vallribera, M. and Bonjoch, J. (2001) *Chemistry – European Journal*, **7**, 3446–3460.
(d) Valls, N., Vallribera, M., Carmeli, S. and Bonjoch, J. (2003) *Organic Letters*, **5**, 447–450.
(e) Ishida, K., Okita, Y., Matsuda, H., Okino, H.T. and Murakami, M. (1999) *Tetrahedron*, **55**, 10971–10988.
(f) Valls, N., Vallribera, M., Lopez-Canet, M. and Bonjoch, J. (2002) *The Journal of Organic Chemistry*, **67**, 4945–4950.
(g) Tao, J. and McGee, K. (2002) *Organic Process Research & Development*, **6**, 520–524.
(h) Dragovich, P.S., Prins, T.J., Zhou, R., Brown, E.L., Maldonado, F.C., Fuhrman, S.A., Zalman, L.S., Tuntland, T., Lee, C.A., Patick, A.K., Matthews, D.A., Hendrickson, T.F., Kosa, M.B., Liu, B., Batugo, M.R., Gleeson, J.-P.R., Sakata, S.K., Chen, L., Guzman, M.C., Meador, J.W., Ferre, R.A. and Worland, S.T. (2002) *Journal of Medicinal Chemistry*, **45**, 1607–1623.
(i) Sheldon, R.A. (1993) *Chirotechnology*, Marcel Dekker, New York, pp. 362–367.
(j) Sheldon, R.A., Zeegers, H.J.M., Houbiers, J.P.M. and Hulshof, L.A. (1991) *Chimica Oggi*, **9**, 35–36.

6 (a) Schmidt, E., Blaser, H.U., Fauquex, P.F., Sedelmeier, G. and Spindler, F. (1992) *Microbial Reagents in Organic Synthesis*, NATO ASI Series C, Vol. 381 (ed. S. Servi), Kluwer Academic, Dordrecht, pp. 377–386.
(b) Azerad, R. and Buisson, D. (1992) *Microbial Reagents in Organic Synthesis*, NATO ASI Series C; Vol. 381 (ed. S. Servi), Kluwer Academic, Dordrecht, pp. 421–430.

7 Huerta, F.F., Laxmi, Y.R.S. and Bäckvall, J.-E. (2000) *Organic Letters*, **2**, 1037–1040.

8 Huerta, F.F. and Bäckvall, J.-E. (2001) *Organic Letters*, **3**, 1209–1212.

9 Strauss, U.T. and Faber, K. (1999) *Tetrahedron, Asymmetry*, **10**, 4079–4081.

10 (a) Fee, J.A., Hegeman, G.D. and Kenyon, G.I. (1974) *Biochemistry*, **13**, 2528–2532.
(b) Neidhart, D.J., Howell, P.L., Petsko, G.A., Powers, V.M., Li, R., Kenyon, G.L. and Gerlt, J.A. (1991) *Biochemistry*, **30**, 9264–9273.
(c) Kenyon, G.L., Gerlt, J.A., Petsko, G.A. and Kozarich, J.W. (1995) *Accounts of Chemical Research*, **28**, 178–186.
(d) Gerlt, J.A., Kozarich, J.W., Kenyon, G.L. and Gassman, P.G. (1991) *Journal of the American Chemical Society*, **113**, 9667–9669.
(e) Gerlt, J.A., Kenyon, G.L., Kozarich, J.W., Neidhart, D.J., Petsko, G.A. and Powers, V.M. (1992) *Current Opinion in Structural Biology*, **2**, 736–742.

11 Felfer, U., Goriup, M., Koegl, M.F., Wagner, U., Larissegger-Schnell, B., Faber, K. and Kroutil, W. (2005) *Advanced Synthesis & Catalysis*, **347**, 951–961.

12 Larissegger-Schnell, B., Glueck, S.M., Kroutil, W. and Faber, K. (2006) *Tetrahedron*, **62**, 2912–2916.

13 (a) Adam, W., Lazarus, M., Boss, B., Saha-Möller, C.R., Humpf, H.U. and Schreier, P. (1997) *The Journal of Organic Chemistry*, **62**, 7841–7843.

(b) Adam, W., Lazarus, M., Saha-Möller, C.R. and Schreier, P. (1998) *Tetrahedron: Asymmetry*, **9**, 351–355.

14 Takahashi, E., Nakamichi, K. and Furui, M. (1995) *Journal of Fermentation and Bioengineering*, **80**, 247–250.

15 Biade, A.E., Bourdillon, C., Laval, J.M., Mairesse, G. and Moiroux, J. (1992) *Journal of the American Chemical Society*, **114**, 893–897.

16 Oikawa, T., Mukoyama, S. and Soda, K. (2001) *Biotechnology and Bioengineering*, **73**, 80–82.

17 Soda, K., Oikawa, T. and Yokoigawa, K. (2001) *Journal of Molecular Catalysis B: Enzymatic*, **11**, 149–153.

18 (a) Takahashi, E., Nakamichi, K. and Furui, M. (1995) *Journal of Fermentation and Bioengineering*, **80**, 247–250.
(b) Xie, S.X., Ogawa, J. and Shimizu, S. (1999) *Bioscience Biotechnology and Biochemistry*, **63**, 1721–1729.
(c) Nakamura, K., Inoue, Y., Matsuda, T. and Ohno, A. (1995) *Tetrahedron Letters*, **36**, 6263–6266.
(d) Nakamura, K., Fujii, M. and Ida, Y. (2001) *Tetrahedron: Asymmetry*, **12**, 3147–3153.
(e) Allan, G.R. and Carnell, A.J. (2001) *The Journal of Organic Chemistry*, **66**, 6495–6497.
(f) Takemoto, M., Matsuoka, Y., Achiwa, K. and Kutney, J.P. (2000) *Tetrahedron Letters*, **41**, 499–502.
(g) Page, P.C., Carnell, A. and McKenzie, M.J. (1998) *Synlett*, **7**, 774–776.
(h) Hasegawa, J., Ogura, M., Tsuda, S., Maemoto, S., Kut-suki, H. and Ohashi, T. (1990) *Agricultural And Biological Chemistry*, **54**, 1819–1827.

19 (a) Chadha, A. and Baskar, B. (2002) *Tetrahedron, Asymmetry*, **13**, 1461–1464.
(b) Padhi, S.K. and Chadha, A. (2005) *Tetrahedron, Asymmetry*, **16**, 2790–2798.
(c) Baskar, B., Pandian, N.G., Priya, K. and Chadha, A. (2005) *Tetrahedron*, **61**, 12296–12306.
(d) Padhi, S.K., Pandian, N.G. and Chadha, A. (2004) *Journal of Molecular Catalysis B: Enzymatic*, **29**, 25–29.

20 (a) Gregory, R.J.H. (1999) *Chemical Reviews*, **99**, 3649–3682.
(b) Johnson, D.V., Zabelinskaja-Mackova, A.A. and Griengl, H. (2000) *Current Opinion in Chemical Biology*, **4**, 103–109.
(c) Effenberger, F., Förster, S. and Wajant, H. (2000) *Current Opinion in Biotechnology*, **11**, 532–539.

21 (a) Gröger, H. (2001) *Chemistry – European Journal*, **7**, 5247–5251.
(b) Brussee, J. and van der Gen, A. (2000) *Stereoselective Biocatalysis* (ed. R.N. Patel), Marcel Dekker, New York, pp. 289–320.

22 Purkarthofer, T., Skranc, W., Schuster, C. and Griengl, H. (2007) *Applied Microbiology and Biotechnology*, **76**, 309–320.

23 (a) Inagaki, M., Hiratake, J., Nishioka, T. and Oda, J. (1991) *Journal of the American Chemical Society*, **113**, 9360–9361.
(b) Inagaki, M., Hiratake, J., Nishioka, T. and Oda, J. (1992) *The Journal of Organic Chemistry*, **57**, 5643–5649.
(c) Inagaki, M., Hatanaka, A., Mimura, M., Hiratake, J., Nishioka, T. and Oda, J. (1992) *Bulletin of the Chemical Society of Japan*, **65**, 111–120.

24 (a) Veum, L. and Hanefeldt, U. (2004) *Tetrahedron, Asymmetry*, **15**, 3707–3709.
(b) Li, Y.-X., Straathof, A.J.J. and Hanefeld, U. (2002) *Tetrahedron, Asymmetry*, **13**, 739–743.

25 (a) Kanerva, L.T., Rahiala, K. and Sundholm, O. (1994) *Biocatalysis*, **10**, 169–180.
(b) Paizs, C., Tosa, M., Majdik, C., Tähtinen, P., Irimie, F.D. and Kanerva, L.T. (2003) *Tetrahedron, Asymmetry*, **14**, 619–627.
(c) Paizs, C., Tähtinen, P., Lundell, K., Poppe, L., Irimie, F.D. and Kanerva, L.T. (2003) *Tetrahedron, Asymmetry*, **14**, 1895–1904.

26 Bommarius, A.S., Kottenhahn, M., Klenk, H. and Drauz, K. (1992) *Microbial Reagents in Organic Synthesis*, NATO ASI Series (ed. S. Servi), Kluwer Academic Publisher, Dordrecht, pp. 161–170.

27 Bommarius, A.S., Drauz, K., Klenk, H. and Wandrey, C. (1992) *Annals of the New York Academy of Sciences*, **672**, 126–136.

28 Caligiuri, A., D'Arrigo, P., Gefflaut, T., Molla, G., Pollegioni, L., Rosini, E., Rossi, C. and Servi, S. (2006) *Biocatalysis and Biotransformation*, **24**, 409–413.

29 (a) Taylor, P.P., Pantaleone, D.P., Senkpeil, R.F. and Fotheringham, I.G. (1998) *Trends in Biotechnology*, **16**, 412–418.
(b) Li, T., Kootstra, A.B. and Fotheringham, I.G. (2002) *Organic Process Research & Development*, **6**, 533–538.

30 Nakajima, N., Esaki, N. and Soda, K. (1990) *Journal of the Chemical Society D: Chemical Communications*, 947–948.

31 Krix, G., Bommarius, A.S., Drauz, K., Kottenham, M., Schwarm, M. and Kula, M.R. (1997) *Journal of Biotechnology*, **53**, 29–39.

32 Peters, K.V., Gunawardana, M., Rozzell, J.D. and Novick, S.J. (2006) *Journal of the American Chemical Society*, **128**, 10923–10929.

33 Mortarino, M., Negri, A., Tedeschi, G., Simonic, T., Duga, S., Gassen, H.G. and Ronchi, S. (1996) *European Journal of Biochemistry*, **239**, 418–426.

34 Fotheringham, I.G., Taylor, P.P. and Ton, J.L. (1998) US Patent, 5,728, 555.

35 (a) Ebbers, E.J., Ariaans, G.J.A., Houbiers, J.P.M., Bruggink, A. and Zwanenburg, B. (1997) *Tetrahedron*, **53**, 9417–9476.
(b) Valcarce, R. and Smith, G.G. (1992) *Chemometrics and Intelligent Laboratory Systems*, **16**, 61–68.
(c) Smith, G.G., Khatib, A. and Reddy, G.S. (1983) *Journal of the American Chemical Society*, **105**, 293–295.
(d) Smith, G.G. and Reddy, G.S. (1989) *The Journal of Organic Chemistry*, **54**, 4529–4535.

36 (a) Pietzsch, M. and Syldatk, C. (2002) *Enzyme Catalysis in Organic Synthesis*, Vol. II (eds K. Drauz and H. Waldmann), Wiley-VCH Verlag GmbH, Weinheim, pp. 761–784.
(b) Ogawa, J., Soong, C.-L., Kishino, S., Li, Q.-S., Horinouchi, N. and Shimizu, S. (2004) *Tetrahedron, Asymmetry*, **70**, 574–578.

37 Olivieri, R., Fascetti, F., Angelini, L. and Degen, L. (1981) *Biotechnology and Bioengineering*, **23**, 2173–2183.

38 (a) Gross, C., Syldatk, C. and Wagner, F. (1987) *Biotechnology Techniques*, **1**, 85–90.
(b) Nishida, Y., Nakamichi, K., Nabe, K. and Tosa, T. (1987) *Enzyme and Microbial Technology*, **9**, 721–725.
(c) Cotoras, D. and Wagner, F. (1984) *III European Congress on Biotechnology, Poster Abstracts*, **1**, 351–356.

39 May, O., Verseck, S., Bommarius A. and Drauz, K. (2002) *Organic Process Research & Development*, **6**, 452–457.

40 Gu, R.L., Lee, I.S. and Sih, C.J. (1992) *Tetrahedron Letters*, **33**, 1953–1956.

41 (a) Chen, S., Fujimoto, Y., Girdaukas, G. and Sih, C.J. (1982) *Journal of the American Chemical Society*, **104**, 7294–7299.
(b) Crich, J.Z., Brieva, R., Marquart, P., Gu, R.L., Flemming, S. and Sih, C.J. (1983) *The Journal of Organic Chemistry*, **58**, 3252–3258.
(c) Bevinakatti, H.S., Newadkar, R.V. and Banerji, A.A. (1990) *Journal of The Chemical Society D: Chemical Communications*, 1091–1092.
(d) Bevinakatti, H.S., Banerji, A.A., Newadkar, R.V. and Mokashi, A.A. (1992) *Tetrahedron, Asymmetry*, **3**, 1505–1508.

42 (a) Turner, N.J., Winterman, J.R., McCague, R., Parrat, J.S. and Taylor, S.J.C. (1995) *Tetrahedron Letters*, **36**, 1113–1146.
(b) Brown, S.A., Parker, M.-C. and Turner, N.J. (2000) *Tetrahedron, Asymmetry*, **11**, 1687–1690.

43 Arosio, D., Caligiuri, A., D'Arrigo, P., Pedrocchi-Fantoni, G., Rossi, C., Saraceno, C., Servi, S. and Tessaro, D. (2007) *Advanced Synthesis & Catalysis*, **349**, 1345–1348.

44 (a) Clark, J.C., Phillips, G.H. and Steer, M.R. (1976) *Journal of The Chemical Society – Perkin Transactions 1*, 475–481.
(b) Honnoraty, A.M., Mion, L., Collet, H., Teissedre, R. and Commeyras, A. (1995) *Bulletin de la Societe Chimique de France*, **132**, 709–720.

45 Wegman, M.A., Rops, M.A.P.J., Hacking, J., Pereira, P., van Rantwijk, F. and Sheldon, R.A. (1999) *Tetrahedron, Asymmetry*, **10**, 1739–1750.

46 Liese, A., Seelbach, K. and Wandrey, C. (2000) *Industrial Biotransformations*, Wiley-VCH Verlag GmbH, Weinheim.

47 May, O., Verseck, S., Bommarius A. and Drauz, K. (2002) *Organic Process Research & Development*, **6**, 452–457.

48 (a) Soda, K., Oikawa, T. and Yokoigawa, K. (2001) *Journal of Molecular Catalysis B: Enzymatic*, **11**, 149–153.

(b) Huh, J.W., Yokoigawa, K., Esaki, N. and Soda, K. (1992) *Journal of Fermentation And Bioengineering*, **74**, 189–190.

49 Huh, J.W., Yokoigawa, K., Esaki, N. and Soda, K. (1992) *Bioscience Biotechnology and Biochemistry*, **56**, 2081–2082.

50 (a) Fotheringham, I., Archer, I., Carr, R., Speight, R. and Turner, N.J. (2006) *Biochemical Society Transactions*, **34**, 287–290.
(b) Turner, N.J. (2004) *Current Opinion in Chemical Biology*, **8**, 114–119.

51 (a) Alexandre, F.R., Pantaleone, D.P., Taylor, P.P., Fotheringham, I.G., Ager, D.J. and Turner, N.J. (2002) *Tetrahedron Letters*, **43**, 707–710.
(b) Beard, T.M. and Turner, N.J. (2002) *Journal of the Chemical Society D: Chemical Communications*, 246–247.

52 Enright, A., Alexandre, F.R., Roff, G., Fotheringham, I.G., Dawson, M.J. and Turner, N.J. (2003) *Journal of the Chemical Society D: Chemical Communications*, 2636–2637.

53 Kato, D.I., Miyamoto, K. and Ohta, H. (2005) *Biocatalysis and Biotransformation*, **23**, 375–379.

54 (a) Curti, B., Ronchi, S. and Pilone, M.S. (1992) *Chemistry and Biochemistry of Flavoenzymes*, Vol. 3 (ed. F. Müller), CRC Press, Boca Raton, pp. 69–94.
(b) Du, X-Y. and Clemetson, K.J. (2002) *Toxicon*, **40**, 659–665.
(c) Pollegioni, L., Piubelli, L., Sacchi, S., Pilone, M.S. and Molla, G. (2007) *Cellular and Molecular Life Sciences*, **64**, 1373–1394.

55 Tishkov, V.I. and Khoronenkova, S.V. (2005) *Biochemistry (Moscow)*, **70**, 40–54.

56 (a) Umhau, S., Pollegioni, L., Molla, G., Diederich, K., Welte, W., Pilone, M.S. and Ghisla, S. (2000) *Proceedings of the National Academy of Sciences of the United States of America*, **97**, 12463–12468.
(b) Pollegioni, L., Diederichs, K., Molla, G., Umhau, S., Welte, W., Ghisla, S. and Pilone, M.S. (2002) *Journal of Molecular Biology*, **324**, 535–546.

57 Pollegioni, L., Sacchi, S., Caldinelli, L., Boselli, A., Pilone, M.S. and Molla, G. (2007) *Current Protein & Peptide Science*, **8**, 600–618.

58 Pawelek, P.D., Cheah, J., Coulombe, R., Macheroux, P., Ghisla, S. and Vrielink, A. (2000) *EMBO Journal*, **19**, 4204–4215.

59 Pantaleone, D.P., Geller, A.M. and Taylor, P.P. (2001) *Journal of Molecular Catalysis B: Enzymatic*, **11**, 795–803.

60 (a) Shaw, J.P., Petsko, G.A. and Ringe, D. (1997) *Biochemistry*, **36**, 1329–1342.
(b) Morollo, A.A., Petsko, G.A. and Ringe, D. (1999) *Biochemistry*, **38**, 3293–3301.
(c) Yoshimura, T. and Esaki, N.J. (2003) *Journal of Bioscience and Bioengineering*, **2**, 103–109.
(d) Yohda, M., Endo, I., Abe, Y., Ohta, T., Iida, T., Maruyama, T. and Kagawa, Y. (1996) *The Journal of Biological Chemistry*, **271**, 22017–22021.
(e) Matsumoto, M., Long, H., Homma, Z., Imai, K., Iida, T., Maruyama, T., Aikawa, Y., Endo, I. and Yohda, M. (1999) *Journal of Bacteriology*, **181**, 6560–6563.

61 Yokoigawa, K., Okubo, Y., Kawai, H., Esaki, N. and Soda, K. (2001) *Journal of Molecular Catalysis B-Enzymatic*, **12**, 27–35.

62 (a) Schnell, B., Faber, K. and Kroutil, W. (2003) *Advanced Synthesis & Catalysis*, **345**, 653–666.
(b) Fukumura, T. (1977) *Agricultural and Biological Chemistry*, **41**, 1509–1510.
(c) Ahmed, S.A., Esaki, N., Tanaka, H. and Soda, K. (1983) *Agricultural and Biological Chemistry*, **47**, 1149–1150.
(d) Ahmed, S.A, Esaki, N., Tanaka, H. and Soda, K. (1984) *FEBS Letters*, **174**, 76–79.
(e) Yamada, M. and Kurahashi, K. (1969) *Journal of Biochemistry*, **66**, 529–540.
(f) Takashashi, H., Sato, E. and Kurahashi, K. (1971) *Journal of Biochemistry*, **69**, 973–976.
(g) Yorifuji, T. and Ogata, K. (1971) *The Journal of Biological Chemistry*, **246**, 5085–5092.
(h) Palmer, D.R.J., Garrett, J.B., Sharma, V., Meganathan, R., Babbitt, P.C. and Gerlt, J.A. (1999) *Biochemistry*, **38**, 4252–4258.
(i) Tokuyama, S. (2001) *Journal of Molecular Catalysis B: Enzymatic*, **12**, 3–14.

63 Glavas, S. and Tanner, M.E. (1999) *Biochemistry*, **38**, 4106–4113.

64 (a) Felfer, U., Goriup, M., Koegl, M.F., Wagner, U., Larissegger-Schnell, B., Faber,

K. and Kroutil, W. (2005) *Advanced Synthesis & Catalysis*, **347**, 951–961.
(b) Tsou, A.J., Ransom, S.C. and Gerlt, J. A. (1990) *Biochemistry*, **29**, 9856–9862.
(c) Fewson, C.A. (1988) *FEMS Microbiology Reviews*, **54**, 85–110.
(d) Kenyon, G.L. and Hegeman, G.D. (1979) *Advances in Enzymology and Related Areas of Molecular Biology*, **50**, 325–360.
(e) Neidhart, D.J., Howell, P.L., Petsko, G.A., Powers, V.M., Li, R., Kenyon, G.L. and Gerlt, J.A. (1991) *Biochemistry*, **30**, 9264–9273.
(f) Kenyon, G.L., Gerlt, J.A., Petsko, G.A. and Kozarich, J.W. (1995) *Accounts of Chemical Research*, **28**, 178–186.
(g) Gerlt, J.A., Kenyon, G.L., Kozarich, J.W., Neidhart, D.J., Petsko, G.A. and Powers, W.M. (1992) *Current Opinion in Structural Biology*, **2**, 736–742.
(h) Stecher, H., Felfer, U. and Faber, K. (1997) *Journal of Biotechnology*, **56**, 33–40.
(i) Tsou, A.Y., Ransom, S.C., Gerlt, J.A., Powers, V.M. and Kenyon, G.L. (1989) *Biochemistry*, **28**, 969–975.

65 (a) Li, T., Kootstra, A.B. and Fotheringham, I.G. (2002) *Organic Process Research & Development*, **6**, 533–538.
(b) Taylor, P.P., Pantaleone, D.P., Senkpeil, R.F. and Fotheringham, I.G. (1998) *Trends in Biotechnology*, **16**, 412–418.

66 (a) Schulz, A., Taggeselle, P., Tripier, D. and Bartsch, K. (1990) *Applied and Environmental Microbiology*, **56**, 1–6.
(b) Dichmann, K., Bartsch, R., Schmitt, P., Uhlmann, E. and Schulz, A. (1990) *Applied and Environmental Microbiology*, **56**, 7–12.
(c) Wiyakrutta, S. and Meevootisom, V. (1997) *Journal of Biotechnology*, **55**, 193–203.

67 Sugio, S., Petsko, G.A., Manning, J.M., Soda, K. and Ringe, D. (1995) *Biochemistry*, **34**, 9661–9669.

14
Nitrilases from Filamentous Fungi

Ludmila Martínková, Vojtěch Vejvoda, Ondřej Kaplan, Vladimír Křen, Karel Bezouška and Maria Cantarella

14.1
Introduction

Nitrilases constitute branch 1 of the nitrilase superfamily, which contains enzymes acting on various non-peptide carbon–nitrogen bonds [1, 2]. The occurrence of nitrilases in microorganisms appears to be rather rare, as indicated by the search for nitrilase genes in sequenced microbial genomes [3]. Nevertheless, culture enrichment techniques have allowed microorganisms with nitrilase activities to be isolated. In addition, uncultured environmental samples were explored by the metagenomic approach, providing over 200 unique nitrilase sequences, out of which 137 were examined for substrate specificity and enantioselectivity [4].

The well-characterized nitrilases listed in recent reviews [5–9] have mostly been of bacterial origin. Few eukaryotic nitrilases have been purified and characterized to date. Among these, the enzymes best characterized at the protein and gene level belong to plant nitrilases, that is primarily AtNIT1-4 enzymes from *Arabidopsis thaliana* (for reviews see [2, 6]). Knowledge of fungal nitrilases was very limited until recently. Databases contain about 40 sequences of fungal nitrilases or cyanide hydratases, which are largely putative proteins disclosed by genome sequencing. Two nitrilases from the genus *Fusarium* were purified and characterized but their amino acid sequences have not been reported [10, 11]. Very recently, our work improved knowledge of fungal nitrilases, in particular those from *Aspergillus niger* and *Fusarium solani*.

On the basis of sequence data and biochemical characterization, this chapter will compare the structure and enzymatic properties of nitrilases from filamentous fungi with those of prokaryotic and plant nitrilases as well as other related enzymes like cyanide hydratases and cyanide dihydratases. The biotechnological potential of fungal nitrilases will be evaluated.

Modern Biocatalysis: Stereoselective and Environmentally Friendly Reactions.
Edited by Wolf-Dieter Fessner and Thorleif Anthonsen

ISBN: 978-3-527-32071-4

14.2
Distribution and Evolutionary Relationship of Fungal Nitrilases

14.2.1
Molecular Genetic Analysis

The enzymes from *A. niger* K10 [12] and *F. solani* O1 [13] were the first biochemically characterized fungal nitrilases, the partial amino acid sequences of which have been reported [14, 15]. Within the time elapsed since the purification and characterization of the nitrilase from *A. niger* K10 [14], the corresponding gene has been sequenced and translated to the amino acid sequence (see Figure 14.1). The search for related proteins using the protein blast program (http://www.ncbi.nlm.nih.gov/BLAST/) revealed high similarity (≥ 85% identities and ≥ 92% positives) of the purified nitrilase from *A. niger* to putative cyanide hydratases/nitrilases encoded in *A. niger, Neosartorya fischeri, Aspergillus fumigatus, Aspergillus terreus, Aspergillus clavatus* and *Aspergillus nidulans*. Out of these proteins, the putative nitrilase from *A. niger* CBS 513.88 (XP_001389844) showed the highest similarity with the purified protein (98.8%).

N-Terminal sequencing and mass spectrometry analysis of the purified nitrilase from *F. solani* O1 revealed four peptide fragments that were identical to those of a putative nitrilase from *Gibberella moniliformis* (anamorph *Fusarium verticillioides*) [15]. A highly homologous gene was also found in *Gibberella zeae* (85% identities and 91% positives).

The multiple sequence alignment of fungal nitrilases and related enzymes (prokaryotic nitrilases and cyanide (di)hydratases) and the level of homology between these proteins are shown in Figures 14.1 and 14.2, respectively. The fungal nitrilases display on average 51% similarity with each other, but the median is only 35%. Conserved sequences were found in all proteins, with the *N*- or *C*-terminal sequences showing low similarities throughout the set of enzymes examined, as expected.

It is notable that the new nitrilases from *F. solani* O1 (presumably highly homologous with the nitrilases from *G. moniliformis* and *G. zeae*) and *A. niger* K10 are relatively distant from each other in evolution (with only about 36% homology). On the other hand, a higher homology level of 53–69% was observed between the nitrilase from strain K10 and the cyanide hydratases from *Gloeocercospora sorghi* [16, 17], *F. solani* [18], *Fusarium lateritium* [19], and *Leptosphaeria maculans* [20]. Cyanide hydratases show a high specifity for HCN, which they transform to formamide. Nevertheless, minor activities for organic nitriles were reported for the enzymes from *F. lateritium* (0.02–0.4% of the HCN activity) and *Fusarium oxysporum* (0.05% of the HCN activity). Cyanide hydratases have been reported in filamentous fungi exclusively and form a much more closely related group of enzymes than nitrilases (see [6] for a review).

Cyanide dihydratases are also HCN specific, but differ in the reaction mechanism, which leads to formate. Their phylogenetic distribution is also different: these enzymes occur in bacteria, for instance, *Bacillus pumilus* [21] or *Pseudomonas*

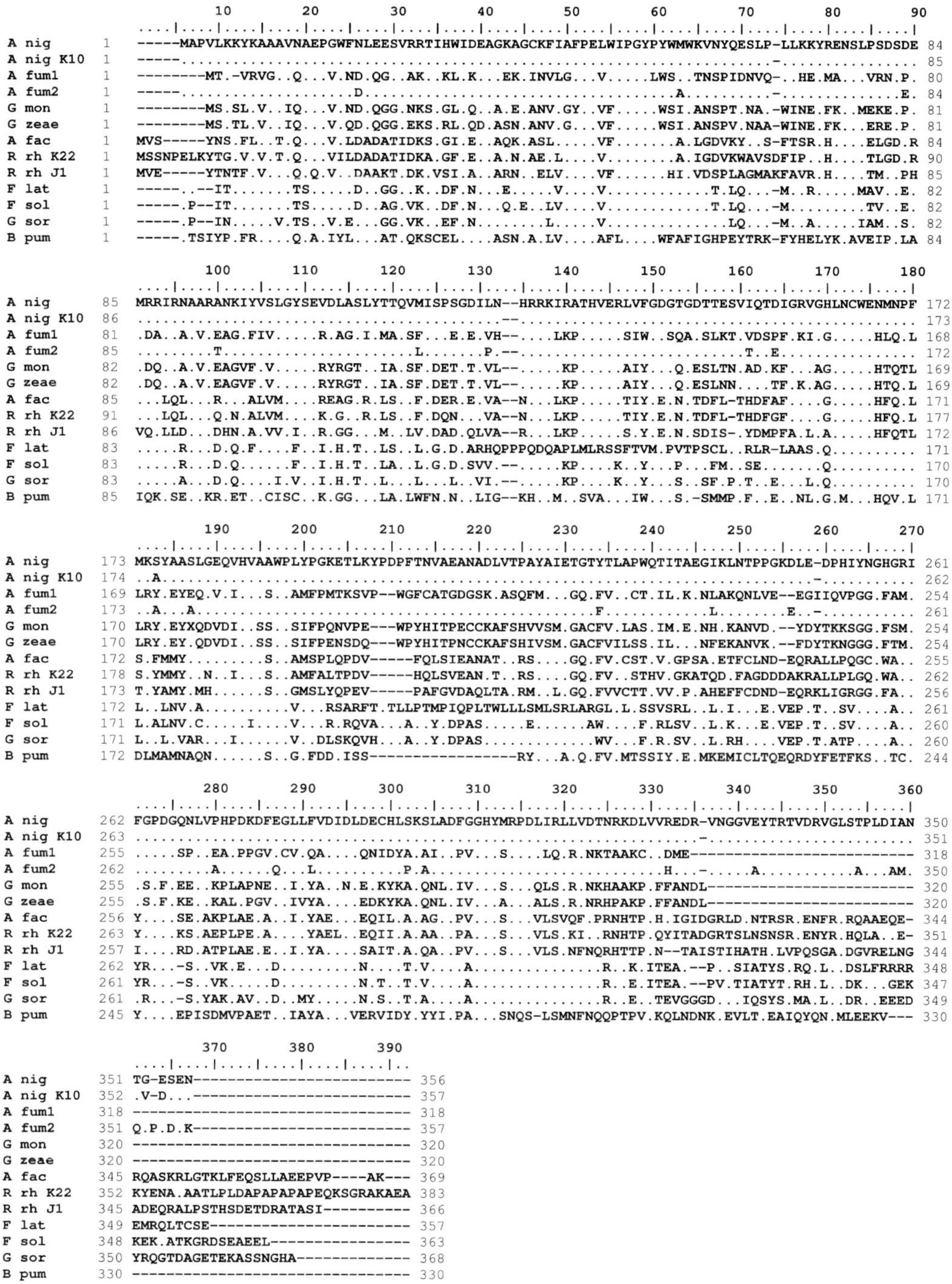

Figure 14.1 Alignment of amino acid sequences of fungal nitrilases and related proteins (using ClustalW software). A nig, *A. niger* CBS 513.88 (putative nitrilase, XP_001389844); A nig K10, *A. niger* K 10 (aromatic nitrilase, ABX75546); A fum1, *A. fumigatus* Af293 (putative cyanide hydrolase/nitrilase, XP_747028); A fum2, *A. fumigatus* Af293 (putative nitrilase, XP_756085); G mon, *G. moniliformis* (putative nitrilase, ABF83489); G zeae, *G. zeae* PH-1 (hypothetical protein, XP_386656); A fac, *Acidovorax facilis* 72W (nitrilase, ABD98457); R rh K22, *R. rhodochrous* K22 (aliphatic nitrilase, BAA02127); R rh J1, *R. rhodochrous* J1 (aliphatic nitrilase, Q03217); F lat, *F. lateritium* (cyanide hydratase, AAA33336); F sol, *F. solani* (cyanide hydratase, CAC69666); G sor – *G. sorghi* (cyanide hydratase, P32964); B pum, *B. pumilus* C1 (cyanide dihydratase, AAN77003).

	A nig	A nig K10	A fum1	A fum2	G mon	G zeae	R rh K22	R rh J1	A fac 72W	F lat	F sol	G sorghi	B pum
A niger CBS 513.88		99,1%	34,7%	92,4%	30,7%	31,0%	33,9%	34,1%	31,2%	51,1%	62,1%	60,3%	29,3%
A niger K10	99,1%		34,7%	92,1%	30,7%	31,0%	33,9%	34,1%	31,2%	50,8%	62,1%	60,1%	29,3%
A fumigatus Af 293 (1)	34,7%	34,7%		35,1%	55,1%	55,7%	36,1%	35,4%	36,8%	28,0%	32,6%	33,5%	29,3%
A fumigatus Af 293 (2)	92,4%	92,1%	35,1%		30,9%	30,6%	33,9%	34,1%	31,4%	51,6%	63,2%	61,7%	29,8%
G moniliformis	30,7%	30,7%	55,1%	30,9%		85,0%	34,9%	31,8%	35,5%	25,9%	29,4%	30,2%	29,7%
G zeae PH-1	31,0%	31,0%	55,7%	30,6%	85,0%		34,4%	31,3%	35,0%	25,1%	29,4%	29,1%	30,9%
R rhodochrous K22	33,9%	33,9%	36,1%	33,9%	34,9%	34,4%		69,1%	52,0%	27,8%	34,5%	35,6%	30,2%
R rhodochrous J1	34,1%	34,1%	35,4%	34,1%	31,8%	31,3%	69,1%		47,6%	27,5%	34,3%	35,1%	28,6%
A facilis 72W	31,2%	31,2%	36,8%	31,4%	35,5%	35,0%	52,0%	47,6%		28,2%	33,4%	33,9%	31,3%
F lateritium	51,1%	50,8%	28,0%	51,6%	25,9%	25,1%	27,8%	27,5%	28,2%		67,3%	58,1%	22,5%
F solani	62,1%	62,1%	32,6%	63,2%	29,4%	29,4%	34,5%	34,3%	33,4%	67,3%		73,9%	26,0%
G sorghi	60,3%	60,1%	33,5%	61,7%	30,2%	29,1%	35,6%	35,1%	33,9%	58,1%	73,9%		29,3%
B pumilus C1	29,3%	29,3%	29,3%	29,8%	29,7%	30,9%	30,2%	28,6%	31,3%	22,5%	26,0%	29,3%	

Figure 14.2 Homology analysis of amino acid sequences of fungal nitrilases and related proteins (see Figure 14.1) using BioEdit software.

stutzeri [22], from which they have been purified. They show a lower level of homology with the nitrilase from *A. niger* (for instance, about 29% for the enzymes from *B. pumilus*) than cyanide hydratases.

The similarity of fungal nitrilases to plant nitrilases or to the Nit domain of the NitFHit protein from *Caenorhabditis elegans*, one of the crystallized members of the nitrilase superfamily, is below 30%, but the sequences flanking the catalytic triad (Glu–Lys–Cys) residues show higher similarities (see Figure 14.3).

Among prokaryotic enzymes, the highest sequence similarities with the nitrilases from *A. niger* and *G. moniliformis* were found for those from *Rhodococcus rhodochrous* J1 and Acidovorax facilis 72W (34.1 and 35.5%). The above results confirm structural differences in fungal nitrilases from those in other organisms as well as among each other and encourage scientists to search for new enzymes with beneficial biocatalytic properties in filamentous fungi.

14.2.2
Selection and Screening of Nitrilase Activity

Most of the well-characterized nitrilases, which are largely of bacterial origin, have been obtained by selection methods allowing only the positive strains to grow on a medium with a nitrile as the sole nitrogen source. The same approach provided nitrile-hydrolyzing fungal isolates belonging to the genus *Fusarium*, among them *F. solani* O1, from environmental samples. 3-Cyanopyridine was used as the sole nitrogen source, while rose Bengal inhibited the growth of contaminating bacteria [23].

Utilization of 3-cyanopyridine for mycelial growth was also the criterion used for identification of nitrilase-producing strains among collection strains of filamentous fungi. Out of approximately 100 strains, 13 strains belonging to the genera *Aspergillus, Fusarium, Penicillium,* and *Tala*romyces [24] grew on this

A nig	GCKFIAFPELWIPGYP	GDILNHRRKIRAT	GRVGHLNCWENMNPFMK
A nig K10	....	....	....
A fum1	.INVLG...V.....L	.E.VH....LKP.	.KI.G.....HLQ.LLR
A fum2	....	...P.........	....
G mon	.ANV.GY..VF.....	.T.VL.....KP.	...AG.....HTQTLLR
G zeae	.ANV.G...VF.....	.T.VL.....KP.	.K.AG.....HTQ.LLR
A fac	.ASL.....VF.....	.E.VAN...LKP.	....G.....HFQ.LS.
P fluo	.ASLV....A.L....	.RVVAT...LKP.	..L.A.C.A.HIQ.LS.
R rh K22	.AE.L....V......	...VAN...LKP.	....G.....HFQ.LS.
R rh J1	..ELV....VF.....	.QLVAR...LKP.	A.L.A.....HFQTLT.
K oza	.AQLV...........	.ITKIR...LKP.	....A...A..LQSLN.
F lat	....V....V......	.ARHQPPPQDQ.P	AAS.Q..........L.
F sol	E..LV....V......	.SVV......KP.	....Q..........L.
G sor	...L.....V......	..VI......KP.	..L.Q..........L.
B pum	.A.LV....AFL....	..LIGKH..M..S	.NL.G.M...HQV.LDL
S cer[a]	DT.LVVL..CFNSP.S	.KLIDKH..VHLF	.KF.VGI.YDMRF.ELA
At nit1	.AELVL...GF.G...	.QF.GKH..LMP.	.KL.AAI....RM.LYR
At nit2	.SELVV...AF.G...	.QF.GKH..LMP.	.KL.AAI....RM.LYR
At nit3	.A.LVL...AF.G...	.QF.GKH..VMP.	.KI.AAI....RM.LYR
At nit4	.SQLVV...AF.G...	.LF.GKH..LMP.	.KI.AAI....RM.SLR
C ele[a]	.AELVL...AF.G...	.-Y.GKH..LLP.	.KI.SAI....YM.LYR
A sp[a]	.AN..V....ALTTFF	.K.VGKY...HLP	AKM.MFI.NDRRW.EAW
H pyl[a]	.VEL.I...YSTQ.LN	.K.ILKY..LFPW	SKLAVCI.HDG.I.ELA
P hor[a]	.A.LVVL...FDT..N	.-YIGKY...HLF	AK..VMI.FDWFF.ESA

Figure 14.3 Alignment of sequences flanking the catalytic triad Cys–Glu–Lys (using ClustalW software). A nig, A nig K10, A fum1, A fum2, G mon, G zeae, A fac, R rh K22, R rh J1, F lat, F sol, G sor, and B pum – see Figure 14.1. P fluo, *P. fluorescens* (nitrilase, Q5EG61); K oza – *Klebsiella pneumoniae* subsp. *ozaenae* (nitrilase, P10045); S cer, *Saccharomyces cerevisiae* (probable hydrolase NIT3, P49954), At nit1, *Arabidopsis thaliana* (nitrilase 1, P32961); At nit2, *A. thaliana* (nitrilase 2, NP_190016); At nit3, *A. thaliana* (nitrilase 3, NP_190018); At nit4, *A. thaliana* (nitrilase 4, NP_197622); C ele, *C. elegans* (hypothetical protein/CN hydrolase, CAA84681); A sp – *Agrobacterium* sp. KNK712 (*N*-carbamoyl-D-amino acid hydrolase (D-*N*-alpha-carbamylase), P60327); H pyl, *Helicobacter pylori* J99 (Formamidase, Q9ZJY8); P hor, *Pyrococcus horikoshii* OT3 (hypothetical protein PH0642, NP_142600).
[a] Proteins, the crystal structures of which were determined.

medium, but considerable levels of nitrilase activity were only found in one strain of *A. niger,* two strains of *F. oxysporum,* and one strain of *Penicillium multicolor* [25].

The expression of nitrilase may be enhanced and, thus, screening for its activity facilitated by the choice of an appropriate nitrilase inducer. For instance, in five strains belonging to three fungal genera 2-cyanopyridine enhanced the nitrilase activity by two to three orders of magnitude. Therefore, this nitrilase inducer appears to be very efficient and of wide use in filamentous fungi [25].

Wide distribution of nitrile-hydrolyzing enzymes (presumably at least partially nitrilases) among different genera of filamentous fungi was suggested by another study [26], the primary aim of which was to examine the distribution of aldoxime dehydratases in microorganisms including 102 fungal strains belonging to 37 genera. Owing to simultaneous expression of nitrile-hydrolyzing enzymes along with aldoxime dehydratases, hydrolytic activities towards 2-phenylacetonitrile and

3-cyanopyridine were detected in 31 and eight strains, respectively, albeit largely at low levels. Obviously, the results of the screening are strongly dependent on the substrate. Out of the total positive strains, eight and four strains belonged to the genera *Fusarium* and *Aspergillus*, respectively, but nitrile-hydrolyzing enzymes were also found in the genera *Schizophyllum, Rhizopus, Talaromyces, Mortierella, Flammulina, Pycnoporus, Keratinomyces, Mucor, Coprinus, Cunninghamella*, and *Phycomyces*, in which no nitrilase or nitrile hydratase genes have been sequenced.

14.3 Structural Properties

In general, nitrilases contain a single type of subunit, the molecular mass of which has been assessed at 32–45 kDa in a number of purified enzymes (for a review see [5, 6]). The size of subunits in fungal nitrilases from *A. niger* K10, *F. solani* O1, and *F. oxysporum* f. sp. *melonis* is in this range (see Table 14.1). In this context, the 76-kDa size reported for the nitrilase subunit in *F. solani* IMI196840 [30] is exceptional [10].

Typically, native nitrilases are oligomers consisting of two to about 16 identical subunits, while active monomers are rare (for a review see [5, 6]). Fungal enzymes belong to the largest molecules among nitrilases (see Table 14.1), the molecular weight of those from *F. oxysporum* f. sp. *melonis, F. solani* IMI196840, *F. solani* O1, and *A. niger* K10 having been assessed at 550, 620, 580, and >650 kDa, respectively. Smaller oligomers can retain activity, as demonstrated for the enzyme from *F. oxysporum* f. sp. *melonis*, which was resolved by non-denaturing polyacrylamide gel electrophoresis into active bands having molecular weights of 170–880 kDa with increments of 70 kDa, that is into protein species made of four to 26 subunits [11].

Co-purification of nitrilases with chaperonin proteins, which was observed for enzymes from *Pseudomonas fluorescens* DSM 7155 [31] and *B. pallidus* Dac521 [32], was recently also reported for the nitrilase from *A. niger* K10 [14]. According to *N*-terminal amino acid sequencing, the proteins co-purified with nitrilases in *P. fluorescens* and *B. pallidus* were found to exhibit a high homology with Cpn60 and GroEL proteins, respectively. The protein co-purified with the nitrilase from *A. niger* was probably an eukaryotic equivalent of these proteins, its *N*-terminal amino acid sequence showing high similarity with that of Hsp60 chaperonine polypeptide. The chaperonins may play role in assemblying subunits into a multimeric structure or in protein stabilization.

Several nitrilase and cyanide dihydratase holoenzymes share the ability to assemble into supramolecular species, which were characterized as helical filaments (see Figure 14.4). This unique property was revealed by electron microscopy selected as a method of choice for the study of high-molecular weight holoenzymes. The nitrilase from *F. solani* formed helical rods up to 500 nm long or aggregates (see Figure 14.4a and b). Long rods were also observed in samples of the *A. niger* nitrilase, which, however, differed in structure and length (up to 250 nm) from those of the *F. solani* enzyme [15]. In addition, the enzyme from

Table 14.1 Subunit and holoenzyme molecular masses of nitrilases from filamentous fungi: comparison with selected prokaryotic nitrilases, cyanide hydratases and cyanide dihydratase.

Enzyme	Organism	Molecular mass (kDa)			Reference
		Subunit experimental	Predicted[c]	Holoenzyme	
Nitrilase	*A. niger* K10	38.5	40.2	> 650	[14]
	A. niger CBS 513.88[a]	–	40.0	–	–
	F. oxysporum f. sp. *melonis*	37	–	550 (170–880[d])	[11]
	F. solani IMI196840	76	–	620	[10]
	F. solani O1	40	–	580	[15]
	G. moniliformis[a,b]	nd	36.0	–	–
	R. rhodochrous J1	40	40.2	480[e]	[27]
	R. rhodochrous K22	41	42.3	650	[28]
Cyanide hydratase	*G. sorghi*	45	40.9	> 300	[16]
	F. solani	45	40.8	> 300	[18]
Cyanide dihydratase	*B. pumilus* C1	37	37.3	417	[21]

Note: no entry indicates no data available.
a Putative nitrilases.
b Presumably highly homologous to the enzyme from *F. solani* O1 [15].
c Calculated by BioEdit sowftware.
d Oligomers of different size produced by non-denaturing electrophoresis [11].
e The 80-kDa dimer is inactive [29].

A. niger seemed to form no aggregates (see Figure 14.4c and d). The cyanide dihydratase from *B. pumilus* formed left-handed helical rods at pH 5.4 and short spirals at pH 8.0 [33]. The long helices were also found in samples of the bacterial nitrilase from *R. rhodochrous* J1 [29]. In its native form (480 kDa) this enzyme formed 'c'-shaped particles, but after 1 month of storage at 4 °C the cleavage of approximately 39 amino acids from the C-terminus led to the formation of active clusters with a molecular weight of >1500 kDa, which were observed under electron microscopy as left-handed helical rods of different length. The same forms were detected in mutants with C-terminus deletions. The C-terminal amino acids sequence seems to make the formation of helical filaments in this nitrilase impossible.

The lack of crystallography data prevented gaining a deep insight into the structure and function of nitrilases, but a considerable homology level between the active sites of these enzymes and proteins, the structures of which have been solved (see Figure 14.3), can be helpful in nitrilase modeling, as demonstrated for the enzyme from *R. rhodochrous* J1 [29]. The members of the nitrilase superfamily are typically α/β proteins with a four-layer $\alpha\beta\beta\alpha$ sandwich architecture. These units associate in a hexamer of formamidase from *Helicobacter pylori* [34], tetramers of NitFhit protein from *C. elegans* [35], and *N*-carbamoyl-D-amino acid amidohydrolase from *Agrobacterium* sp. strain KNK712 [36] or dimers of hypothetical protein from *Pyrococcus horikoshii* OT3 [37] and CN hydrolase from *Saccharomyces cerevisiae* [38].

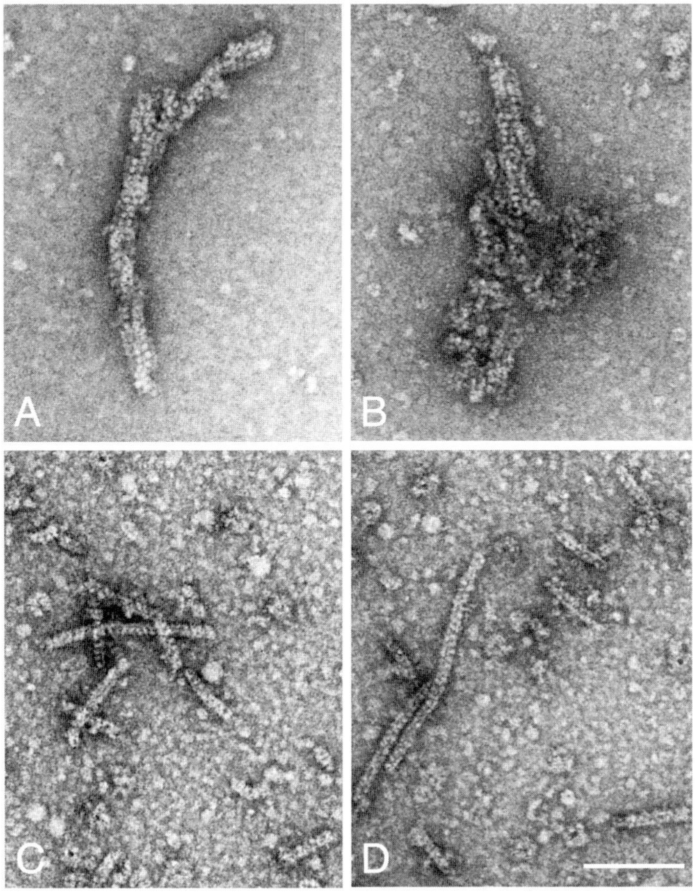

Figure 14.4 Electron micrographs of negatively stained nitrilases from (a and b) *F. solani* O1 and (c and d) *A. niger* K10. The scale bar represents 100 nm. From [15] with permission.

14.4
Catalytic Properties

14.4.1
Reaction Mechanism

The hypothetical nitrilase catalytic mechanism based on previous schemes (for a review see [6]), which has been recently refined to bring more understanding to the formation of two possible end-products [39], is discussed in Chapter 16. According to a generally accepted hypothesis, the catalytic mechanism involves a nucleophilic attack on the cyano carbon by the sulfhydryl group of the conserved

Table 14.2 Ratio of amide : acid produced by nitrilases from different organisms.

Organism	Amide : Acid ratio (substrate)[a]	Reference
A. thaliana[b]	19 (3-nitroacrylonitrile)	[41]
	13.3 (3-cyanoacrylonitrile)	–
	≈4.5–5.7 (α-fluoroarylacetonitriles)	–
A. thaliana[c]	≈1.5 (β-cyano-L-alanine)	[42]
A. niger K10	5.2 (2-cyanopyridine)	[14]
	1.1 (4-chlorobenzonitrile)	–
	0.5 (4-cyanopyridine)	–
	0.4 (1,4-dicyanobenzene)	–
F. oxysporum f. sp. *melonis*	0.04–0.06 (benzonitrile, propionitrile, acrylonitrile)	[11]
F. solani O1	0.03 (3-cyanopyridine, 4-cyanopyridine)	[15]
	0.02 (benzonitrile)	–
Pseudomonas sp.	≈0.1 (*N*-methyl-3-cyano-4-methoxy-2-pyridone)	[43]
P. fluorescens EBC 191	10 (*R*,*S*)-2-chloro-2-phenylacetonitrile	[39, 44]
	1.7 (*S*)-*O*-acetylmandelonitrile[d]	–
	1.25 (*S*)-mandelonitrile[d]	–
	0.5 (*R*)-*O*-acetylmandelonitrile[d]	–
	0.12 (*R*)-mandelonitrile[d]	–
Rhodococcus sp. ATCC39484	0.02 2-phenylacetonitrile	[45]

a Only substrates from which the maximum amide amounts were produced are shown for each enzyme.
b AtNIT1 enzyme.
c AtNIT4 enzyme.
d At pH 6 (amide : acid ratio pH-dependent).

cysteine, leading to a covalent enzyme–thioimidate complex that is hydrolyzed to give a tetrahedral intermediate. Two pathways of intermediate decomposition involve either ammonia or enzyme as the leaving group, thus releasing acyl-enzyme or amide, respectively (see Figure 16.10). Data from experiments with bacterial and plant nitrilases have indicated the importance of the electronic or steric effect of the substrate, amides being preferably formed from nitriles with electron-withdrawing or bulky substituents [40–45]. Reactions catalyzed by fungal nitrilases support this view (see Table 14.2). The nitrilase from *A. niger* K10 belongs to enzymes that are very prone to amide formation, as it forms high amide/acid molar ratios from nitriles with an electron-withdrawing heteroatom or substituent (up to about 5 : 1 for 2-cyanopyridine) [14]. The nitrilase from *F. solani* O1 forms only low amounts of amide by-products (≤3% of the total product) [15]. This is in agreement with amide production by the nitrilase from *F. oxysporum*, which produced 4–6% of amides from some nitriles [11]. The chemoselectivity of the third purified nitrilase from the genus *Fusarium* is unknown as the enzyme assay was based on ammonia determination [10].

14.4.2
Substrate Specificity

All known fungal nitrilases exhibit high relative activities towards benzonitrile and its *m*- and *p*-substituted derivatives. Therefore, according to the nitrilase classification [46] they belong to aromatic nitrilases. However, almost all fungal nitrilases also hydrolyze aliphatic nitriles, albeit at lower relative rates.

The substrate specificities of all four biochemically characterized fungal nitrilases and two enzymes from rhodococci (an aromatic and an aliphatic nitrilase) are compared in Table 14.3. Only a few nitriles were examined as substrates of all these enzymes. Moreover, the comparison of substrate specificity is difficult in some cases as different activity assays have been used for different enzymes. For instance, the determination of activity by ammonia measurement did not reflect potential amide formation. Nevertheless, common patterns can be found in substrate preferences of aromatic nitrilases.

In general, substitution of the benzene ring decreases the reactivity of the compounds with a few exceptions. Halobenzonitriles are generally among the best substrates due to a strong electron-withdrawing effect of the substituents, which is favorable for the nucleophilic attack of the enzyme on the cyano group. On the other hand, the derivatives with electron-donating groups (3-hydroxybenzonitrile and tolunitriles) are largely hydrolyzed at low relative rates, with a few exceptions. For instance, the nitrilases from *F. solani* O1 [15] and *R. rhodochrous* J1 [27] have high relative activities for 3-hydroxybenzonitrile and tolunitriles, respectively.

Differences between the reactivities of *meta*- and *para*-isomers also seem to reflect the electron densities on the cyano group [14]. For instance, a lower electron density may be expected on the cyano group of 1,4-dicyanobenzene in comparison with 1,3-dicyanobenzene and the much greater reactivity of the former compound is in accordance with this expectation.

Nitrilases are largely very sensitive to steric hindrances by the substituent or heteroatom at the *ortho*-position, 2-substituted benzonitriles or 2-cyanopyridine being rarely hydrolyzed at acceptable rates.

4-Cyanopyridine is a better substrate than 3-cyanopyridine for all the examined nitrilases. The difference between the reactivities of these compounds can also be rationalized by a lower electron density on the cyano group of 4-cyanopyridine. 2,4- and 2,6-Pyridinedicarbonitrile are also accepted as substrates by nitrilases from *A. niger* K10 and *F. solani* O1, the major products being the corresponding cyanocarboxylic acids [47].

Among aliphatic nitriles, unbranched saturatured or unsaturated compounds with a medium chain length (for instance, propionitrile, butyronitrile, hexanenitrile, and acrylonitrile) are hydrolyzed with the highest relative rate (about 20–35% of that for benzonitrile) by aromatic nitrilases. Branched nitriles are in general poor substrates of these enzymes but, for instance, isobutyronitrile is one of the best substrates of the aliphatic nitrilase from *R. rhodochrous* K22 [28].

Table 14.3 Substrate specificity of purified nitrilases from filamentous fungi : comparison with prokaryotic nitrilases.

Substrate[a]	Relative rate[b]					
	A. niger K10[c]	*F. oxysporum* f. sp. *melonis*[c]	*F. solani* IMI196840[d]	*F. solani* O1[c]	*R. rhodochrous* J1[c]	*R. rhodochrous* K22[c]
Benzonitrile	100	100	100	100	100	27.1
3-Hydroxybenzonitrile	5.8	0	21.0	80	22.2	–
3-Fluorobenzonitrile	–	–	74.3	–	24.8	–
4-Fluorobenzonitrile	–	–	150.2	–	26.2	–
3-Chlorobenzonitrile	41.0	–	40.4	87	137	–
4-Chlorobenzonitrile	29.8	–	85.6	40	114	–
3-Bromobenzonitrile	–	–	34.4	–	26.3	–
4-Bromobenzonitrile	–	–	37.6	–	23.6	–
3-Nitrobenzonitrile	–	–	7.0	–	74.2	74.5
4-Nitrobenzonitrile	–	–	27.6	–	74.2	–
3-Tolunitrile	5.5	0	17.9	33	99.6	–
4-Tolunitrile	3.4	9.0	9.9	16	116	–
1,4-Benzodinitrile	79.5	–	213.3	–	–	–
3-Cyanopyridine	32.4	25	31.0	28	22.1	7.3
4-Cyanopyridine	410.7	–	124.7	130	25.6	9.7
2-Phenylacetonitrile	10.8	8	0	–	1.8	27.3
Crotononitrile	–	17	–	–	1.7/11.5[e]	100
Acetonitrile	0	0.3	–	–	–	28.3
Propionitrile	6.9	20	–	18	–	12.7
Butyronitrile	17.6	33	–	20	–	18
Isobutyronitrile	0	15	–	4	–	271
Valeronitrile	19.6	–	–	26	–	40.8
Hexanenitrile	–	24	–	–	–	–
Adiponitrile[e]	–	11	–	–	–	110[f]
Acrylonitrile	–	35	–	–	0/58.8[e]	348
Methacrylonitrile	–	19	–	14	–	143
Allylonitrile	–	34	–	–	–	–
2-Thiophenacetonitrile	56.1	–	–	–	10.7	73.5[g]

Note: no entry indicates no data available.

a This table shows only those substrates that were transformed by at least one of the nitrilases at ≥20% of the reaction rate for the reference substrate (crotononitrile for *R. rhodochrous* K22 and benzonitrile for other organisms).

b The relative activities of enzymes from *F. oxysporum* f. sp. *melonis*, *F. solani* IMI196840, *R rhodochrous* J1, and *R. rhodochrous* K22 have been taken from the previous review [6], while those of enzymes from *A. niger* K10 and *F. solani* O1 were taken from works [14] and [15], respectively.

c Relative rates at a given substrate concentration are compared to that for benzonitrile.

d Maximum reaction velocities (V_{max}) are compared to that of benzonitrile.

e Activity for crotononitrile and acrylonitrile was recovered by subunit aggregation in presence of 10% saturated ammonium sulfate.

f The enzyme also hydrolyzed other dinitriles, that is glutaronitrile, succinonitrile, fumaronitrile, malononitrile, and sebaconitrile at relative rates of 345, 271, 27.4, 45.1, and 15.7%.

g The enzyme also hydrolyzed 3-thiophenacetonitrile at a relative rate of 66.6%.

14.4.3
Activity and Stability

Out of the three purified nitrilases from the genus *Fusarium*, two enzymes have high specific activities towards benzonitrile (143 and 156 U/mg of protein), the nitrilase from *F. solani* IMI196840 being exceptional regarding not only its subunit size (see Section 14.3) but also its specific activity, which is two orders of magnitude lower (see Table 14.4). The specific activity of the nitrilase from *A. niger* (91.6 U/mg of protein) is probably underestimated, the enzyme having not yet been purified to homogeneity. Fungal cyanide hydratases are also characterized by high specific activities, which may be up to one order of magnitude greater than those of fungal nitrilases [17] and, in general, also markedly higher than those of bacterial cyanide dihydratases (see Table 14.4). The specific activites of bacterial nitrilases are generally lower than those of fungal nitrilases, as illustrated by two examples of enzymes from rhodococci in Table 14.4.

The nitrilase from *F. solani* IMI196840 also differs from other fungal nitrilases in terms of its K_m value for benzonitrile (0.039 mM) [10], which is markedly lower than those of other fungal nitrilases (0.2–1.5 mM) [11, 14, 15]. In general, even higher K_m-values, that is in the range of 2.6–12 mM for wild-type enzymes and 5.9–90 mM for hexahistidine-tagged enzymes, have been determined for cyanide hydratases/dihydratases [17].

Typically, nitrilases exhibit the highest activities at neutral or slightly alkaline pH values except for the aliphatic nitrilase from *R. rhodochrous* K22 with a slightly acidic pH optimum (for a review see [5]; Table 14.4). Nevertheless, examination of the activity profiles of fungal nitrilases revealed some differences among these enzymes. Contrary to the nitrilase from *F. oxysporum* f. sp. *melonis*, which was highly active irrespective of pH value in the range of pH 6–11, the nitrilases from *F. solani* O1 and *A. niger* K10 required a relatively narrow range of neutral and slightly alkaline pH values. In general, similar conditions are also beneficial for nitrilase stability (see Table 14.4).

The temperature optima of fungal nitrilases (see Table 14.4) are similar to those of bacterial nitrilases, which are largely highly active at 40–50 °C except for the rare nitrilases from moderate thermophiles (for a review see [5]) and the highly thermostable nitrilase from *Pyrococcus abyssi* [48]. Apparently, there is no marked difference between the fungal nitrilases in this respect, though accurate comparison is not possible due to the different protocols used. The enzyme from *A. niger* K10 was fairly stable at 30 °C with a half-life of 11 h, but its stability decreased sharply at 35 and 40 °C, as indicated by enzyme half-lives of 6.2 and 2.8 h, respectively [14]. The half-life of the purified nitrilase from *F. solani* O1 was about 20 and 11 h at 35 and 45 °C, respectively, indicating higher stability of this enzyme (unpublished results).

In order to assess the potential applications of these new nitrilases in biocatalytic processes, data on their operational stability were required. To this end, we investigated the kinetic behavior of enzymes from *F. solani* and *A. niger* either immobilized on solid supports or retained in stirred ultrafiltration membrane reactors in continuous experiments.

Table 14.4 Specific activity, reaction optima, and stability of purified nitrilases from filamentous fungi: comparison with prokaryotic nitrilases, cyanide hydratase, and cyanide dihydratase.

Enzyme	Organism	Specific activity (U/mg protein)	Optimum		Stability		Reference
			pH	T (°C)	pH	T (°C)	
Aromatic nitrilase	*A. niger* K10	91.6 (benzonitrile)	8	45	7.2–9.0	≤ 30	[14]
	F. solani f. sp. *melonis*	143 (benzonitrile)	6–11	40	6–11	≤ 40	[11]
	F. solani IMI196840	1.66 (benzonitrile)	7.8–9.1	50	–	–	[10]
	F. solani O1	156.0 (benzonitrile)	7–9	40–45	6–8[a]	≤ 35[a]	[15]
	R. rhodochrous J1	15.9 (benzonitrile)	7.6	45	–	≤ 45	[27]
Aliphatic nitrilase	*R. rhodochrous* K22	0.737 (crotononitrile)	5.5	50	6–8	≤ 40	[28]
Cyanide hydratase	*G. sorghi*	555 (KCN)	7.8	–	–	23	[16, 17]
Cyanide dihydratase	*B. pumilus* C1	50.9 (KCN)	7.8–8.0	37	–	–	[17, 21]

Note: no entry indicates no data available.
a Unpublished results.

The nitrilases were immobilized with 70–95% activity recovery on hydrophobic carriers (for instance, Butyl Sepharose) packed in columns [49–51]. The benefit of this immobilization method is its simplicity and quickness, as well as the possibility of compensating biocatalyst deactivation by reloading the column with fresh enzyme. On the other hand, the high salt concentration (for instance, 0.8 M ammonium sulfate), which is necessary for enzyme binding onto the carrier, complicates the purification of the reaction products. This is avoided by using ion exchangers, for instance Q Sepharose [49], as the solid supports, but this method is not compatible with high concentrations of substrates, as the corresponding products (acids) interfere with enzyme binding.

Fungal nitrilases immobilized on these columns were applicable to continuous biotransformation of heteroaromatic nitriles like 3- and 4-cyanopyridine, the products of which, nicotinic and isonicotinic acid, respectively, are of commercial interest. The enzyme from *F. solani* exhibited a higher stability than that from *A. niger* at 35 °C. The conversion of 3-cyanopyridine by the former enzyme was nearly quantitative within 24 h [50], while it decreased by 30% within 15 h in the case of the latter [49]. Similar differences in operational stabilities were observed during conversion of 4-cyanopyridine. The stability of the enzymes depended on the substrate used, both nitrilases being more stable during the conversion of 4-cyanopyridine in comparison with 3-cyanopyridine.

Continuous stirred membrane reactors proved to be useful in laboratory-scale studies of continuous biocatalytic processes based on reactions catalyzed by nitrile hydratase and amidase (see Chapter 17). The retention of enzymes in ultrafiltration membrane reactors can be regarded as a special type of immobilization with a number of advantages resulting from the absence of enzyme–carrier interactions. At temperatures below the reaction optimum the enzymes from both *F. solani* and *A. niger* showed fair stability during long-term runs in ultrafiltration membrane reactors. Similarly, as in the above experiments, however, the former enzyme was more stable. During conversions of 50 mM of 4-cyanopyridine, *A. niger* showed a half-life of 50 h at 30 °C (unpublished results) and *F. solani* a half-life of 277 h at 35 °C [15]. However, as a very small amount of the enzyme (<0.05 mg protein) was used for reactor loading, mechanical stress possibly occurred. Therefore, further experiments with an increased enzyme load are needed to confirm the reaction parameters. Obviously, the substrate has a stabilizing effect on the nitrilase, as its operational stabilities are generally higher than storage stabilities. It is possible that the substrate (3-cyanopyridine or 4-cyanopyridine) supports the multimeric structure of the enzyme, as reported, for instance, for the nitrilase from *R. rhodochrous* J1, the subunits of which associated into an active form in the presence of benzonitrile [52].

14.5
Conclusions and Outlook

Filamentous fungi seem to be a rich but so far little exploited source of nitrilases. There is an evident discrepancy between the large number of putative nitrilase genes resulting from database searches and the limited number of biochemically characterized nitrilases. Moreover, screening of culture collection suggests a wide phylogenetic distribution of fungal nitrilases with sequences that are still unknown.

The relatively low homology among fungal nitrilases may suggest their different enzymatic properties. This expectation was partly confirmed by biochemical characterization of the enzymes from *A. niger, F. solani*, and *F. oxysporum*. In particular, their distinct chemoselectivity is notable. Sequence database mining, on one hand and screening and selection of enzymatically active strains, on the other hand, appear to be viable and complementary approaches towards the finding of new nitrilases in filamentous fungi.

A method of efficient induction of fungal nitrilases in wild strains was established and seems to be of wide application throughout this enzyme group. However, heterologous expression of fungal nitrilases remains to be solved in order to facilitate the use of these enzymes in industrial biotechnology. To what extent the desirable properties (broad substrate specificity, thermostability, chemoselectivity and enantioselectivity) can be improved by mutagenesis of these enzymes should also be examined.

Acknowledgment

The ESF/COST (Action D25), the Czech Science Foundation (project 203/05/2267), the Grant Agency of the Academy of Sciences of the Czech Republic (project IAA 500200708) and the institutional research concept AV0Z50200510 (Institute of Microbiology) supported this work.

References

1 Pace, H.C. and Brenner, C. (2001) *Genome Biology*, **2**, 0001.1–0001.9.
2 Brenner, C. (2002) *Current Opinion in Structural Biology*, **12**, 775–782.
3 Podar, M., Eads, J.R. and Richardson, T.H. (2005) *BMC Evolutionary Biology*, **5**, 42–54.
4 Robertson, D.E., Chaplin, J.A., DeSantis, G., Podar, M., Madden, M., Chi, E., Richardson, T., Milan, A., Miller, M., Weiner, D.P., Wong, K., McQuaid, J., Farwell, B., Preston, L.A., Tan, X., Snead, M.A., Keller, M., Mathur, E., Kretz, P.L., Burk, M.J. and Short, J.M. (2004) *Applied and Environmental Microbiology*, **70**, 2429–2436.
5 Banerjee, A., Sharma, R. and Banerjee, U.C. (2002) *Applied Microbiology and Biotechnology*, **60**, 33–44.
6 O'Reilly, C. and Turner, P.D. (2003) *Journal of Applied Microbiology*, **95**, 1161–1174.
7 Martínková, L. and Mylerová, V. (2003) *Current Organic Chemistry*, **7**, 1279–1295.
8 Martínková, L. and Kren, V. (2002) *Biocatalysis and Biotransformation*, **20**, 73–93.
9 Singh, R., Sharma, R., Tewari, N., Geetanjali and Rawat, D.S. (2006) *Chemistry and Biodiversity*, **3**, 1279–1287.
10 Harper, D.B. (1977) *The Biochemical Journal*, **167**, 685–692.
11 Goldlust, A. and Bohak, Z. (1989) *Biotechnology and Applied Biochemistry*, **11**, 581–601.
12 Deposited in the Culture Collection of Fungi of the Charles University Prague (accession number CCF 3411).

13 Deposited in the Culture Collection of Fungi of the Charles University Prague (accession number CCF 3635).

14 Kaplan, O., Vejvoda, V., Plíhal, O., Pompach, P., Kavan, D., Bojarová, P., Bezouška, K., Macková, M., Cantarella, M., Jirků, V., Kren, V. and Martínková, L. (2006) *Journal of Applied Microbiology and Biotechnology*, **73**, 567–575.

15 Vejvoda, V., Kaplan, O., Bezouška, K., Pompach, P., Šulc, M., Cantarella, M., Benada, O., Uhnáková, B., Rinágelová, A., Lutz-Wahl, S., Fischer, L., Kren, V. and Martínková, L. (2008) *Journal of Molecular Catalysis B-Enzymatic*, **50**, 99–106.

16 Wang, P., Matthews, D.E. and VanEtten, H.D. (1992) *Archives of Biochemistry and Biophysics*, **298**, 569–575.

17 Jandhyala, D.M., Willson, R.C., Sewell, B.T. and Benedik, M.J. (2005) *Applied Microbiology and Biotechnology*, **68**, 327–335.

18 Barclay, M., Day, J.C., Thompson, I.P., Knowles, C.J. and Bailey, M.J. (2002) *Environmental Microbiology*, **4**, 183–189.

19 Cluness, M.J., Turner, P.D., Clements, E., Brown, D.T. and O'Reilly, C. (1993) *Journal of General Microbiology*, **139**, 1807–1815.

20 Sexton, A.C. and Howlett, B.J. (2000) *Molecular & General Genetics*, **263**, 463–470.

21 Meyers, P.R., Rawlings, D.E., Woods, D.R. and Lindsey, G.G. (1993) *Journal of Bacteriology*, **175**, 6105–6112.

22 Sewell, T., Berman, M.N., Meyers, P.R., Jandhyala, D. and Benedik, M.J. (2003) *Structure*, **11**, 1413–1422.

23 Kaplan, O., Nikolaou, K., Pišvejcová, A. and Martínková, L. (2006) *Enzyme and Microbial Technology*, **38**, 260–264.

24 Šnajdrová, R., Kristová-Mylerová, V., Crestia, D., Nikolaou, K., Kuzma, M., Lemaire, M., Gallienne, E., Bolte, J., Bezouška, K., Kren, V. and Martínková, L. (2004) *Journal of Molecular Catalysis B-Enzymatic*, **29**, 227–232.

25 Kaplan, O., Vejvoda, V., Charvátová-Pišvejcová, A. and Martínková, L. (2006) *Journal of Industrial Microbiology & Biotechnology*, **33**, 891–896.

26 Kato, Y., Ooi, R. and Asano, Y. (2000) *Applied and Environmental Microbiology*, **66**, 2290–2296.

27 Kobayashi, M., Nagasawa, T. and Yamada, H. (1989) *European Journal of Biochemistry*, **182**, 349–356.

28 Kobayashi, M., Yanaka, N., Nagasawa, T. and Yamada, H. (1990) *Journal of Bacteriology*, **72**, 4807–4815.

29 Thuku, R.N., Weber, B.W., Varsani, A. and Sewell, B.T. (2007) *FEBS Journal*, **274**, 2099–2108.

30 Deposited in the International Mycological Institute, Egham, Surrey, UK.

31 Layh, N., Parratt, J. and Willetts, A. (1998) *Journal of Molecular Catalysis B: Enzymatic*, **5**, 467–474.

32 Almatawah, Q.A., Cramp, R. and Cowan, D.A. (1999) *Extremophiles*, **3**, 283–291.

33 Jandhyala, D., Berman, M., Meyers, P.R., Sewell, B.T., Wilson, R.C. and Benedik, M. J. (2003) *Applied and Environmental Microbiology*, **69**, 4794–4805.

34 Hung, C.-L., Liu, J.-H., Chiu, W.-C., Huang, S.-W., Hwang, J.-K. and Wang, W.-C. (2007) *Journal of Biological Chemistry*, **282**, 12220–12229.

35 Pace, H.C., Hodawadekar, S.C., Draganescu, A., Huang, J., Bieganowski, P., Pekarsky, Y., Croce, C.M. and Brenner, C. (2000) *Current Biology*, **10**, 907–917.

36 Nakai, T., Hasegawa, T., Yamashita, E., Yamamoto, M., Kumasaka, T., Ueki, T., Nanba, H., Ikenaka, Y., Takahashi, S., Sato, M. and Tsukihara, T. (2000) *Structure*, **8**, 729–737.

37 Sakai, N., Tajika, Y., Yao, M., Watanabe, N. and Tanaka, I. (2004) *Proteins*, **57**, 869–873.

38 Kumaran, D., Eswaramoorthy, S., Gerchman, S.E., Kycia, H., Studier, F.W. and Swaminathan, S. (2003) *Proteins*, **52**, 283–291.

39 Fernandes, B.C.M., Mateo, C., Kiziak, C., Chmura, A., Wacker, J., van Rantwijk, F., Stolz, A. and Sheldon, R.A. (2006) *Advanced Synthesis & Catalysis*, **348**, 2597–2603.

40 Effenberger, F. and Oßwald, S. (2001) *Tetrahedron: Asymmetry*, **12**, 279–285.

41 Oßwald, S., Wajant, H. and Effenberger, F. (2002) *European Journal of Biochemistry*, **269**, 680–687.

42 Piotrowski, M., Schönfelder, S. and Weiler, E.W. (2001) *Journal of Biological Chemistry*, **276**, 2616–2621.

43 Hook, R.H. and Robinson, W.G. (1964) *Journal of Biological Chemistry*, **239**, 4263–4267.

44 Kiziak, C., Conradt, D., Stolz, A., Mattes, R. and Klein, J. (2005) *Microbiology*, **151**, 3639–3648.

45 Stevenson, D.E., Feng, R., Dumas, F., Groleau, D., Mihoc, A. and Storer, A.C. (1992) *Biotechnology and Applied Biochemistry*, **15**, 283–302.

46 Kobayashi, M. and Shimizu, S. (1994) *FEMS Microbiology Letters*, **120**, 217–223.

47 Vejvoda, V., Šveda, O., Kaplan, O., Prikrylová, V., Elišáková, V., Himl, M., Kubáč, D., Pelantová, H., Kuzma, M. and Martínková, L. (2007) *Biotechnology Letters*, **29**, 1119–1124.

48 Müller, P., Egorova, K., Vorgias, C.E., Boutou, E., Trauthwein, H., Verseck, S. and Antranikian, G. (2006) *Protein Expression and Purification*, **47**, 672–681.

49 Vejvoda, V., Kaplan, O., Bezouška, K. and Martínková, L. (2006) *Journal of Molecular Catalysis B-Enzymatic*, **39**, 55–58.

50 Vejvoda, V., Kaplan, O., Klozová, J., Masák, J., Cejková, A., Jirků, V., Stloukal, R. and Martínková, L. (2006) *Folia Microbiologica*, **51**, 251–256.

51 Vejvoda, V., Kaplan, O., Kubáč, D., Kren, V. and Martínková, L. (2006) *Biocatalysis and Biotransformation*, **24**, 414–418.

52 Nagasawa, T., Wieser, M., Nakamura, T., Iwahara, N., Yoshida, T. and Gekko, K. (2000) *European Journal of Biochemistry*, **267**, 138–144.

15
Nitrilase- and Nitrile Hydratase-catalyzed Enantioselective Preparation of Non-proteinogenic Amino Acids

Norbert Klempier and Margit Winkler

15.1
Introduction

Nitriles are versatile precursors of carboxylic acids. Their non-enzymatic chemical synthesis is generally straightforward, allowing the introduction of the carboxylic acid functionality into the molecule at any desired stage of the synthesis. Since non-enzymatic nitrile hydrolyzes sometimes require harsh reaction conditions, enzymatic procedures have been established as efficient alternatives. Much more important, however, are the benefits from enzymatic enantioselective hydrolysis of nitriles. The respective enzymes involved in preparative nitrile hydrolysis, nitrile hydratases (EC 4.2.1.84), amidases (EC 3.5.1.4), and nitrilases (EC 3.5.5.1), are frequently used in whole cell systems, although few researchers have applied them in purified form. The enzymes of the nitrile hydratase/amidase pathway allow the isolation of accumulated amide, depending on the relative activity of both associated enzymes. Nitrilases, in contrast, offer the advantage of directly transforming a nitrile to the acid without the release of an intermediate hydrolysis product (amide) from the catalytic site. However, novel insights into the nitrilase reaction, where amides were isolated as products, revealed that this is not necessarily the case. An inconvenient restriction for the organic chemist using whole cells is the high effort of cell cultivation. Recently, nitrilases became commercially available as ready for use catalysts, thus simplifying the reaction protocol considerably.

Enantiomerically pure non-proteinogenic amino acids have attracted recent attention due to their antibiotic [1], antifungal [2], cytotoxic [3], and other important pharmacological properties [4]. Frequently, they also occur incorporated in natural products, such as peptides, depsipeptides, and other macrocyclic compounds. Other important applications are to serve as building blocks in asymmetric synthesis. A specifically prominent class of them are cyclic β- and γ-amino acids, the subject to which the present chapter is dedicated.

Our novel contributions to the enzymatic (microbial) synthesis of some relevant β-amino acids is summarized in the succeeding part of this chapter (Section 15.2). Such β-amino acids occur as key components in many natural products [4] and

Modern Biocatalysis: Stereoselective and Environmentally Friendly Reactions.
Edited by Wolf-Dieter Fessner and Thorleif Anthonsen

ISBN: 978-3-527-32071-4

are key components of many naturally occurring peptides [6]: the most well known example is (2*R*,3*S*)-phenylisoserine, an essential constituent of the potent antitumor agent paclitaxel (Taxol®) and analogs thereof (Taxotere®) [5]. They also exhibit pharmacological properties *per se*, such as (1*R*,2*S*)-2-aminocyclopentanecarboxylic acid (−)-**2c**, the antifungal antibiotic cispentacin (Figure 15.1) [7]. Replacement of α-amino acids in biologically active peptides by β-amino acids can alter the secondary structure formation distinctly, resulting in modified biological properties of the unnatural analogs [8]. Currently, the synthesis of oligopeptide chains of β-amino acids is attracting much interest because of their ability to fold into defined three-dimensional structures [9]. Accordingly, the efforts being made to synthesize these compounds enantiomerically pure have been reflected by several recent reviews [4, 10, 11]. Besides, certain heterocyclic non-proteinogenic amino acids, such as β-proline (pyrrolidine-3-carboxylic acid (−)-**10c**, depicted in Figure 15.1), pipecolic acid (piperidine-2-carboxylic acid) and nipecotic acid (piperidine-3-carboxylic acid), are common substructures in natural and synthetic bioactive compounds and also deserve attention due to numerous other applications [12]. Some successful attempts have been made to synthesize such acids enantioselectively by chemical means, though they generally include multistep procedures [13]. Recent approaches have used biocatalysts for introducing asymmetry, such as the synthesis of both enantiomers of β-proline by microbial Baeyer–Villiger oxidation [14], (*S*)-pipecolic acid by (*R*)-hydroxynitrile lyase [15], and amides by stereospecific amidases from whole cells [16] and acylase [17].

The enzymatic synthesis and relevance of carbocyclic γ-amino acids will be addressed in Section 15.3 of this chapter and is mainly related to their potential to mimic the structure of γ-amino butyric acid (GABA) on specific receptor sites. GABA is the major inhibitory neurotransmitter in the mammalian central nervous system (CNS) [18]. As a highly flexible molecule it can participate in many low-energy conformation-binding processes. In recent times, the application of conformationally restricted GABA mimics, typically cyclic compounds containing a rigid carbon backbone [19], has contributed to a better understanding in GABA neuroreceptor research [20]. For example, 3-amino cyclopentanecarboxylic acid isomers were recognized to be particularly efficient stereomeric probes for GABA binding site topography. 3-Aminocyclohexanecarboxylic acids, however, were found to act selectively as GABA uptake inhibitors [21]. Moreover, analogs there-

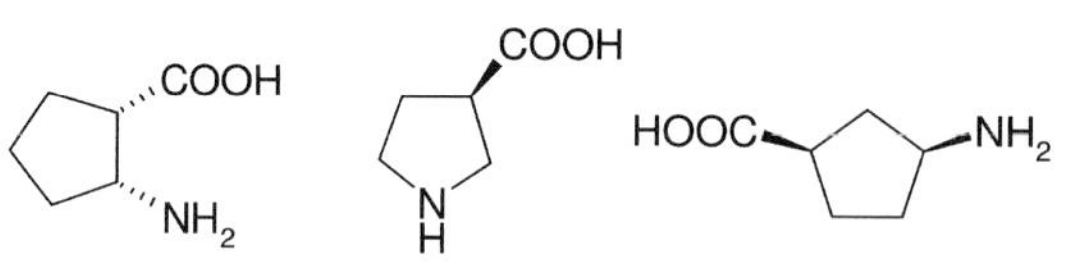

(-)-**2c** (-)-**10c** (-)-**13c**

Figure 15.1 (1*R*,2*S*)-2-Aminocyclopentanecarboxylic acid (cispentacin), (*R*)-pyrrolidine-3-carboxylic acid (β-proline), and (1*R*,3*S*)-3-aminocyclopentanecarboxylic acid.

from were investigated as therapeutic agents for a range of CNS disorders [22]. All these facts account for the need to elaborate synthetic procedures for producing enantiomerically pure cyclic GABA analogs.

Surprisingly, methods for preparing enantioenriched 3-aminocyclopentanecarboxylic acids and 3-aminocyclohexanecarboxylic acids have rarely been reported. The preparation of both enantiomers of *cis*-3-aminocyclohexanecarboxylic acids was achieved via classical fractional crystallization of the diastereomeric *L*-ornithine and brucine salts [23]. Both isomers of 3-oxo-cyclopentane carboxylic acid were obtained by resolution of their brucine adducts, following several transformation steps to either enantiomer of *cis*- and *trans*-3-aminocyclopentanecarboxylic acid [24]. The *cis*-enantiomers were also prepared by resolution of their (−)-1-phenylethylammonium salts [25]. *cis*-(−)-3-Aminocyclopentanecarboxylic acid (−)-**13c** (Figure 15.1) was isolated as a degradation product from the antiviral antibiotic amidinomycin (−)-**17c** (Figure 15.8) [26, 27]. Additional to these long-established methods, approaches to *cis*-3-aminocyclopentanecarboxylic acids comprise multiple-step asymmetric syntheses [28–30]. In fact, few methods have taken advantage of the benefits of biocatalysis. Thus, only the enantiomers of *cis*-3-aminocyclopentanecarboxylic acid were obtained by esterase- and lipase-catalyzed desymmetrization of meso *cis*-1,3-cyclopentanedicarboxylic esters [27, 31] and by lactamase-catalyzed kinetic resolution of a bicyclic lactam [32]. However, the enantioselective synthesis of *trans*-3-aminocyclopentanecarboxylic acid as well as either enantiomer of *cis*- and *trans*-3-aminocyclohexanecarboxylic acids has not yet been accomplished using the great resolution potential of biocatalysts.

15.2 Nitrile Hydratase/Amidase Biotransformations

15.2.1 Protecting Groups for Amino Nitriles

The transformation of unprotected amino nitriles appears to be practicable in a straightforward manner in aqueous buffer, as far as the water solubility of substrates is concerned. However, the whole procedure is severely hindered by insufficient reaction monitoring because of poor ultraviolet (UV) sensitivity on both thin layer chromatograpy (TLC) and reverse-phase high-performance liquid chromatography (RP-HPLC), in particular on a screening scale, as well as a lack of a precise distinction of strongly polar reaction products such as amides and carboxylic acids on TLC.

Another drawback is the unexpectedly high water solubility of all carbocyclic amides and acids, thus complicating the isolation by extraction into an organic solvent.

Suitable amino protecting groups can circumvent most of these drawbacks, though not all requirements can be met in a single protecting group. Our experiments revealed that *N*-toluenesulfonamides suffice most of the requirements

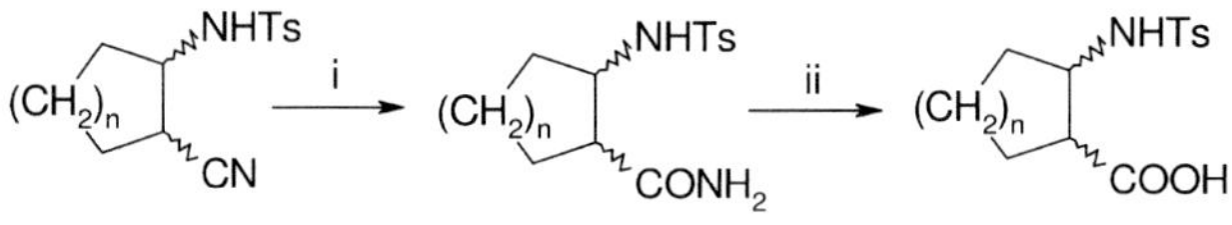

Scheme 15.1 *Rhodococcus* nitrile hydratase (i) – amidase (ii) whole cell transformation of alicyclic nitriles (±)-**1a**–(±)-**4a** to amides **1b–4b** and carboxylic acids **1c–4c**; $n = 1, 2$.

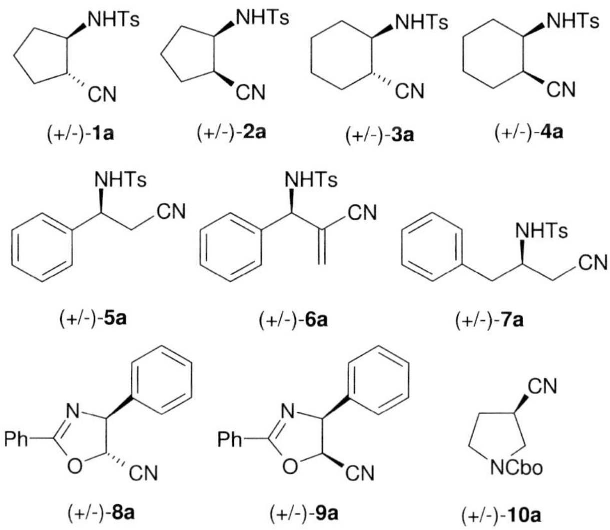

Figure 15.2 β-Amino nitriles for whole cell transformations with *Rhodococci* (only one enantiomer is depicted).

mentioned above, such as excellent extractability and UV sensitivity. Yet more importantly, they do not undergo undesired enzymatic hydrolysis by the strains investigated. Alternatively, butylcarbamates are easier deprotected than toluene-sulfonyl groups: however, their UV sensitivity is poor.

15.2.2
Enantioselective Hydrolysis of β-Amino Nitriles

During our longstanding interest in the biohydrolysis of nitriles, we found that whole cell preparations of certain *Rhodococci*, such as *R. erythropolis* A4 (formerly *R. equi* A4), *R.* sp. R312, and *R. erythropolis* NCIMB 11540, containing the nitrile hydratase/amidase enzyme system, are efficient catalysts for stereoselective microbial hydrolysis of *N*-protected carbocyclic β-amino nitriles (±)-**1a**–(±)-**4a**, to β-amino acids **1c–4c** and amides **1b–4b**, respectively (Scheme 15.1) [33, 34].

The structures of *N*-tosylated amino nitriles for enantioselective hydrolysis are depicted in Figure 15.2. The results listed in Tables 15.1–15.3 are isolated yields of the whole cell transformations after extraction and chromatographic purification. The enantiomeric excesses (e.e.s) are listed in parentheses. All reactions were

Table 15.1 *Rhodococcus erythropolis* A4 whole cell biohydrolysis: isolated products.

	(±)-1a	(±)-2a	(±)-3a	(±)-4a	(±)-6a	(±)-10a
Nitrile[a]	40 (47)	71 (5)	26 (78)	47 (8)	84 (0)	5[b]
Amide[a]	14 (>99)	14 (51)	54 (65)	48 (6)	10 (11)	0
Acid[a]	44 (2)	0	13 (>99)	0	0	95[b]

a Percentage isolated yield after chromatographic purification with percentage e.e. in parentheses.
b The e.e. was not determined.

Table 15.2 *Rhodococcus erythropolis* NCIMB 11540 whole cell biohydrolysis: isolated products.

	(±)-1a	(±)-2a	(±)-3a	(±)-4a	(±)-6a	(±)-10a
Nitrile[a]	0	50 (16)	24 (98)	50 (10)	91 (1)	0
Amide[a]	13 (>99)	49 (15)	56 (59)	41 (8)	2 (32)	0
Acid[a]	86 (5)	0	15 (97)	0	0	100

a Percentage isolated yield after chromatographic purification with percentage e.e. in parentheses.

Table 15.3 *Rhodococcus sp.* R312 whole cell biohydrolysis: isolated products.

	(±)-1a	(±)-2a	(±)-3a	(±)-4a	(±)-6a	(±)-10a
Nitrile[a]	46 (30)	11 (51)	33 (47)	44 (10)	73 (0)	9[b]
Amide[a]	10 (>99)	75 (7)	42 (77)	43 (4)	25 (6)	8[b]
Acid[a]	34 (14)	0	16 (87)	0	0	83[b]

a Percentage isolated yield after chromatographic purification with percentage e.e. in parentheses.
b The e.e. was not determined.

run at standardized conditions (amount of wet cell weight per amount of substrate, substrate concentration, and so on) and were stopped after 24 h in order to compare the particular product distribution [34]. Generally, the five-ring amino nitriles reacted significantly faster than the six-ring compounds. Notably, the relative configuration of the ring substituents had a distinct impact on the product selectivity of the whole cell transformations: *trans*-2-aminocyclopentane carboxylic acid and *trans*-2-aminocyclohexanecarboxylic acid could be synthesized from the respective amino nitriles *trans*-(±)-**1a** and *trans*-(±)-**3a** exclusively. Whereas *trans*-2-amino nitriles reacted smoothly to acids and amides, accumulating the acid with the preceding reaction time, the *cis*-amino nitriles (±)-**2a** and (±)-**4a** reacted more slowly to the amides. A similar diastereodifferentiation was also observed for structurally analogous carbocyclic β-hydroxy nitriles.

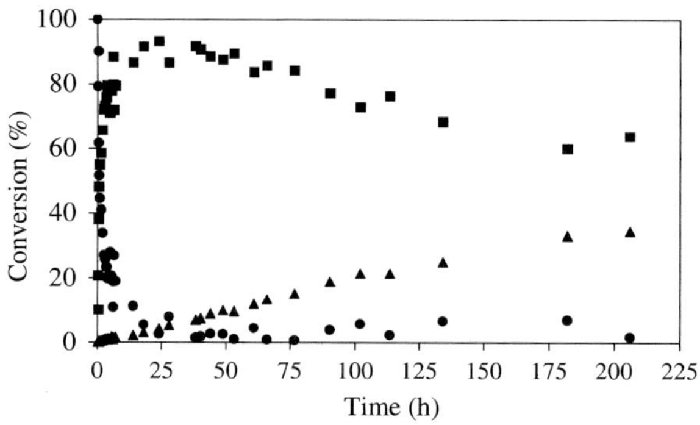

Figure 15.3 Transformation of (±)-**2a** by *R. erythropolis* A4 with dependence on the reaction time; ● nitrile, ■ amide, and ▲ acid.

Surprisingly, acyclic vinylogous amino nitrile (±)-**6a**, prepared by the Aza–Baylis–Hillman reaction, only reacted to the corresponding α-methylene-β-amino amide [34]. In agreement with this, the related oxo-analog was recently reported to give the amide as well upon incubation with *Rhodococcus* sp. AJ270 cells [35].

The enantioselective synthesis of 2-phenylisoserine derivatives is an attractive goal and several attempts, such as the application of acylase [36], lipase [37], and, more recently, bakers yeast reductase enzymes [38], were made to prepare the taxol side chain (see Section 15.1). No nitrile-converting enzymes have been investigated so far. The chemical instability of 2-hydroxy-3-amino-3-phenylpropionitrile (a cyanohydrin) in aqueous solution afforded the preparation of the cyclic protection products *trans*-(±)-**8a** and *cis*-(±)-**9a** (Figure 15.2).

Incubation with whole cells of *Rhodococcus* sp. R 312 yielded (i.e. 50%) the amide *trans*-**8b** in 98% e.e. quantitatively, with not even traces of the acid, leaving the *cis*-nitrile untouched [34].

Acyclic amino nitriles (±)-**5a** and (±)-**7a** reacted sluggishly and no clear trend in the formation of a preferred product was observed, thus their results have been omitted. *N*-Carbobenzoxylated pyrrolidine-3-carbonitrile, a precursor for β-proline, was the fastest reacting five-membered compound. The substrate was entirely consumed within a few minutes yielding β-proline in low enantiomeric purity [39].

The hydrolysis of *cis*-2-NHTs-cyclopentane nitrile (±)-**2a** over a prolonged reaction time was investigated to check whether it could be extended over the amide stage to formation of acid **2c**, a derivative of cispentacin. The conversion–time diagram of (±)-**2a** is depicted in Figure 15.3, suggesting that the acid **2c** could be accumulated in an amount of 34% after 206 h [34].

Similar to diastereoselectivities, the enantioselectivities also strongly correlate with the relative stereochemistry of the 1,2-substituents and with the ring size of amino nitriles (±)-**1a**–(±)-**4a**: *trans*-cyclopentane carbonitriles gave the amides in excellent enantiopurity, whereas *trans*-cyclohexane carbonitriles resulted in the

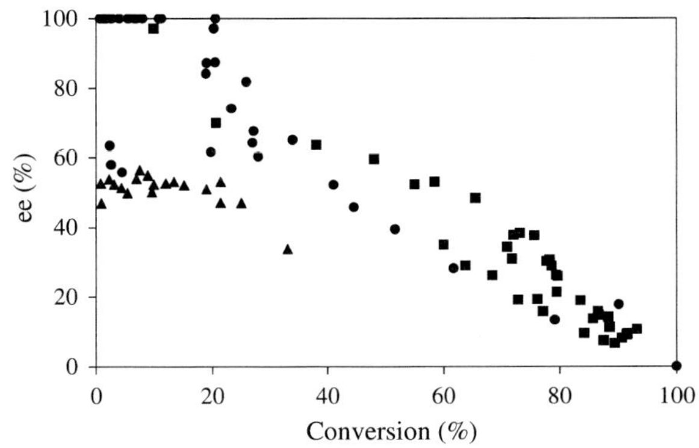

Figure 15.4 Enantiomeric excess of **2a–2c** with dependence on conversion by *R. erythropolis* A4; ● nitrile, ■ amide, and ▲ acid.

formation of the acids in excellent enantiopurities, though at lower conversion. The e.e.s of *cis*-compounds, however, were lower throughout [34].

Thus, the *trans*-aminocyclohexanecarboxylic acid **3c** was prepared by all three microorganisms in very high e.e. (87–99%), whereas the intermediate amides were formed in moderate e.e. In contrast, five-membered *trans*-amide **1b** was obtained in very high enantiopurity, but not so its acid [34]. The high enantiomeric purity of the remaining *trans*-nitriles **1a** and **3a** is remarkable in view of the extent of conversion (38 and 24%, respectively).This can be attributed to the reportedly high enantioselectivity of the nitrile hydratase from *R. erythropolis* A4 for the transformation of (±)-**1a** to **1b** at the least [40].

The conversion of *cis*-2-NHTs-cyclopentane carbonitrile (±)-**2a** to amide **2b** and acid **2c** is plotted versus their e.e.s in Figure 15.4. The transformation was monitored over a reaction time of 206 h. The curve suggests the typical enantiopurity conversion dependence expected from a kinetic resolution. This is in contrast to the results in Tables 15.1–15.3, where no acid was formed within the reaction time of 24 h: 33% acid could be obtained in 34% e.e. after 206 h [34]. Unfortunately, the e.e. of the acid did not exceed 56%, even at low conversion. The reasons for that are not clear.

15.3 Nitrilase Biotransformations

15.3.1 Enantioselective Hydrolysis of β-Amino Nitriles

The availability of ready for use nitrilase preparations [41] with the advantage of avoiding the laborious handling of whole cell biotransformation systems has simplified the reaction protocol substantially and prompted the screening of nitrilases

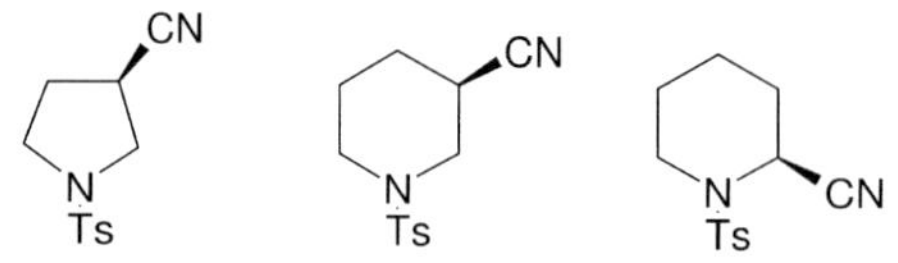

(+/-)-**10a** (+/-)-**11a** (+/-)-**12a**

Figure 15.5 Heterocyclic amino nitriles for enzymatic transformations (only one enantiomer is depicted).

Table 15.4 Nitrilase hydrolysis to heterocyclic amino acids **10c**–**12c**.

	(±)-**10c**[a]	(±)-**11c**[a]	(±)-**12c**[a]
NIT-101	56[b]	–	–
NIT-104	–	15 (75)	–
NIT-105	44 (73)	–	–
NIT-106	44 (76)	–	5[b]
NIT-107	–	50 (93)	–

a Percentage conversion determined by HPLC with the percentage e.e. in parentheses.
b The e.e. was not determined.

with regard to novel substrates. Surprisingly, none of the carbocyclic β-amino nitriles (±)-**1a**–(±)-**4a** depicted in Figure 15.2 were accepted by the enzymes [39]. This situation changed when the nitrogen atom was incorporated into the ring, as in case of the *N*-heterocyclic nitriles (±)-**10a**–(±)-**12a** shown in Figure 15.5, which were found to be highly applicable substrates. The results from the transformations are given in Table 15.4 [42].

A significantly different enzyme activity was noticed here depending on the ring size. Five-membered pyrrolidine-3-carbonitriles were formed close to their theoretical yields within a shorter reaction time (a maximum of 24h) than six-membered piperidine-3- and 4-carbonitriles, which required transformation times within days. Nevertheless, the enzyme activities remained nearly unchanged over this time period. Nipecotic acid **11c** could be prepared and isolated in 93% e.e., thus allowing an efficient and decisively short enantioselective synthesis of this heterocyclic amino acid [42], compared to existing literature procedures (Section 15.1). *N*-Toluenesulfonyl protected acids were formed in enantioselectivities superior to *N*-carbobenzoxylated derivatives, a fact also observed throughout the work on carbocyclic γ-amino acids.

Although some product from (±)-**12a** was formed by NIT-106, pipecolic acid **12c** could not be synthesized in a preparatively satisfying way, in particular because the amide formation is twice as high as the acid formation. A structural comparison of heterocyclic amino nitriles (±)-**10a**–(±)-**12a** with carbocyclic β-amino nitriles

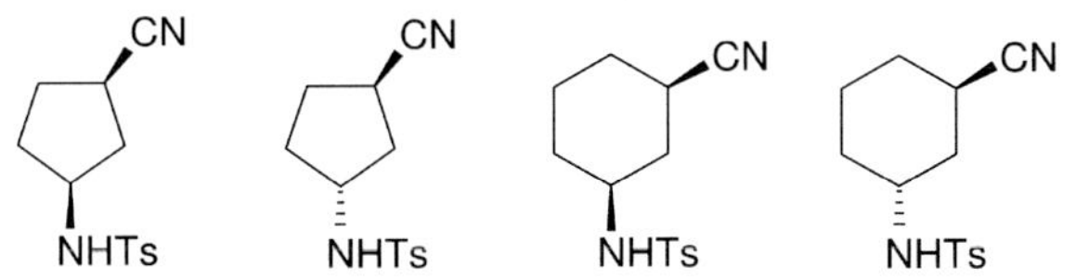

(+/-)-**13a** (+/-)-**14a** (+/-)-**15a** (+/-)-**16a**

Figure 15.6 Novel (±)-γ-amino nitriles for enzymatic transformations to enantioenriched γ-amino carboxylic acids (only one enantiomer is depicted).

(±)-**1a**–(±)-**4a** provides an explanation for the differing reactivities of nitrilases versus nitrile hydratases with regard to these compounds. None of them, carbocyclic substrates (±)-**1a**–(±)-**4a** and pipecolic carbonitrile (±)-**12a**, react with nitrilases. This can likely be attributed to a sterically unfavorable 1,2-substitution pattern in the carbocycle and the piperidine-2-carbonitrile (±)-**12a**, where the protecting group of the ring nitrogen atom is taking up a similar position to the 1,2-carbocyclic substituents, thus resulting in substantial hindrance for the enzyme. On the other hand, nitriles (±)-**10a** and (±)-**11a** are structurally more closely related to γ-amino nitriles (±)-**13a**–(±)-**16a** (Section 15.3.2 and Figure 15.6) where the protecting group attached to the ring nitrogen atom represents the exocyclic amino substituent in the 3-position of the carbocycles. The carbocyclic γ-amino nitriles (±)-**13a**–(±)-**16a** were found to react consistently in excellent manner with the nitrilases throughout (see Section 15.3.2 and Figure 15.6).

Not surprisingly, some nitrilase reactions were accompanied by the formation of the corresponding amides, such as pipecolic amide **12b** (up to 10%) and pyrrolidine-3-carboxamide **10b** (for a discussion of nitrile hydratase activity of nitrilases see Section 15.3.3).

15.3.2
Enantioselective Hydrolysis of γ-Amino Nitriles

During our work on enzymatic nitrile transformation, carbocyclic γ-amino nitriles have emerged from a comprehensive screening among structurally diverse amino nitriles as being well-suited substrates for nitrilase-mediated hydrolysis. In contrast, as stated in the previous section, the analogous carbocyclic β-amino nitriles are strictly non-substrates for nitrilases.

For reasons already outlined in the introduction (Section 15.1), we were especially interested in the enantioselective synthesis of cyclic γ-amino acids serving as conformational GABA mimics in neuroreceptor research. The structures of amino nitriles (±)-**13a**–(±)-**16a** are depicted in Figure 15.6 [43].

All nitrilases examined [41] that exhibited sufficient activity from a screening [43] turned out to be sensitive towards changes in the ring size and relative configuration of the 1,3-substituents. The results of preparative transformations are shown in Table 15.5. Generally, cyclopentane carbonitriles reacted much faster

Table 15.5 Synthesis of enantioenriched γ-amino acids **13c–16c** from (±)-γ-amino nitriles.

	(±)-13c[a]	**(±)-14c**[a]	**(±)-15c**[a]	**(±)-16c**[a]
NIT-106	45 (97)	36 (55)	29 (99)	–
NIT-107	–	–	–	46 (86)

a Percentage isolated yield after chromatographic purification with the percentage e.e. determined by HPLC in parentheses.

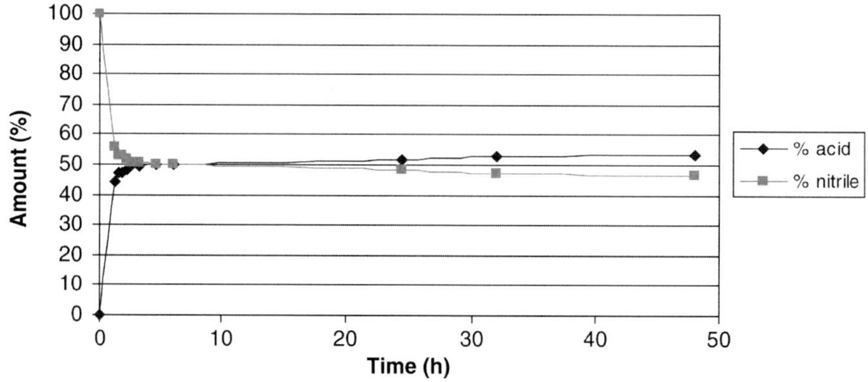

Figure 15.7 Conversion–time diagram of *cis*-NHTs-cyclopentane carbonitrile ((±)-**13a**) hydrolysis by NIT-106.

than the cyclohexane carbonitriles. The *cis*-acids (+)- or (–)-**13c** and (+)- or (–)-**15c** are formed in excellent e.e. (97 and 99%, respectively) but not so the corresponding *trans*-acids **14c** and **16c** [43].

Essentially, nitrilases NIT-106 and NIT-107 were the most efficient catalysts, whereas NIT-101 and NIT-105 turned out to be less suitable for the synthesis of γ-amino acids. *cis*-3-Aminocyclopentanecarboxylic acid (+)-**13c** was prepared by NIT-104, whereas (–)-**13c** was produced by NIT-106 in an enantiocomplementary manner in a high e.e. (97%) close to the theoretical yield of a kinetic resolution. The respective *trans*-isomer (–)-**14c** was obtained in only 55% e.e. by the same enzyme. All other nitrilases examined could not enhance this result. NIT-106 revealed similar outstanding selectivities in the transformation of six-membered aminonitrile *cis*-(±)-**15a** to (–)-**15c** in almost optical purity (>99% e.e.) and in a 29% isolated yield [43].

The conversion–time diagram of (±)-**13a** to **13c** in Figure 15.7 shows that the kinetic resolution of this substrate proceeds nearly quantitatively within 2 h.

The enzyme's prerequisite for both high catalytic activity and enantioselectivity is fulfilled at best by a 1,3-diequatorial position of the substituents, as present in

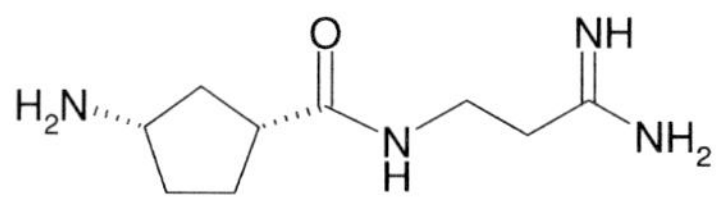

Figure 15.8 Structure of amidinomycin (–)-**17c**.

cis-(±)-**13a** and *cis*-(±)-**15a** (Figure 15.6). Notably, the remaining nitriles were also recovered in up to 98% e. In contrast, the respective *trans*-isomers **14c** and **16c** were obtained in good yields exclusively by NIT-107, though in diminished e.e. (86 and 74%, respectively). In this case, nitrilase NIT-106 completely failed to catalyze the transformation of *trans*-(±)-**16a.**

The absolute configuration of *cis*-**13c** was assigned to be (1*R*,3*S*) by comparison of the HPLC elution order with the available reference acid *cis*-(1*R*,3*S*)-3-ACPA and was found to coincide serendipitously with the configuration of natural amidinomycin (Figure 15.8). The absolute configuration of *cis*-**15c** could be assigned analogously. For an assignment of the *trans*-amino acids **14c** and **16c**, the *trans*-carbonitriles were chemically epimerized by base, though with concomitant nitrile hydrolysis, to the thermodynamically more stable 1,3-*cis*-isomers following the assignment by the HPLC elution order again.

Interestingly, NIT-106 acted highly *R*-selective in contrast to all other nitrilases applied in this study, which were *S*-selective with regard to the ring carbon at C-1 [43].

15.3.3
Nitrile Hydratase Activity of Nitrilases

The formation of amides during nitrilase reactions was observed as early as 1964 [44]. However, the rise of more sophisticated analytical techniques has triggered communications on this subject substantially in recent times. This 'nitrile hydratase activity' of nitrilases was frequently observed when substrates activated in the α-position to the nitrile group were used [45, 46] . Most recently, a study appeared on this subject suggesting a mechanistic rationale of amide formation [47].

Throughout the work on amino nitriles discussed in this chapter, amide formation was observed in agreement with the structural features mentioned above. Thus, α-activated nitrile (±)-**12a** (Figure 15.5), a poor substrate for all nitrilases except NIT-106, gave pipecolic amide **12b** (in an amount of up to 10% even as the predominant product) and (±)-**10a** (Figure 15.6) unexpectedly gave pyrrolidine-3-carboxamide **10b** in amounts of up to 31% depending on the nitrilase employed (the 31% value was achieved by NIT-104), although its nitrile moiety is not in agreement with an α-activating position [42]. On the other hand, there was no formation of piperidine-3-carboxamide. Both protected α-hydroxy-activated amino nitriles (±)-*trans*-**8a** and (±)-*cis*-**9a** (Figure 15.2) gave rise to the amides **8b** and **9b** as the sole products [48]. Consistent with these findings, neither carbocyclic β-amino amides nor γ-amino amides were detected at any stage of the nitrilase transformations.

References

1 Garner, P. and Ramakanth, S. (1986) *The Journal of Organic Chemistry*, **51**, 2609–2612.
2 Davies, S.G., Ichihara, O., Lenoir, I. and Walters, I.A.S. (1994) *Journal of the Chemical Society-Perkin Transactions*, **1**, 1411–1415.
3 Sone, H., Nemoto, T., Ishiwata, H., Ojika, M. and Yamada, K. (1993) *Tetrahedron Letters*, **34**, 8449–8452.
4 Juaristi, E. (ed.) (1997) *Enantioselective Synthesis of β-Amino Acids*, Wiley-VCH Verlag GmbH, Weinheim.
5 Denis, J.N., Correa, A. and Greene, A.E. (1990) *The Journal of Organic Chemistry*, **55**, 1957–1959.
6 Cardillo, G. and Tomasini, C. (1996) *Chemical Society Reviews*, **25**, 117–128.
7 Konishi, M., Nishio, M., Saitoh, K., Miyaki, T., Oki, T. and Kawaguchi, H. (1989) *Journal of Antibiotics*, **42**, 1749–1755.
8 Gellman, S.H. (1998) *Accounts of Chemical Research*, **31**, 173–180.
9 Cheng, R.P., Gellman, S.H. and DeGrado, W.F. (2001) *Chemistry Reviews*, **101**, 3219–3232.
10 Juaristi, E., Quintana, D. and Escalante, J. (1994) *Aldrichimica Acta*, **27**, 3–11.
11 Cole, D.C. (1994) *Tetrahedron*, **50**, 9517–9582.
12 see references 1-3 in [42]
13 see references 4 in [42]
14 Mazzini, C., Lebreton, J., Alphand, V. and Furstoss, R. (1997) *The Journal of Organic Chemistry*, **62**, 5215–5218.
15 Nazabadioko, S., Pérez, R.J., Brieva, R. and Gotor, V. (1998) *Tetrahedron: Asymmetry*, **9**, 1597–1604 and references therein.
16 Eichhorn, E., Roduit, J.-P., Shaw, N., Heinzmann, K. and Kiener, A. (1997) *Tetrahedron: Asymmetry*, **8**, 2533–2536 and references therein.
17 Sánchez-Sancho, F. and Herradón, B. (1998) *Tetrahedron: Asymmetry*, **9**, 1951–1965.
18 Andersen, K.E., Sorensen, J.L., Lau, J., Lundt, B.F., Petersen, H., Huusfeldt, P.O., Suzdak, P.D. and Swedberg, M.D. (2001) *Journal of Medicinal Chemistry*, **44**, 2152–2163 and references therein.
19 Simonyi, M. (1996) *Enantiomer*, **1**, 403–414.
20 Chebib, M. and Johnston, G.A.R. (1999) *Clinical and Experimental Pharmacology & Physiology*, **26**, 937–940.
21 Krogsgaard-Larsen, P., Froelund, B. and Frydenvang, K. (2000) *Current Pharmaceutical Design*, **6**, 1193–1209.
22 Chebib, M., Duke, R.K., Allan, R.D. and Johnston, G.A.R. (2001) *European Journal of Pharmacology*, **430**, 185–192.
23 Allan, R.D., Johnston, G.A.R. and Twitchin, B. (1981) *Australian Journal of Chemistry*, **34**, 2231–2236.
24 Allan, R.D., Johnston, G.A.R. and Twitchin, B. (1979) *Australian Journal of Chemistry*, **32**, 2517–2521.
25 Milewska, M.J. and Polonski, T. (1994) *Tetrahedron: Asymmetry*, **5**, 359–362.
26 Nakamura, S., Karasawa, K., Yonehara, H., Tanaka, N. and Umezawa, H. (1961) *The Journal of Antibiotics*, **14** (Ser. A), 103–106.
27 Chênevert, R., Lavoie, M., Courchesne, G. and Martin, R. (1994) *Chemistry Letters*, **23**(1), 93–96.
28 Bergmeier, S.C., Cobás, A.A. and Rapoport, H.J. (1993) *Bioorganic Chemistry*, **58**, 2369–2376.
29 Trost, B.M., Stenkamp, D. and Pulley, S.R. (1995) *Chemistry – European Journal*, **1**, 568–572.
30 Sung, S.-Y. and Frahm, A.W. (1996) *Archiv der Pharmazie*, **329**, 291–300.
31 Chênevert, R. and Martin, R. (1992) *Tetrahedron: Asymmetry*, **3**, 199–200.
32 Evans, C., McCague, R., Roberts, S.M. and Sutherland, A.G. (1991) *Journal of the Chemical Society-Perkin Transactions*, **1**, 656–657.
33 Preiml, M., Hillmayer, K. and Klempier, N. (2003) *Tetrahedron Letters*, **44**, 5057–5059.
34 Winkler, M., Martínková, L., Knall, A.C., Krahulec, S. and Klempier, N. (2005) *Tetrahedron*, **61**, 4249–4260.
35 Wang, M.-X. and Wu, Y. (2003) *Organic & Biomolecular Chemistry*, **1**, 535–540.
36 Soloshonok, V.A. (1997) Biocatalytic entry to enantiomerically pure β-amino acids, in *Enantioselective Synthesis of β-Amino Acids* (ed. E. Juaristi), Wiley-VCH Verlag GmbH, Weinheim, pp. 443–464.

37 Sih, C.J. (1997) Chemoenzymatic synthesis of the side chain of taxol, in *Enantioselective Synthesis of -Amino Acids* (ed. E. Juaristi), Wiley-VCH Verlag GmbH, Weinheim, pp. 433–442.

38 Feske, B.D., Kaluzna, I.A. and Steward, J.D. (2005) *The Journal of Organic Chemistry*, **70**, 9654–9657.

39 Knall, A.C. (2005) TU Graz, Diploma Thesis.

40 Prepechalová, I., Martínková, L., Stolz, A., Ovesná, M., Bezouska, K., Kopecký, J. and Kren, V. (2001) *Applied Microbiology and Biotechnology*, **55**, 150–156.

41 *Nitrilase NIT-101-NIT-108*, Codexis Inc., Pasadena, CA.

42 Winkler, M., Meischler, D. and Klempier, N. (2007) *Advanced Synthesis & Catalysis*, **349**, 1475–1480.

43 Winkler, M., Knall, A.C., Kulterer, M.R. and Klempier, N. (2007) *The Journal of Organic Chemistry*, **72**, 7423–7426.

44 Hook, R. and Robinson, W. (1964) *The Journal of Biological Chemistry*, **239**, 4263–4267.

45 Effenberger, F. and Osswald, S. (2001) *Tetrahedron: Asymmetry*, **12**, 279–285.

46 Osswald, S., Wajan, H. and Effenberger, F. (2002) *European Journal of Biochemistry*, **269**, 680–687.

47 Fernandes, B.C.M., Mateo, C., Kiziak, C., Chmura, A., Wacker, J., van Rantwijk, F., Stolz, A. and Sheldon, R.A. (2006) *Advanced Synthesis & Catalysis*, **348**, 2597–2603.

48 Winkler, M., Glieder, A. and Klempier, N. (2006) *Chemical Communications*, **12**, 1298–1300.

16
Nitrilases in the Enantioselective Synthesis of α-Hydroxycarboxylic Acids

Fred van Rantwijk, Cesar Mateo, Andrzej Chmura, Bruno C. M. Fernandes and Roger A. Sheldon

16.1
Routes to Enantiomerically Pure α-Hydroxycarboxylic Acids

Interest in the synthesis of enantiopure 2-hydroxycarboxylic acids via asymmetric enzymatic transformations is still increasing and two pathways have risen into prominence recently. The first is based on enantioselective hydrocyanation of the appropriate aldehyde in the presence of an oxynitrilase (hydroxynitrile lyase, EC 4.1.2.10), which gives rise to the corresponding enantiomerically pure cyanohydrin, followed by chemical hydrolysis in the presence of strong acid (Figure 16.1, route a). This latter step generates copious quantities of salt and is not compatible with sensitive functional groups, which is a serious limitation.

Alternatively, enantiopure 2-hydroxycarboxylic acids can be obtained via a dynamic kinetic resolution of the (chemically synthesized) cyanohydrin in the presence of an enantioselective nitrilase (EC 3.5.5.1) (see Figure 16.1, route b). Racemization of the cyanohydrin, via reversible dehydrocyanation, takes place readily at pH 7 or above. The methodology [1] is attractive on account of the mild reaction conditions and is industrially applied in the multiton-scale synthesis of (*R*)-mandelic acid [2].

This nitrilase dynamic kinetic resolution (DKR) methodology depends on the availability of highly enantioselective biocatalysts that generate a minimum amount of amide. This latter issue may seem trivial and has long been disregarded somewhat, but reports of modest amounts of amide co-products date back to the early days of nitrilase enzymology. Only recently has the subject come under more intense scrutiny [3–5] and has a relationship with the stereochemistry of the nitrile been demonstrated [3, 5]. Hence, we set out to investigate the enantiomer and chemical selectivity of nitrilases in the hydrolysis of a representative set of cyanohydrins.

Modern Biocatalysis: Stereoselective and Environmentally Friendly Reactions.
Edited by Wolf-Dieter Fessner and Thorleif Anthonsen

ISBN: 978-3-527-32071-4

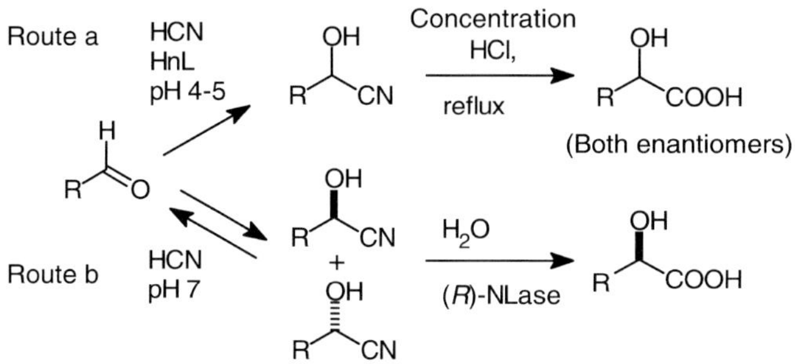

Figure 16.1 Synthetic routes to enantiomerically pure 2-hydroxycarboxylic acids, via oxynitrilase (hydroxynitrile lyase) catalysed enantioselective hydrocyanation (route A) and (*R*)-nitrilase (nitrilase) mediated dynamic kinetic resolution (route B).

Table 16.1 Nitrilase selectivity in the hydrolysis of 2-hydroxynitriles.[a]

Cyanohydrin	Nitrilase[b]	Initial rate (mmol mg^{-1} min^{-1})	ee_{acid} (3, % *R*)	Amide (4, %)
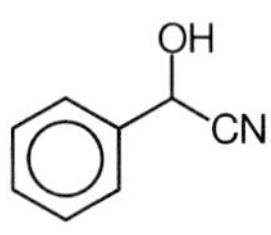	NIT-104	0.2	98	<0.5
	NIT-106	0.5	91	3
	NIT-107	0.014	95	<0.5
	PfNLase	4.3	40	15
1a				
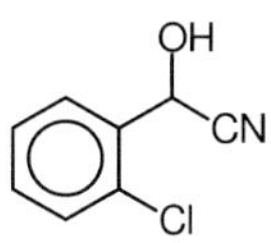	NIT-104	0.03	n.d.	12
	NIT-106	1.8	91	3
	NIT-107	0.023	n.d.	9
	PfNLase	3.3	22	8
1b				
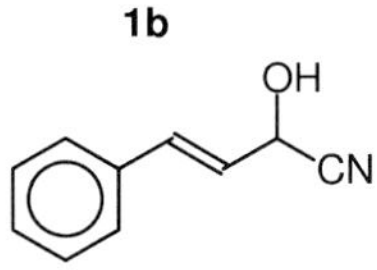	NIT-102	1×10^{-3}	n.d.	<0.5
	NIT-104	18×10^{-3}	92	3
	NIT-105	0.035×10^{-3}	n.d.	14
	NIT-106	0.54×10^{-3}	n.d.	54
1c	NIT-107	15×10^{-3}	88	4
	PfNLase	0.15×10^{-3}		14

a Reactions at pH 7 and 20 °C.
b Nitrilases NIT-102-107 are from BioCatalytics, Inc. (now Codexis, Inc.).

16.2 Nitrilase-mediated Hydrolysis of Cyanohydrins

Three cyanohydrins (**1a–c**) (see Table 16.1) were subjected to hydrolysis in the presence of a number of nitrilases (Figure 16.2). The reactants included the standard substrate mandelonitrile (**1a**) and its *o*-chloro derivative (**1b**), which is of

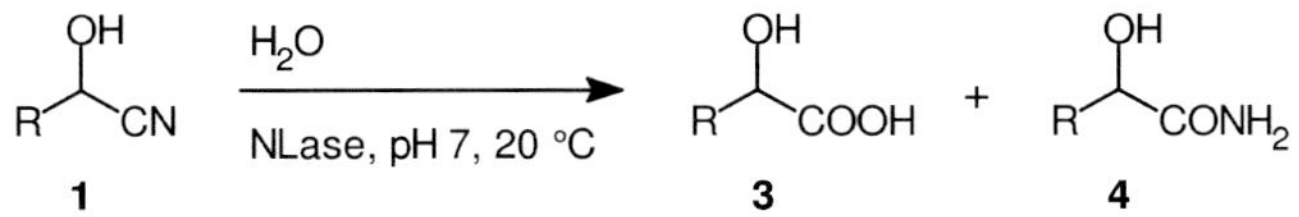

(a) R = C_6H_5
(b) R = 2-Cl-C_6H_5
(c) R = *trans*-C_6H_5–CH=CH

Figure 16.2 Nitrilase mediated hydrolysis of **1a–c**.

interest as the (*R*)-acid is a building block for the antithrombotic agent clopidogrel [6]. Furthermore, the enzymatic hydrolysis of 4-phenyl-2-hydroxy-(*E*)-but-3-enenitrile (**1c**), the HCN adduct of cinnamic aldehyde, was investigated. The enzymatic hydrocyanation of this latter aldehyde is slow and does not readily proceed to completion [7, 8]; hence a DKR of **1c** could be an attractive option.

The biocatalysts that we employed (see Table 16.1) included the set of recombinant nitrilases from Biocatalytics, Inc. (now Codexis, Inc.), as well as the nitrilase originating from *Pseudomonas fluorescens* EBC 191 (PfNLase) [3, 9]. These biocatalysts were compared with regard to activity, selectivity (acid versus amide) and enantioselectivity. Only a few nitrilases, generally speaking NIT-104, 106, 107 and PfNLase, converted **1a–c** at a useful rate. When comparing the hydrolysis of **1a** and **1b** it is noteworthy that the *o*-chloro substituent in **1b** reduced the activity of NIT-104 by nearly an order of magnitude but, in contrast, exerted a small activating effect on NIT-106 and 107. Nitrile **1c** proved a particularly difficult substrate that reacted, in the presence of NIT-104, 10 times slower than **1a** and useful rates were only obtained with NIT-104 and 107. PfNLase hydrolysed **1c** at only a minute rate (30×10^3 times slower than **1a**).

The extent of amide production varied erratically, depending on the nitrile and the biocatalyst. It was negligible in the hydrolysis of **1a,** except when PfNLase was employed as the biocatalyst [3, 5]. The *o*-chloro substituent in **1b** caused NIT-104 and 107 to produce considerable amounts of amide. NIT-106, which acted selectively in the hydrolysis of **1a** and **1b,** sluggishly mediated the hydrolysis of **1c** into a nearly equimolar mixture of acid and amide. In conclusion, there is no 'best' nitrilase with regard to acid/amide selectivity.

The nitrilases from Biocatalytics were (*R*)-selective in all cases where measurements were feasible, but not always to the same extent. PfNLase, in contrast, showed little enantiomeric bias with these nitriles. It is worth noting that in a recent survey the majority of nitrilases that hydrolyzed **1a** were biased towards the (*R*)-enantiomer [10].

From these results we concluded that the preferred biocatalysts for the selective hydrolysis of **1a, 1b** and **1c** into the (*R*)-acids are, respectively, NIT-104, NIT-106 and NIT-104, although the hydrolysis of **1b** and **1c** is rather less enantioselective than would be desirable. Thus, in a preparative experiment, NIT-104 converted **1a** (0.1 M starting concentration) into (*R*)-mandelic acid (98% conversion, 98% *ee*), attesting that under the prevailing conditions the racemisation of the nitrile is fast compared with hydrolysis. The hydrolysis of **1b** in the presence of NIT-106 likewise proceeded to complete conversion into the (*R*)-acid but with incomplete enantioselectivity, as would be expected. In conclusion, the hydrolysis of **1a** under DKR conditions provides access to enantiomerically pure (*R*)-mandelic acid but no sufficiently selective biocatalyst is available for **1b** and **1c**.

16.3
A Bienzymatic Approach to Enantiopure 2-Hydroxycarboxylic Acids

The nitrilase mediated DKR route to enantiomerically pure 2-hydroxycarboxylic acids is restricted to the (*R*)-enantiomers because, to our knowledge, no (*S*)-selective nitrilases for cyanohydrin substrates are commonly available [11]. We reasoned that a fully enzymatic route to the (*S*)-acids should be possible by combining an (*S*)-selective oxynitrilase (hydroxynitrile lyase, EC 4.1.2.10, (*S*)-hydroxynitrile lyase) and a non-selective nitrilase in a bienzymatic cascade (see Figure 16.3). Besides being more environmentally acceptable than chemical hydrolysis, the mild reaction conditions of the combined enzymatic reaction would be compatible with a wide range of hydrolysable groups.

Although the methodology seems neat and straightforward, there are potential incompatibilities to be be resolved with regard to the pH and the reaction medium. Hydroxynitrile lyase-mediated hydrocyanations are preferably carried out at pH < 5 to suppress the competing uncatalysed hydrocyanation [12]. The background reaction is usually further reduced by the use of a biphasic aqueous-organic [12] or micro-aqueous reaction medium [13, 14]. Such reaction conditions are tolerated rather well by the readily available (*S*)-selective hydroxynitrile lyase from *Manihot esculenta* (MeHnL) [12, 15]. Nitrilases, in contrast, are preferentially used at pH close to 7 and are readily deactivated by organic solvents. Rendering the nitrilase compatible with the conditions of enzymatic hydrocyanation is obviously an issue on which the success of the methodology depends and proper immobilization could present a solution to this latter problem. Hence, we undertook to extend the cross-linked enzyme aggregate (CLEA) methodology [16] to nitrilases.

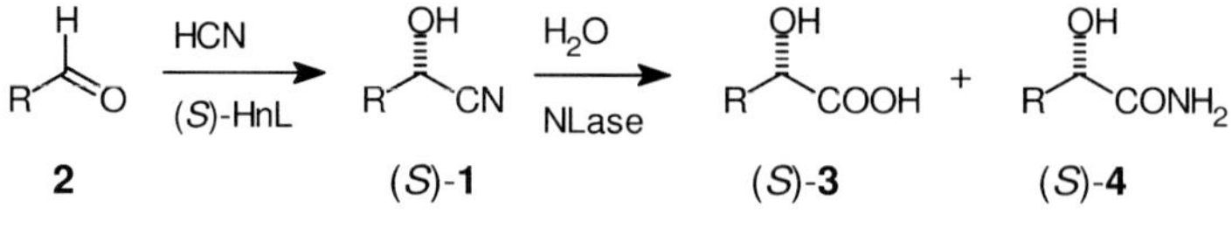

(a) R = C_6H_5

Figure 16.3 Bienzymatic procedure for the synthesis of (*S*)-2-hydroxycarboxylic acids, using an (*S*)-specific hydroxynitrile lyase and a non-specific nitrilase in tandem.

16.4
Stabilization of NLases as Cross-linked Enzyme Aggregates

The preparation of a cross-linked enzyme aggregate (CLEA) involves precipitation and aggregation, using a precipitant, followed by cross-linking with a bifunctional reagent, usually glutaraldehyde [16]. Applying this standard procedure to nitrilases did not result in an active preparation, however [17]. We surmised that glutaraldehyde, which is supposed to react with Lys residues on the enzyme's surface, could easily penetrate the active site and react with the catalytic Lys [11], resulting in loss of activity.

The solution is to employ a high molecular mass cross-linking reagent that is prevented from entering the active site on account of its size. Indeed we found that upon cross-linking with dextran polyaldehyde (MW 100 kDa), 50–60% of the original activity was recovered. Subsequent to cross-linking, the Schiff's base linkages were permanently stabilised by reduction with borohydride [18]. The methodology was demonstrated with NIT-104 and PfNLase [16], later also with NIT-106. Presumably, it is generally applicable to nitrilases.

16.5
Hydrocyanation and Hydrolysis in a Bienzymatic Cascade

As noted above, enzymatic hydrocyanations are preferably performed at pH < 5, to suppress the non-enzymatic background reaction whereas the pH optimum of the common nitrilases is 7. A compromise pH is obviously required and we accordingly assessed the effects of the pH on the MeHnL-mediated hydrocyanation of benzaldehyde (**2a**, see Figure 16.3) in a biphasic aqueous-diisopropyl ether (DIPE) medium. We found that enantioselectivity was maintained at pH 5.5, which we adopted as a compromise pH for the bienzymatic reactions, provided that the aqueous buffer phase accounted for <10% of the reaction volume. PfNLase was the obvious choice for the second step as it stayed active at pH 5.5 and converted (*S*)- and (*R*)-**1a** at comparable rates.

Accordingly, we conducted experiments with CLEAs of MeHnL and PfNLase in tandem in a 90 : 10 DIPE-buffer pH 5.5 medium [19]. The reaction proceeded to nearly full conversion (see Figure 16.4a) and the product *ee* was 94%. Combining both enzymes in a bienzymatic catalyst (combi-CLEA, Figure 16.4b) resulted in further improvement and 98% enantiomerically pure (*S*)-**3a** was obtained. It would seem that the nitrile intermediate is immediately hydrolysed in the combiCLEA particles, which supresses diffusion into the water phase and possible racemisation. The amount of (*S*)-mandelic amide ((*S*)-**4a**, see Figures 16.3 and 16.4, approx. 40%) that accompanied the formation of (*S*)-**3a** was more than would be expcctcd from the data in Table 16.1 and made us aware of possible stereochemical effects on the acid/amide mechanistic switch [5] in PfNLase as will be discussed later.

We conclude from the formation of nearly enantiomerically pure (*S*)-**3a** that the bienzymatic methodology is basically sound and that racemisation of the interme-

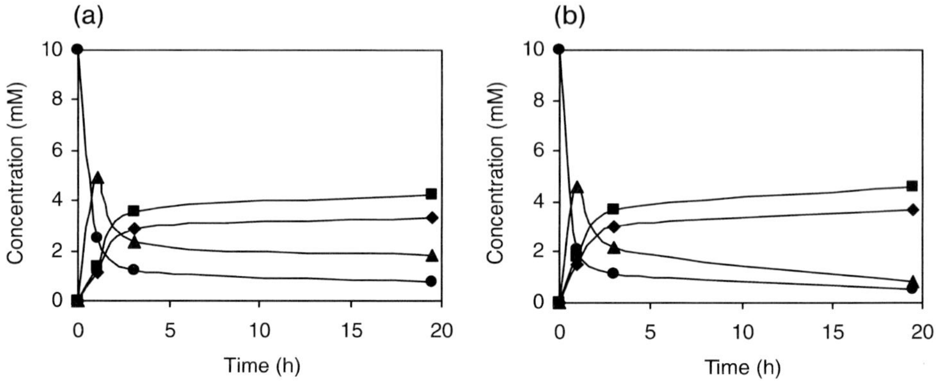

Figure 16.4 Bienzymatic synthesis of (*S*)-mandelic acid from benzaldehyde (10 mM) and HCN (50 mM) in the presence of MeHnL and PfNLase in 90 : 10 DIPE-buffer pH 5.5 [19], (a) two CLEAs, (b) combiCLEA; legend: **1a: ▲**; **2a: ●**; **3a: ■**, **4a: ◆**.

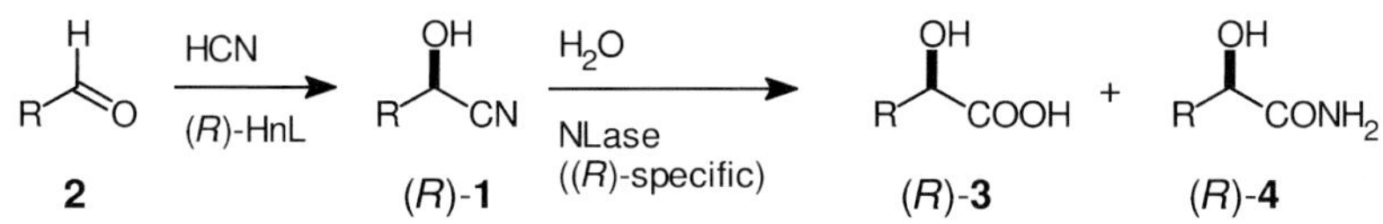

(b) R = 2-Cl-C_6H_5
(c) R = *trans*-C_6H_5–CH=CH

Figure 16.5 Bienzymatic synthesis of (*R*)-2-hydroxycarboxylic acids **3b** and **3c.**

diate cyanohydrin can be avoided. The starting concentrations could be increased to 0.25 M **2a** and 0.75 M HCN while maintaining complete conversion. Some loss of enantioselectivity was observed, presumably due to a slight upward pH drift as the reaction progressed.

The bienzymatic approach described above could also be advantageously applied to the synthesis of (*R*)-2-hydroxycarboxylic acids in cases where no satisfactorily enantioselective nitrilase is available (Figure 16.5). The best enantioselectivity in the hydrolysis of **1b**, for example, was 92% *ee*. The enantioselectivity of the hydroxynitrile lyase from almonds (PaHnL) in the synthesis of **1b** is also less then perfect [13], but we found that a combiCLEA of PaHnL and NIT-106 quantitatively converted **2b** (0.1 M starting concentration) into **3b** with *ee* > 99% *R* (reaction in 90 : 10 DIPE-buffer pH 5.5, as before) with very little (>3%) amide formation.

Furthermore, we considered that the hydrocyanation of cinnamic aldehyde (**2c**) suffers from an unfavorable equilibrium [7, 8] and (apparent) mediocre enantioselectivity [7], which could be increased to >96% only upon careful optimization [8]. We reasoned that the equilibrium would be improved by removing the cyanohydrin **1c** by *in-situ* hydrolysis. Moreover, such a bienzymatic procedure would also obviate the erosion of the cyanohydrin *ee* which tends to become apparent at the

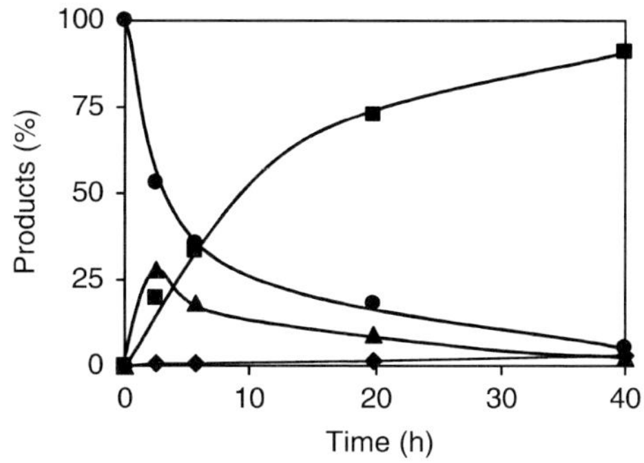

Figure 16.6 Bienzymatic synthesis of (*R*)-4-phenyl-2-hydroxy-(*E*)-but-3-enoic acid from cinnamic aldehyde (0.1 M) and HCN (0.5 M) in the presence of a combi-CLEA of PaHnL and NIT-104, in 50:50 DIPE-buffer pH 5.5 at r.t; legend: **1c**: ▲; **2c**: ●; **3c**: ■, **4c**: ◆.

approach of equilibrium [15]. The nitrilase of choice for such a procedure is NIT-104, as discussed above. Indeed we found that **2c** could be converted into **3c** (96% *ee*) in an acceptable yield (Figure 16.6). The amide (**4c**) was formed in minute amounts only, as would be expected from the results in Table 16.1.

In conclusion, the bienzymatic transformation of aldehydes and HCN into the enantiomerically pure 2-hydroxycarboxylic acids is feasible. The stereochemistry can be steered either by the hydroxynitrile lyase or by both enzymes in combination and the hydrocyanation equilibrium is no longer an issue because it can be shifted to complete conversion. The formation of large amounts of amide, in particular (*S*)-**4a**, somewhat reduces the immediate practical value of our procedure. Ways to obviate this unwanted side-reaction will be discussed later.

16.6 Nitrilases Acting as Nitrile Hydratases

It has been noted already that reports can be found in the literature, as far back as the 1960s, of modest amounts of amides being formed in the presence of nitrilases [20–22]. This side activity (see Figure 16.7) could not be accounted for by the commonly accepted nitrilase mechanism [21] and was largely neglected until quite recently. It was then shown, for example, that nitrile hydration in the presence of recombinantly expressed and purified nitrilases from *Arabidopsis thaliana* was a major pathway with some nitriles, in particular electron-deficient ones [23–25].

We found that nitrile hydration in the presence of PfNLase is likewise subject to electronic effects. Hardly any amide was formed from 2-phenylpropionitrile (**1d**) in the presence of this latter enzyme but the hydration pathway predominated, in contrast, with the electron-deficient 2-chloro-2-phenylacetonitrile (**1e**); the extent

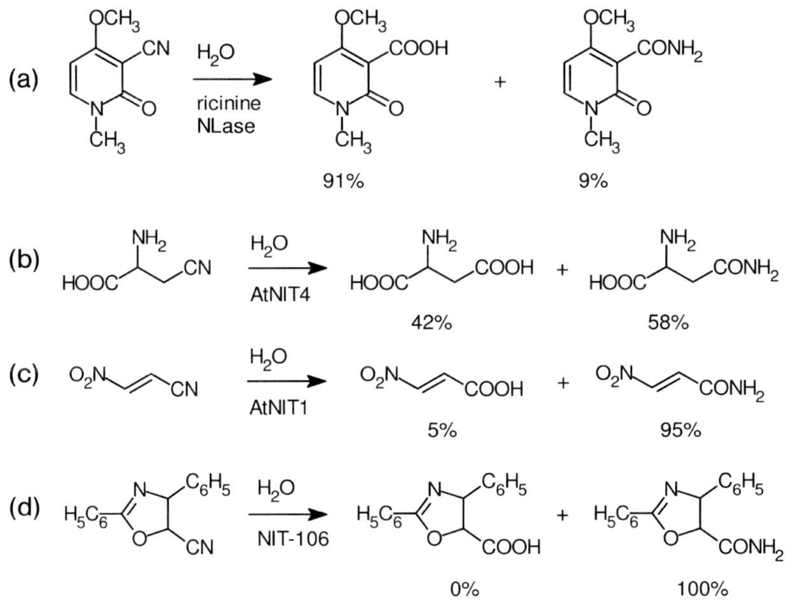

Figure 16.7 Examples of amide formation in the presence of nitrilases. References: a, [20]; b, [23]; c, [24]; d, [4].

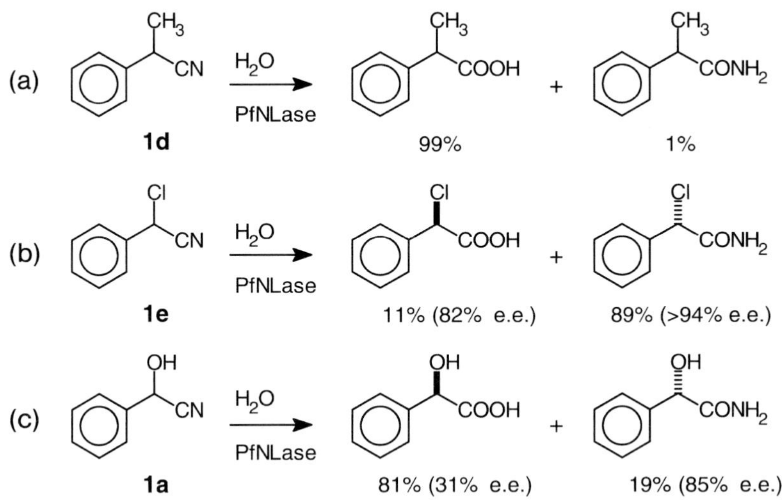

Figure 16.8 Substrate effects on the extent of amide formation by PfNLase. References: a, b, [5]; c, [3].

of amide formation from **1a** was between these two extremes (see Figure 16.8). It became apparent, moreover, that the hydrolysis pathway was biased towards the (*R*)-acid, whereas the (*S*)-amide predominated as hydration product [3, 5]. A very similar observation had been made by the Effenberger group with regard to the AtNIT catalyzed hydrolysis of 2-fluoro-2-phenylacetonitrile [25].

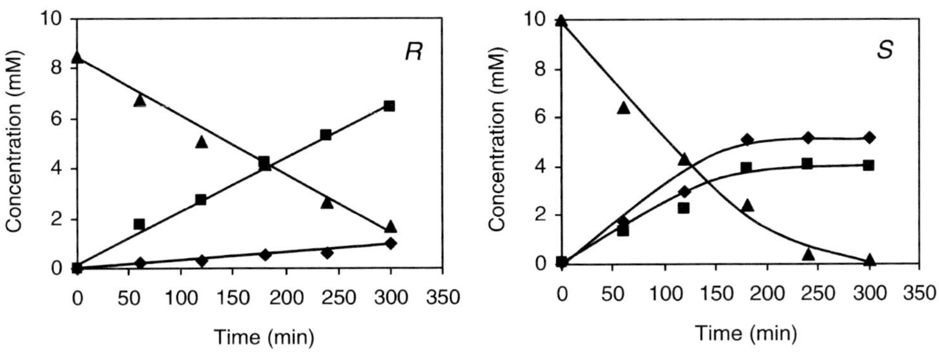

Figure 16.9 Hydrolysis of (*R*)- and (*S*)-mandelonitrile in the presence of *P. fluorescens* nitrilase at pH 6, 0 °C; legend: **1a:** ▲; **3a:** ■; **4a:** ◆; figure taken from [19].

To aqcuire more insight in amide formation we undertook to study the hydrolysis of enantiomerically pure cyanohydrins. Nitrile **1e** racemises too readily but the stereochemical integrity of (*R*)- and (*S*)-**1a** could be maintained with careful adjustment of the reaction conditions (pH 6, 0 °C). Thus, enantiomerically pure (*R*)- and (*S*)-**1a** were subjected to hydrolysis in the presence of PfNLase (see Figure 16.9) [5]. It became clear that only 11% of amide was formed from the (*R*)-enantiomer, whereas the (*S*)-enantiomer was hydrolyzed into 55% amide and 45% acid under otherwise identical conditions (see Figure 16.9). Stereochemical integrity was fully maintained under the reaction conditions and **3a** and **4a** were formed with complete retention of configuration, as would be expected.

Summarising, the nitrile hydration pathway is often prominent with nitrilases and should be mechanistically accomodated. It is commonly accepted that nitrilases harbour a Cys-Glu-Lys triad in their active sites [26] and that the reaction takes place via a thioimidate intermediate (**I** in Figure 16.10) [20, 21, 27]. Elimination of ammonia from the tetrahedral intermediate (pathway A), with formation of the acylenzyme intermediate **III**, requires a positive charge on the nitrogen atom in the reactant that is stabilized by the Glu residue. If, in contrast, the positive charge is not on the reactant but, for example, on the Lys residue (tetrahedral intermediate **IIb**), formal thiol elimination is expected to prevail (pathway B). Such a charge distribution would be expected, for example, when an electron-demanding R group destabilizes the the positive charge on the reactant N. Otherwise, steric interactions could force the N atom away from the stabilizing Glu (see Figure 16.10, pathway **B**). Summarizing, nitrilase mediated nitrile hydrolysis into the carboxylic acid and hydration to give the amide are two branches of the nitrilase mechanism, as originally proposed by Hook and Robinson [20]. It is worth noting that the active site does not contain bound amide at any time, according to our mechanism, and amide, once formed, is not hydrolyzed any further.

The formation of amide by nitrilase somewhat reduces the synthetic value of our hydroxynitrile lyase-nitrilase based bienzymatic procedure for (*S*)-**3a**, as noted

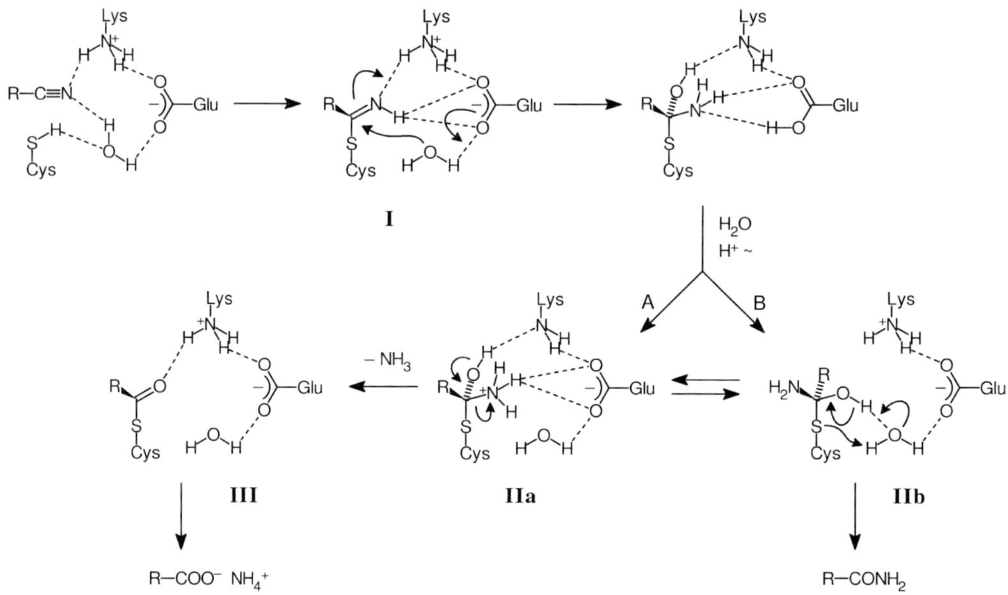

Figure 16.10 Proposed nitrilase mechanism for the formation of acid (**a**) and amide (**b**) [5].

above. Obvious ways to avoid or remedy amide formation would be to obtain a more selective nitrilase via screening or mutagenesis. Alternatively, a suitable amidase could be employed to hydrolyze any amide into the desired acid. Indeed, a nearly quantitative yield of (*S*)-**3a** was obtained upon hydrocyanation of **2a** in the presence of MeHnL, PfNLase and penicillin G acylase. This procedure only applies to mandelic acid and simple derivatives, owing to the strict substrate specificity of penicillin acylase, which is an undesirable limitation. We are now extending the three enzyme methodology to non-specific amidases, such as, for example, the one from *Rhodococcus erythopolis* MP50 [28].

16.7 Conclusion

Commonly available nitrilases hydrolyze cyanohydrins with modest to excellent selectivity for the (*R*)-enantiomer. The formation of amide coproducts, once thought to be an insignificant side-reaction, varies erratically in cyanohydrin hydrolysis between minute and copious, depending on the enzyme and the steric character of the nitrile.

The bienzymatic transformation of aldehydes and HCN in the presence of an oxynitrilase and a nitrilase is a useful addition to the synthetic repertoire, provided that the coproduction of amide can be avoided.

Acknowledgments

This research was supported by the Netherlands Research Council NWO and the Deutsche Forschungsgemeinschaft in the framework of the CERC-3 programme. Support by the Spanish CICYT (project PPQ 2002-01231), by a Ramon Y Cajal Contract from M.E. y C. (Spain) and by COST D25/0002/02 is gratefully acknowledged.

References

1 (a) Yamamoto, K., Oishi, K., Fujimatsu, I. and Komatsu, K.-I. (1991) *Applied and Environmental Microbiology*, **57**, 3028–3032. (b) Endo, T. and Tamura, K. (1991) (Nitto Chem. Ind.), EP 449648, [*Chem. Abstr.* 1992, **116**, 5338].

2 (a) Yamaguchi, Y., Ushigome, M. and Kato, T. (1997) (Nitto Chem. Ind.), EP 773297, [*Chem. Abstr.* 1997, **127**, 4190]. (b) Ress-Löschke, M., Friedrich, T., Hauer, B., Mattes, R. and Engels, D. (2000) (BASF AG), DE 19848129, [*Chem. Abstr.* 2000, **132**, 292813].

3 Kiziak, C., Conradt, D., Stolz, A., Mattes, R. and Klein, J. (2005) *Microbiology*, **151**, 3639–3648.

4 Winkler, M., Glieder, A. and Klempier, N. (2006) *Chemical Communications*, 1298–1300.

5 Fernandes, B.C.M., Mateo, C., Kiziak, C., Chmura, A., Wacker, J., Van Rantwijk, F., Stolz, A. and Sheldon, R.A. (2006) *Advanced Synthesis Catalysis*, **348**, 2597–2603.

6 Bousquet, A. and Musolino, A. (1999) PCT Int. Appl. WO 9918110, [*Chem. Abstr.* 1999, **130**, 296510e].

7 Warmerdam, E.G.J.C., Van den Nieuwendijk, A.M.C.H., Kruse, C.G., Brussee, J. and Van der Gen, A. (1996) *Recueil des Travaux Chimiques des Pays-Bas*, **115**, 20–24.

8 Gerrits, P.-J., Willeman, W.F., Straathof, A.J.J., Heijnen, J.J., Brussee, J. and Van der Gen, A. (2001) *Journal of Molecular Catalysis B: Enzymatic*, **15**, 111–121.

9 Layh, N., Parratt, J. and Willets, A. (1998) *Journal of Molecular Catalysis B: Enzymatic*, **5**, 467–474.

10 Robertson, D.E., Chaplin, J.A., DeSantis, G., Podar, M., Madden, M., Chi, E., Richardson, T., Milan, A., Miller, M., Weiner, D.P., Wong, K., McQuaid, J., Farwell, B., Preston, L.A., Tan, X., Snead, M.A., Keller, M., Mathur, E., Kretz, P.L., Burk, M.J. and Short, J.M. (2004) *Applied and Environmental Microbiology*, **70**, 2429–2436.

11 Pace, H.C., Hodawadekar, S.C., Draganescu, A., Huang, J., Bieganowski, P., Pekarsky, Y. and Brenner, C. (2000) *Current Biology*, **10**, 907–917.

12 Schmidt, M. and Griengl, H. (1999) *Topics in Current Chemistry*, **200**, 193–226.

13 Han, S., Lin, G. and Li, Z. (1998) *Tetrahedron: Asymmetry*, **9**, 1835–1838.

14 Van Langen, L.M., Van Rantwijk, F. and Sheldon, R.A. (2003) *Organic Process Research and Development*, **7**, 828–831.

15 Chmura, A., Van der Kraan, G.M., Kielar, F., Van Langen, L.M., Van Rantwijk, F. and Sheldon, R.A. (2006) *Advanced Synthesis Catalysis*, **348**, 1655–1661.

16 (a) Cao, L., van Rantwijk, F. and Sheldon, R.A. (2000) *Organic Letters*, **2**, 1361–1364. (b) Mateo, C., Palomo, J.M., van Langen, L.M., van Rantwijk, F. and Sheldon, R.A. (2004) *Biotechnology and Bioengineering*, **86**, 273–276. (c) Van Langen, L.M., Selassa, R.P., Van Rantwijk, F. and Sheldon, R.A. (2005) *Organic Letters*, **7**, 327–329.

17 Mateo, C., Palomo, J.M., Van Langen, L.M., Van Rantwijk, F. and Sheldon, R.A. (2004) *Biotechnology and Bioengineering*, **86**, 273–276.

18 Fernández-Lafuente, R., Rodriguez, V., Mateo, C., Penzol, G., Hernández-Justiz, O., Irazoqui, G., Villarino, A., Ovsejevi, K., Batista, F. and Guisán, J.M. (1999) *Journal*

of Molecular Catalysis B: Enzymatic, **7**, 181–189.

19 Mateo, C., Chmura, A., Rustler, S., Van Rantwijk, F., Stolz, A. and Sheldon, R.A. (2006) *Tetrahedron: Asymmetry*, **17**, 320–323.

20 Hook, R.H. and Robinson, W.G. (1964) *The Journal of Biological Chemistry*, **239**, 4263–4267.

21 Goldlust, A. and Bohak, Z. (1989) *Biotechnology and Applied Biochemistry*, **11**, 581–601.

22 Stevenson, D.E., Feng, R., Dumas, F., Groleau, D., Mihoc, A. and Storer, A.C. (1992) *Biotechnology and Applied Biochemistry*, **15**, 283–302.

23 Piotrowski, M., Schönfelder, S. and Weiler, E.W. (2001) *The Journal of Biological Chemistry*, **276**, 2616–2621.

24 Osswald, S., Wajant, H. and Effenberger, F. (2002) *European Journal of Biochemistry*, **269**, 680–687.

25 Effenberger, F. and Oßwald, S. (2001) *Tetrahedron: Asymmetry*, **12**, 279–285.

26 Brenner, C. (2002) *Current Opinion in Structural Biology*, **12**, 775–782.

27 Harper, D.B. (1977) *The Biochemical Journal*, **165**, 309–319.

28 Trott, S., Bürger, S., Calaminus, C. and Stolz, A. (2002) *Applied and Environmental Microbiology*, **68**, 3279–3286.

17
UF-Membrane Bioreactors for Kinetics Characterization of Nitrile Hydratase–Amidase-catalyzed Reactions: a Short Survey

Maria Cantarella, Alberto Gallifuoco, Agata Spera, Laura Cantarella, Ondřej Kaplan and Ludmila Martínková

17.1
Introduction

Nitrile biodegradation is performed by a variety of microorganisms and proceeds through two different enzymatic pathways: direct transformation to carboxylic acids and ammonia, with some exceptions, catalyzed by a nitrilase (EC 3.5.5.1) [1–3] or a two-step reaction, the first catalyzed by nitrile hydratase (EC 4.2.1.84) that produces an amide intermediate, which is further hydrolyzed to the acid and ammonia by an amidase (EC 3.5.1.4) [4, 5].

As far as the two-step reaction is concerned, a wide variety of bacteria possess the nitrile hydratase enzyme able to convert nitriles to corresponding amides: *Arthrobacter* [2, 6–8], *Bacillus* [9, 10], *Brevibacterium* [11], *Comamonas* [12], *Corynebacterium* [13], *Nocardia* [4], *Rhodococcus* [14, 15], *Pseudonocardia* [16], and *Pseudomonas* [17].

Some filamentous fungi, mainly representatives of the genera *Fusarium, Aspergillus,* and *Penicillium,* degrade nitriles preferentially through nitrilase activity [18, 19] (see Chapter 14).

Yeast strains belonging to the species of *Candida fabianii, Candida guilliermondii, Candida tropicalis, Debaryomyces hansenii, Saccharomyces cerevisiae, Torulopsis candida* (*Candida fumata*), and *Williopsis saturnus* also exhibit the nitrile hydratase/amidase system able to use the series of aliphatic mono- and dinitriles as well as their matching amides as the sole N-source [20].

Rhodococcus sp. N-774 and *Pseudomonas chlororaphis* B23 resting cells have been used at industrial scale (as first- and second-generation biocatalysts) for the biological production of acrylamide from acrylonitrile since the 1980s [21]. Currently *Rhodococcus rhodochrous* J1 is being adopted as a third-generation biocatalyst (Mitsubishi Rayon Co.). The industrial production of nicotinamide from 3-cyanopyridine is also operated with this strain (Lonza AG). However, despite the enormous potentiality of nitrile-hydrolyzing biocatalysts for industrial applications, only a few commercial processes have been realized [22].

Modern Biocatalysis: Stereoselective and Environmentally Friendly Reactions.
Edited by Wolf-Dieter Fessner and Thorleif Anthonsen

ISBN: 978-3-527-32071-4

The nitrile-converting enzymes in *R. rhodochrous* J1 strain exhibit high versatility against a variety of aromatic and heteroaromatic nitriles [5, 15, 23] with the only formation of amide, this strain being deficient of amidase activity toward aromatic amides.

So far from the numerous results achieved with a variety of strains generated in different culture conditions a broad substrate tolerance or stereoselectivity of the enzyme has emerged, even though interpertation of the results has been rather complicated. Nitrile hydratases are considered to be active preferentially on aliphatic nitriles with only a marginal activity on aromatic ones [14, 15, 24–26].

Moreover, nitrile hydratases have been found to be thermally unstable, substrate inhibited, and also inhibited by the amides or acids produced during bioconversion.

The control of reaction kinetics (mainly substrate and/or product inhibition) through the correct choice of bioreactor configuration has also gained attention. Recycle fed-batch reactors, packed-bed reactors, and dual hollow fiber bioreactors have been used for producing acrylamide using *Brevibacterium* sp. CH1 [27–30]. The resting cells of *Microbacterium imperiale* (previously named *Brevibacterium imperialis*) confined in a membrane reactor were also investigated by us for acrylamide production [31].

In the last decade our research has focused on nitrile and amide biotransformation using the enzymatic system of *M. imperiale* CBS 498-74. In this strain both nitrile hydratase and amidase are present and nitriles are transformed in the corresponding acid in a two-step reaction with an amide as an intermediate. Aliphatic and aromatic substrates have been tested using resting cells as a biocatalyst. In particular, nitrile hydratase activity was characterized in the presence of acrylonitrile [31, 32], propionitrile [33], benzonitrile [34], and amidase with nicotinamide [35]. Amidase activity was investigated either independently, in the presence of the exogenous matching amide, or as part of the cascade system, where the matching amide was generated by the nitrile hydratase activity. Figure 17.1 illustrates the reaction scheme with the substrates tested.

The investigation into the kinetic behavior of both enzymes was in general performed in both batch and continuous stirred ultrafiltration (UF)-membrane reac-

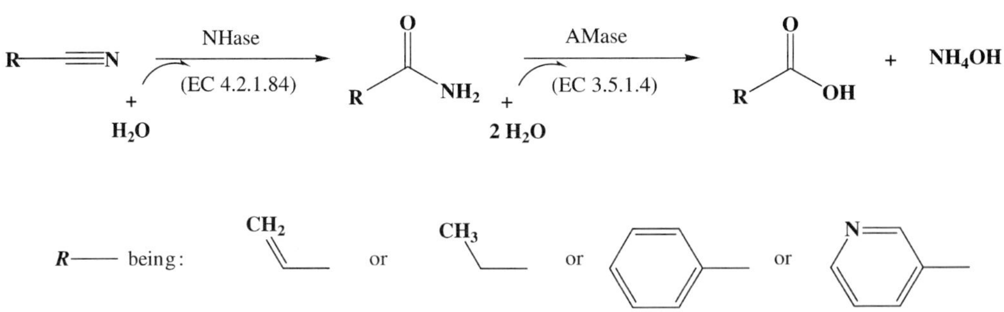

Figure 17.1 Nitrile conversion catalyzed by nitrile hydratase (NHase)–amidase (AMase) system from *Microbacterium imperiale CBS 498-74* here illustrated.

tors (CSMRs). The body of the experimental results collected so far by our research group [31–39] on the effect on the kinetics of parameters such as the substrate and enzyme concentrations, temperature, pH, stirring velocity, and space velocity suggested an emphasis on the preliminary investigation being carried out in batch reactors by a more in-depth kinetic characterization of the system in a continuous reactor. Characterizing the biocatalyst kinetic behavior in defined operational conditions that are essential for increasing the substrate conversion and reaction yield could be a quite useful investigation strategy. This approach could be adopted as a pre-industrial study before its transfer to a larger scale.

In this paper, several examples are summarized where the systematic approach with CSMR investigations allowed the acquisition of experimental data useful for the evaluation of key parameter values for current kinetic models. The temperature dependence of catalyzed reactions was studied in depth and from the fitting of CSMR experimental data the activation energies of both catalyzed reactions were confirmed and those of the inactivation processes evaluated. A high irreversible inactivation by the substrate was elucidated in the CSMR investigation. Finally, the effect of the substrate concentration, cell loading, and mean residence time on the reactor capacity and percentage of conversion are presented. The microorganism *M. imperiale* CBS 498-74, cultivated as fully described previously [36], was employed throughout as resting cells.

17.2
Experiment Design

Continuous runs were carried out in a stirred cell ultrafiltration module using the fluoropolymer membrane FS61PP with a nominal molecular weight cut-off of 20 kDa. The reactor, loaded with an appropriate amount of resting cells, was fed with a buffered substrate solution by a peristaltic pump with the flow rate set at 12–14 ml/h in most runs or as indicated in the experiments at different τ. The stirred cell module was submerged in a thermostated water bath for temperature control (±0.1 °C). A constant stirrer speed (250 rev/min) guaranteed mixing and absence of cell polarization phenomena on the membrane surface. The chemicals used in the runs did not affect the membrane flow characteristics and no rejection of solutes was determined. The permeate was collected by means of an automatic fraction collector, the volume measured, and the samples analyzed every 2 h for reaction product(s) and unreacted substrate.

17.3
Temperature Dependence of the Nitrile Hydratase–Amidase Cascade System

Initially the effect of temperature on both cascade system enzymes was investigated in batch reactors in the presence of the proper substrate for each enzyme using resting cells as the biocatalyst. Figure 17.2a shows the initial reaction rate

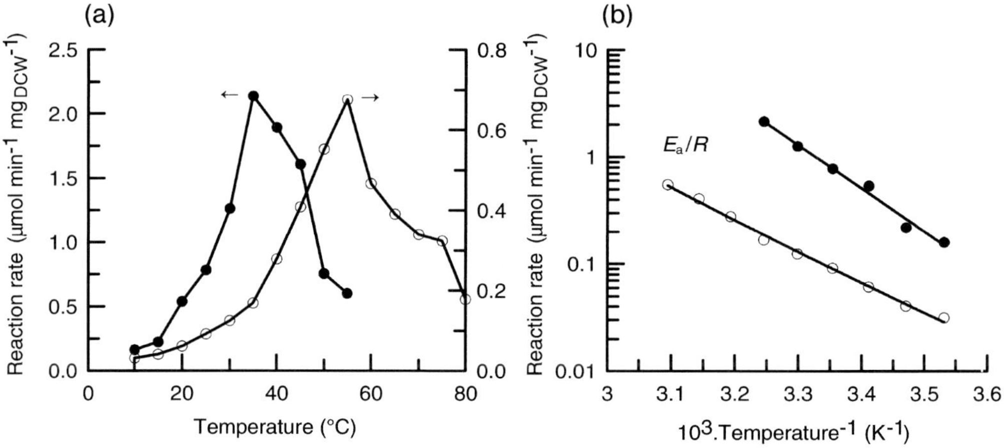

Figure 17.2 (a) Effect of temperature on nitrile hydratase (●) and amidase (○) initial reaction rate. Batch reactor: resting cell load 0.16 $mg_{DCW}\,ml^{-1}$, 10 mM of benzonitrile and benzamide respectively. (b) Arrhenius plot of initial reaction rate catalysed by nitrile hydratase (●) and amidase (○).

of the benzonitrile nitrile hydratase-catalyzed reaction and benzamide amidase-catalyzed reaction as a function of the temperature. The dependence on temperature of the activity of the two enzymes was quite different, the optimal temperatures being 35 and 55 °C for the nitrile hydratase- and amidase-catalyzed reactions, respectively. Interestingly, at 35 °C the nitrile hydratase specific activity was approximately 12.6 times that of amidase, while it dropped to approximately comparable levels at 55 °C, thus indicating a significant thermal instability of nitrile hydratase. Another indication emerged that either the expression or the activity of amidase for the substrates tested in this study appeared considerably lower than that of nitrile hydratase, as also quoted elsewhere [40]. The data for the ascending part of the curve fitted the Arrhenius law well as shown in Figure 17.2b. The slopes allow evaluation of the activation energy of the reaction (regression coefficient approximately 0.99 for both series of data). The apparent activation energies of the nitrile hydratase- and amidase-catalyzed reactions were 77.06 and 55.61 KJ/mol, respectively. A major dependence on temperature of the first reaction in the cascade system was clearly indicated. Interestingly, a similar investigation cited elsewhere [35] evidenced that the activation energy of amidase-catalyzed biotransformation of nicotinamide into nicotinic acid was about 52.6 ÷ 53.5 KJ/mol, as detected in the experiments carried out in both batch reactors and the CSMR. As these E_a-values are quite similar, they suggest that the nature of the aromatic ring in benzamide and nicotinamide has a small or no influence on the amidase-catalyzed reaction.

The thermal stability of enzymes also has to be investigated in order to characterize the temperature dependence of the reaction kinetics fully. The rate at which the enzyme activity declines is a critical characteristic since the economical feasibility of the process depends on the useful lifetime of the enzyme. It is well known that, for many enzymes, the rate of deactivation depends not only on the temperature, but also on the substrate concentration and protein concentration. Hence, fairly high concentrations of both the substrate and protein can drastically reduce the rate of enzyme denaturation and consequently increase the enzyme half-life. It is therefore crucial to test the enzyme stability under conditions as close as possible to those encountered during reactor operation.

17.4 CSMR Investigations

The operational thermal stability of enzymes can be easily evaluated in experiments carried out in a CSMR fed with a saturating substrate concentration, while varying the temperature but maintaining all the other parameters constant. Each enzyme of the cascade system was tested by feeding the CSMR with the appropriate substrate. The kinetic characterization of amidase-catalyzed reactions in runs fed with a nitrile was hampered by the fact that the intracellular enzyme works in cascade with nitrile hydratase. The concentration of amide, produced *in situ* in the first step, varied with the time and reaction conditions and did not assure the differential conditions needed for an accurate analysis, the amide being completely converted by amidase in some runs. Hence, amidase activity was characterized independently by feeding the reactor with amide as the substrate [35].

The flow rate across the membrane, the total mass of resting cells, and the concentration of the product(s) formed in the effluent stream collected at time intervals allowed the calculation of the specific reaction rate during process time [32]. These data may be illustrated in a semi-logarithmic graph as a function of the process time. Figure 17.3a shows a few representative examples of the time course of the reaction rate at different temperatures: that of the benzonitrile nitrile hydratase-catalyzed biotrasformation carried out at 15 °C and the benzamide amidase-catalyzed one performed at 40 and 45 °C. The thermal behavior of the two enzymes is visibly different and can be quantified in the following manner. The data are correlated by a straight line (dotted lines in the graph) once the steady state of the reactor is attained, thus indicating that the decay in enzyme activity with time follows a first-order deactivation mechanism. The slope represents the inactivation constant k_d and the extrapolated *y*-axis intercept the value of the initial velocity r_0. The r_0-values are also well correlated in an Arrhenius plot and allow evaluation of the activation energy of the catalyzed reaction: the E_a-values are in good agreement with those obtained in batch reactors, as also indicated for the nicotinamide amidase-catalyzed reaction [35]. On the other hand, the inactivation constants, which were calculated for all the series of data (not shown) obtained for nitrile hydratase and amidase in the temperature range 5 ÷ 45 °C, were arranged in the Arrhenius plot and are illustrated in Figure 17.3b. Both series of data (those referring to benzonitrile and benzamide transformation) fit a straight line well, the

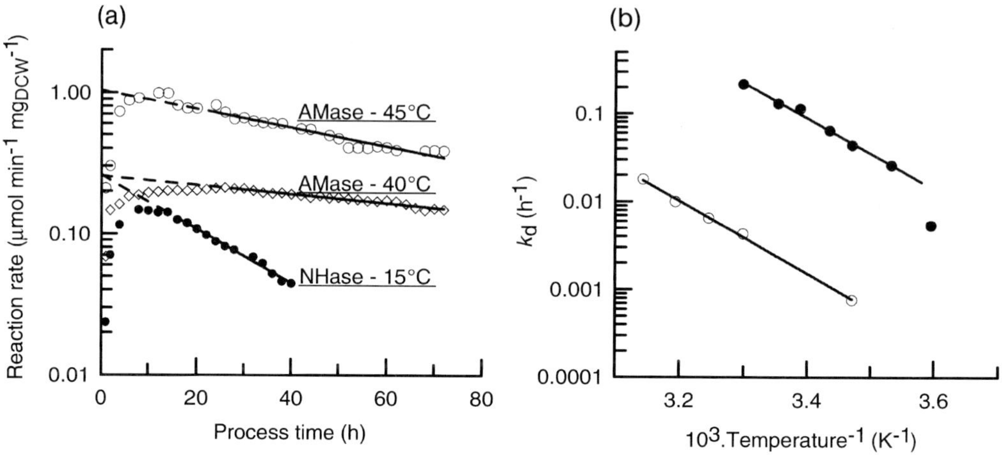

Figure 17.3 (a) Time course of reaction rate in UF-membrane bioreactor at different temperature. Appropriate substrate feed solution (benzonitrile or benzamide) 10 mM, resting cell load 2 mg_{DCW}, flow-rate 12 ml h^{-1}. Filled symbols for nitrile hydratase activity; empty symbols for amidase activity. (b) Arrhenius plot of inactivation constant obtained from UF-runs. Filled symbols for nitrile hydratase activity; empty symbols for amidase activity.

activation energies of the inactivation process being 91.63 ($R = 0.997$) and 89.21 ($R = 0.991$) KJ/mol for nitrile hydratase and amidase, respectively. Although the activation energy values are similar, they were evaluated in different temperature ranges, that is 5–30 °C and 15–45 °C for nitrile hydratase and amidase, respectively and, thus, at any temperature nitrile hydratase will be inactivated more rapidly than amidase. Accordingly, the k_d-values for nitrile hydratase are approximately 57 times higher than that for amidase at 15 °C.

Similar data were obtained for 3-cyanopyridine biotransformation into nicotinamide and nicotinic acid (unpublished data). The higher dependence of the nitrile hydratase deactivation process on temperature has already been observed with other substrates, such as in acrylonitrile bioconversion into acrylamide where the nitrile hydratase half-life dropped from 33 h to approximately 7 h when the temperature was varied from 4 to 10 °C [37].

The above procedure allows the thermal kinetic behavior of both enzymes involved in the nitrile bioconversion to be characterized fully and helps obtain the complete kinetic equation.

As demonstrated elsewhere [34], it is possible to utilize such a different thermal dependence of the enzymes involved in the cascade system for process control. The appropriate choice of operational conditions allowed the enzyme activities to be controlled and directed the selectively of the process to the first reaction product, the amide or to the second reaction product, the acid. A higher process temperature and residence time (20 °C and $\tau = 10$ h) favor production of the acid (amidase activity prevails), while a lower temperature and residence time (5 °C and $\tau = 5$ h) allow the production of the intermediate amide (amidase activity negligible) [34].

17.5 Substrate Concentration Effects on the Reaction Rate, Enzyme Stability, Substrate Conversion, and Reactor Capacity

During the course of our research on nitrile hydratase- and amidase-catalyzed reactions we had to deal with different aspects of substrate concentration effects on the enzyme kinetics and all were suitably investigated making use of a CSMR, as shown by the following case studies.

The effect of the substrate (benzonitrile) concentration on nitrile hydratase activity was investigated in a CSMR in the range of 2–10 mM at low temperature (10 °C) in order to reduce the enzyme thermal inactivation as much as possible. The experimental runs are illustrated in Figure 17.4a as a semi-logarithmic plot of the specific reaction rate against process time. Both the r_0- and the k_d-values were calculated (as summarized in the previous paragraph) and are quoted in Table 17.1.

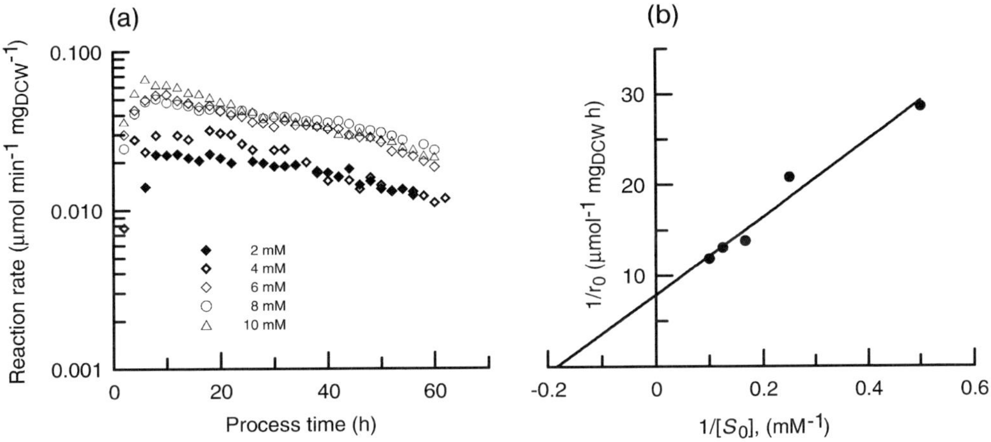

Figure 17.4 (a) Time course of product concentration in UF-membrane bioreactor at various substrate concentration (benzonitrile in 50 mM sodium phosphate buffer, pH 7.0). Cell load 10 mg_{DCW}, temperature 10 °C, flow-rate 12 ml h^{-1}. (b) Linewaever-Burk plot of data obtained from UF-runs.

Table 17.1 r_0- and k_d-values obtained form steady-state data of runs performed in CSMR at different substrate concentration feed.

[S] mM	2	4	6	8	10
r_0 nitrile hydratase µmol/min.mg_{DCW}	0.0349	0.0480	0.0723	0.0764	0.0842
k_d h^{-1}	0.020	0.023	0.020	0.018	0.021

The k_d-value is approximately the same (0.020/h) in all runs, thus indicating a half-life of nitrile hydratase equal to 34.66 h. The invariance of the k_d-value suggests that at least this benzonitrile concentration range has no effect on nitrile hydratase inactivation. The plot of the initial reaction rate versus substrate concentration obeys Michaelis–Menten kinetics, as illustrated in the Lineweaver–Burk plot of Figure 17.4b. The following apparent kinetic parameters were evaluated: $K_m^{app} = 5.43\,mM$ and $V_{max}^{app} = 0.127\,\mu mol\,min^{-1}\,mg_{DCW}^{-1}$. These values are in agreement with those determined using the classic differential batch reactor technique. The kinetic parameters are quoted in Table 17.2, together with the substrate range and reaction temperature for some nitrile hydratase- and amidase-catalyzed reactions. It is interesting to note that different culture conditions to those compared in Table 17.2 (0.5% glucose on line 1, 0.5% glucose plus acrylonitrile on lines 2 and 3, and 0.5% glucose plus acrylamide on line 4) apparently did not alter the substrate (acrylonitrile) transport through the cell wall and the enzyme specificity towards the substrate (see reference [36] for details). The difference in the levels of V_{max}^{app} are due to the different amount of enzyme units expressed by the resting cells.

The propionitrile nitrile hydratase-catalyzed reaction presented kinetics inhibited by high substrate concentrations that were clearly evidenced in the CSMR, as revealed in [33]. The kinetic parameters for nitrile hydratase-catalyzed reactions of

Table 17.2 Kinetic parameters of nitrile hydratase and amidase catalysed reactions.

	Activity	Substrate	Range mM	T °C	K_m mM	V_{max} $U.mg_{DCW}^{-1}$	Reference
1.	nitrile hydratase	acrylonitrile	2–50	20	9.8	11.48	[36]
2.	nitrile hydratase	acrylonitrile	2–50	20	9.8	16.16	[36]
3.	nitrile hydratase	acrylonitrile	2–50	20	9.4	19.52	[36]
4.	nitrile hydratase	acrylonitrile	2–50	20	9.4	25.80	[36]
5.	nitrile hydratase	propionitrile	10–500	10	21.6	11.04	[33]
6.	nitrile hydratase	3-cyanopyridine	0.1–10	20	11.4	3.62	–
7.	amidase	nicotinamide	0.5–10	20	2.6	0.17	–
8.	amidase[a]	nicotinamide	10–50	20	8.2	0.22	–

a Crude enzymes obtained from precipitation by addition of ammonium sulfate of cell crude extract.
– data not published previously.
Notes:

1. Nitrile hydratase and amidase activities were assayed determining the amount of product formed per unit weight of dry cell under the following conditions: reaction mixture 2 ml, stirring rate 250 rpm, temperature 20 °C, reaction time 20 min, pH 7.0 (50 mM Na-phosphate buffer). The specific nitrile hydratase activity of the resting cells freshly collected from broth was roughly 15.5 U/mg_{DCW}. The nitrile hydratase unit (U) was defined as the μmol of acrylamide formed per minute at 20 °C during incubation of 50 mM acrylonitrile. One unit of amidase activity was defined as the amount of enzyme required to release 1 $\mu mol\,min^{-1}$ of product from 100 mM substrate concentration.
2. Products were quantitatively determined with HPLC analyses. The analytical conditions adopted for benzonitrile, benzamide and benzoic acid determination are fully described in [34], and those for nicotinic acid and nicotinamide in [35]. Propionitrile and propionamide were identified by their retention time in gas chromatography analyses [33].

3-cyanopyridine have also been summarized (personal communication). These data are homogeneous, as far as the K_m^{app}-value is concerned, with the other nitrile hydratase-catalyzed reactions (entries 1–5). The amidase activity on nicotinamide shows a lower K_m^{app}-value (entry 7 in Table 17.2) when compared with that obtained with crude enzyme, thus indicating probable mass transfer limitations (personal communication).

Many of the nitriles investigated are classified as toxic agents and could inactivate enzymes at high concentration. The acrylonitrile nitrile hydratase-catalyzed biotransformation presents this limitation [31] and the CSMR proved to be a useful tool for assessing the reversibility of inactivation induced by the high substrate concentration. In that study, two runs performed in UF-reactors were compared and Figure 17.5 shows schematically how inverting the substrate concentration feed (100–800 mM of acrylonitrile) influences the enzyme inactivation level. The first run was fed with 100 mM of acrylonitrile and the reaction rate was recorded for a certain period after the steady-state conditions were achieved. Data analysis and interpretation (described above) allowed an unequivocal evaluation of the nitrile hydratase inactivation constant, $k_d^{100\,mM}$. The substrate concentration change in the feed stream (from 100 to 800 mM of acrylonitrile) generated a new transient period due to substrate and product accumulation inside the reactor, until a new steady state was established. From that moment onwards the recorded enzyme activity showed a marked decline (due to the high substrate concentration) measurable through the new nitrile hydratase inactivation constant $k_d^{800\,mM}$.

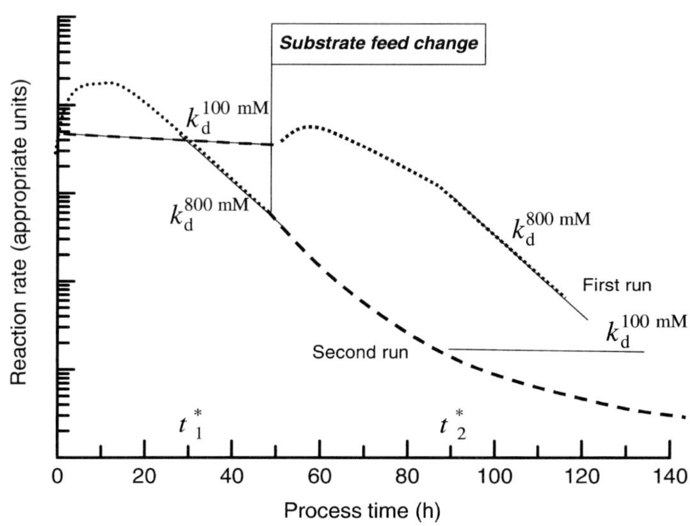

Figure 17.5 Scheme of the reaction runs performed in UF-membrane bioreactor in the presence of high substrate concentration inactivating the nitrile hydratase activity (for details see reference [31]). (– – – –) 100 mM and ( ) 800 mM acrylonitrile concentration in the feed stream.

In the second experiment the reactor feeds were inverted, that is the first feed and the second feed contained 800 and 100 mM of acrylonitrile, respectively. The dynamics of the reactor remained identical (constant flow rate) and the times needed to achieve the first and second steady states in the reactor (t_1^*, t_2^* in the scheme) were similar. Interestingly, the inactivation constant $k_d^{800\,mM}$ evaluated in both runs was the same. The productivity in the second part of the run decreased because of the drastic activity loss from exposure of the enzyme to the high substrate concentration in the first part of the run. However, the expected new steady-state conditions in the second part of the run were not attained. Conversely, the activity decay continued at a rate higher than that recorded (in the first run) in the presence of 100 mM of substrate. The absence of initial activity recovery suggested that the enzymes were irreversibly damaged and not reversibly inhibited by the high substrate concentration.

Figure 17.6a shows the effect of the substrate concentration on the regime conversion and the UF-reactor capacity. A 50% substrate conversion was the maximum attained with 2–4 mM of substrate, while increasing the substrate concentration to over 4 mM led to a decreased conversion. The reactor capacity increased with the substrate concentration and reached a plateau. Since the goal might be to increase the product concentration in the reactor permeate, the enzyme load in the reactor was doubled and, as expected, the steady-state conversion increased from 50 to 62.5% (run fed with 4 mM of substrate). However, the reactor capacity dropped from 2.17 to 1.24 μmol mg_{DCW}^{-1} h^{-1}.

An increase in the flow rate Q(l/h) at constant reactor volume V_R (l) led to adecrease in the space velocity τ (h), the latter being defined as $\tau = V_R/Q$.

The space velocity effect was investigated and data analysis allowed evaluation of the steady-state conversion of the reactor loaded with 10 mg and a substrate feed of 4 mM in buffer (50 mM of sodium phosphate buffer at pH 7.0) solution. These data are illustrated in Figure 17.6b. An increase in τ gives a higher product conversion but the reactor capacity is lower.

Similar behavior was obtained with experiments performed in UF-membrane bioreactors at 50 °C and fed with 100 mM of nicotinamide at constant τ (5.83 h) at varying cell loadings inside the reactor [35]. The steady-state values of the product concentration elaborated in terms of percentage conversion and reactor capacity are illustrated in Figure 17.7. The reactor capacity reaches the maximum value at 20 mg_{DCW}: for higher cell concentrations inside the reactor the capacity drops, while the conversion remains approximately constant. This latter diagram shows that the optimal cell loading was reached and that further biocatalyst increase did not produce any process benefit.

Collectively, the experiments clarify the optimal mean residence time (see Figure 17.6b) and cell loading (see Figure 17.7). The optima occur whilst the conversions attained in the CSMR are 30 and 50%, respectively. Varying these operational parameters, the CSMR could produce well-concentrated streams at a still acceptable reactor capacity. Due to the complex effect of the substrate concentration on the reaction rate, the highest conversion is attained at a reactor capacity approximately half that of the maximum (see Figure 17.6a).

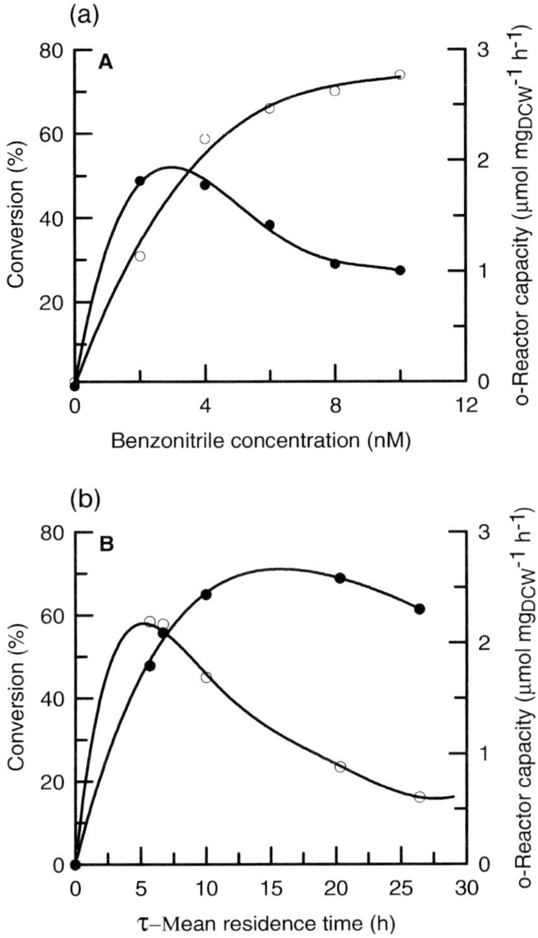

Figure 17.6 UF-membrane bioreactor loaded with 10 mg_{DCW}, and run at 10 °C. The reactor was fed with benzonitrile buffered (50 mM sodium phosphate buffer, pH 7.0) solution. (a) % Conversion at steady state (●) and reactor capacity (○) as a function of benzonitrile concentration fed at flow-rate of 12 ml h^{-1}. (b) % Conversion at steady state (●) and reactor capacity (○) as a function of τ-mean residence time in runs at 4 mM benzonitrile.

17.6 Concluding Remarks

The microorganism *Microbacterium imperiale* CBS 498-74, used as resting cells, is able to hydrate nitriles via two steps by nitrile hydratase and amidase catalysed reactions. Both aliphatic (acrylonitrile, propionitrile) and aromatic (benzonitrile, 3-cyanopyridine) nitriles were converted to the corresponding amide or acid.

The use of a CSMR allowed to investigate the dependence of reaction rate on key parameters, such as temperature, substrate concentration and cell loading.

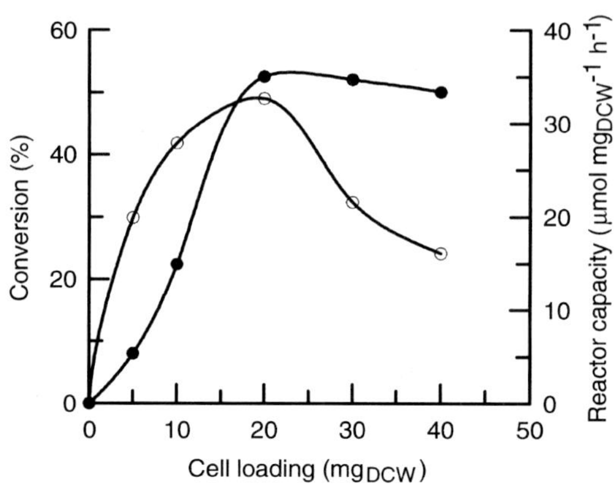

Figure 17.7 % Conversion at steady state (●) and reactor capacity (○) as a function of cell loading in UF-membrane reactors operated at 50 °C with substrate feed 100 mM Nicotinamide in buffered solution.

This was performed for each enzyme independently, feeding the reactor with the appropriate substrate (nitrile for the cascade reaction, amide for the sole amidase). The activation energies of both catalysed reactions were evaluated together with those of the inactivation process that inevitably takes place even under the most suitable operational conditions. In the nitrile hydratase/amidase cascade system nitrile hydratase is the more labile enzyme that imposes process temperature choice. These findings make accessible the complete kinetic expression of the dependence from temperature of reaction rate, allowing accurate prediction on reactor performances for process scale-up.

As far as substrate concentration is concerned, most of the investigated reactions were substrate or product inhibited, and in the case of acrylonitrile bioconversion nitrile hydratase is irreversibly inactivated by substrate concentration as high as 800 mM. However, we proved [31] that using high biocatalyst loading in multi-cycle CSMR experiments performed at 4 °C it was possible to control nitrile hydratase inactivation during 500 hours operation. In these conditions, feeding CSMR with inactivating acrylonitrile concentration (300 mM) almost complete conversion (93.7%) was attainable. Acrylamide in the effluent could be concentrated, restoring the initial substrate concentration in the product stream after each cycle and feeding it to the subsequent reactor. The apparent lack of amidase activity that we also observed in the propionamide production [33] could be attributed either to the low temperature chosen to perform the CSMR runs or to amidase substrate (acrylamide, propionamide) inhibition. In benzonitrile biotransformation in CSMR, the correct choice of operational conditions (mainly temperature and residence time) permitted to arrest the nitrile hydratase/amidase reaction either at the amide intermediate or at the final corresponding acid stage [34].

The continuous reactor proved to be very flexible, in that varying the operational parameters one can get a wide spectrum of product stream quality, that is high productivity or high concentration.

Acknowledgments

The authors would like to acknowledge the support of the University of L'Aquila Research Fund, COST action D25/0002/02 'Nitrile- and Amide-hydrolysing Enzymes as Tools in Organic Chemistry', and the institutional research concept AV0Z50200510 (Institute of Microbiology).

References

1 Osswald, S., Wajant, H. and Effenberger, F. (2002) *European Journal of Biochemistry*, **269**, 680–688.
2 Asano, Y., Tani, Y. and Yamada, H. (1980) *Agricultural and Biological Chemistry*, **44**, 2251–2252.
3 Fernandes, B.C.M., Mateo, C., Kiziak, C., Chmura, A., Wacker, J., van Rantwijk, F., Stolz, A. and Sheldon, R.A. (2006) *Advanced Synthesis Catalysis*, **348**, 2597–2603.
4 Harper, D. (1977) *The Biochemical Journal*, **165**, 309–319.
5 Kobayashi, M. and Shimizu, S. (1994) *FEMS Microbiology Letters*, **120**, 217–224.
6 Asano, Y., Fujishiro, K., Tani, Y. and Yamada, H. (1982) *Agricultural and Biological Chemistry*, **46**, 1165–1174.
7 Asano, Y., Tachibana, M., Tani, Y. and Yamada, H. (1982) *Agricultural and Biological Chemistry*, **46**, 1175–1181.
8 Ramakrishna, C. and Desai, J.D. (1992) *Biotechnology Letters*, **14**, 827–830.
9 Cramp, R.A. and Cowan, D.A. (1999) *Biochimica et Biophysica Acta*, **1431**, 249–260.
10 Pereira, R.A., Graham, D., Rainey, F.A. and Cowan, D.A. (1998) *Extremophiles*, **2**, 347–357.
11 Mayaux, J.F., Cerberlaud, E., Soubrier, F., Faucher, D. and Pétré, D. (1990) *Journal of Bacteriology*, **172**, 6764–6773.
12 Cerbelaud, E., Levy-Schil, S., Petre, D. and Soubrier, F. (1995) PCT Int. Appl. WO 9504828 A1 950216 (Patent), Rhone–Poulenc Chemie.
13 Tani, Y., Kurihara, M. and Nishise, H. (1989) *Agricultural and Biological Chemistry*, **53**, 3151–3158.
14 Endo, T. and Wantanabe, I. (1989) *FEBS Letters*, **243**, 61–64.
15 Nagasawa, T. and Yamada, H. (1989) *Trends in Biotechnology*, **7**, 153–158.
16 Yamaki, T., Oikawa, T., Ito, K. and Nakamura, T. (1997) *Journal of Fermentation and Bioengineering*, **5**, 474–477.
17 Nawaz, M.S. and Chapatwala, K.D. (1991) *Canadian Journal of Microbiology*, **37**, 411–418.
18 Mylerová, V. and Martínková, L. (2003) *Current Organic Chemistry*, **7**, 1–17.
19 Kaplan, O., Vejvoda, V., Plíhal, O., Pompach, P., Kavan, D., Fialová, P., Bezouška, K., Macková, M., Cantarella, M., Jirků, V., Křen, V. and Martínková, L. (2006) *Applied Microbiology and Biotechnology*, **73**, 567–575.
20 Van der Walt, J.P., Brewis, E.A. and Prior, B.A. (1993) *Systematic and Applied Microbiology*, **16**, 330–332.
21 Yamada, H. and Kobayashi, M.B. (1996) *Bioscience Biotechnology and Biochemistry*, **60**, 1391–1400.
22 Martínková, L. and Křen, V. (2002) *Biocatalysis and Biotransformation*, **20**, 73–93.
23 Ogawa, J. and Shimizu, S. (1999) *Trends in Biotechnology*, **17**, 13–20.
24 Watanabe, I., Satoh, Y. and Enomoto, K. (1987) *Agricultural and Biological Chemistry*, **51**, 3193–3199.

25 Watanabe, I., Satoh, Y., Enomoto, K., Seki, S. and Sakashita, K. (1987) *Agricultural and Biological Chemistry*, **51**, 3201–3206.

26 Tani, Y., Kurihara, M. and Nishise, H. (1989) *Agricultural and Biological Chemistry*, **53**, 3151–3158.

27 Lee, C.Y. and Chang, H.N. (1990) *Biotechnology Letters*, **12**, 23–28.

28 Lee, C.Y., Sang, K.C. and Chang, H.N. (1993) *Journal of Microbiology and Biotechnology*, **3**, 36–45.

29 Hwang, J.S. and Chang, H.N. (1987) *Biotechnology Letters*, **9**, 237–242.

30 Hwang, J.S. and Chang, H.N. (1989) *Biotechnology and Bioengineering*, **34**, 380–386.

31 Cantarella, M., Cantarella, L., Spera, A. and Alfani, F. (1998) *Journal of Membrane Science*, **147**, 279–290.

32 Cantarella, M., Spera, A. and Alfani, F. (1998) *Annals of the New York Academy of Sciences*, **864**, 224–227.

33 Cantarella, M., Cantarella, L., Gallifuoco, A., Frezzini, R., Spera, A. and Alfani, F. (2004) *Journal of Molecular Catalysis B: Enzymatic*, **29**, 105–113.

34 Cantarella, M., Cantarella, L., Gallifuoco, A. and Spera, A. (2006) *Enzyme and Microbial Technology*, **38**, 126–134.

35 Cantarella, M., Cantarella, L., Gallifuoco, A., Intellini, R., Kaplan, O., Spera, A. and Martínková, L. (2008) *Enzyme and Microbial Technology*, **42**, 222–229.

36 Cantarella, M., Spera, A., Leonetti, P. and Alfani, F. (2002) *Journal of Molecular Catalysis B: Enzymatic*, **19–20C**, 405–414.

37 Alfani, F., Cantarella, M., Spera, A. and Viparelli, P. (2001) *Journal of Molecular Catalysis B: Enzymatic*, **11**, 687–697.

38 Alfani, F., Cantarella, M. and Vitolo, M. (1999) *Revista de Farmácia e Química*, **32**, 17–23.

39 Cantarella, M., Cantarella, L., Gallifuoco, A. and Spera, A. (2006) *Journal of Industrial Microbiology and Biotechnology*, **33**, 208–214.

40 Wegman, M.A., Heinemann, U., Stolz, A., van Rantwijk, F. and Sheldon, R.A. (2000) *Organic Process Research and Development*, **4**, 318–322.

18
Enzymes Catalyzing C—C Bond Formation for the Synthesis of Monosaccharide Analogs

Laurence Hecquet, Virgil Hélaine, Franck Charmantray and Marielle Lemaire

18.1
Introduction

Recent advances in chemoenzymatic synthesis have made possible the stereocontrolled synthesis of a large number of organic substances, in particular by carbon–carbon bond-forming reactions. One area is the synthesis of water-soluble multifunctional organic compounds, such as carbohydrates and carbohydrate-like chemical entities. Our group is interested in using enzymes catalyzing C—C bond formation for the synthesis of monosaccharides and analogs, specifically a lyase such as fructose-1,6-bisphosphate aldolase (EC 4.1.2.13) and a transferase such as transketolase (E.C 2.2.1.1). Here, we present recent syntheses using both enzymes. In addition, to extend the range of monosaccharides and analogs produced, the modification of the substrate specificity of transketolase is discussed, together with the corresponding screening and selection tests.

18.2
Recent Syntheses Involving Transketolase and Fructose-1,6-*bis*phosphate Aldolase

The use of aldolases and transketolase has opened the way to many highly multifunctional organic compounds [1]. In organic synthesis, the most widely used dihydroxyacetonephosphate (DHAP) aldolase is the commercially available fructose-1,6-*bis*phosphate aldolase from rabbit muscle (FruA). This enzyme is a key enzyme of the glycolytic pathway, reversibly catalyzing the cleavage of fructose-1,6-*bis*phosphate into two three-carbon units, DHAP and D-glyceraldehyde-3-phosphate. *In vitro*, the equilibrium is in favor of the addition reaction and the enzyme forms the C3–C4 bond with the (3*S*,4*R*)-configuration. This enzyme is very specific towards its donor substrate, DHAP, tolerating only very close DHAP analogs, but accepting a large number of aldehydes as acceptors. Many strategies for obtaining sugars and analogs using an FruA-catalyzed reaction for stereochemical control have been described [1–5]. For synthetic applications, DHAP is required and

Modern Biocatalysis: Stereoselective and Environmentally Friendly Reactions.
Edited by Wolf-Dieter Fessner and Thorleif Anthonsen

ISBN: 978-3-527-32071-4

various strategies have been reported [6], together with our own procedures [7b, 8–10]. Transketolase *in vivo* is the key enzyme in the non-oxidative pentose phosphate pathway. Transketolase catalyzes the reversible transfer of a two-carbon ketol unit (hydroxyacetyl group) from a ketose phosphate donor to an aldose phosphate acceptor, creating a new asymmetric center (C3) with the *S*-configuration. Transketolase requires a divalent cation (Mg^{2+} or Ca^{2+}) and thiamine pyrophosphate (ThDP) as co-factors. We carried out the first systematic explorations on the substrate specificity and synthetic usefulness of transketolase. Our group showed [11a] that hydroxypyruvate (HPA), which is easily prepared from bromopyruvate, could be used as a donor substrate, rendering the reaction irreversible owing to its decarboxylation catalyzed by transketolase. Transketolase appears highly specific for ketol donor substrates and for hydroxyaldehyde acceptor substrates with the (*R*)-configuration leading to D-*threo* (3*S*,4*R*)-ketoses. For synthetic purposes, a wide range of aldehyde acceptors have been used with transketolase isolated from spinach leaves [11] and then from *Escherichia coli* [12] and *Saccharomyces cerevisiae* [13]. Transketolase and FruA can yield the same ketose if the aldehyde acceptor of FruA (especially an α-hydroxylated aldehyde as shown in Figure 18.1) has one carbon less than the aldehyde acceptor used for transketolase. Hence for a particular synthesis, the choice of the enzyme will depend strongly on the availability of the acceptor substrate.

In this section we describe recent syntheses of monosaccharide analogs we have developed with FruA and transketolase (e.g. aminocyclitols, D-xylulose and D-xylose analogs) including the recent syntheses of DHAP, the donor substrate of FruA.

18.2.1
DHAP Syntheses

In this section we will describe our own recent contributions: a chemical route from dihydroxyacetone based on monofunctionalization catalyzed by a lipase [9]

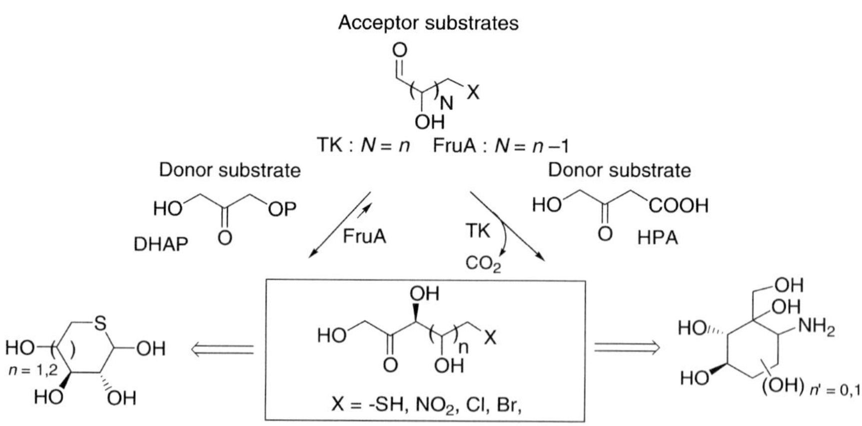

Figure 18.1 FruA and transketolase as complementary tools for the synthesis of monosaccharide analogs and derivatives.

and an enzymatic route from glycidol [10]. The DHAP preparations by chemical and enzymatic routes have been recently reviewed by Schümperli *et al.* [6].

18.2.1.1 DHAP Synthesis from Dihydroxyacetone

With the aim of generating DHAP of high purity on a large scale and with high reproducibility for further applications in synthesis, we developed the route depicted in Figure 18.2. The dimethyl acetal precursor **4** used by Ballou *et al.* and Valentin *et al.* [7] appeared best suited for achieving the last steps in high yield. It offers the advantage of producing highly pure DHAP, which can greatly simplify the purification steps of sugars and analogs after FruA aldolization. In addition, compound **4** is stable and can be stored for several months and DHAP can be prepared just before use. This work thus focused on improving the chemical route to supply this stock compound in a satisfactory total yield and with short purification steps. The use of a lipase to control the monoesterification of the diol **2** formed by treatment of **1** with trimethylorthoformate was our key step. The lipase AK from Amano proved to be efficient in the transesterification with vinylacetate in diisopropylether. The choice of benzyl over phenyl for phosphate protection was based on the ready hydrogenolysis of the former with no need for pressure or a costly PtO_2 catalyst. The phosphorylation of alcohol **3** was then performed with dibenzyl phosphoroiodidate generated in *situ* from tribenzylphosphite and iodine [8] to yield compound **4**.

From **4**, three reactions were carried out consecutively without isolation of intermediates. First, a basic hydrolysis of the ester group gave the crude alcohol **5**. Next, this compound was hydrogenated in the presence of Pd/C (10%) in MeOH to remove the benzyl groups quantitatively. Finally, we took advantage of the acidity of the free phosphate monoester to catalyze the hydrolysis of the dimethoxyacetal of **6**. This reaction was carried out in distilled water at 45 °C. Hydrolysis of **6** gave a 424 mM DHAP solution in 50 min (84% yield calculated from **4**). Under acidic conditions DHAP is unstable, with low pH and high temperature inducing phosphate hydrolysis. Under our conditions, the ketal hydrolysis rate was high enough and the phosphate hydrolysis rate low enough to give DHAP in high yield and with excellent purity. This short and easy procedure needed only two chromatography runs on silica gel and used inexpensive reagents. Overall the yield of DHAP from dihydroxyacetone (DHA) was 47%. A DHAP solution of high purity

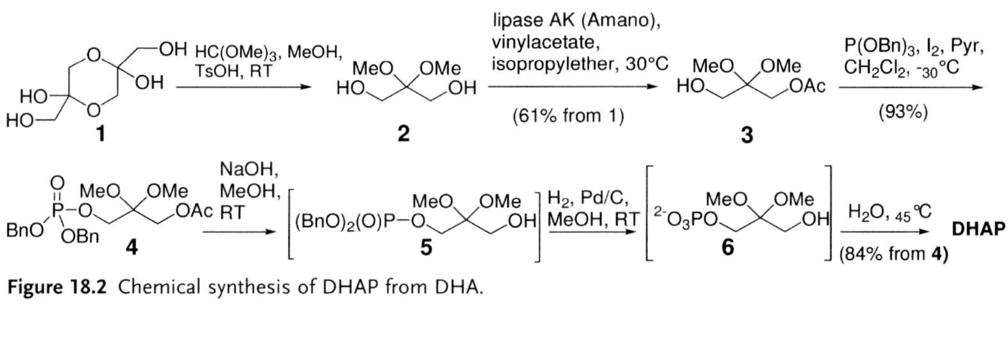

Figure 18.2 Chemical synthesis of DHAP from DHA.

and concentration could thus be obtained on a scale of several grams and was stored at −18 °C for several months without noticeable decomposition.

18.2.1.2 DHAP Synthesis from Rac-Glycidol

However, wider practical applications of aldolases require cheap and ready access to DHAP. Expensive or toxic reagents, multistep purification procedures, and functional group protection complicate the chemical strategies. A promising alternative would combine a short and inexpensive DHAP synthesis followed by an *in situ* aldolization reaction catalyzed by aldolase. By this approach, DHAP was recently obtained by DHA phosphorylation catalyzed by DHA kinase with ATP [14]. One drawback of this procedure was the need for a system of ATP regeneration from acetyl phosphate. Sheldon *et al.* [15, 16] also described a coupled multienzymatic system for preparing carbohydrates from glycerol. The key step of this strategy was an efficient DHAP synthesis from L-glycerol-3-phosphate (L-G3P) catalyzed by L-glycerophosphate oxidase (L-GPO) in the presence of catalase [17]. The authors elegantly prepared D,L-G3P by phosphorylation of glycerol with inexpensive pyrophosphate catalyzed by phytase. However, the quantitative conversion of pyrophosphate into D,L-G3P required a glycerol concentration of 95% (v/v), restricting the synthesis to hydrophobic carbohydrate analogs. We described a short, practical, and efficient chemoenzymatic synthesis of DHAP from inexpensive *rac*-glycidol **7** [10] (Figure 18.3). Our route had two steps: regioselective opening of the *rac*-glycidol epoxide ring **7** with phosphate sources to generate D,L-G3P and oxidation of L-G3P **8** to DHAP under oxygen mediated by L-GPO, coupled with hydrogen peroxide decomposition by catalase.

In the first step, we showed that, with various phosphate sources, the opening of *rac*-glycidol **7** was pH dependent with an optimum pH above 10 obtained with $Na_2(K_2)HPO_4$ and Na_3PO_4. In these cases, D,L-G3P was obtained in a 50–60% yield. Similar results were observed when Na^+ replaced K^+. Surprisingly, the yield of D,L-G3P decreased when the concentrations of the two reagents increased. For these reasons, we decided to use stoichiometric amounts of *rac*-glycidol and Na_2HPO_4 or Na_3PO_4 as phosphate sources. In the second step, after conversion of glycidol **7** into D,L-G3P, the pH was adjusted to 6.8 in order to optimize GPO activity and then catalase and L-GPO were added to the reaction mixture. The behavior of L-GPO was different when Na_2HPO_4 or Na_3PO_4 were used in the previous step. With Na_3PO_4, whatever the quantity of L-GPO, the oxidation reaction stopped after a few minutes. In contrast, when Na_2HPO_4 was used, L-G3P was fully converted into DHAP after oxidation by L-GPO. Finally, this two-step proce-

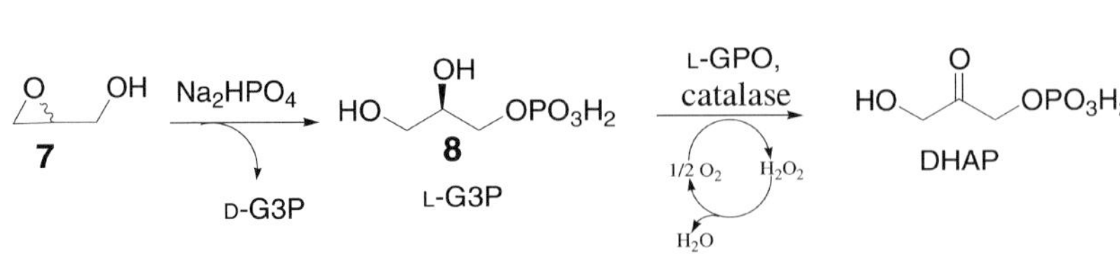

Figure 18.3 Enzymatic synthesis of DHAP from *rac*-glycidol.

dure from *rac*-glycidol **7** gave DHAP in a 28% overall yield (the maximum theoretical yield was 50%). To conclude, we performed an attractive two-step 'one-pot' synthesis of DHAP from *rac*-glycidol, a cheap commercially available starting material. The controlled opening of the *rac*-glycidol epoxide ring with Na_2HPO_4 in water gave D,L-G3P in a 55% yield. L-G3P conversion to DHAP by means of L-GPO and catalase was found to be quantitative. We noted that L-GPO and catalase could be co-immobilized as described earlier [18]. DHAP was thus readily available for an *in situ* coupled FruA reaction as we showed with the preparation of 5-halo-D-xylulose (see Section 18.2.3.1).

18.2.2
Synthesis of Aminocyclitols

Recently, we developed a highly stereoselective methodology for nitro and aminocyclitol preparation using FruA, which catalyzed the addition of DHAP to nitrobutyraldehydes [19, 20] (Figure 18.4). These, with or without hydroxyl groups, are good substrates for FruA. In this procedure, the key step is based on a one-pot, two-enzyme process, where three reactions take place: aldolization catalyzed by FruA, phosphate hydrolysis catalyzed by a phosphatase, and an intramolecular Henry reaction (nitroaldolization). Two families of nitrocyclitols were prepared according to the configuration of the carbon β to the nitro group.

In a first exploration [19], starting from **9**, the linear ketose usually obtained after aldol condensation was never obtained: the nitromethylene reacted instantaneously with the ketone to give nitrocyclitol **11** in a highly stereoselective process. Only one major isomer was isolated after hydrolysis of the phosphate with phytase (Figure 18.5).

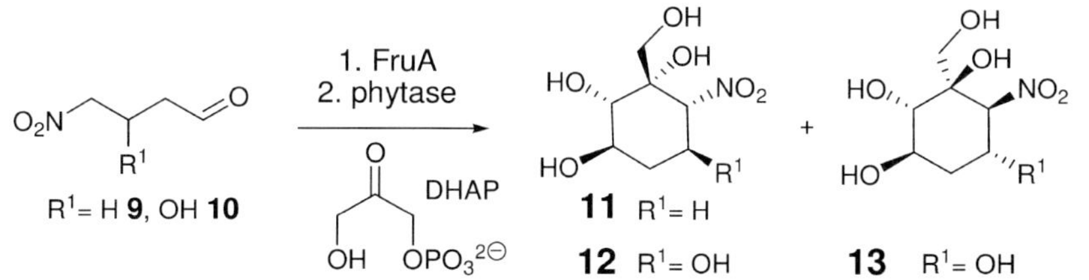

Figure 18.4 Enzymatic strategy for nitrocyclitol synthesis.

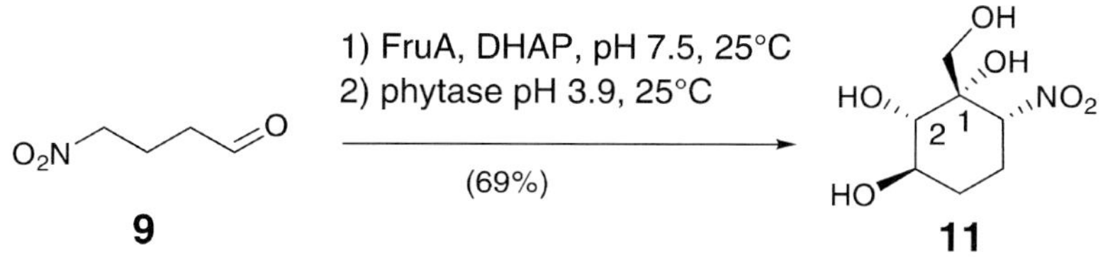

Figure 18.5 Isomer obtained after aldolization with FruA.

FruA is stereoselective and stereocenter **2** is created with the *S*-configuration. Taking into account this stereoselectivity, the Nuclear Overhauser Effects (NOEs) observed, and the coupling constants measured, the configurations found for compound **11** were 1*S*, 2*S*, 3*R*, and 6*R*. All the substituents are equatorially positioned on the cyclohexane ring. The (2*S*,3*R*)-configurations are consistent with the usually observed diastereoselectivity of FruA. In a second exploration [20], the racemic aldehyde **10** was condensed with DHAP after ketal hydrolysis of *rac*-**14** under acidic conditions. In this case, two major isomers **12** and **13** were isolated in 35 and 29% yields after flash chromatography purification.

Interestingly, the configuration of the alcohol function on **14** influenced the intramolecular nitrocyclization. When the alcohol was *S*, the nitrocyclitol **12**, possessing the same (1*S*,6*R*)-configurations as compound **11** (Figure 18.6), was isolated. When the alcohol was *R*, the configuration 1*R*,6*S* was obtained for compound **13**. When an optically pure compound (*R*)-**14** was used only one isomer **13** was isolated in a 50% yield.

The nitro group of **11**, **12**, and **13** was then reduced on PtO_2 under 50 psi of hydrogen (Figure 18.7). The three aminocyclitols **15**, **16**, and **17** were isolated in 80, 76, and 76% yields, respectively. These new aminocyclitols are analogs of valiolamine, a natural inhibitor of α-glucosidases.

Opening a new route to valiolamine analogs, this work broadens the scope of the enzymatic aldol reaction combining in one pot the formation of two carbon–carbon bonds with high stereoselectivity. We demonstrate a short efficient synthesis of new nitro- and aminocyclitols that considerably reduces the laborious

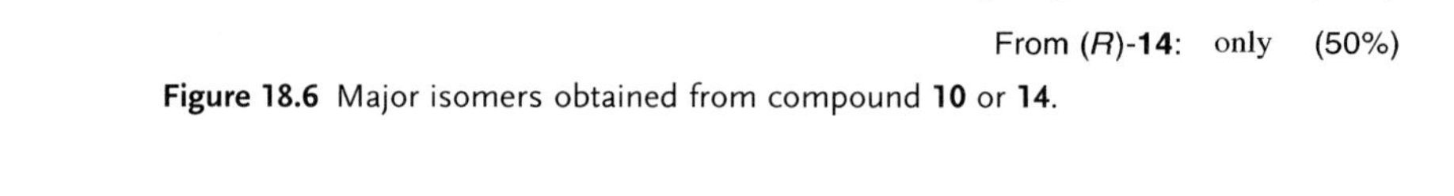

Figure 18.6 Major isomers obtained from compound **10** or **14**.

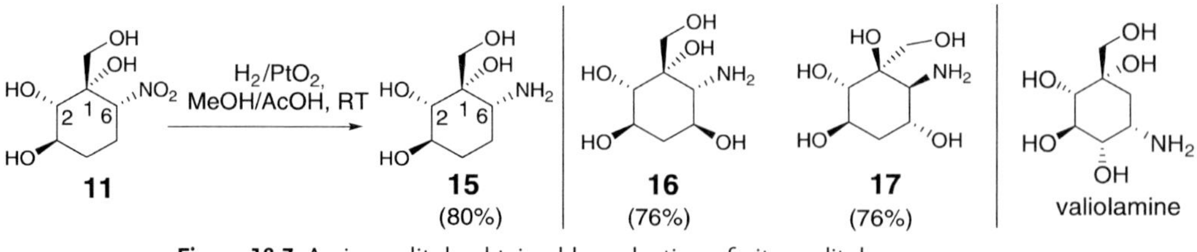

Figure 18.7 Aminocyclitols obtained by reduction of nitrocyclitols.

protection–deprotection steps classically necessary when sugars are used as starting materials.

18.2.3
Synthesis of 5-D-Xylulose and 5-D-Xylose Analogs

Halo-sugars are very useful building blocks for designing further reactions by substitution of halogen atoms. From these compounds, we have developed a chemoenzymatic strategy giving 5-thio-D-xylose in four steps (the key step being catalyzed by FruA).

18.2.3.1 Synthesis of 5-halo-D-xylulose

We have developed a two-step procedure, from DHAP produced *in situ* from glycidol **7** (see Section 18.2.1.2), for the synthesis of 5-halo-D-xylulose **19** mediated by FruA [10] (Figure 18.8).

*Fru*A was added to the reaction mixture when the oxidation of L-glycerol-3-phosphate catalyzed by L-GPO was complete. FruA was used for catalyzing the aldol addition of DHAP onto 2-chloro- or 2-bromoacetaldehydes **20** to give the aldol adduct 5-chloro or 5-bromo-D-xylulose-1-phosphate. The yield was quantitative. The dephosphorylation of 5-chloro and 5-bromo-D-xylulose-1-phosphate was carried out by the addition of acid phosphatase. After purification, 5-chloro-D-xylulose and 5-bromo-D-xylulose were recovered as pure compounds in 47 and 12% yields, respectively, from DHAP. In this study, we have shown that DHAP generated from glycidol **7** can be used *in situ* as a donor substrate of *Fru*A in the presence of 2-halo-acetaldehydes **20** as acceptor substrates for the synthesis of 5-halo-D-xylulose **19**. Given that DHAP aldolases display a broad specificity towards acceptor substrates, this strategy can be applied generally to the synthesis of various analogs of monosaccharides.

18.2.3.2 Synthesis of 5-thio-D-xylopyranose 21

We investigated the use of the 5-halo-D-xylulose sugars **19** obtained above to design further reactions by substitution of the halogen atom and in particular by a thiol group for the preparation of 5-thio-D-xylopyranose **21** [13f]. This compound is an inhibitor of β-D-xylosidase and a useful chiral building block for the synthesis of D-xylopyranosides displaying antithrombotic activities [21a].

Clinical studies with some of these compounds are in progress. For instance, Odiparcil [21b] (**22** X = *S*, Y = *O*, *R* = methylcoumarin, Figure 18.9) is currently

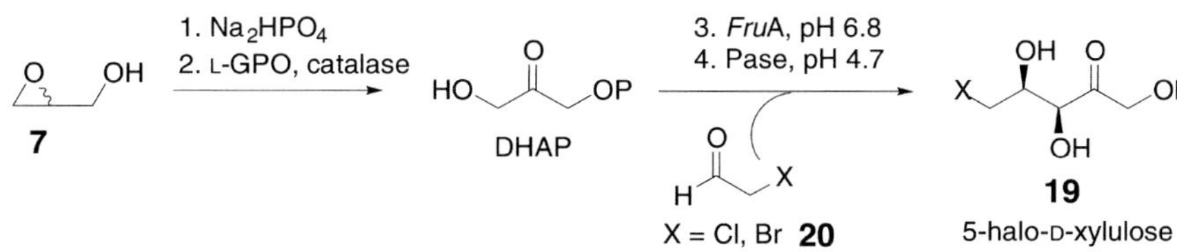

Figure 18.8 Enzymatic strategy for the synthesis of 5-halo-D-xylulose from glycidol.

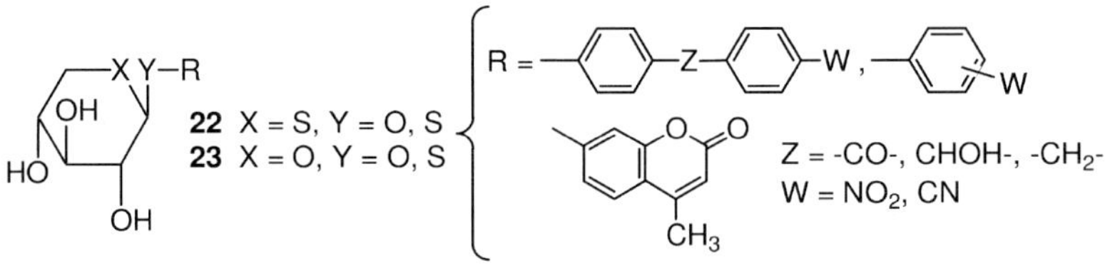

Figure 18.9 D-Xylopyranosides displaying antithrombotic activities.

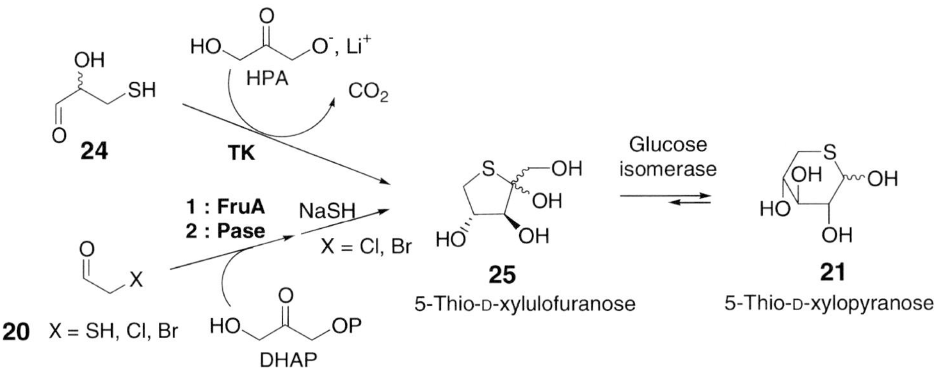

Figure 18.10 Routes catalyzed by transketolase or FruA to give 5-thio-D-xylopyranose **21**.

undergoing phase II clinical trials. Various methods for producing these glycosides in a highly stereoselective manner have been described [22]. The synthesis of the aldose moiety of this compound, 5-thio-D-xylopyranose **21** has been obtained from D-xylose involving protection–deprotection and replacement with nucleophilic sulfur-containing reagents [21b, 23]. Our interest was in preparing the aldose moiety by enzymatic means based on the isomerization of the corresponding ketose catalyzed by glucose isomerase (EC 5.3.1.5) GlcI. This strategy was proposed by Wong *et al.* [24a]. The efficiency of this rather general approach was initially reported for the synthesis of D-glucose derivatives [24b] and also by Fessner *et al.* for the synthesis of L-fucose analogs [24c] using L-fucose isomerase. First of all, we investigated chemoenzymatic strategies based on transketolase or FruA reactions to obtain the appropriate ketose **25**: a route catalyzed by transketolase with HPA as a donor substrate and thioglyceralehyde **24** as acceptor substrate and a route catalyzed by FruA with DHAP as donor substrate and acetaldehyde analogs **20** as acceptor substrates (Figure 18.10). We showed that the use of an acceptor substrate with a free thiol function was rate limiting for these enzymes.

Taking this result into account, we have developed an efficient enzymatic route catalyzed by FruA in which the acceptor substrate is a commercially available 2-haloacetaldehydes (chloro- or bromo-) **19** yielding 5-halo-D-xylulose after dephosphorylation catalyzed by acid phosphatase (see Section 18.2.3.1). The thiol was introduced after the enzymatic step by displacement of halogen with NaSH, to give **25**. After isomerization of **25** catalyzed by glucose isomerase, the final yield of 5-thio-D-xylopyranose **21** was 23% from commercially available 2-chloro- or 2-bromoacetaldehyde **20**. To our knowledge, enzymatic isomerization of thioketose into thioaldose on a preparative scale had never been described before. The yield of this chemoenzymatic way to 5-thio-D-xylopyranose **21** is lower than with common chemical synthesis, but this strategy offers an attractive alternative to conventional chemical methods because of its stereochemical control, mild conditions, and non-requirement of a protecting group. To improve the yield of isomerization of thioketose **25** into thioaldose **21** recycling of unreacted **25** could be considered, as in industrial production of D-fructose from D-glucose.

18.3 Modification of Substrate Specificity of Yeast Transketolase

In 1998, our group was the first to investigate a project on modification of the substrate specificity of yeast transketolase [25]. First of all, in collaboration with Profssor Gunter Schneider on the basis of the three-dimensional structure of yeast transketolase, we showed the role of the residues of the active site by directed mutagenesis and particularly the role of Asp477 in the control of the enantioselectivity of the transketolase reaction [25]. The replacement of this residue by an alanine caused loss of enantioselectivity. Our goal being to create libraries of mutants in order to modify the substrate specificity of transketolase, we looked for a screening or selection system, an absolute prerequisite for identifying evolved enzyme variants that display improved properties. Only screening tests have been developed for this type of enzyme. To detect modifications of the stereospecificity of mutant transketolase, in 2003 we developed the first stereospecific and fluorogenic screening test for this enzyme based on the detection of coumarine from a stereochemical probe [26] (Figure 18.11). From this compound, the transketolase reaction released an aldehyde, leading to β-elimination of umbelliferone (highly fluorescent) catalyzed by bovine serum albumin as reported previously by Reymond with other enzymes as transaldolase [27]. We showed that this assay can be used for discriminating the stereoselectivity of transketolase using fluorogenic probes with various configurations on C3 and C4 [28].

However, these *in vitro* assays have a major disadvantage for screening large libraries of mutant enzymes: each clone has to be disrupted separately to

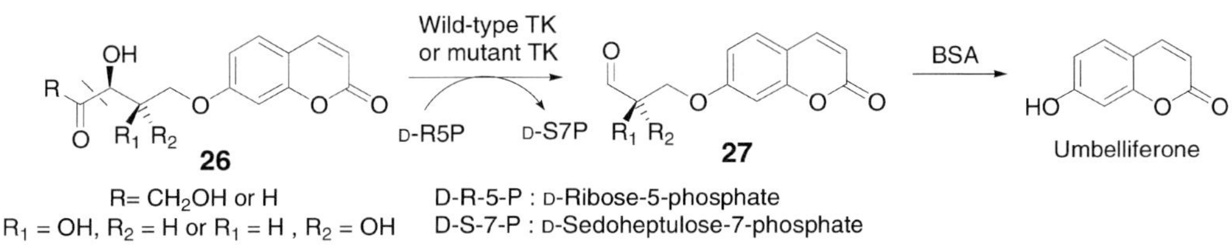

Figure 18.11 Fluorogenic assay for transketolase activity.

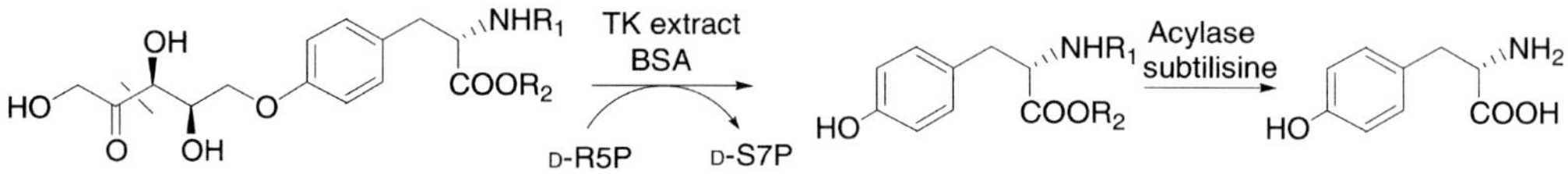

Figure 18.12 Enzymatic assay for transketolase based on L-tyrosine release.

determine the catalytic properties of the mutant enzyme. Our purpose is now to develop an *in vivo* selection test, never before described for transketolase, that allows the selection of large libraries of microorganisms. This approach needs a system for linking catalytic activity to a survival factor, a growth advantage for the microorganisms. Our strategy is based on the use of substrates that can release an amino acid required for the growth of cells, auxotrophic for this amino acid and containing a sugar moiety with various stereochemical features specifically recognized by a desired mutant transketolase. We turned to the system used previously for the detection of transketolase activity with a fluorogenic substrate [26, 28] (Figure 18.12). To mimic this strategy, we replaced coumarine by L-tyrosine, the only amino acid bearing a phenolate moiety.

In the first step, we showed by analytical studies that compound **28** was a donor substrate for transketolase in the presence of D-ribose-5-phosphate as acceptor substrate and that in the second step, the hydroxylated aldehyde released **29** led to the β-elimination of protected L-tyrosine. We showed that the free L-tyrosine can thus be obtained by enzymatic deprotection of *N*-acetyl-L-tyrosine ethyl ester using acylase and subtilisine. In this conditions, it should be possible to carry out this assay *in vivo* in the presence of host cells overexpressing transketolase and auxotrophic for L-tyrosin. This strategy should offer the first stereospecific selection test of transketolase mutants. The principle of this assay may be extended to other enzymes that can release aldehydes β-substituted by L-tyrosine.

18.4 Conclusion

FruA and transketolase offer complementary tools for the enzymatic synthesis of monosaccharides and analogs. FruA creates two new asymmetric centers, against only one asymmetric center in the case of transketolase. However, transketolase being enantioselective, a D-*threo* (3*S*,4*R*) ketose is finally obtained in both cases. To create other configurations on C3 and C4, other aldolases can be used to catalyze the synthesis of the three other stereoisomers. These are tagatose-1,6-*bis*phosphate aldolase, rhamnulose-1-phosphate aldolase and fuculose-1-phosphate aldolase. Another route is the modification of the substrate specificity by mutagenesis, in which we are engaged with transketolase. The general goal is now to extend the synthetic potential of these enzymes and also to seek new ones catalyzing C—C bond formation.

References

1 (a) Machajewski, T.D. and Wong, C.H. (2000) *Angewandte Chemie – International Edition*, **39**, 1352–1374.
(b) Dean, S.M., Greenberg, W.A. and Wong, C.-H. (2007) *Advanced Synthesis Catalysis*, **349**, 1308–1320 (and ref cited therein).
(c) Samland, A.K. and Sprenger, G.A. (2006) *Applied Microbiology and Biotechnology*, **71**, 253–264.

2 Wong, C.H., Mazenod, F.P. and Whitesides, G.M. (1983) *The Journal of Organic Chemistry*, **48**, 3493–3497.

3 Straub, A., Effenberger, F. and Fischer, P. (1990) *The Journal of Organic Chemistry*, **55**, 3926–3932.

4 (a) Bednarski, M.D., Waldmann, H.J. and Whitesides, G.M. (1986) *Tetrahedron Letters*, **27**, 5807–5810.
(b) Bednarski, M.D., Simon, E.S., Bischofberger, N., Fessner, W.D., Kim, M.J., Lees, W., Saito, T., Waldmann, H. and Whitesides, G.M. (1989) *Journal of the American Chemical Society*, **111**, 627–635.

5 (a) Lemaire, M., Valentin, M.L., Hecquet, L., Demuynck, C. and Bolte, J. (1995) *Tetrahedron: Asymmetry*, **6**, 67–70.
(b) Andre, C., Guérard, C., Hecquet, L., Demuynck, C. and Bolte, J. (1998) *Journal of Molecular Catalysis B: Enzymatic*, **5**, 459–466.
(c) Crestia, D., Demuynck, C. and Bolte, J. (2004) *Tetrahedron*, **60**, 2417–2425.

6 Schümperli, M., Pellaux, R. and Panke, S. (2007) *Applied Microbiology and Biotechnology*, **75**, 33–45.

7 (a) Ballou, C.E. and Fischer, H.O.L. (1956) *Journal of the American Chemical Society*, **78**, 1659.
(b) Valentin, M.-L. and Bolte, J. (1995) *Bulletin de la Societe Chimique de France*, **132**, 1167–1171.

8 Gefflaut, T., Lemaire, M., Valentin, M.L. and Bolte, J. (1997) *The Journal of Organic Chemistry*, **62**, 5920–5922.

9 Charmantray, F., El Blidi, L., Gefflaut, T., Hecquet, L., Bolte, J. and Lemaire, M. (2004) *The Journal of Organic Chemistry*, **69**, 9310–9312.

10 Charmantray, F., Dellis, P., Samreth, S. and Hecquet, L. (2006) *Tetrahedron Letters*, **47**, 3261–3263.

11 (a) Bolte, J., Demuynck, C. and Samaki, H. (1987) *Tetrahedron Letters*, **27**, 5525–5528.
(b) Bolte, J., Demuynck, C., Hecquet, L. and Samaki, H. (1990) *Carbohydrate Research*, **206**, 79–85.
(c) Hecquet, L., Bolte, J. and Demuynck, C. (1993) *Bioscience, Biotechnology, and Biochemistry*, **57**, 2174–2176.
(d) Hecquet, L., Bolte, J. and Demuynck, C. (1996) *Tetrahedron*, **52**, 8223–8232.

12 (a) Morris, K.G., Smith, M.E.B. and Turner, N.J. (1996) *Tetrahedron: Asymmetry*, **7**, 2185–2188.
(b) Zimmermann, F.T., Schneider, A., Schörken, Y., Sprenger, G.A. and Fessner, W.D. (1999) *Tetrahedron: Asymmetry*, **10**, 1643–1646.

13 (a) Ziegler, T., Straub, A. and Effenberger, F. (1988) *Angewandte Chemie – International Edition in English*, **27**, 716–721.
(b) Kobori, Y., Myles, D.C. and Whitesides, G.M. (1991) *The Journal of Organic Chemistry*, **22**, 5899–5907.
(c) André, C., Guérard, C., Hecquet, L., Demuynck, C. and Bolte, J. (1998) *Journal of Molecular Catalysis B: Enzymatic*, **5**, 459–466.
(d) Guérard, C., Alphand, V., Archelas, A., Demuynck, C., Hecquet, L., Furstoss, R. and Bolte, J. (1999) *European Journal of Organic Chemistry*, **1**, 3399–3402.
(e) Crestia, D., Guérard, C., Veschambre, H., Hecquet, L., Demuynck, C. and Bolte, J. (2001) *Tetrahedron: Asymmetry*, **12**, 869–876.
(f) Charmantray, F., Dellis, P., Samreth, S. and Hecquet, L. (2006) *European Journal of Organic Chemistry*, **24**, 5526–5532.

14 Sánchez-Moreno, I., García-García, J.F., Bastida, A. and García-Junceda, E. (2004) *Chemical Communications*, **14**, 1634–1635.

15 Schoevaart, R., van Rantwijk, F. and Sheldon, R.A. (1999) *Chemical Communications*, **24**, 2465–2466.

16 Schoevaart, R., van Rantwijk, F. and Sheldon, R.A. (2000) *The Journal of Organic Chemistry*, **65**, 6940–6943.

17 Fessner, W.-D. and Sinerius, G. (1994) *Angewandte Chemie – International Edition in English*, **33**, 209–212.

18 Krämer, L. and Steckhan, E. (1997) *Tetrahedron*, **53**, 14645–14650.
19 El Blidi, L., Crestia, D., Gallienne, E., Demuynck, C., Bolte, J. and Lemaire, M. (2004) *Tetrahedron: Asymmetry*, **15**, 2951–2954.
20 E1 Blidi, L., Ahbala, M., Bolte, J. and Lemaire, M. (2006) *Tetrahedron Asymmetry*, **17**, 2684–2688.
21 (a) Samreth, S., Barberousse, V., Bellamy, F., Horton, D., Masson, P., Millet, J., Renaut, P. and Sepulchre, C. (1994) Theveniaux. *Actualites de Chimie Therapeutique*, **21**, 23–33.
(b) Bellamy, F., Horton, D., Millet, J., Picard, F., Samreth, S. and Chazan, J.B. (1993) *Journal of Medicinal Chemistry*, **36**, 898–903.
22 (a) Paulsen, H. (1982) *Angewandte Chemie – International Edition in English*, **21**, 155–173.
(b) Schmidt, R.R. (1986) *Angewandte Chemie – International Edition in English*, **25**, 212–235.
(c) Kunz, H. (1987) *Angewandte Chemie – International Edition in English*, **26**, 294–308.
23 Lalot, J., Stasik, I., Demailly, G. and Beaupère, D. (2003) *Carbohydrate Research*, **338**, 2241–2245.
24 (a) Durrwachter, J.R., Drueckhammer, D.G., Nozaki, K., Sweers, H.M. and Wong, C.-H. (1986) *Journal of the American Chemical Society*, **108**, 7812–7818.
(b) Chou, W.-C., Chen, L., Fang, J.-M. and Wong, C.-H. (1994) *Journal of the American Chemical Society*, **116**, 6191–6194.
(c) Fessner, W.-D., Gosse, C., Jaeschke, G. and Eyrisch, O. (2000) *European Journal of Organic Chemistry*, **1**, 125–132.
25 (a) Wikner, C., Meshalkina, L., Nilsson, U., Nikkola, M., Lindqvist, Y. and Schneider, G. (1994) *The Journal of Biological Chemistry*, **269**, 32144–32150.
(b) Nilsson, U., Hecquet, L., Gefflaut, T., Guérard, C. and Schneider, G. (1998) *FEBS Letters*, **424**, 49–52.
26 Sevestre, A., Helaine, V., Guyot, G., Martin, C. and Hecquet, L. (2003) *Tetrahedron Letters*, **44**, 827–830.
27 (a) Klein, G. and Reymond, J.L. (1998) *Bioorganic and Medicinal Chemistry Letters*, **8**, 1113–1116.
(b) Jourdain, N., Carlon, R.P. and Reymond, J.L. (1998) *Tetrahedron Letters*, **39**, 9415–9418.
28 Sevestre, A., Charmantray, F., Helaine, V., Lasikova, A. and Hecquet, L. (2006) *Tetrahedron*, **62**, 3969–3976.

19
Novel Strategies in Aldolase-catalyzed Synthesis of Iminosugars

Pere Clapés, Georg A. Sprenger and Jesús Joglar

19.1
Introduction

Oligosaccharides and glycoconjugates (i.e. with lipids and proteins) are molecules of paramount importance in biochemical recognition processes such as cellular adhesion, viral infections, cellular differentiation, metastasis, and numerous signal transduction events [1]. It has been shown that disorders in their metabolism have important consequences as they are involved in diseases such as type 2 diabetes, hepatitis B and C, glycosphingolipid storage disorders (e.g. Gaucher, Fabry, and Tay–Sachs diseases), cystic fibrosis, rheumatoid arthritis (e.g. chronic polyarthritis), colon–rectal cancer, or viral infections, including AIDS [2, 3]. Glycosyltransferases and glycosidases, being responsible for the biosynthesis of these complex carbohydrate structures, are thus specific targets for inhibition or activation (chemical chaperones) [4].

Iminosugars are naturally occurring polyhydroxylated alkaloids,[5] many of them are potent inhibitors of glycosidases and glycosyltransferases (Figure 19.1) [6]. Because of that they have attracted much interest with respect to potential therapeutic applications [2, 7]. Furthermore, they are useful probes in fundamental biochemical studies of glycosidases mechanism [8].

The synthesis of iminocyclitols both natural and/or their analogs has been accomplished either using chemical and/or chemo-enzymatic methodologies [9, 10]. Importantly, generation of configurational diversity on the hydroxyl groups appears to be of paramount importance to optimize their activity and selectivity [11]. Consequently, novel methodologies for the preparation of these compounds with a broad structural and configurational diversity with minimal steps are of increasing interest.

In this sense, enzymatic catalysis is especially attractive for syntheses demanding highly regio- and stereoselectivity. Particularly dihydroxyacetone phosphate (DHAP) dependent aldolases are among the most suitable biocatalyst for iminosugar synthesis due to their high stereoselectivity and chiral induction capacity [10, 12, 13]. The aldol addition of DHAP (**1**) to a synthetic equivalent of an

Modern Biocatalysis: Stereoselective and Environmentally Friendly Reactions.
Edited by Wolf-Dieter Fessner and Thorleif Anthonsen

ISBN: 978-3-527-32071-4

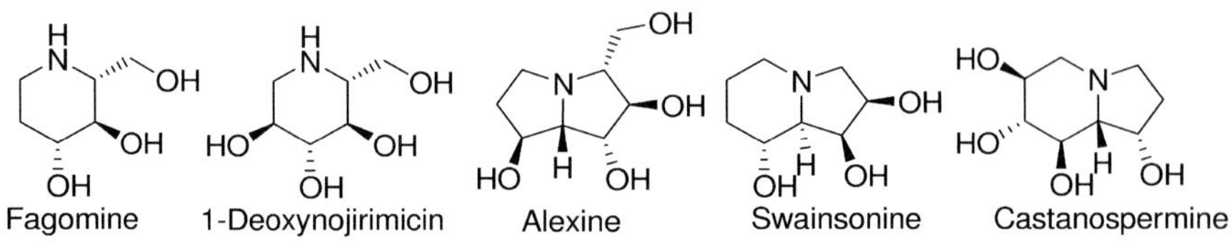

Figure 19.1 Examples of important iminosugars found in nature.

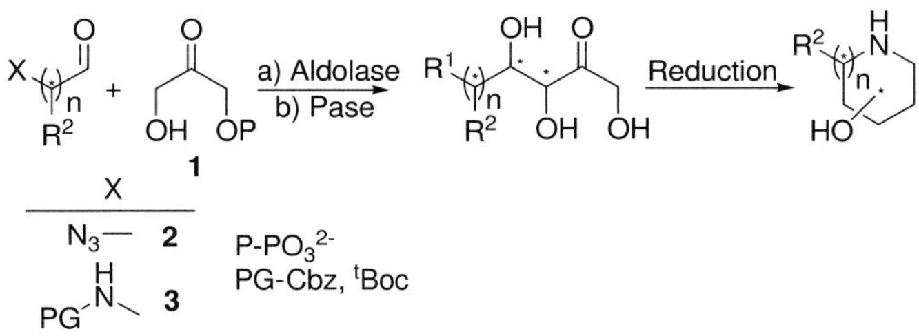

Figure 19.2 Chemo-enzymatic synthesis of iminosugars. * = stereocenters; Pase (Acid phosphatase).

aminoaldehyde catalyzed by aldolases constitutes the synthetic key step of chemo-enzymatic strategies of iminosugars (Figure 19.2).

Among synthetic equivalents of aminoaldehydes it has been shown that azido aldehydes (**2**) are well accepted for many aldolases at synthetically practical rates and yields [14, 15]. Drawbacks are the limited commercial availability, low stability, toxicity and that they can cause violent explosions [16]. Convenient protection of the amino group of amino aldehydes (**3**) resulted most attractive due to their simplicity. Most importantly, *N*-protected amino aldehydes have the tremendous advantage that they can be easily obtained from the wide structural variety of readily available optically pure α- or β-amino acids or alcohols and their derivatives [17]. Moreover, the use of well-known benzyloxycarbonyl (Cbz) or *tert*-butyloxycarbonyl (Boc), as *N*-protecting groups for amino aldehydes may be an excellent complement for orthogonal protection schemes, in particular, when the aldol adducts obtained are to be used as chiral building blocks. *N*-Cbz amino aldehydes were found to be poor substrates for fructose 1,6-diphosphate aldolase [14, 15, 18–20]. Only the products formed from *N*-(Cbz)aminoacetaldehyde (i.e. Cbz-glycinal) and *N*-(Cbz)-3-aminopropanal and DHAP have been characterized although obtained in low yields [18, 20].

Here we account on our research about a novel chemo-enzymatic approach to the synthesis of iminosugars from *N*-protected-aminoaldehydes as donor substrates and aldolases as biocatalysts for the key aldol addition reaction.

19.2
DHAP-Aldolase-mediated Synthesis of Iminosugars from *N*-Cbz-amino Aldehydes

19.2.1
Reaction Media

We faced the problem of the poor solubility of most *N*-protected amino aldehydes in water, which might account for the low reactivity observed with D-fructose-1,6-diphosphate aldolase from rabbit muscle (RAMA) [14, 15, 19–21]. Increasing the percentage of organic co-solvent (e.g. dimethylformamide) in the medium to make the aldehyde soluble may lead to either a dramatic enzyme deactivation [22] or an insolubilization of the donor (e.g. dihydroxyacetone (DHA) and DHAP sodium salt). As a result, no reaction or insufficient product yields are often obtained.

We have developed new reaction systems based on colloidal dispersions [23, 24], namely highly concentrated water-in-oil (gel) emulsions, which could overcome most of the disadvantages of the aqueous-co-solvent mixtures such as inactivation of the aldolase and incomplete aldehyde solubilization in the medium. These emulsions are characterized by volume fractions of dispersed phase higher than 0.73 [25]: therefore, the droplets are deformed and/or polydisperse, separated by a thin film of continuous phase. Water-in-oil gel emulsions of water/$C_{14}E_4$/oil 90/4/6 wt%, where $C_{14}E_4$ is a technical grade poly(oxyethylene) tetradecyl ether surfactant, with an average of four moles of ethylene oxide per surfactant molecule and oil can be octane, decane, dodecane, tetradecane, hexadecane, or squalane, were typically chosen as reaction media [23, 26].

Gel emulsions were applied successfully for the first time in aldol additions of DHAP to phenylacetaldehyde and benzyloxyacetaldehyde as model aldehydes catalyzed by RAMA [24]. The first interesting observation was that the stability of RAMA in water-in-oil gel emulsions improved by 25-fold compared to that in dimethylformamide/water 1/4 v/v co-solvent mixture. The reported experimental data concluded that both the highest enzymatic activities and equilibrium yields were observed in water-in-oil gel emulsion systems with the lowest water–oil interfacial tension attained with the most hydrophobic oil component (i.e. tetradecane, hexadecane, and squalane).

19.2.2
Aldolase-catalyzed Aldol Additions of DHAP to *N*-Cbz-Amino Aldehydes

Water-in-oil gel emulsions were tested in enzymatic aldolization of selected *N*-Cbz-amino aldehydes (Figure 19.3), *N*-Cbz-3-amino propanal (**4**), *N*-Cbz-glycinal, (**5**), (*S*)-*N*-Cbz-alaninal (**6**), and (*R*)-*N*-Cbz-alaninal (**7**) catalyzed by RAMA and L-rhamnulose-1-phosphate aldolase (RhuA) and L-fuculose-1-phosphate aldolase (FucA) from *Escherichia coli* [27, 28]. The largest differences between conventional dimethyl formamide (DMF)/water co-solvent systems and gel emulsions were observed with RAMA and FucA catalysts (Figure 19.3). The emulsion media enhanced the catalytic efficiency of RAMA towards the *N*-Cbz amino aldehydes tested three, five,

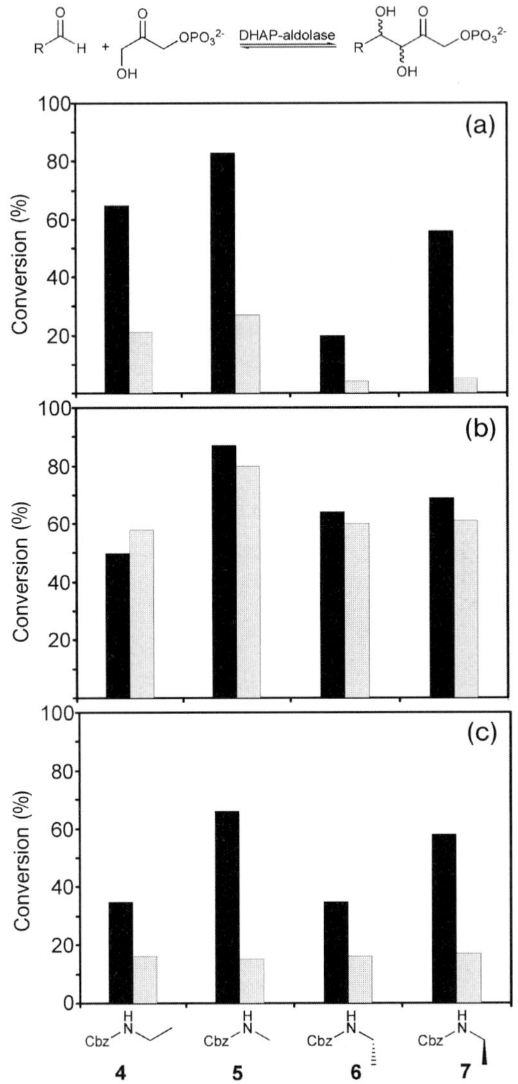

Figure 19.3 DHAP-aldolase catalyzed aldol addition of DHAP to *N*-Cbz-aminoaldehydes. Reaction conversion to aldol adduct in gel emulsion (black bars) and dimethylformamide/water 1:4 system (gray bars) for reactions catalyzed by (a) RAMA, (b) RhuA and (c) FucA.

and even 10-fold, allowing the synthesis of the corresponding products at the preparative level from moderate to good yields. Good results were achieved regardless of the reaction media using RhuA, due to both its higher tolerance towards hydrophobic substrates and its high stability in the presence of organic polar solvents [29]. These results show that it is difficult to generalize and each case must be analyzed in detail, bearing in mind that the reaction performance may be affected not only by the structure of the amino aldehyde but also by the reaction medium. It is important to note that the results presented should reflect the reaction conversion to product close to a quasi-stationary situation in which no changes in substrates and products composition with the time were observed.

The stereochemistry outcome of the aldol additions is an issue of paramount importance in the synthesis of iminosugars. Based on mechanistic considerations of the DHAP aldolases [29, 30] it can be assumed that the absolute configuration at C-3 (i.e. the stereocenter arising from DHAP) is independent of the acceptor used in the reaction. Analysis of the stereochemistry at C-4 (i.e. the one generated from the aldehyde) can be used to infer the kinetic stereoselectivity of the aldolases towards each of the *N*-protected amino aldehydes (Figure 19.4). For the selected

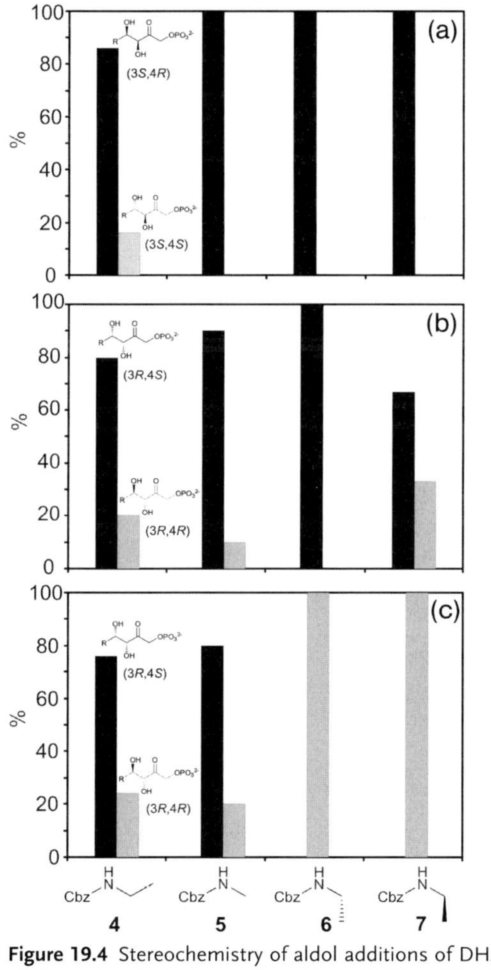

Figure 19.4 Stereochemistry of aldol additions of DHAP to *N*-Cbz-aminoaldehydes catalyzed by (a) RAMA, (b) RhuA and (c) FucA.

examples, RAMA exhibited high stereoselectivity forming single diastereoisomers (3*S*,4*R*) from aldehydes **5**, **6**, and **7**. Reaction with aldehyde **4** afforded 14% of a minor diastereoisomer (3*S*,4*S*). This stereochemical inversion has also been observed with a series of pyridine carbaldehydes and diethylamino acetaldehyde [31]. Class II RhuA and FucA diastereoselectivity depended on both the structure and the stereochemistry of the *N*-Cbz amino aldehyde. RhuA catalysis gave 10–33% of the epimeric aldol adducts at C-4 for *N*-amino aldehydes **4**, **5**, and **7** while a single diasteromer was observed for **6** ((*S*)-Cbz-*N*-alaninal). FucA showed low stereoselectivity towards the *N*-Cbz amino aldehydes **4** and **5**, the configuration at C-4 of the major adduct being inverse (i.e. 3*R*,4*S*) to that found for its natural acceptor, that is L-lactaldehyde (i.e. 3*R*,4*R*), and other non-natural aldehydes. [32] Experiments performed with similar substrates revealed that the (4*R*) epimer was kinetically favored while the (4*S*) one was the major product obtained and, presumably, was thermodynamically more stable [27]. On the other hand, FucA was highly stereoselective with both enantiomers of Cbz-alaninal (**6** and **7**) with an approximately 99:1 diasteromeric ratio of the expected 4*R* diastereoisomer (Figure 19.4). Formation of diastereoisomers, epimeric at C-4, using class II RhuA and FucA catalysis has also been reported with a series of non polar aliphatic aldehydes [29]. From computational model studies, it appears that aldehydes **1** and **2** can adopt similar conformations in the enzyme pocket, independently of their reactive orientation (*si* or *re*). These studies concluded that the conformational differences due to the two possible orientations (*si* or *re*) of the aldehydes or between the two corresponding epimeric adducts are rather small. Therefore, it seems plausible to extrapolate that the differences between the geometries and, by extension, the energies of the corresponding transition states should be small, which may explain the relatively small kinetic preference shown by the enzyme [27].

19.2.3
Effect of *N*-Protecting Groups

Encouraged by the results obtained using *N*-Cbz-amino aldehydes for the enzymatic aldol addition we investigated whether other *N*-blocking groups could also be tolerated by DHAP–aldolases and could be a suitable alternative for the synthesis of iminosugars. Two additional purposes stimulated this study: first to determine the effect of the *N*-protecting group structure on the selectivity and conversion of the enzymatic aldol addition and, second, to have alternative *N*-blocking groups to suit any further synthetic strategies upon the *N*-protected amino-2-keto-3,4-diols. Three *N*-protecting groups were selected (Figure 19.5) on 3-aminopropanal, which was chosen as the model aldehyde for the sake of experimental simplicity: phenylacetyl (PhAc) (**8**), which is structurally similar to Cbz and can be removed by penicillin amidase (PGA)-catalyzed hydrolysis under mild and selective conditions [33], *tert*-Boc (**9**), which is cleavable under acidic conditions but less strenuous than simple amides like acetyl; [14], and fluoren-9-ylmethoxycarbonyl (Fmoc) (**9**), which is removable in the presence of secondary amines, such as piperidine, by base-induced β-elimination.

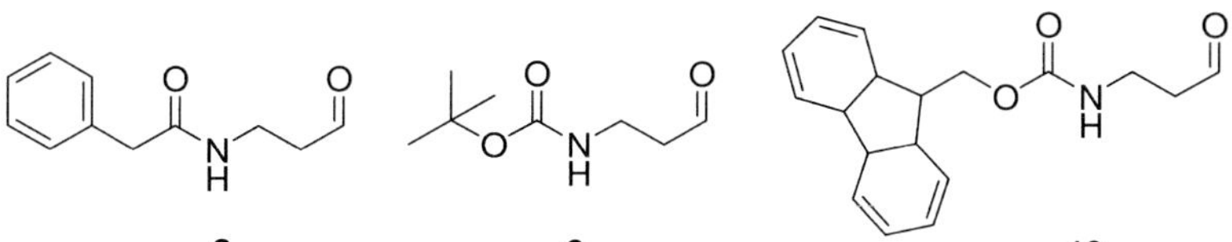

8 9 10

Figure 19.5 Structure of the *N*-protected-3-aminopropanal for the enzymatic aldol addition.

Both PhAc and *tert*-Boc derivatives (**8** and **9**) were tolerated as substrates, the conversions to aldol adduct being similar to those achieved with the Cbz group: for PhAc they were 66% using RAMA, 47% using RhuA and 71% using FucA, and for *tert*-Boc they were 70% using RAMA, 63% using RhuA and 70% using FucA [27, 28]. The most hydrophobic Fmoc derivative, **10**, was also tolerated for the aldolases, but gave the lowest reaction conversions: 25, 15, and 20% using RAMA, RhuA and FucA, respectively. The RAMA catalyst was the most stereoselective towards the *N*-protected 3-aminopropanal derivatives, both *tert*-Boc and Fmoc giving the highest diasteromeric ratios (92 : 8 (3*S*,4*R*) : (3*S*,4*S*)). The RhuA enzyme gave diasteromeric ratio values of 19 : 81, 30 : 70 and 23 : 77 for (3*R*,4*S*) : (3*R*,4*R*) for PhAc, *tert*-Boc, and Fmoc, respectively, which were lower than those of RAMA. Similarly to *N*-Cbz-3-amino propanal, [27] the major diasteromers obtained by the FucA catalyst were (3*R*,4*S*) with the stereochemistry at C-4 inverted in comparison with the one of L-fuculose-1-phosphate. In this case the diasteromeric ratio values were around 30 : 70 for (3*R*,4*S*) : (3*R*,4*R*), which were similar to those achieved by RhuA catalysis.

Docking simulations carried out with the aldol adducts bound into the active center of RAMA and RhuA suggested that in all cases the bulky *N*-protecting group cannot get into the catalytic site of the protein. That would explain the small effect of the *N*-protecting groups with different sizes and shapes on the stereochemical outcome of the reactions, since those groups would remain far from the reactive atoms.

N-PhAc and *N*-*tert*-Boc adduct product derivatives were successfully deprotected by PGA-catalyzed hydrolysis at pH 7 and with aqueous trifluoroacetic acid (6% v/v), respectively. However, the corresponding six-membered imine sugar could not be isolated. The product isolated upon *N*-PhAc removal was identified as a five-membered iminosugar. Boc derivatives gave the expected imine sugar in solution, but after work-up or simple lyophilization complex nuclear magnetic resonance (NMR) spectra were recorded with signals that presumably belong to a number of decomposition products. This behavior may be due to the simultaneous presence of a highly reactive primary amine and a keto group in the molecule, which may lead to a number of undesired reactions. Treatment of the Cbz derivative with hydrogen in the presence of Pd/C promoted the deprotection and the reductive amination successively in one pot reaction, safely transforming the imine intermediate into the iminosugar. Therefore, the Cbz group was the protecting group of choice for the chemoenzymatic synthesis of iminosugars. Protecting groups such as PhAc, *tert*-Boc, and Cbz also provide a range of removal conditions for fulfilling most of the required orthogonality for other types of functional group manipulation on the 2-ketoaminodiols.

19.2.4
Synthesis of Iminosugars: Reductive Amination

Following the aforementioned chemoenzymatic strategy the aldol adducts from the enzymatic addition were first dephosphorylated and then treated with H_2 in the presence of Pd/C. Under these conditions the Cbz group removal and the reductive amination took place in one pot, the iminosugars being obtained as hydrochlorides upon lyophilization at pH 6.5. Figure 19.6 depicts the structure of the iminosugars obtained, either as diasteromeric mixtures or pure compounds, from the *N*-Cbz-amino aldehydes described in this account.

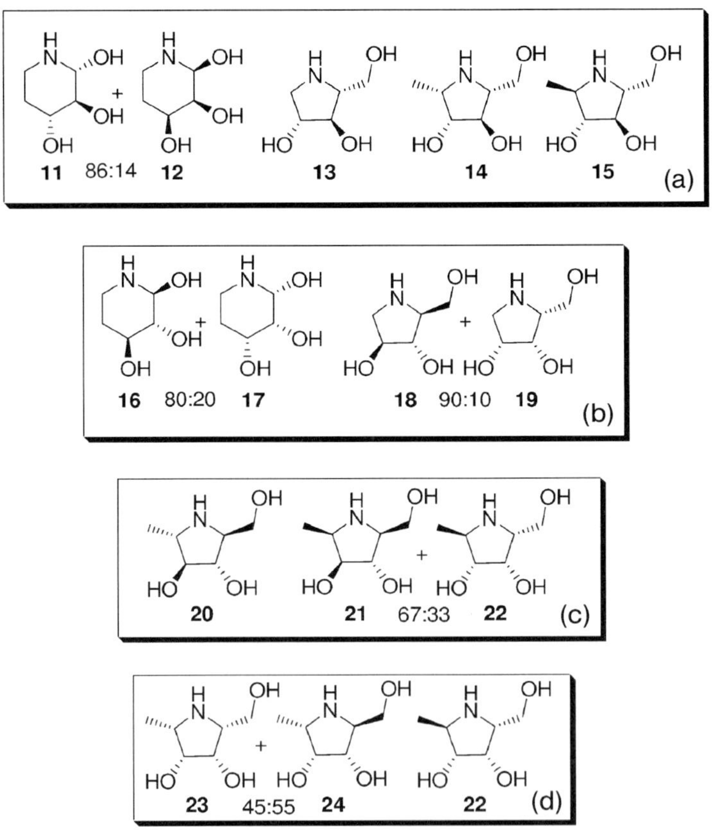

Figure 19.6 Piperidine and pyrrolidine iminosugars obtained by catalysis with (a) RAMA, (b) RhuA and FucA, (c) RhuA and (d) FucA.

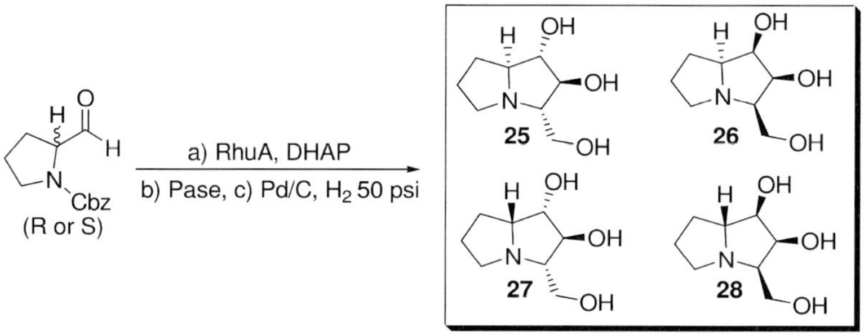

Figure 19.7 Structure of the polyhydroxylated pyrrolizidines obtained by the chemo-enzymatic scheme. Pase-(Acid phosphatase).

An inspection of the stereochemistry at C-2 for the iminosugars revealed that the reductive amination with Pd/C was highly diastereoselective. Interestingly, as already stated [15, 18, 20, 34], we found that the hydrogen was added to the face opposite to the C-4 hydroxyl group, regardless of the relative stereochemistry of the other substituents. Hence, the stereochemistry observed at C-2 was controlled exclusively by the configuration at C-4. An exception was found for the reductive amination of compound **7**. In this case, there was no face selectivity and a circa 1 : 1 diastereomeric mixture was obtained.

Recently, the chemoenzymatic synthetic strategy described thus far has been extended to the synthesis of polyhydroxylated pyrrolizidines using (*R*) and (*S*)-Cbz-prolinal as the amino aldehyde [35]. In this case only the RhuA catalyst tolerated the prolinal derivatives as the substrate (Figure 19.7), no aldol adduct being detected by either RAMA or FucA aldolases. The chemoenzymatic process produced four polyhydroxylated pyrrolizidine diastereomers **25–28** which were efficiently separated by simple cation exchange chromatography. Interestingly, (−)-hyacinthacine A_2 (**25**) was synthesized for the first time [35] and was a good inhibitor of α-D-glucosidase from rice whereas the natural enantiomer, (+)-hyacinthacine A_2 [36], was not.

19.3
D-Fructose-6-Phosphate Aldolase as Catalyst for Iminosugar Synthesis

The chemoenzymatic strategies described heretofore are based on DHAP-dependent aldolases. One of the main drawbacks of DHAP aldolases is their strict specificity toward DHAP and a negligible activity with the unphosphorylated DHA analog (**29**) (Figure 19.8). The chemical synthesis of DHAP involves several steps in an approximately 70% overall yield. [37] Alternatively, multienzymatic methods coupled with the aldol reaction have also been described [38]. However, drawbacks such as the different and often incompatible optimal reaction conditions of

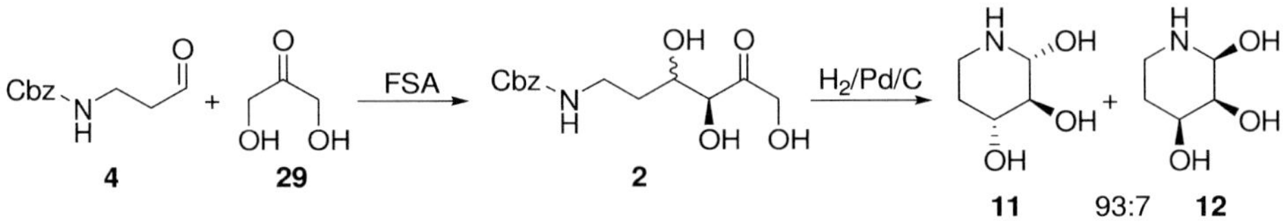

Figure 19.8 D-fructose-6-bisphosphate aldolase-catalyzed synthesis of D-fagomine (**11**).

the enzymes involved and the generation of elusive purification mixtures were detected [31].

Hence, the use of DHA as a donor substrate continues to generate enormous research expectations because it opens the possibility for a wide practical industrial application of this methodology [13].

Recently, it was been found that RhuA accepts DHA in the presence of boric-borate buffer, presumably by reversible *in situ* formation of DHA borate ester [39]. Most recently, the group of Wong selected an RhuA mutant enzyme that accepts DHA as a donor [40].

Most interestingly, two fructose 6-phosphate aldolase isozymes (transaldolase C (TALC) and D-fructose-6-phosphate aldolase (FSA)) were reported from the *E. coli* genome and characterized in recombinant form [41].

FSA is a novel class I aldolase from *E. coli* related to a novel group of bacterial transaldolases that catalyzes the aldol addition of DHA to glyceraldehyde-3-phosphate [41, 42]. The most interesting feature of FSA is that it utilizes DHA instead of either DHAP or DHA/esters, which greatly simplifies the chemoenzymatic strategies to α,β-dihydroxyketones. Furthermore, the enzyme displays a remarkable thermostability (half-life of 16 h at 75 °C), which facilitates its purification from recombinant *E. coli* strains by a simple heat step [41]. The enzyme also accepts hydroxyacetone as a donor compound, which allows access to 1-deoxysugars [42]. We have recently reported [43] on the first example of the use of FSA in iminosugar synthesis: a straightforward procedure for the stereoselective synthesis of D-fagomine (**11**), a piperidine-type iminosugar first isolated from buckwheat seeds of *Fagopyrum esculentum* Moench, [44] (Figure 19.8). The key step was the stereoselective aldol addition of DHA to *N*-Cbz-3-amino propanal (**4**). D-Fagomine (**11**) was then obtained by selective catalytic reductive amination of **2** (Pd/C, H_2 and 50 psi) in an 89% isolated yield without further purification and a 93:7 diastereomeric ratio by NMR. The minor diastereoisomer was identified as D-2,4-di-epi-fagomine (**12**). This compound arose from the *re* face attack of the DHA–FSA complex on the aldehyde, similar to that found with FSA. Further examples using azido and *N*-Cbz aldehydes and DHA, hydroxyacetone and 1-hydroxybutanone as donors have been recently reported [45].

Furthermore, FSA tolerates organic co-solvents such as DMF (10–20% v/v), acetonitrile (10–20% v/v), ethanol (20%), and even a water/ethyl acetate two-phase system.

Investigations of the synthetic abilities of FSA in organic synthesis are currently in progress. We have found that this enzyme accepts, among others, *N*-Cbz glycinal and *N*-Cbz-3-amino 2-hydroxypropanal, but α-alkyl branched *N*-Cbz-amino aldehydes were not tolerated as substrates. Recently, an improved FSA mutant (Ala129Ser) was described, which displays lowered K_m values for DHA as a donor and a doubled V_{max} value [46]. Further analysis of this mutant enzyme is in progress.

19.4 Summary and Outlook

Chemo-enzymatic approaches to the synthesis of iminosugars have been demonstrated to be a powerful tool to generate a variety of both naturally occurring and analog structures with a wide structural and stereochemical diversity. The key step of this methodology is the enzymic aldolization of simple starting materials DHAP, DHA or hydroxyacetone to synthetic equivalents of amino aldehydes. The aldolase controls formally the stereochemistry of the two newly created stereogenic centers. Moreover, in most instances the aldolases tolerate well both enantiomers of a chiral acceptor, and the reductive amination is highly stereoselective. Thus, this methodology yields a variety of stereoisomers which is of paramount importance because the orientation of polar hydroxyl groups provides the precise hydrogen-bond interactions that are essential for biological activity and specificity.

Advances to novel enzymes and molecular biology to obtain both new and engineered aldolases with broad substrate tolerance and improved or altered stereoselectivities are being one of the goals and challenges of this field. Moreover, full exploitation of the synthetic abilities of the available aldolases is important to access new promising structures with interesting biological activities.

Last but not least aldolases are fully reactive in aqueous media, they are selective and biodegradable and, therefore, in accordance to the principles of Green Chemistry.

References

1 Koeller, K.M. and Wong, C.H. (2000) *Nature Biotechnology*, **18**, 835–841.

2 Asano, N. (2000) *Journal of Enzyme Inhibition*, **15**, 215–234.

3 (a) Asano, N. (2003) *Glycobiology*, **13**, 93R–104R.
(b) Wong, C.-H. (ed.) (2003) *Carbohydrate-Based Drug Discovery*, Vol. **1**, **2**, Wiley-VCH Verlag GmbH & Co. KGaA, Weinheim.
(c) Fiaux, H., Popowycz, F., Favre, S., Schutz, C., Vogel, P., Gerber-Lemaire, S. and Juillerat-Jeanneret, L. (2005) *Journal of Medicinal Chemistry*, **48**, 4237–4246.

4 Kolter, T. and Wendeler, M. (2003) *Chembiochem*, **4**, 260–264.

5 Asano, N., Nash, R.J., Molyneux, R.J. and Fleet, G.W.J. (2000) *Tetrahedron: Asymmetry*, **11**, 1645–1680.

6 Compain, P. and Martin, O.R. (2001) *Bioorganic and Medicinal Chemistry*, **9**, 3077–3092.

7 (a) Watson, A.A., Fleet, G.W.J., Asano, N., Molyneux, R.J. and Nash, R.J. (2001) *Phytochemistry*, **56**, 265–295.
(b) Butters, T.D., Dwek, R.A. and Platt, F.M. (2003) *Current Topics in Medicinal Chemistry*, **3**, 561–574.

8 (a) Heightman, T.D. and Vasella, A.T. (1999) *Angewandte Chemie – International Edition*, **38**, 750–770.
(b) Vasella, A., Davies, G.J. and Bohm, M. (2002) *Current Opinion in Chemical Biology*, **6**, 619–629.

9 (a) Cipolla, L., La Ferla, B. and Gregori, M. (2006) *Combinatorial Chemistry and High Throughput Screening*, **9**, 571–582.
(b) Wrodnigg, T.M. and Sprenger, F.K. (2004) *Mini Reviews in Medicinal Chemistry*, **4**, 437–459.
(c) Ayad, T., Genisson, Y. and Baltas, M. (2004) *Current Organic Chemistry*, **8**, 1211–33.
(d) Cipolla, L., La Ferla, B. and Nicotra F. (2003) *Current Topics in Medicinal Chemistry*, **3**, 485–511.

10 Whalen, L.J. and Wong, C.H. (2006) *Aldrichimica Acta*, **39**, 63–71.

11 (a) Kato, A., Kato, N., Kano, E., Adachi, I., Ikeda, K., Yu, L., Okamoto, T., Banba, Y., Ouchi, H., Takahata, H. and Asano, N. (2005) *Journal of Medicinal Chemistry*, **48**, 2036–2044.
(b) Ouchi, H., Mihara, Y., Watanabe, H. and Takahata, H. (2004) *Tetrahedron Letters*, **45**, 7053–7056.
(c) Asano, N., Ikeda, K., Yu, L., Kato, A., Takebayashi, K., Adachi, I., Kato, I., Ouchi, H., Takahata, H. and Fleet, G.W.J. (2005) *Tetrahedron: Asymmetry*, **16**, 223–229.
(d) Takahata, H., Banba, Y., Ouchi, H., Nemoto, H., Kato, A. and Adachi, I. (2003) *The Journal of Organic Chemistry*, **68**, 3603–3607.

12 Fessner, W.D. (2007) *Asymmetric Synthesis with Chemical and Biological Methods* (eds D. Enders and K.-E. Jaeger), Wiley-VCH Verlag GmbH & Co KGaA, Weinheim, pp. 351–375.

13 (a) Dean, S.M., Greenberg, W.A. and Wong, C.-H. (2007) *Advanced Synthesis Catalysis*, **349**, 1308–1320.
(b) Samland, A.K. and Sprenger, G.A. (2006) *Applied Microbiology and Biotechnology*, **71**, 253–264.

14 Hung, R.R., Straub, J.A. and Whitesides, G.M. (1991) *The Journal of Organic Chemistry*, **56**, 3849–3855.

15 Look, G.C., Fotsch, C.H. and Wong, C.H. (1993) *Accounts of Chemical Research*, **26**, 182–190.

16 Bräse, S., Gil, C., Knepper, K. and Zimmermann, V. (2005) *Angewandte Chemie – International Edition*, **44**, 5188–5240.

17 (a) Jurczak, J. and Golebiowski, A. (1989) *Chemical Reviews*, **89**, 149–164.
(b) Jurczak, J., Gryko, D., Kobrzycka, E., Gruza, H. and Prokopowicz, P. (1998) *Tetrahedron*, **54**, 6051–6064.

18 Von der Osten, C.H., Sinskey, A.J., Barbas, C.F. III, Pederson, R.L., Wang, Y.F. and Wong, C.H. (1989) *Journal of the American Chemical Society*, **111**, 3924–3927.

19 Romero, A. and Wong, C.H. (2000) *The Journal of Organic Chemistry*, **65**, 8264–8268.

20 Pederson, R.L. and Wong, C.H. (1989) *Heterocycles*, **28**, 477–480.

21 Azema, L., Bringaud, F., Blonski, C. and Perie, J. (2000) *Bioorganic and Medicinal Chemistry*, **8**, 717–722.

22 Sobolov, S.B., Bartoszko-Malik, A., Oeschger, T.R. and Montelbano, M.M. (1994) *Tetrahedron Letters*, **35**, 7751–7754.

23 Clapés, P., Espelt, L., Navarro, M.A. and Solans, C. (2001) *Journal of the Chemical Society. Perkin Transactions*, **2**, 1394–1399.

24 Espelt, L., Clapes, P., Esquena, J., Manich, A. and Solans, C. (2003) *Langmuir*, **19**, 1337–1346.

25 (a) Ostwald, W. (1910) *Wilmersdorf Z Chem Ind Kolloide*, **8**, 103–109.
(b) Lissant, K.J. (1966) *Journal of Colloid and Interface Science*, **22**, 462–468.
(c) Princen, H.M. (1979) *Journal of Colloid and Interface Science*, **71**, 55–66.

26 Solans, C., Pons, R. and Kuniedain, H. (1998) in *Modern Aspects of Emulsion Science* (eds B.P. Binks), The Royal Society of Chemistry, Cambridge, U.K, pp. 367–394.

27 Espelt, L., Bujons, J., Parella, T., Calveras, J., Joglar, J., Delgado, A. and Clapés, P. (2005) *Chemistry – A European Journal*, **11**, 1392–1401.

28 Espelt, L., Parella, T., Bujons, J., Solans, C., Joglar, J., Delgado, A. and Clapés, P. (2003) *Chemistry – A European Journal*, **9**, 4887–4899.

29 Fessner, W.D., Sinerius, G., Schneider, A., Dreyer, M., Schulz, G.E., Badia, J. and Aguilar, J. (1991) *Angewandte Chemie*, **103**, 596–599.

30 (a) Gefflaut, T., Blonski, C., Perie, J. and Willson, M. (1995) *Progress in Biophysics and Molecular Biology*, **63**, 301–340.
(b) Fessner, W.-D., Schneider, A., Held, H., Sinerius, G., Walter, C., Hixon, M. and Schloss, J.V. (1996) *Angewandte Chemie–International Edition*, **35**, 2219–2221.
(c) Dalby, A., Dauter, Z. and Littlechild, J.A. (1999) *Protein Science*, **8**, 291–297.
(d) Hall, D.R., Leonard, G.A., Reed, C.D., Watt, C.I., Berry, A. and Hunter, W.N. (1999) *Journal of Molecular Biology*, **287**, 383–394.

31 Fessner, W.-D. and Walter, C. (1996) *Topics in Current Chemistry*, **184**, 97–194.

32 (a) Wong, C.-H., Alajarin, R., Moris-Varas, F., Blanco, O. and Garcia-Junceda, E. (1995) *The Journal of Organic Chemistry*, **60**, 7360–7363.
(b) Mitchell, M., Qaio, L. and Wong, C.H. (2001) *Advanced Synthesis Catalysis*, **343**, 596–599.
(c) Joerger, A.C., Gosse, C., Fessner, W.-D. and Schulz, G.E. (2000) *Biochemistry*, **39**, 6033–6041.

33 Waldmann, H. and Reidel, A. (1997) *Angewandte Chemie–International Edition*, **36**, 647–649.

34 (a) Kajimoto, T., Chen, L., Liu, K.K.C. and Wong, C.H. (1991) *Journal of the American Chemical Society*, **113**, 6678–6680.
(b) Liu, K.K.C., Kajimoto, T., Chen, L., Zhong, Z., Ichikawa, Y. and Wong, C.H. (1991) *The Journal of Organic Chemistry*, **56**, 6280–6289.
(c) Takayama, S., Martin, R., Wu, J., Laslo, K., Siuzdak, G. and Wong, C.-H. (1997) *Journal of the American Chemical Society*, **119**, 8146–8151.

35 Calveras, J., Casas, J., Parella, T., Joglar, J. and Clapés, P. (2007) *Advanced Synthesis Catalysis*, **349**, 1661–1666.

36 Asano, N., Kuroi, H., Ikeda, K., Kizu, H., Kameda, Y., Kato, A., Adachi, I., Watson, A.A., Nash, R.J. and Fleet, G.W.J. (2000) *Tetrahedron: Asymmetry*, **11**, 1–8.

37 (a) Jung, S.-H., Jeong, J.-H., Miller, P. and Wong, C.-H. (1994) *The Journal of Organic Chemistry*, **59**, 7182–7184.
(b) Gefflaut, T., Lemaire, M., Valentin, M.-L. and Bolte, J. (1997) *The Journal of Organic Chemistry*, **62**, 5920–5922.
(c) Ferroni, E.L., Ditella, V., Ghanayem, N., Jeske, R., Jodlowski, C., Oconnell, M., Styrsky, J., Svoboda, R., Venkataraman, A. and Winkler, B.M. (1999) *The Journal of Organic Chemistry*, **64**, 4943–4945.
(d) Charmantray, F., El Blidi, L., Gefflaut, T., Hecquet, L., Bolte, J. and Lemaire, M. (2004) *The Journal of Organic Chemistry*, **69**, 9310–9312.
(e) Meyer, O., Ponaire, S., Rohmer, M. and Grosdemange-Billiard, C. (2006) *Organic Letters*, **8**, 4347–4350.

38 (a) Fessner, W.D. and Sinerius, G. (1994) *Angewandte Chemie–International Edition*, **33**, 209–212.
(b) Sanchez-Moreno, I., Francisco Garcia-Garcia, J., Bastida, A. and Garcia-Junceda, E. (2004) *Chemical Communications*, 1634–1635.
(c) van Herk, T., Hartog, A.F., Schoemaker, H.E. and Wever, R. (2006) *The Journal of Organic Chemistry*, **71**, 6244–6247.

39 Sugiyama, M., Hong, Z.Y., Whalen, L.J., Greenberg, W.A. and Wong, C.H. (2006) *Advanced Synthesis Catalysis*, **348**, 2555–2559.

40 Sugiyama, M., Hong, Z., Greenberg, W.A. and Wong, C.-H. (2007) *Bioorganic and Medicinal Chemistry*, **15**, 5905–5911.

41 Schürmann, M. and Sprenger, G.A. (2001) *The Journal of Biological Chemistry*, **276**, 11055–11061.

42 Schürmann, M., Schürmann, M. and Sprenger, G.A. (2002) *Journal of Molecular Catalysis B: Enzymatic*, **19**, 247–252.

43 Castillo, J.A., Calveras, J., Casas, J., Mitjans, M., Vinardell, M.P., Parella, T., Inoue, T., Sprenger, G.A., Joglar, J. and Clapés, P. (2006) *Organic Letters*, **8**, 6067–6070.

44 Koyama, M. and Sakamura, S. (1974) *Agricultural and Biological Chemistry*, **38**, 1111–1112.

45 Sugiyama, M., Hong, Z., Liang, P.H., Dean, S.M., Whalen, L.J., Greenberg, W.A. and Wong, C.H. (2007) *Journal of the American Chemical Society*, **129**, 14811–14817.

46 Sprenger, G.A., Schürmann, M., Schürmann, M., Johnen, S., Sprenger, G., Sahm, H., Inoue, T. and Schörken, U. (2007) *Asymmetric Synthesis with Chemical and Biological Methods* (eds D. Enders and K.-E. Jaeger), Wiley-VCH Verlag GmbH & Co KGaA, Weinheim, pp. 312–326.

20
Biocatalytic Asymmetric Oxidations with Oxygen

Roland Wohlgemuth

20.1
Introduction

Oxidation reactions in general describe electron transfer reactions between atoms, ions, and molecules that can exist in different oxidation states and may involve nearly every element of the periodic system with the exception of the noble gases. The subclass of oxidation reactions in organic chemistry is probably among the most widely investigated process class of all, since it contains a wide variety of reactions with a range of different oxidants as the electron acceptors. The oxidation reactions of fossil resources are fundamental to raw materials and intermediates in our global chemical supply chain. Large-scale oxidation reactions are powerful tools for creating new functional groups at non-functional positions or transforming a protected functional group into the desired functional group in a higher oxidation state. It has been estimated that only about 3% of the large-scale reactions in the pharmaceutical industry are oxidation reactions [1]. Catalytic converters in fuel production and refining have to withstand harsh process conditions and have become the most known and broadly used catalysts [2]. The conversion of fossil fuels to basic chemicals and industrial organic oxidations depend on robust catalysts, which are able to catalyze selective oxidations with high activity and stability. Whatever the targets of the oxidation reactions may be, whether carbon dioxide as the ultimate end-product of oxidations at carbon atoms or organic compounds with intermediate oxidation states of the carbon atoms like alcohols, aldehydes, ketones, lactones, or carboxylic acids, the tremendous success of catalysts in directing the reaction pathway to the intended products with high efficiency and economy has led to numerous industrial applications.

Oxidation reactions in flammable solvents are critical to exothermic, fire, and explosion events. The simultaneous presence of reactive oxidants and flammable organic solvents represent operating conditions that require substantial safety precautions and have the disadvantage of generating stoichiometric amounts of waste from the oxidant. Therefore the replacement of solvents, oxidants, and reaction methodology by environmentally compatible and safe reagents is of prime

Modern Biocatalysis: Stereoselective and Environmentally Friendly Reactions.
Edited by Wolf-Dieter Fessner and Thorleif Anthonsen

ISBN: 978-3-527-32071-4

importance, not only in production but also in research-scale laboratories. The bonus of selective and orthogonal oxidation methodology without the need to introduce additional protection–deprotection loops for other labile functional groups is attractive.

The introduction of catalytic oxidations using oxygen from air is of much interest and breakthrough discoveries and spectacular progress have been achieved in research on catalytic asymmetric oxidations since the early part of the last century [3–5]. The increased number of publications per year in this area, as extracted from the electronically accessible literature in a SciFinder search, is shown in Figure 20.1 and is due to the importance of catalysis and chirality for oxidation methods in organic synthesis. Directing asymmetric insertion of oxygen into carbon backbones by catalytic methods and with high selectivity is a major challenge and continues to be of relevance for direct routes to oxygenated chiral compounds. The development of safe and environment-friendly oxidants like molecular oxygen has become a major research area in modern oxidation reactions [5], because many of the currently used oxidants require special safety, health, and environment precautions. Organic peroxides and peroxide-containing materials including hydrogen peroxide represent major global safety and security risk factors due to their explosive properties, which are quite difficult to detect by standard methods [6]. Oxidation reactions for alcohols include transition metal compounds like chromiumtrioxide pyridine, pyridinium chlorochromate, potassium permanganate, and ruthenium-tetroxide as oxidants, which may lead to unwanted oxidation of other functional groups due to their strong oxidation power and may also cause additional negative

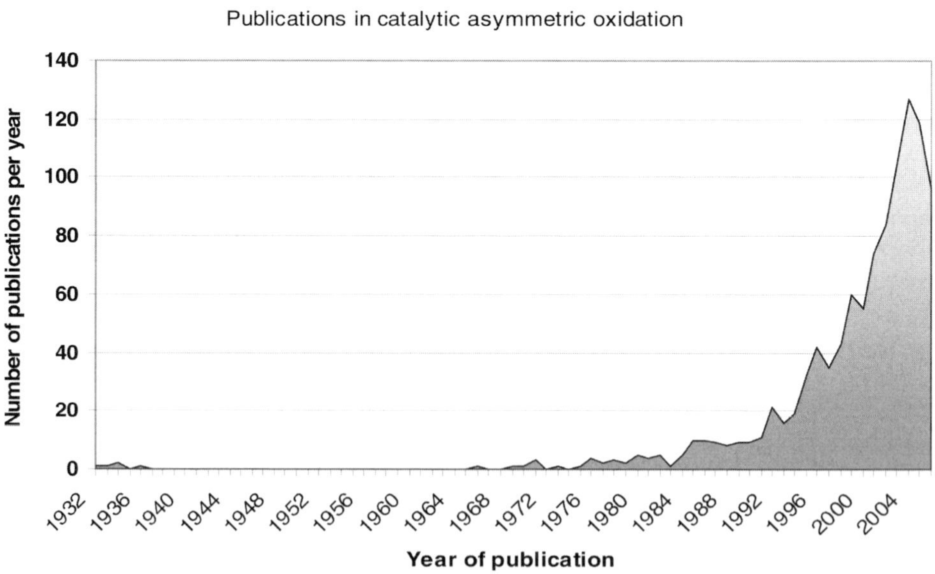

Figure 20.1 Historical development of the number of publications on catalytic asymmetric oxidation.

global health and environmental effects. A range of alcohol oxidation methods make use of dimethylsulfoxide in conjunction with electrophilic reagents like dicyclohexylcarbodiimide, sulfurtrioxide or oxalylchloride, with the disadvantage of accumulating stoichiometric waste. Oxidative decarboxylation reactions of carboxylic acids with lead tetraacetate or mercury oxide are highly problematic. Oxidation of alkenes with ozone is an important reaction for the cleavage of C=C double bonds, but requires special equipment and the products isolated depend on the work-up conditions. The difficulties in performing direct selective oxidations as one of the key steps in organic synthesis have led to the detour of halogenated intermediates for the synthesis of oxygenated compounds, with the associated accumulation of stoichiometric waste. It is therefore of global importance to re-engineer our vulnerable oxidation processes and to turn to nature in order to determine safe and secure direct oxidation reactions compatible with life in a more complex and densely populated world.

The challenge for oxidation reactions in life processes is the balance of the beneficial oxidation of food-providing energy in the process of oxidative phosphorylation in the mitochondria with the toxic effects of oxygen and its products on the life cycle of the biological cell. Many oxidation reactions in life processes are performed by biocatalysts in a highly stereoselective way with oxygen, which is abundantly available in various ecological niches from the photosynthetic splitting of water. Thereby oxygen can act either as an oxidant with no oxygen atom incorporated into the substrate molecule or as an oxygen supplier with one or both oxygen atoms of molecular oxygen ending up in the product. Although molecular oxygen is the most common electron acceptor in biological systems, not only oxygen but also sulfate [7] and substrates [8] can act as electron acceptors for oxidative enzymes.

Biocatalytic asymmetric oxidations were developed very early for key steps in the production of vitamin C [9] and steroid hormones [10] and for a series of applications in organic synthesis [8], as illustrated in Figure 20.2. The use of biocatalysts in oxidation reactions is growing [11–15] and the inherent chirality of the enzymes enables a wide variety of biocatalytic asymmetric oxidations, an overview of which is given in the next six sections.

The use of oxygen as an oxidant is common to a variety of enzymes. As the large-scale preparation of these enzymes is a prerequisite for the development of biocatalytic oxidations with oxygen, an overview of their production is given. The application of oxidases towards selected amino acids has been useful for the resolution of non-natural racemic amino acids. Monooxygenases are versatile biocatalysts for a range of asymmetric oxidations like sulfur and nitrogen oxidations, epoxidations, and Baeyer–Villiger oxidations. Classical Baeyer–Villiger oxidations often give rise to non-specific side reactions, overoxidation, and rearrangements, which are disadvantages in chemo- and enantioselectivity. In contrast, biocatalytic asymmetric Baeyer–Villiger oxidations show excellent selectivity over a broad substrate range and the Baeyer–Villiger monooxygenases have been prepared on a large scale. Optimized productivity has been achieved with a simple adsorber technique. Dioxygenases are no less interesting than monooxygenases and their

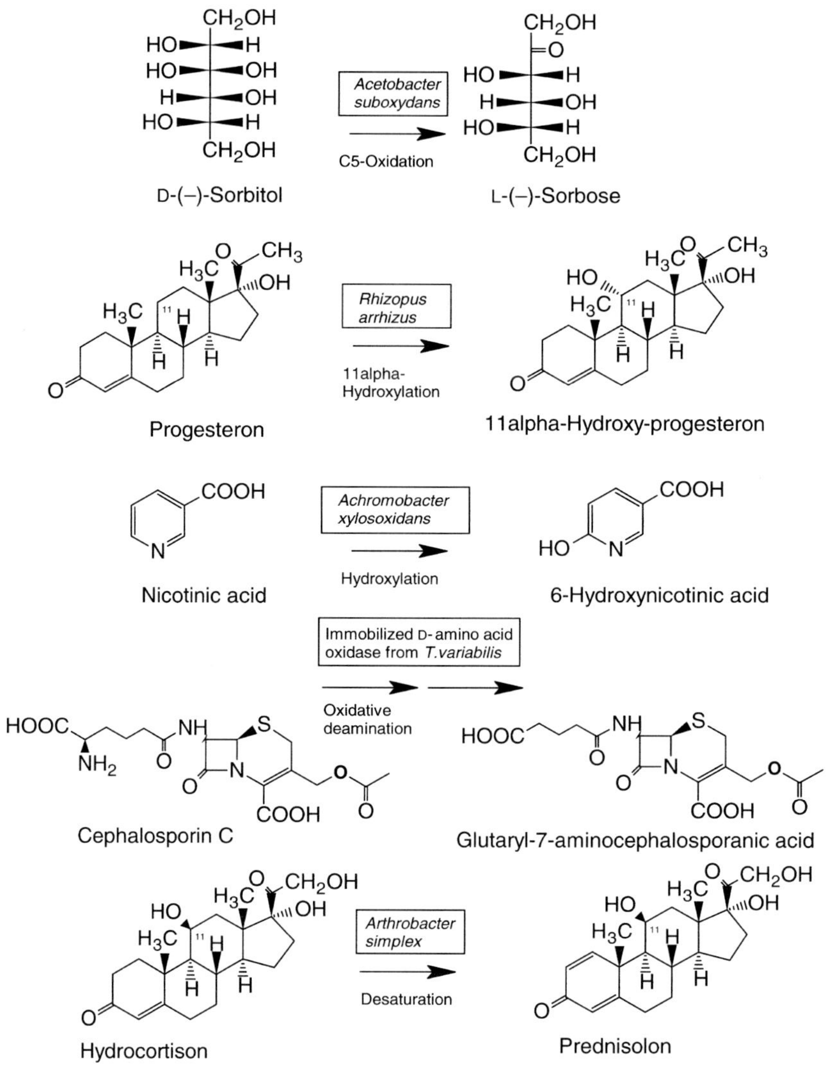

Figure 20.2 Historical industrial examples of biocatalytic asymmetric oxidations with oxygen.

application in the regioselective asymmetric dihydroxylation of aromatic nitriles gives access to the interesting class of dihydrodiolnitriles. An outlook of the potential of biocatalytic asymmetric oxidations using oxidoreductases or other enzyme classes in an oxidative reaction environment with oxygen will be given.

The focus is on proteins with redox functionalities like oxidases, mono- and dioxygenases, and other enzymes able to utilize molecular oxygen as an oxidant, but dehydrogenases and peroxidases will also be discussed. The use of common proteins such as bovine serum albumin with no distinctive redox functionalities as chiral templates for asymmetric oxidations has been reviewed elsewhere [16]. As the oxidations require the transfer of electrons from a substrate to an electron acceptor and the side chains of natural amino acids do not contain redox-active groups able to support electron transfer, it is necessary for redox enzymes to have the assistance of cofactors like transition metals or organic molecules at the active site in order to achieve efficient electron transfer. The growing knowledge of redox enzymes is also finding increased applications in the pharmaceutical and biotechnology industries [17] and the chemical industry [14].

20.2 Biocatalytic Asymmetric Oxidations with Oxidases

The wide variety of known oxidases reflects the complexities of the different substrate classes such as carbohydrates, amino acids, lipids, amines, metabolites, alcohols, acids, and other chiral building blocks. An overview of synthetic applications of oxidases is given in Figure 20.3.

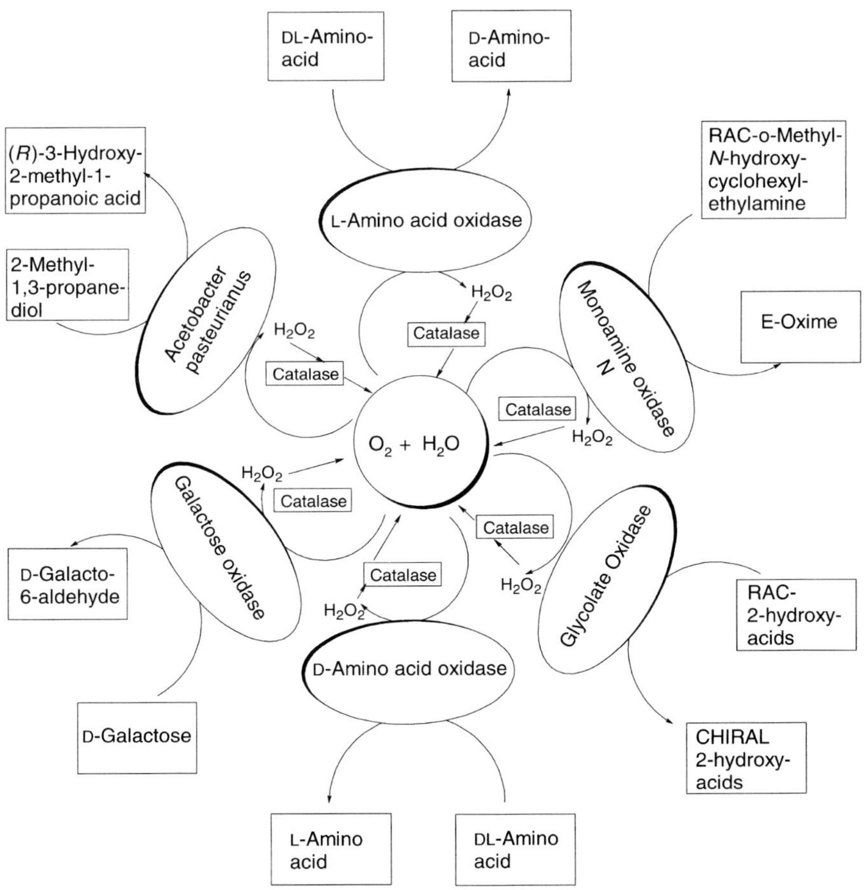

Figure 20.3 Overview of oxidase reactions in synthetic chemistry.

The selective oxidation of D- or L-amino acids by the corresponding amino acid oxidases is well established and convenient to perform for the resolution of racemic amino acids, if the peroxide formed is continuously destroyed by catalase [13]. Both natural as well as new mutant D- and L-specific amino acid oxidases are available for preparing both the chiral amino acid as well as the α-ketoacid from the racemic amino acid substrate [18]. The rapid access to racemic unnatural aminoa cids makes resolutions with amino acid oxidases a reliable step in the planning of synthetic routes [19]. A product recycling method can enhance the yield further. Enantiospecific oxidation of a racemic amino acid substrate with D-amino acid oxidase and chemical non-enantiospecific reduction of the product with sodium borohydride in one pot (see Figure 20.4) was successfully applied to a variety of racemic amino acids to give the corresponding L-amino acid in high enantiomeric purity and yield [20, 21]. The fact that amino acid oxidase, like other members of the oxidase family, evolves hydrogen peroxide as a reaction by-product, which can be detected colorimetrically directly upon individual bacterial colonies, has been successfully used for evolving highly process-suitable amino acid or amine oxidases from the wild-type enzymes [22]. Other oxidases like L-lactate oxidase have been applied in such a deracemization or enantiomerization of a racemic substrate [20].

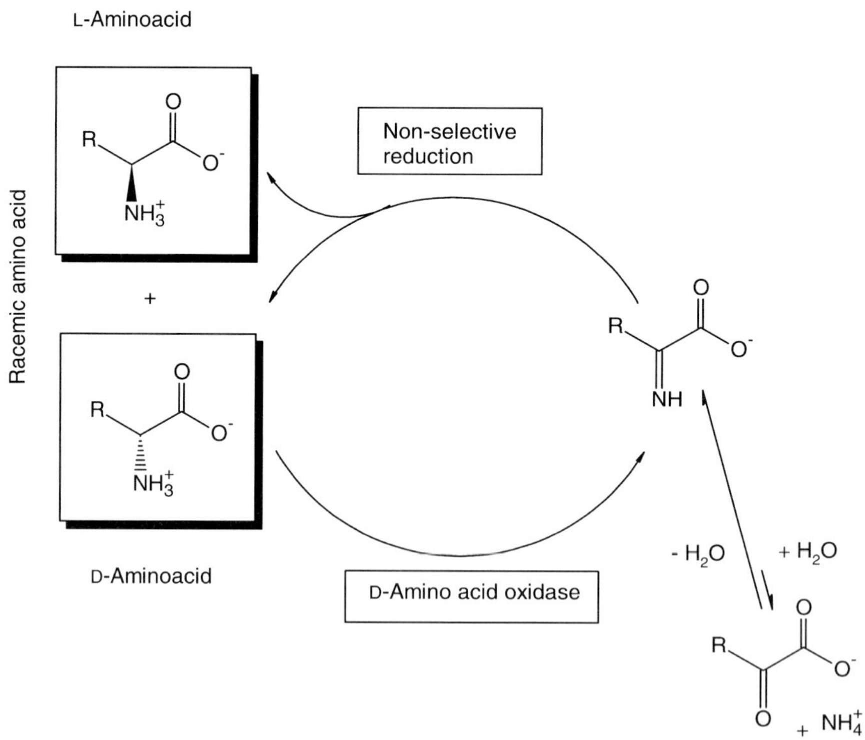

Figure 20.4 Resolution of racemic amino acids with enantioselective amino acid oxidase.

Enantioselective oxidation of racemic *O*-methyl-*N*-hydroxycyclohexylethylamine using monoamine oxidase N from *Aspergillus niger* yields unreacted (*R*)-enantiomer together with the oxime exclusively in the (*E*)-configuration [23].

The selective oxidation of a single hydroxyl group in complex molecules carrying several hydroxyl groups requires a number of protection–deprotection steps when conventional chemical means are to be utilized. In contrast, the biocatalytic single-step oxidation of specific hydroxyl groups in carbohydrates by monosaccharide oxidases has been achieved successfully for a number of sugars and polyols, leading to the corresponding keto polyols or keto acids. Therefore the biocatalytic oxidation is of both interest to the preparation of unusual sugars [24], nucleosides, and chiral building blocks as well as for the synthesis of glycoconjugates.

(*R*)-3-Hydroxy-2-methyl propionic acid, an important building block for the synthesis of the widely used antihypertensive drug captopril, was obtained with 97% enantiomeric excess (e.e.) and 100% molar conversion by microbial oxidation of prochiral 2-methyl -1,3-propandiol with *Acetobacter pasteurianus* [25].

Enantioselective oxidation of 2-hydroxyacids with oxygen by the action of glycolate oxidase from spinach has been used for resolving racemic 2-hydroxyacids [26].

20.3 Biocatalytic Asymmetric Oxidations with Peroxidases

Although peroxidases require peroxides as oxidants, the low operational stability of peroxidases towards peroxides favors *in situ* generation of hydrogen peroxide from oxygen and a sacrificial reductant leading to an overall oxidation with molecular oxygen. The use of peroxidases has been documented for nearly two centuries and knowledge about their use in selective oxidations of electron-rich substrates [27] and asymmetric oxidations is growing [28]. These enzymes are able to preserve high oxidation states of metal centers in a protein environment and yet achieve a strikingly diverse range of chemistries by similar high-oxidation state intermediates. The high oxidation state intermediates in a redox enzyme have been shown for the first time with the catalytic pathway of horseradish peroxidase [29].

As enantiomerically pure sulfoxides are excellent chiral auxiliaries for asymmetric synthesis, different approaches for biocatalytic asymmetric oxidations at the *S*-atom have been explored [30, 31]. Asymmetric peroxidase-catalyzed sulfoxidation of organic sulfides to sulfoxides in organic solvents opens up attractive opportunities by increased substrate solubility and diminished side reactions [32]. Plant peroxidases located in the cell wall are capable of oxidizing a broad range of structurally different substrates to products with antioxidant, antibacterial, antifungal, antiviral, and antitumor activities [33]. Hydroperoxides and their alcohols have been obtained in excellent e.e. in the biocatalytic kinetic resolution of secondary hydroperoxides with horseradish and *Coprinus* peroxidase [34].

The first example of an asymmetric electroenzymatic synthesis with a peroxidase was carried out with thioanisole and chloroperoxidase from *Caldariomyces fumago*.

This catalytic asymmetric oxidation yielded *R*-methylphenylsulfoxide with a productivity of 30 g/l/day and an e.e. >98% [35]. Chloroperoxidase is the most versatile peroxidase with better stability compared to other peroxidases, because spontaneous oxidation can be suppressed in the presence of ascorbic acid or dihydroxyfumaric acid, and with better enantioselectivity because substrate access to the heme iron and ferryl oxygen favors stereoselective oxygen transfer [36]. Chloroperoxidase has been used for catalyzing the oxidation of *cis*-cyclopropylmethanols with much higher enantioselectivity than *trans*-isomers [37].

An interesting oxidative coupling mechanism has been proposed for the biotransformation of *p*-coumaric acid with peroxidase, leading to dimer and trimer structures with stronger antioxidant activities [38].

20.4
Biocatalytic Asymmetric Oxidations with Dehydrogenases

The broad range of enantioselective dehydrogenases also shows interesting applications in the direction of oxidations. In the production of (*R*)-1,3-butandiol, the enzymatic resolution by the enantioselective oxidation of the undesired (*S*)-1,3-butanediol in the racemate has been found to be the most practical process [39]. The responsible enzyme was (*S*)-1,3-butanediol dehydrogenase, a novel secondary alcoholdehydrogenase, from *Candida parapsilosis*. The use of D-sorbitol dehydrogenase in the oxidation of D-sorbitol to L-sorbose in vitamin C synthesis [9] or the oxidation of *N*-protected 1-amino-D-sorbitol to the key intermediate in miglitol synthesis [40] are classical examples of the industrial utility of polyol dehydrogenases in the selective oxidation of a single hydroxyl group in complex molecules with various other hydroxyl groups and other oxidizable functional groups. Bypassing a number of the chemical protection–deprotection steps required in conventional non-selective oxidations has been the key success factor for the application of these dehydrogenases in selective single-step oxidations. The selective oxidation of each of the hydroxyl groups of the steroid backbone in cholic acid has been achieved by various hydroxysteroid dehydrogenases [41]. The oxidation of meso-diols with secondary hydroxyl functions to (*S*)-α-hydroxyketones by glyceroldehydrogenase is an interesting direct route to chiral hydroxyketones [42]. The desymmetrization of meso-diols or prochiral diols by alcohol dehydrogenase has been used as a valuable and selective two-step one-pot method for the synthesis of chiral lactones [43]. A wide variety of prochiral and meso-diols, including sterically demanding bicyclic meso-diols [44–45], have been oxidized by alcohol dehydrogenase to the corresponding chiral lactones with excellent enantiomeric purity. As the chemical resolution of racemic lactols suffers from low enantioselectivity, alcohol dehydrogenases provide the method of choice.

20.5
Biocatalytic Asymmetric Oxidations with Monooxygenases

It is clear that biosynthesis and biodegradation are crucially dependent on selective asymmetric oxidations with easily available natural catalysts and oxidants.Oxygen is available within many environments as an abundant natural oxidant, thereby acting as an electron acceptor. The reactions where one of the atoms of molecular oxygen is introduced into an organic substrate and the other into a molecule of water are catalyzed by monooxygenases. An overview of useful reactions, such as hydroxylations, epoxidations, Baeyer–Villiger oxidations, heteroatom oxidations, and heteroatom dealkylations, is illustrated in Figure 20.5. The oxygen is acting as both an electron acceptor as well as an oxygen transfer agent in these reactions. The repertoire of enzymes for all kinds of asymmetric oxidations is therefore of much interest to the synthetic chemist, because many reaction types, like asymmetric hydroxylations at a saturated carbon atom or oxidative desymmetrizations, are perfectly achieved with the aid of biocatalysts, while being a problem with stoichiometric oxidations and organic or inorganic catalysis.

Both chemical and enzymatic synthetic methods for the asymmetric oxidation of the carbon–carbon double bond have been developed [46], but the area of carbon–carbon double bond oxidations has been shaped by the breakthrough discovery of asymmetric epoxidation of allylic alcohols with the Katsuki–Sharpless method [47]. Catalytic asymmetric synthesis of epoxides from alkenes by Jacobsen [48], the catalytic enantioselective epoxidation of olefins and the desymmetrization of meso-tetrahydrofurans by Katsuki [49], and the development of versatile new catalysts [50] have broadened the scope of these reactions. The excellent enantioselectivity of biocatalysts in the epoxidation of terminal olefins to give optically pure *R*-(+)-epoxides was discovered in 1978 [51] and the selective oxidation of one out of several C—C double bonds in an asymmetric way is a competence of biocatalytic systems [52]. In asymmetric Baeyer–Villiger oxidation and asymmetric heteroatom oxidation, biocatalytic procedures are well ahead of their chemical counterparts [53].

When considering the safety, health, and environment aspects of many types of oxidation reactions [1], the replacement of nitric acid oxidation of carbon feedstocks by more environmentally friendly procedures is highly desirable [54]. The road to the elimination of stoichiometric waste, which can contribute to global warming, ozone depletion, acid rain, and smog, needs catalytic process reengineering and refining and the quest for sustainable chemistry puts the monooxygenases in a central role for a variety of established asymmetric oxidations as well as towards new asymmetric oxidation processes. The cofactors utilized by monooxygenases for the generation of the oxygen-transferring molecules contain either a heteroaromatic molecule like flavin or pterin or a transition metal like iron or copper in a heme or non-heme complex.

Heme-dependent monooxygenases contain ferric protoporphyrin IX, the heme, as the cofactor and protect us from the xenobiotics, toxins, food, and chemicals we take up. As these proteins exhibit a very strong absorption band maximum at

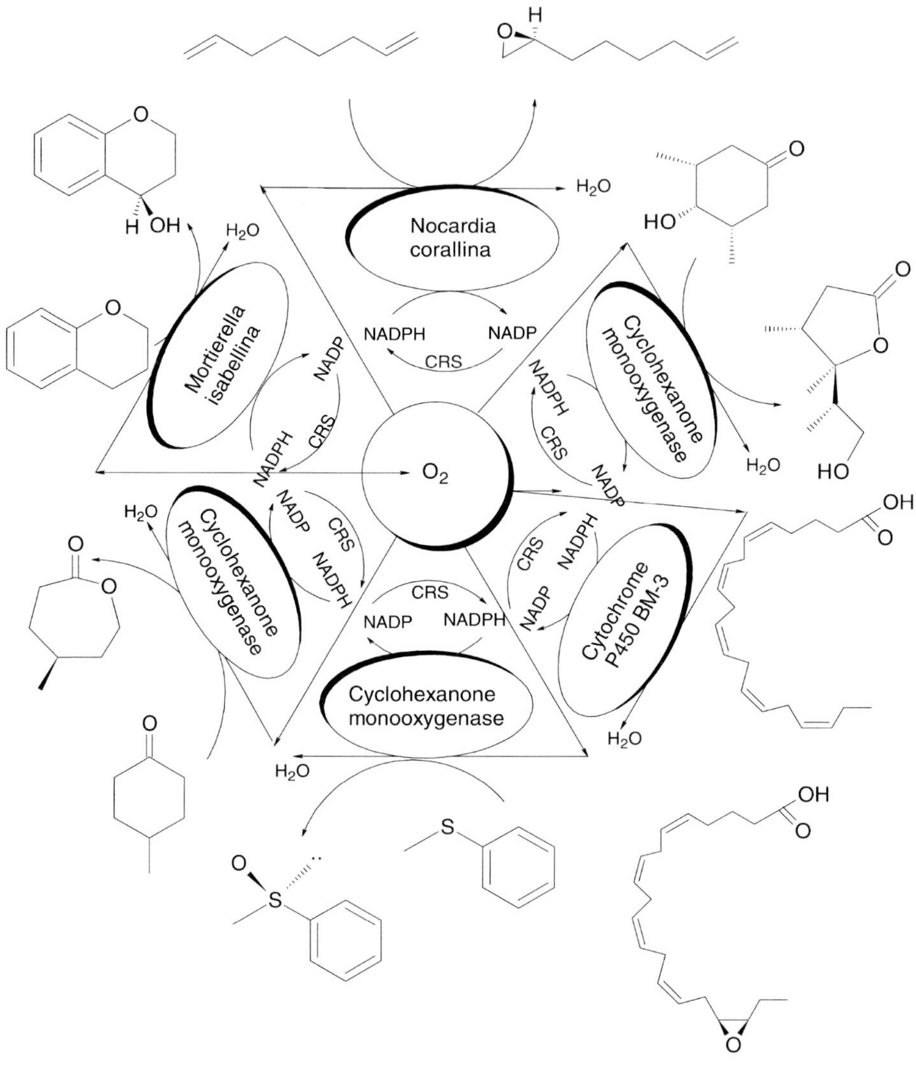

CRS = cofactor regeneration system

Figure 20.5 Overview of monooxygenase reactions in synthetic chemistry.

450 nm in a carbon monoxide difference spectrum, when carbon monoxide is added, they are also referred to as the cytochrome P450 superfamily. With more than 500 isozymes they are widespread in nature and and catalyze a variety of asymmetric oxidations (see Figure 20.5), which are difficult to achieve by classical organic chemistry. Stereospecific hydroxylations at non-activated carbon atoms under physiological conditions, epoxidations, and heteroatom dealkylations in organisms ranging from bacteria to humans are among the specialty oxidation reactions the heme-dependent monooxygenases are able to catalyze with remark-

able selectivity. The structures of three intermediates in the camphor hydroxylation reaction have revealed insight into the reaction pathway and the mechanism of oxygen activation [55]. An interesting selectivity of cytochrome P450 monoxygenase BM-3 from *Bacillus megaterium* is the enantioselective epoxidation of the *E*-isomer geranylacetone to the 9,10-epoxygeranyl-acetone exclusively, while the oxidation of the *Z*-isomer nerylacetone resulted in a product distribution [56]. This well-studied cytochrome P450 monooxygenase BM-3 has been engineered to perform selective hydroxylation at the 2-position of 2-arylacetic acid derivatives and the 6-position of the anti-anxiety agent buspirone with excellent enantiomeric excess [57].

Non-heme iron-containing monooxygenases are able to catalyze the oxidation of hydrocarbons to the corresponding hydrocarbon alcohols and have been found in methanotrophic bacteria [58].

Dopamine hydroxylase belongs to the copper-dependent monooxygenases and is important in the control of the neurotransmitter concentrations of dopamine and norepinephrine [59].

The important metabolism of the neurotransmitters norepinephrine, epinephrine, dopa, and serotonin involves pterin-dependent monooxygenases. The direct biocatalytic hydroxylation of the aromatic amino acids phenylalanine, tyrosine, and tryptophane requires tetrahydrobiopterin and Fe^{2+} as the cofactors [60]. The cleavage of unsaturated glyceryl ethers by glyceryl ether monooxygenase also requires tetrahydrobioterin as the cofactor [61].

Flavin-dependent monooxygenases are of increasing interest [62] and have been useful in preparative biocatalytic asymmetric Baeyer–Villiger oxidations due to their excellent enantioselectivity [63–65]. Baeyer–Villiger monoxygenases [66] are a group of enzymes that can not only perform asymmetric oxidations of linear and cyclic ketones to their corresponding esters and lactones, but also achieve *N*- or *S*-heteroatom oxidations in a highly enantioselective way. As these chiral compounds are difficult to obtain by direct classical oxidations and are useful intermediates in the production of pharmaceuticals, flavors, fragrances, and agrochemicals, the development of robust and scalable processes is a key goal [66]. The production of recombinant Baeyer–Villiger monooxygenases and the identification of reaction and downstream processing bottlenecks has led to the development of large-scale processes [67–69]. The hurdles to further large-scale oxygenase applications have been described to include, among other factors, substrate uptake, substrate toxicity and also oxygen mass transfer [70]. Technical improvements using sinter-metal spargers and the proper balance of cell concentration and oxygen mass transfer [71] have been shown as key success factors for an optimized biocatalytic asymmetric Baeyer–Villiger oxidation process. Symmetric prochiral ketones are of special interest as a biocatalytic Baeyer–Villiger oxidation yields only one chiral product with complete conversion. Resembling this desymmetrization of a prochiral ketone and of similar high synthetic value is the asymmetric oxidation of a thioether to a chiral sulfoxide [72]. Non-symmetric racemic ketones (see Figure 20.6) can be either kinetically resolved by the selective oxidation of only one ketone enantiomer, leaving the opposite enantiomer unchanged and then separating the

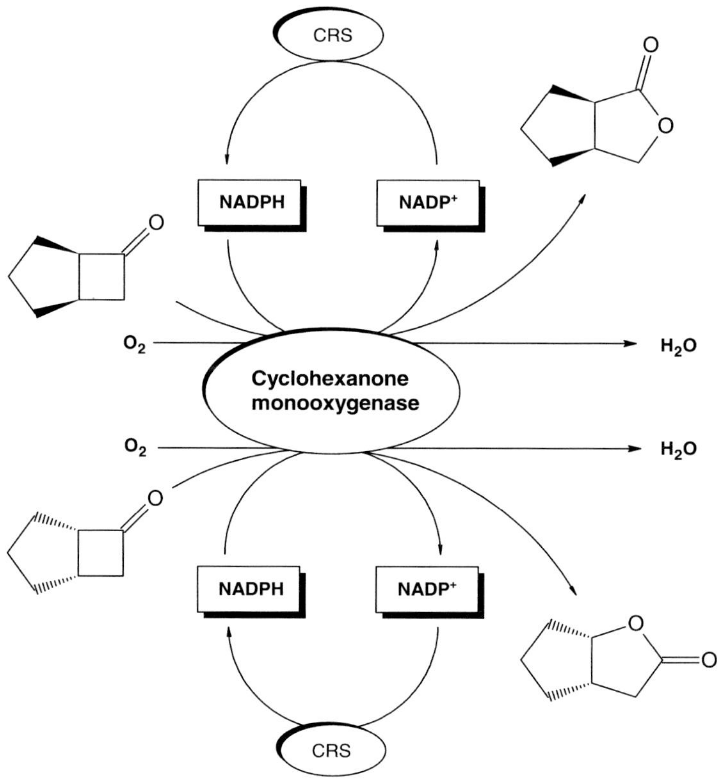

CRS = cofactor regeneration system

Figure 20.6 Large-scale biocatalytic asymmetric Baeyer–Villiger oxidations.

product lactone from the remaining ketone or by enantiodivergent oxidation of both enantiomers of the racemic ketone and the separation of the resulting regioisomeric lactones. The ability to oxidize racemic bicyclic diketones selectively, such as the Wieland–Miescher and the Hajos–Parrish diketones to the corresponding lactones in a highly regio- and enantioselective manner [73] or the ability to perform a Baeyer-Villiger oxidation of a glycosylated substrate [74], further emphasize the exquisite selectivity of these biocatalysts.

The first example of a catalytic enantioselective epoxidation by cyclohexanone monooxygenase was shown with a fosfomycin-related model compound [75]. The efficient asymmetric epoxidation of styrene to (*S*)-styrene oxide by recombinant styrene monooxygenase has been achieved by increasing biocatalyst concentrations and reducing the exposure time of the biocatalyst to the product [76].

A large range of microbial hydroxylations of aliphatic and aromatic hydrocarbons, terpenes, and steroids is of flavor and fragrance, pharmaceutical, and fundamental interest, because the hydroxylation of non-activated carbon centers has only very few counterparts in classical organic chemistry [10, 11, 13, 18, 77].

20.6 Biocatalytic Asymmetric Oxidations with Dioxygenases

The most important discovery of the Sharpless asymmetric *cis*-vicinal dihydroxylation [78] reaction in twentieth-century organic chemistry has its counterpart in nature where the microbial dioxygenation of aromatic compounds became known through the characterization of the metabolic pathway of aromatic compound degradation by Gibson and coworkers [79–82]. Mutant bacterial strains containing dioxygenases but lacking the corresponding diol dehydrogenase activity [80] have been developed for the asymmetric dihydroxylation of a remarkable range of aromatic compounds to *cis*-dihydrodiols with excellent stereoselectivity [81–83]. These reactions, where both of the atoms of molecular oxygen are introduced into an organic substrate, are catalyzed by dioxygenases and represent highly interesting, safe, and environment-friendly transformations to synthons for organic chemistry [84–86]. The further extension of biocatalytic dihydroxylations (see Figure 20.7) has been facilitated by biodiversity [87], the recombinant protein expression of dioxygenase genes in *Escherichia coli* [88–90], and directed evolution [91]. Recombinant dioxygenases were applied in the determination of the absolute configuration of metabolites from the toluene dioxygenase-catalyzed dihydroxylation of bromo-(methylsulfanyl)benzenes [92] and from the chlorobenzene dioxygenase-catalyzed asymmetric dihydroxylation of cinammonitrile [93].

Recombinant toluene dioxygenase and chlorobenzene dioxygenase whole cell biocatalysts can perform *cis*-vicinal dihydroxylation of aromatic nitriles to nitriledihydrodiols [94], the isolated yield of which can be improved by *in situ* product recovery and which can also be hydrolyzed by nitrilase to the corresponding carboxylic acid [95]. More than 300 *cis*-vicinal diol-metabolites have now been prepared with dioxygenases [96] from aromatic substrates, a selection of which is shown in Figure 20.8. Recent evidence for the synthesis of tetraoxygenated chiral metabolites [97] by the action of toluene dioxygenase and biphenyl dioxygenase shows that *cis*-dihydrodiols and acetonide derivatives can themselves be substrates for these enzymes.

The class of intramolecular dioxygenases mentioned above incorporates both oxygen atoms into the same molecule and can perform selective oxidations not only on aromatic but also aliphatic substrates. The activation and incorporation of oxygen into the substrates can thereby occur via different mechanisms to highly reactive intermediates like hydroperoxides, endoperoxides, or dioxetanes. Lipoxygenases and cyclooxygenases can incorporate oxygen into an achiral polyunsaturated fatty acid to form a chiral product with high regio- and stereochemical purity [98].

The biocatalytic asymmetric incorporation of oxygen by soybean lipoxygenase into polyunsaturated fatty acids with a non-conjugated 1,4-diene unit like linoleic

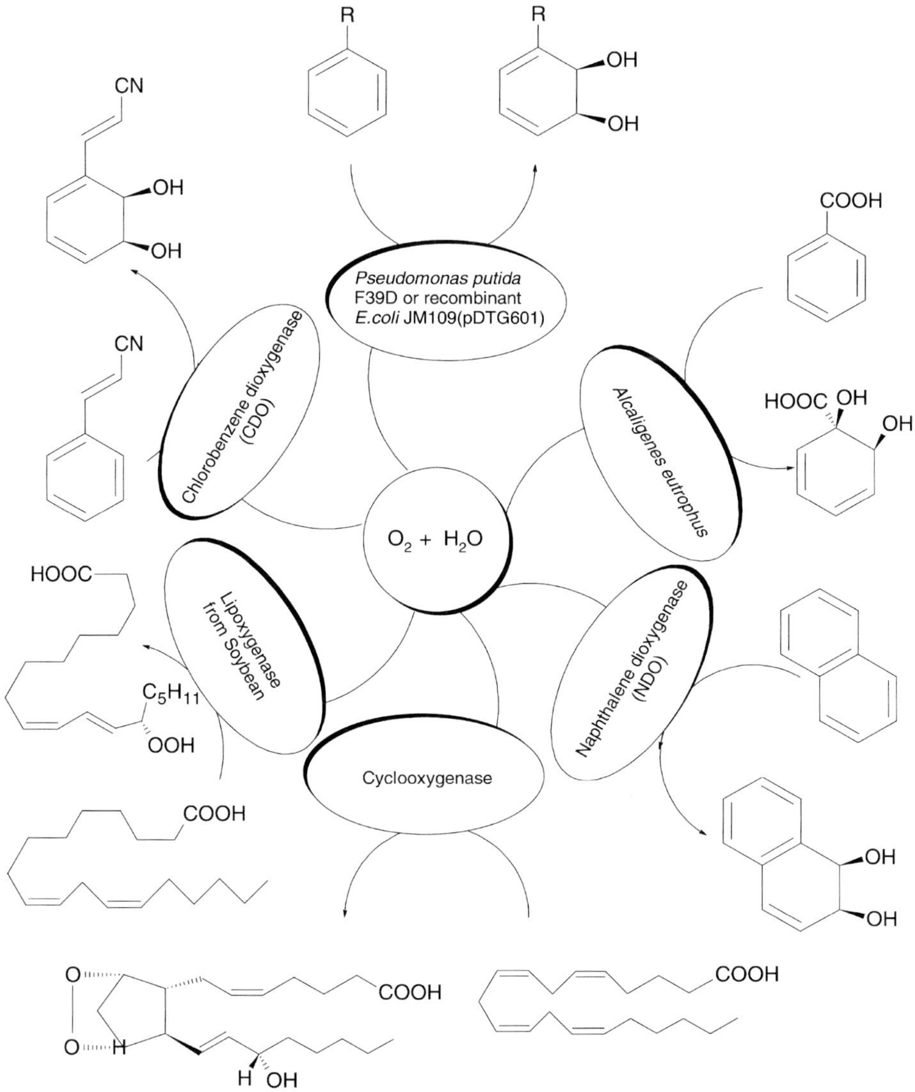

Figure 20.7 Overview dioxygenase reactions in synthetic chemistry.

acid yields chiral hydroperoxides [99]. Cyclooxygenases catalyze the oxidation of arachidonic acid to interesting cyclic endoperoxides and prostaglandins [100]. Dioxygenase-catalyzed oxidative cleavage of carbon–carbon double bonds was discovered by Osamu Hayaishi in his pioneering studies of tryptophan metabolism in 1957 and continues to be of fundamental biochemical interest [101]. Whether the central cleaving enzyme of β-carotene into two molecules of retinal is a monooxygenase [102, 103] or a dioxygenase [104] has been discussed. Isotopic labeling

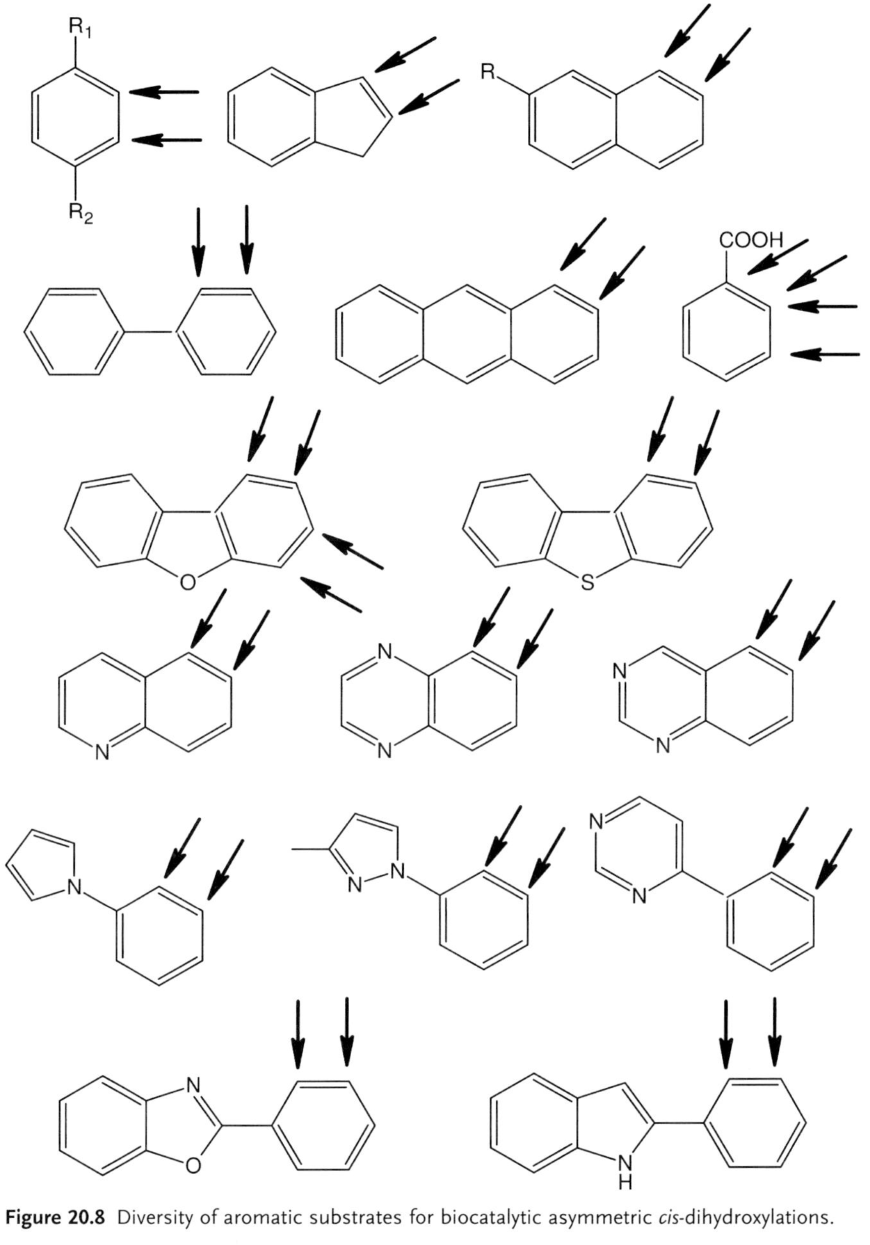

Figure 20.8 Diversity of aromatic substrates for biocatalytic asymmetric *cis*-dihydroxylations.

experiments using $H_2^{18}O$ or $^{18}O_2$ showed that, in the excentral biocatalytic cleavage, the oxygen in the keto-group of β-ionone is derived solely from molecular dioxygen [104]. In any case, selective biocatalytic cleavage of double bonds in polyenes in a central or excentral manner is an important path to central or remote metabolites.

The biocatalytic single-step cleavage of arylalkenes is an interesting synthetic tool and an equivalent to reductive ozonization [105].

The intermolecular dioxygenases incorporate the two oxygen atoms into two molecules and the α-keto acid-dependent enzymes are important in the biosynthesis of a variety of biological compounds like penicillins, cephalosporins, and clavulanic acid. Usually the α-keto acid is α-ketoglutarate, which is oxidatively decarboxylated and one of the atoms from molecular oxygen is incorporated into the product succinate. The decarboxylation of α-ketoglutarate can thereby be coupled by these enzymes with the formation of a high-energy ferryl-oxo intermediate acting as a hydrogen-abstracting species for performing hydroxylation or halogenation [106].

20.7
Biocatalytic Asymmetric Oxidations with Other Enzymes

Many unusual biocatalytic asymmetric oxidation reactions like oxidative cyclization, oxidative ring expansion, oxidative deamination, or oxidative decarboxylation were discovered in the course of studies in natural product biosynthesis and the involved enzyme functions continue to be of great interest.

In addition to the already mentioned dehydrogenases, biocatalytic oxidative cyclizations have been achieved by a variety of enzyme classes like synthases or tyrosinases as shown in Figure 20.9. Isopenicillin N synthase catalyzes the two iron-dioxygen-mediated oxidative ring closure of the tripeptide δ-(L-α-aminoadipoyl)- L-cysteinyl-D-valine to isopenicillin N, the precursor to all penicillins and cephalosporins [107]. The four-electron oxidation of δ-(L-α-aminoadipoyl)- L-cysteinyl-D-valine by isopenicillin N synthase is unusual, requiring no 2-oxoglutarate cofactor. In contrast, clavaminate synthase is a remarkable Fe(II)- and 2-oxo-glutarate-dependent oxygenase, which catalyzes the construction of the bicyclic clavam ring system of clavaminic acid from a monocyclic β-lactam. Clavaminate synthase thereby catalyzes the three quite different oxidation reactions of hydroxylation, oxidative cyclization to dihydroclavaminic acid, and the final desaturation reaction to clavaminic acid [108]. Tyrosinase-catalyzed oxidative cyclization of *N,N*-dimethyltyramine with oxygen has been used for the synthesis of the anachelin chromophore [109].

Selected examples of other biocatalytic asymmetric oxidations are shown in Figure 20.10. In the area of the polyether ionophore monensin a recently proposed mechanism of oxidative cyclization of a linear polyketide intermediate by four enzymes, the products of *monBI, monBII, monCI,* and *monCII,* has been supported experimentally by analysis of a biosynthetic gene cluster [110] and the accumulation of an *E,E,E*-triene, when oxidative cyclization was blocked [111].

Oxidative ring expansion by an additional carbon atom is performed by expandases different from the Baeyer–Villiger monooxygenases capable of introducing oxygen into cyclic ketones. The Fe(II)- and 2-oxoglutarate-dependent oxygenase

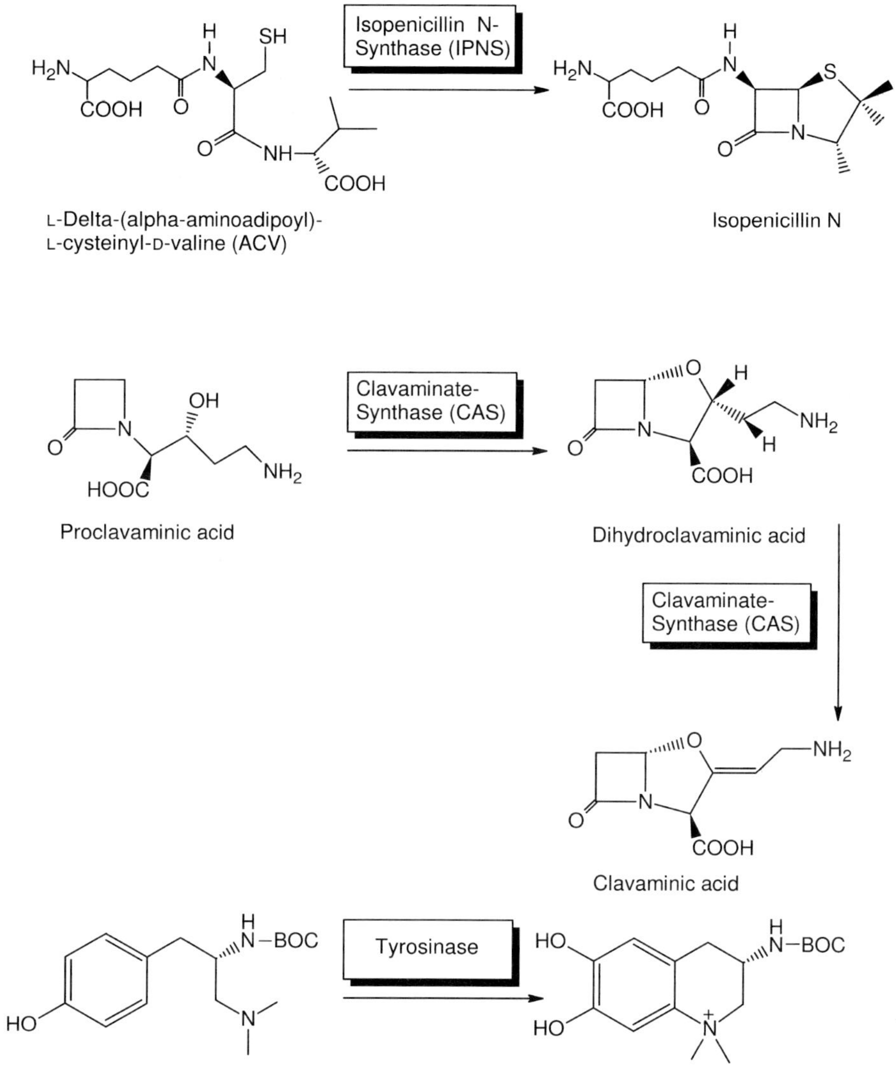

Figure 20.9 Biocatalytic oxidative cyclization reactions.

deacetoxy-cephalosporin C synthase [112] catalyzes the unusual oxidative ring expansion of penicillin N to deacetyoxycephalosporin C.

Useful whole cell asymmetric oxidations of methyl groups to carboxyl groups have been found by classical microbiological screening as for the microbial oxidation of prochiral 2-methyl-1,3-propandiol to (*R*)-3-hydroxy-2-methyl propionic acid with *A. pasteurianus* with 97% e.e. and 100% molar conversion [113].

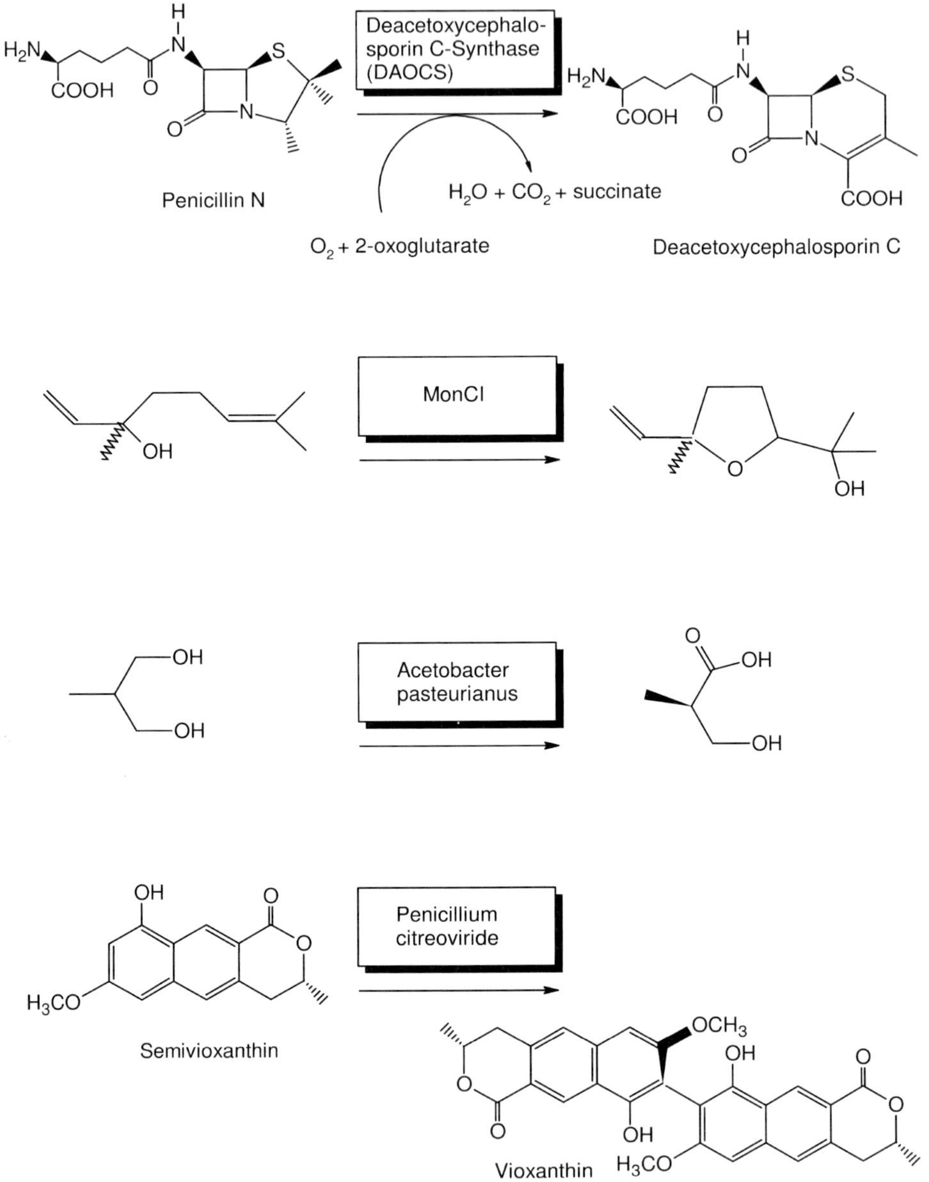

Figure 20.10 Selected examples of other enzymes for asymmetric oxidative transformations.

20.8 Outlook

The organic chemistry of enzyme-catalyzed asymmetric oxidations shows a remarkable selectivity and versatility that is best used in an early phase of preceding and subsequent synthesis steps [114]. It will be interesting to establish which complexity of the catalyst is needed for a given asymmetric oxidation, that is whether a multienzyme complex or a multifunctional enzyme, a monofunctional enzyme, or an organocatalyst like an amino acid [115] will be best. As large-scale industrial oxidations of organic compounds with oxidants in flammable solvents pose not only safety, health, and environmental challenges, but often suffer from insufficient selectivity, the development of selective catalysts is of prime importance. Asymmetric Baeyer–Villiger oxidations by biocatalysts have been established on a large scale with a selectivity not reached by the use of inorganic, organometallic, or organocatalysts [116]. The practical process improvements at the macroenvironmental level go in parallel with the improved compatibility of labile functional groups and increased selectivity at the molecular level. Biological cells make use of the required selectivity in many asymmetric oxidation reaction steps of biosynthetic pathways to central and remote metabolites. The elucidation of novel reaction steps in the biosynthesis of known natural products is therefore an interesting entry point for finding the novel biocatalytic reaction methodologies of a wide variety of compounds. An example of such a reaction step is the eight-electron oxidation of chlorinated chromopyrrolic acid to the rebeccamycin aglycone performed by the enzyme pair RebP and RebC [117]. The mode of action has been investigated by crystallography and suggested an interesting interaction: while the first enzyme RebP produces low levels of a reactive intermediate by an eight-electron oxidation, the second enzyme RebC is needed for accelerating rebeccamycine aglycone formation and eliminating side reactions [118]. The combination of molecular biology, chemistry, and process design will be a major driving force for the implementation of biocatalytic asymmetric oxidations [119, 120]. The discovery and understanding of novel biocatalytic oxidations shows great promise for the synthesis of more complex molecules.

Acknowledgments

I would like to thank all co-workers and members of the EU projects on Baeyer–Villiger oxidation, the Swiss project on asymmetric dihydroxylation for their dedicated work and excellent cooperation, and the following organizations for their support: the COST Action D25 and the COST Office, the Swiss State Secretariat for Education and Research, the program SPP-Biotech of the Swiss National Science Foundation, and the EU Research Directorate General.

References

1 Caron, S., Dugger, R.W., Ruggeri, S.G., Ragan, J.A. and Ripin, B.D.H. (2006) Large-scale oxidations in the pharmaceutical industry. *Chemical Reviews*, **106**, 2943–2989.

2 Ertl, G., Knözinger, H. and Weitkamp, J. (eds) (1999) *Environmental Catalysis*, Wiley-VCH Verlag GmbH, Weinheim.

3 Sharpless, K.B. (2002) Searching for new reactivity. *Angewandte Chemie – International Edition*, **41**, 2024–2032.

4 Katsuki, T. (ed.) (2001) *Asymmetric Oxidation Reactions*, Oxford University Press, Oxford, UK.

5 Bäckvall, J.-E. (ed.) (2006) *Modern Oxidation Methods*, Wiley-VCH Verlag GmbH & Co. KGaA, Weinheim.

6 Dubnikova, F., Kosloff, R., Almog, J., Zeiri, Y., Boese, R., Itzhaky, H., Alt, A. and Keinan, E. (2005) Decomposition of triacetone triperoxide is an entropic explosion. *Journal of the American Chemical Society*, **127**, 1146–1157.

7 Kniemeyer, O., Musat, F., Sievert, S.M., Knittel, K., Wilkes, H., Blumenberg, M., Michaelis, W., Classen, A., Bolm, C., Joye, S.B. and Widdel, F. (2007) Anaerobic oxidation of short-chain hydrocarbons by marine sulphate-reducing bacteria. *Nature*, **449**, 898–901.

8 Holland, H.L. (1992) *Organic Synthesis with Oxidative Enzymes*, VCH Publishers, New York.

9 Reichstein, T. and Grüssner, A. (1934) Eine ergiebige Synthese der L-Ascorbinsäure (vitamin C). *Helvetica Chimica Acta*, **17**, 311–328.

10 Kieslich, K. (1980) Industrial aspects of the biotechnological production of steroids. *Biotechnology Letters*, **2**, 211–217.

11 Schmid, R.D. and Urlacher, V. (eds) (2007) *Modern Biooxidation. Enzymes, Reactions and Applications*, Wiley-VCH Verlag GmbH, & Co. KGaA, Weinheim.

12 Kroutil, W., Mang, H., Edegger, K. and Faber, K. (2004) Biocatalytic oxidation of primary and secondary alcohols. *Advanced Synthesis Catalysis*, **346**, 125–142.

13 Drauz, K. and Waldmann, H. (eds) (2002) *Enzyme Catalysis in Organic Synthesis: a Comprehensive Handbook*, Vol. **1–3**, 2nd edn, completely revised and extended edition, Wiley-VCH Verlag GmbH, Weinheim.

14 Blaser, H.U. and Schmidt, E. (eds) (2004) *Asymmetric Catalysis on Industrial Scale – Challenges, Approaches and Solutions*, Wiley-VCH Verlag GmbH, & Co. KGaA, Weinheim.

15 Liese, A., Seelbach, K. and Wandrey, C. (eds) (2006) *Industrial Biotransformations*, 2nd edn, completely revised and extended edition, Wiley-VCH Verlag GmbH, Weinheim.

16 Dzyuba, S.V. and Klibanov, A.M. (2004) Stereoselective oxidations and reductions catalyzed by nonredox proteins. *Tetrahedron: Asymmetry*, **15**, 2771–2777.

17 Patel, R.N. (ed.) (2007) *Biocatalysis in the Pharmaceutical and Biotechnology Industries*, CRC Taylor and Francis, Boca Raton – London – New York.

18 Caligiuri, A., D'Arrigo, P., Rosini, E., Tessaro, D., Molla, G., Servi, S. and Pollegioni, L. (2006) Enzymatic conversion of unnatural amino acids by yeast D-amino acid oxidase. *Advanced Synthesis Catalysis*, **348**, 2183–2190.

19 Benz, P. and Wohlgemuth, R. (2007) Amino acid oxidase-catalysed resolution and Pictet–Spengler reaction towards chiral and rigid unnatural amino acids. *Journal of Chemical Technology and Biotechnology (Oxford, Oxfordshire: 1986)*, **87**, 1082–1086.

20 Soda, K., Oikawa, T. and Yokoigawa, K. (2001) One-pot chemo-enzymatic enantiomerization of racemates. *Journal of Molecular Catalysis B: Enzymatic*, **11**, 149–153.

21 Trost, E.M. and Fischer, L. (2002) Minimization of by-product formation during D-aminoacid oxidase catalyzed racemate resolution of D/L-amino acids. *Journal of Molecular Catalysis B: Enzymatic*, **19–20**, 189–195.

22 Fotheringham, I., Archer, I., Carr, R., Speight, R. and Turner, N.J. (2006) Preparative deracemization of unnatural amino acids. *Biochemical Society Transactions*, **34**, 287–290.

23 Eve, T.S.C., Wells, A. and Turner, N.J. (2007) Enantioselective oxidation of *O*-methyl-hydroxylamines using monoamine oxidase N as catalyst. *Chemical Communications*, 1530–1531.

24 Root, R.L., Durrwachter, J.R. and Wong, C.-H. (1985) Enzymatic synthesis of unusual sugars: galactose oxidase catalyzed stereospecific oxidation of polyols. *Journal of the American Chemical Society*, **107**, 2997–2999.

25 Molinari, F., Gandolfi, R., Villa, R., Urban, E. and Kiener, A. (2003) Enantioselective oxidation of prochiral 2-methyl-1,3-propandiol by *Acetobacter pasteurianus*. *Tetrahedron: Asymmetry*, **14**, 2041–2043.

26 Adam, W., Lazarus, M., Boss, B., Saha-Möller, C.R., Humpf, H.-U. and Schreier, P. (1997) Enzymatic resolution of chiral 2-hydroxy carboxylic acids by enantio-selective oxidation with molecular oxygen catalyzed by the glycolate oxidase from spinach (*Spinacia oleracea*). *The Journal of Organic Chemistry*, **62**, 7841–7843.

27 Adam, W., Lazarus, M., Saha-Moller, C.R., Weichold, O., Hoch, U. Haring, D. and Schreier, P. (1999) Biotransformations with peroxidases. *Advances in Biochemical Engineering/ Biotechnology*, **63**, 73–108.

28 van de Velde, F., van Rantwijk, F. and Sheldon, R.A. (2001) Improving the catalytic performance of peroxidases in organic synthesis. *Trends in Biotechnology*, **19**, 73–80.

29 Berglund, G.I., Carlsson, G.H., Smith, A.T., Szöke, H., Henriksen, A. and Hajdu, J. (2002) The catalytic pathway of horseradish peroxidase at high resolution. *Nature*, **417**, 463–468.

30 Colonna, S., Del Sordo, S., Gaggero, N., Carrea, G. and Pasta, P. (2002) Enzyme-mediated catalytic asymmetric oxidations. *Heteroatom Chemistry*, **13**, 467–473.

31 Legros, J., Dehli, J.R. and Bolm, C. (2005) Applications of catalytic asymmetric sulfide oxidations to the syntheses of biologically active sulfoxides. *Advanced Synthesis Catalysis*, **347**, 19–31.

32 Klibanov, A.M. (2003) Asymmetric enzymic oxidoreductions in organic solvents. *Current Opinion in Biotechnology*, **14**, 427–431.

33 Ros Barcelo, A. and Pomar, F. (2002) Plant peroxidases: versatile catalysts in the synthesis of bioactive natural products. *Studies in Natural Products Chemistry*, **27**, 735–791.

34 Adam, W., Heckel, F., Saha-Moller, C.R. and Schreier, P. (2002) Biocatalytic synthesis of optically active oxyfunctionalized building blocks with enzymes, chemoenzymes and microorganisms. *Journal of Organometallic Chemistry*, **661**, 17–29.

35 Lutz, S., Steckhan, E. and Liese, A. (2004) First asymmetric electroenzymatic oxidation catalyzed by peroxidase. *Electrochemistry Communications*, **6**, 583–587.

36 Colonna, S., Gaggero, N., Richelmi, C. and Pasta, P. (1999) Recent biotechnological developments in the use of peroxidases. *Trends in Biotechnology*, **17**, 163–168.

37 Hui, S. and Dordick, J.S. (2002) Highly enantioselective oxidation of *cis*-cyclopropyl-methanols to corresponding aldehydes catalyzed by chloroperoxidase. *The Journal of Organic Chemistry*, **67**, 314–317.

38 Liu, H.-L., Huang, X.-F., Wan, X. and Kong, L.-Y. (2007) Biotransfromation of *p*-coumaric acid (= (2*E*)-3-(4-hydroxyphenyl)prop-2-enoic Acid) by *Momor-Dica charantia* peroxidase. *Helvetica Chimica Acta*, **90**, 1117–1132.

39 Matsuyama, A., Yamamoto, H., Kawada, N. and Kobayashi, Y. (2001) Industrial Production of (*R*)-1,3-butanediol by new biocatalysts. *Journal of Molecular Catalysis B: Enzymatic*, **11**, 513–521.

40 Kinast, G. and Schedel, M. (1981) A four-step synthesis of 1-deoxynojirimycin with a biotransformation as cardinal reaction step. *Angewandte Chemie – International Edition*, **20**, 805–806.

41 Riva, S., Bovara, R., Pasta, P. and Carrea, G. (1986) Preparative-scale regio- and stereospecific oxidoreduction of cholic acid and dehydrocholic acid catalyzed by hydroxysteroid dehydrogenases. *The*

Journal of Organic Chemistry, **51**, 2902–2906.

42 Lee, L.G. and Whitesides, G.M. (1986) Preparation of optically active 1,2-diols and a-hydroxy ketones using glycerol dehydrogenase as catalyst: limits to enzyme-catalyzed synthesis due to noncompetitive and mixed inhibition by product. *The Journal of Organic Chemistry*, **51**, 25–36.

43 Irwin, A.J. and Jones, J.B. (1977) Regiospecific and enantioselective horse liver alcohol dehydrogenase catalyzed oxidations of some hydroxycyclopentanes. *Journal of the American Chemical Society*, **99**, 1625–1630.

44 Irwin, A.J. and Jones, J.B. (1977) Asymmetric syntheses via enantiotopically selective horse liver alcohol dehydrogenase catalyzed oxidations of diols containning a prochiral center. *Journal of the American Chemical Society*, **99**, 556–561.

45 Lok, K.P., Jakovac, I.J. and Jones, J.B. (1985) Enzymes in organic synthesis. 34. preparations of enantiomerically pure exo-and endo-bridged bicyclic [2.2.1] and [2.2.2] chiral lactones via stereospecific horse liver alcohol dehydrogenase catalyzed oxidations of meso-diols. *Journal of the American Chemical Society*, **107**, 2521–2526.

46 Chang, D., Zhang, J., Witholt, B. and Li, Z. (2004) Chemical and enzymatic synthetic methods for asymmetric oxidation of the C—C double bond. *Biocatalysis and Biotransformation*, **22**, 113–131.

47 Katsuki, T. and Sharpless, K.B. (1980) The first practical method for asymmetric epoxidation. *Journal of the American Chemical Society*, **102**, 5974–5976.

48 Zhang, W., Loebach, J.L., Wilson, S.R. and Jacobsen, E.N. (1990) Enantioselective epoxidation of unfunctionalized olefins catalyzed by (salen)manganese complexes. *Journal of the American Chemical Society*, **112**, 2801–2803.

49 Katsuki, T. (2001) The catalytic enantioselective synthesis of optically active epoxides and tetrahydrofurans. Asymmetric epoxidation, the desymmetrization of meso-tetrahydrofurans, and enantiospecific ring-enlargement. *Current Organic Chemistry*, **5**, 663–678.

50 Lorenz, J.C., Frohn, M., Zhou, X., Zhang, J.-R., Tang, Y., Burke, C. and Shi, Y. (2005) Transition state studies on the dioxirane-mediated asymmetric epoxidation via kinetic resolution and desymmetrization. *The Journal of Organic Chemistry*, **70**, 2904–2911.

51 Ohta, H. and Tetsukawa, H. (1978) Microbial epoxidation of long-chain terminal olefins. *Chemical Communications*, 849–850.

52 Fourneron, J.D., Archelas, A. and Furstoss, R. (1989) Microbial transformations. 12. Regiospecific and asymmetric oxidation of the remote double bond of geraniol. *The Journal of Organic Chemistry*, **84**, 4686–4689.

53 Mihovilovic, M.D. (2006) Enzyme mediated Baeyer–Villiger oxidations. *Current Organic Chemistry*, **10**, 1265–1287.

54 Sato, K., Aoki, M. and Noyori, R. (1998) A 'green' route to adipic acid: direct oxidation of cyclohexenes with 30 percent hydrogen peroxide. *Science*, **281**, 1646–1647.

55 Schlichting, I., Berndzen, J., Chu, K., Stock, A.M., Maves, S.A., Benson, D.E., Sweet, R.M., Ringe, D., Petsko, G.A. and Sligar, S.G. (2000) The catalytic pathway of cytochrome P450cam at atomic resolution. *Science*, **287**, 1615–1622.

56 Watanabe, Y., Laschat, S., Budde, M., Affolter, O., Shimada, Y. and Urlacher, V.B. (2007) Oxidation of acyclic monoterpenes by P450 BM-3 monooxygenase: influence of the substrate *E*/*Z*-isomerism on enzyme chemo- and regio-selectivity. *Tetrahedron*, **63**, 9413–9422.

57 Landwehr, M., Hochrein, L., Otey, C.R., Kasrayan, A., Bäckvall, J.-E. and Arnold, F.H. (2006) Enantioselective hydroxylation of 2-arylacetic acid derivatives and buspirone catalyzed by engineered cytochrome P450 BM-3. *Journal of the American Chemical Society*, **128**, 6058–6059.

58 Lipscomb, J.D. (1994) Biochemistry of the soluble methane monooxygenase. *Annual Review of Microbiology*, **48**, 371–399.

59 Klinman, J.P., Krueger, M., Brenner, M. and Edmondson, D.E. (1984) Evidence for two copper atoms/subunit in dopamine beta-monooxygenase catalysis. *The Journal of Biological Chemistry*, **259**, 3399–402.

60 Kappock, T.J. and Caradonna, J.P. (1996) Pterin-dependent amino acid hydroxylases. *Chemical Reviews*, **96**, 2659–2756.

61 Werner, E.R., Hermetter, A., Prast, H., Golderer, G. and Werner-Felmayer, G. (2007) Widespread occurrence of glyceryl ether monooxygenase activity in rat tissues detected by a novel assay. *Journal of Lipid Research*, **48**, 1422–1427.

62 Joosten, V. and van Berkel, W.J.H. (2007) Flavoenzymes. *Current Opinion in Chemical Biology*, **11**, 195–202.

63 Roberts, S.M. and Wan, P.W.H. (1998) Enzyme-catalyzed Baeyer–Villiger oxidations. *Journal of Molecular Catalysis B: Enzymatic*, **4**, 111–136.

64 Mihovilovic, M.D. (2006) Enzyme-mediated Baeyer–Villiger oxidations. *Current Organic Chemistry*, **10**, 1265–1287.

65 Doig, S.D., Avenell, P.J., Bird, P.A., Gallati, P., Lander, K.S., Lye, G.J., Wohlgemuth, R. and Woodley, J.M. (2002) Reactor operation and scale-up of whole-cell Baeyer–Villiger catalyzed lactone synthesis. *Biotechnology Progress*, **18**, 1039–1046.

66 Alphand, V., Carrea, G., Furstoss, R., Woodley, J.M. and Wohlgemuth, R. (2003) Towards large-scale synthetic applications of Baeyer–Villiger monooxygenases. *Trends in Biotechnology*, **21**, 318–323.

67 Hilker, I., Alphand, V., Wohlgemuth, R. and Furstoss, R. (2004) Microbial transformations 56. Preparative scale asymmetric Baeyer–Villiger oxidation using a highly productive 'two-in-one' *in situ* SFPR concept. *Advanced Synthesis Catalysis*, **346**, 203–214.

68 Hilker, I., Gutierrez, M.C., Alphand, V., Wohlgemuth, R. and Furstoss, R. (2004) Microbiological transformations 57. Facile and efficient resin-based *in situ* SFPR preparative scale synthesis of an enantiopure 'unexpected' lactone regioisomer via a Baeyer–Villiger oxidation process. *Organic Letters*, **6**, 1955–1958.

69 Hilker, I., Wohlgemuth, R., Alphand, V. and Furstoss, R. (2005) Microbial transformations 59: first kilogram scale asymmetric microbial Baeyer–Villiger oxidation with optimized productivity using a resin-based *in situ* SFPR strategy. *Biotechnology and Bioengineering*, **92**, 702–710.

70 van Beilen, J.B., Duetz, W.A., Schmid, A. and Witholt, B. (2003) Practical issues in the application of oxygenases. *Trends in Biotechnology*, **21**, 170–177.

71 Hilker, I., Baldwin, C., Alphand, V., Furstoss, R., Woodley, J. and Wohlgemuth, R. (2006) On the influence of oxygen and cell concentration in an SFPR whole cell biocatalytic Baeyer–Villiger oxidation process. *Biotechnology and Bioengineering*, **93**, 1138–1144.

72 Holland, H.L. (1988) Chiral sulfoxidation by biotransformation of organic sulfides. *Chemical Reviews*, **88**, 473–485.

73 Ottolina, G., de Gonzalo, G., Carrea, G. and Danieli, B. (2005) Enzymatic Baeyer–Villiger oxidation of bicyclic diketones. *Advanced Synthesis Catalysis*, **347**, 1035–1040.

74 Wang, C., Gibson, M., Rohr, J. and Oliveira, M.A. (2005) Crystallization and X-ray diffraction properties of Baeyer–Villiger monooxygenase MtmOIV from the mithramycin biosynthetic pathway in *Streptomyces argillaceus*. *Acta Crystallographica Section F, Structural Biology and Crystallization Communications*, **61**, 1023–1026.

75 Colonna, S., Gaggero, N., Carrea, G., Ottolina, G., Pasta, P. and Zambianchi, F. (2002) First asymmetric epoxidation catalysed by cyclohexanone monooxygenase. *Tetrahedron Letters*, **43**, 1797–1799.

76 Park, J.B., Bühler, B., Habicher, T., Hauer, B., Panke, S., Witholt, B. and Schmid, A. (2006) The efficiency of recombinant *Escherichia coli* as biocatalyst for stereospecific epoxidation.

Biotechnology and Bioengineering, **95**, 501–512.

77 Faber, K. (2004) *Biotransformations in Organic Chemistry*, 5th revised and corrected edn, Springer-Verlag, Berlin, Heidelberg.

78 Kolb, H.C., Van Nieuwenhenhze, M.S. and Sharpless, K.B. (1994) Catalytic asymmetric dihydroxylation. *Chemical Reviews*, **94**, 2483–2547.

79 Gibson, D.T., Hemsley, M., Yoshioka, H. and Mabry, T.J. (1970) Formation of (+)-*cis*-2,3-dihydroxy-1-methylcyclohexa-4,6-diene from toluene by *Pseudomonas putida*. *Biochemistry*, **9**, 1626–1630.

80 Gibson, D.T., Koch, J.R. and Kallio, R.E. (1968) Oxidative degradation of aromatic hydrocarbons by microorganisms. I. Enzymatic formation of catechol from benzene. *Biochemistry*, **7**, 2653–2662.

81 Hudlicky, T., Gonzalez, D. and Gibson, D.T. (1999) Enzymatic dihydroxylation of aromatics in enantioselective synthesis: expanding asymmetric methodology. *Aldrichimica Acta*, **32**, 35–62.

82 Resnick, S.M., Lee, K. and Gibson, D.T. (1996) Diverse reactions catalyzed by naphthalene dioxygenase from *Pseudomonas* sp. strain NCIB9816. *Journal of Industrial Microbiology*, **17**, 438–457.

83 Boyd, D.R., McMordie, R.A.S., Porter, H.P., Dalton, H., Jenkins, R.O. and Howarth, O.W. (1987) Metabolism of bicyclic aza-arenes by *Pseudomonas putida* to yield vicinal *cis*-dihydrodiols and phenols. *Chemical Communications*, **22**, 1722–1724.

84 Ley, S.V. and Sternfeld, F. (1989) Microbial oxidation in synthesis: preparation of (+)- and (–)-pinitol from benzene. *Tetrahedron*, **45**, 3463–3476.

85 Boyd, D.R. and Sheldrake, G.N. (1998) The dioxygenase-catalyzed formation of vicinal *cis*-diols. *Natural Product Reports*, **15**, 309–324.

86 Johnson, R.A. (2004) Microbial arene oxidations. *Organic Reactions*, **63**, 117.

87 Reiner, A.M. and Hegemann, G.D. (1971) Metabolism of benzoic acid by bacteria accumulation of (–)-3,5-cyclohexadiene-1,2-diol-1-carboxylic acid by a mutant strain of *Alcaligenes eutrophus*. *Biochemistry*, **10**, 2530–2536.

88 Urlacher, V.B. and Schmid, R.D. (2006) Recent advances in oxygenase-catalyzed biotransformations. *Current Opinion in Chemical Biology*, **10**, 156–161.

89 Bui, V.B., Hudlicky, T., Hansen, T.V. and Stenstrom, Y. (2002) Direct biooxidation of arenes to corresponding catechols with *E. coli* JM109(pDTG602). Application to synthesis of combretastatins A-1 and B-1. *Tetrahedron Letters*, **43**, 2839–2841.

90 Parales, R.E. and Resnick, S.M. (2007) Applications of aromatic dioxygenases, in *Biocatalysis in the Pharmaceutical and Biotechnology Industries* (ed. R.N. Patel), CRC Taylor and Francis, Boca Raton–London–New York, pp. 299–332.

91 Cirini, P.C. and Arnold, F.H. (2002) Protein engineering of oxygenases for biocatalysis. *Current Opinion in Chemical Biology*, **6**, 130–135.

92 Finn, K.J., Pavlyuk, O. and Hudlicky, T. (2005) Toluene dioxygenase-mediated oxidation of bromo(methylsulfanyl) benzenes. Absolute configuration of metabolites and evaluation of chemo- and regioselectivity trends. *Collection of Czechoslovak Chemical Communications*, **70**, 1709–1726.

93 Yildirim, S., Zezula, J., Hudlicky, T., Witholt, B. and Schmid, A. (2004) Asymmetric dihydroxylation of cinnamonitrile to *trans*-3-[(5S,6R)-5,6-dihydroxy-cyclohexa-1,3-dienyl]-acrylonitrile using chlorobenzene dioxygenase. *Advanced Synthesis Catalysis*, **346**, 933–942.

94 Yildirim, S., Franco, T., Wohlgemuth, R., Kohler, H.P., Witholt, B. and Schmid, A. (2005) Recombinant chlorobenzene dioxygenase from *Pseudomonas* sp. P51: a biocatalyst for regioselective oxidation of aromatic nitriles. *Advanced Synthesis Catalysis*, **347**, 1060–1073.

95 Yildirim, S., Ruinatscha, R., Gross, R., Wohlgemuth, R., Kohler, H.P., Witholt, B. and Schmid, A. (2006) Selective hydrolysis of the nitrile group of *cis*-dihydrodiols. *Journal of Molecular Catalysis B: Enzymatic*, **38**, 76–83.

96 Boyd, D.R. and Bugg, T.D.H. (2006) Arene *cis*-dihydrodiol formation: from biology to application, organic and

biomolecular chemistry. *Organic and Biomolecular Chemistry*, **4**, 181–192.

97 Boyd, D.R., Sharma, N.D., Belhocine, T., Malone, J.F., McGregor, S. and Allen, C.C.R. (2006) Dioxygenase-catalyzed dihydroxylation of arene *cis*-dihydrodiols and acetonide derivatives: a new approach to the synthesis of enantio-pure tetraoxygenated bioproducts from arenes. *Chemical Communications*, 4934–4936.

98 Schneider, C., Pratt, D.A., Porter, N.A. and Brash, A.R. (2007) Control of oxygenation in lipoxygenase and cyclooxygenase catalysis. *Chemistry and Biology*, **14**, 473–488.

99 Corey, E.J., Albright, J.O., Burton, A.E. and Hashimoto, S. (1980) Chemical and enzymic syntheses of 5-HPETE, a key biological precursor of slow-reacting substance of anaphylaxis (SRS), and 5-HETE. *Journal of the American Chemical Society*, **102**, 1435–1436.

100 Samuelsson, B. (1965) On the incorporation of oxygen in the conversion of 8,11,14-eicosatrienoic acid to prostaglandin E_1. *Journal of the American Chemical Society*, **87**, 3011–3013.

101 Hayaishi, O. (1993) My life with tryptophan – never a dull moment. *Protein Science*, **2**, 472–475.

102 Leuenberger, M.G., Engeloch-Jarret, C. and Woggon, W.D. (2001) The reaction mechanism of the enzyme-catalyzed central cleavage of β-carotene to retinal. *Angewandte Chemie – International Edition*, **40**, 2613–2617.

103 Kloer, D.P., Ruch, S., Al-Babili, S., Beyer, P. and Schulz, G.E. (2005) The structure of a retinal-forming carotenoid oxygenase. *Science*, **308**, 267–269.

104 Schmidt, H., Kurtzer, R., Eisenreich, W. and Schwab, W. (2006) The carotenase AtCCD1 from *Arabidopsis thaliana* is a dioxygenase. *The Journal of Biological Chemistry*, **281**, 9845–9851.

105 Mang, H., Gross, J., Lara, M., Goessler, C., Schoemaker, H.E., Guebitz, G.M. and Kroutil, W. (2006) Biocatalytic single-step alkene cleavage from aryl alkenes: an enzymatic equivalent to reductive ozonization. *Angewandte Chemie – International Edition*, **45**, 2501–2503.

106 Blasiak, L.C., Vaillancourt, F.H., Walsh, C.T. and Drennan, C.L. (2006) Crystal structure of the non-haem iron halogenase SyrB2 in syringomycin biosynthesis. *Nature*, **440**, 368–371.

107 Burzlaff, N.I., Rutledge, P.J., Clifton, I.J., Hensgens, C.M., Pickford, M., Adlington, R.M., Roach, P.L. and Baldwin, J.E. (1999) The reaction cycle of isopenicillin N synthase observed by X-ray diffraction. *Nature*, **401**, 721–724.

108 Zhang, Z., Ren, J., Stammers, D.K., Baldwin, J.E., Harlos, K. and Schofield, C.J. (2000) Structural origins of the selectivity of the trifunctional oxygenase clavaminic acid synthase. *Nature Structural and Molecular Biology*, **7**, 127–133.

109 Gademann, K., Bethuel, Y., Locher, H.H. and Hubschwerlen, C. (2007) Biomimetic total synthesis and antimicrobial evaluation of anachelin H. *The Journal of Organic Chemistry*, **72**, 8361–8370.

110 Oliynyk, M., Stark, C.B.W., Bhatt, A., Jones, M.A., Hughes-Thomas, A., Wilkinson, C., Oliynyk, Z., Demydchuk, Y., Staunton, J. and Leadlay, P.F. (2003) Analysis of the biosynthetic gene cluster for the polyether antibiotic monensin in *Streptomyces cinnamonensis* and evidence for the role of *monB* and *monC* genes in oxidative cyclization. *Molecular Microbiology*, **49**, 1179–1190.

111 Bhatt, A., Stark, C.B.W., Harvey, B.M., Gallimore, A.R., Demydchuk, Y.A., Spencer, J.B., Staunton, J. and Leadlay, P.F. (2005) Accumulation of an *E,E,E*-triene by the monensin-producing polyketide synthase when oxidative cyclization is blocked. *Angewandte Chemie – International Edition*, **44**, 7075–7078.

112 Valegard, K., Terwisscha van Scheltinga, A.C., Lloyd, M.D., Hara, T., Ramaswamy, S., Perrakis, A., Thompson, A., Lee, H.J., Baldwin, J.E., Schofield, C.J., Hajdu, J. and Andersson, I. (1998) Structure of a cephalosporin synthase. *Nature*, **394**, 805–809.

113 Molinari, F., Gandolfi, R., Villa, R., Urban, E. and Kiener, A. (2003) Enantioselective oxidation of prochiral

2-methyl-1,3-propandiol by *Acetobacter pasteuria-nus*. *Tetrahedron: Asymmetry*, **14**, 2041–2043.

114 Wohlgemuth, R. (2007) Interfacing biocatalysis and organic synthesis. *Journal of Chemical Technology and Biotechnology (Oxford, Oxfordshire: 1986)*, **82**, 1055–1062.

115 Cordova, A., Sunden, H., Engqvist, M., Ibrahem, I. and Casas, J. (2004) Direct amino-acid-catalyzed asymmetric incorporation of molecular oxygen to organic compounds. *Journal of the American Chemical Society*, **126**, 8914–8915.

116 Wohlgemuth, R. (2007) Modular and scalable biocatalytic tools for practical safety, health and environmental improvements in the production of speciality chemicals. *Biocatalysis and Biotransformation*, **25**, 178–185.

117 Sanchez, C., Zhu, L., Brana, A.F., Salas, A.P., Rohr, J., Mendez, C. and Salas, J. A. (2005) Combinatorial biosynthesis of antitumor indolocarbazole compounds. *Proceedings of the National Academy of Sciences of the United States of America*, **102**, 461–466.

118 Ryan, K.S., Howard-Jones, A.R., Hamill, M.J., Elliott, S.J., Walsh, C.T. and Drennan, C.L. (2007) Crystallographic trapping in the rebeccamycin biosynthetic enzyme RebC. *Proceedings of the National Academy of Sciences of the United States of America*, **104**, 15311–15316.

119 Wohlgemuth, R. (2006) Tools for selective enzyme reaction steps in the synthesis of laboratory chemicals. *Engineering in Life Sciences*, **6**, 1–8.

120 Kamerbeek, N.M., Janssen, D.B., van Berkel, W.J.H. and Fraaije, M.W. (2003) Baeyer–Villiger monooxygenases, an emerging family of flavin-dependent biocatalysts. *Advanced Synthesis Catalysis*, **345**, 667–678.

21
Second Generation Baeyer–Villiger Biocatalysts

Veronique Alphand, Marco W. Fraaije, Marko D. Mihovilovic and Gianluca Ottolina

21.1
Introduction

The oxidative conversion of a ketone into a lactone or ester functionality was first discovered in 1899 by Adolf von Baeyer and Victor Villiger [1]. The transformation turned out to be particularly valuable, as the rearrangement process commences with strict retention of the configuration at the migrating carbon center and the effects governing the regiochemistry of this reaction (conformational, steric, and electronic effects) have been well understood and became highly predictable [2, 3]. In present day organic chemistry, peroxyacids are used as reagents for this oxygen insertion into a less activated carbon-carbon bond and also asymmetric variants have been proposed [4].

The enzyme mediated equivalent was first discovered in 1948 in the biocatalytic degradation of steroids [5] and later proved to be abundant in fungi and prokaryotic organisms. The transformation was established to be dependent on oxygen and a nicotinamide reduction equivalent [6]. Approximately two decades after the biocatalytic Baeyer–Villiger reaction was observed, first Baeyer–Villiger monooxygenases (BVMOs) were isolated and characterized as responsible catalytic entities [7, 8]. In these proteins, a flavin co-factor replaces the peracid of the chemical transformation while NADH or NADPH is required as an electron donor. It is noteworthy that most reported BVMOs are soluble proteins, in contrast to many other types of monooxygenases, which tend to be associated to membranes. Depending on the nature of the flavin co-factor, at least two types of BVMOs can be classified [9]. Type 1 BVMOs consist of a single polypeptide chain, contain flavin adenine dinucleotide (FAD) as a tightly bound co-factor, and are NADPH dependent: the two co-factors are bound at separate dinucleotide binding domains as indicated by the presence of two Rossmann sequence motifs (GxGxxG). Type 2 BVMOs contain flavin mononucleotide (FMN) and are dependent on NADH: they are composed of two different subunits and display some relationship with flavin-dependent luciferases.

The generally accepted mechanism of BVMOs has been exemplified using a BVMO from *Acinetobacter* [10, 11]. Initially, the NADPH-mediated reduction of the FAD moiety leads to formation of an enamine-type intermediate (Figure 21.1).

Modern Biocatalysis: Stereoselective and Environmentally Friendly Reactions.
Edited by Wolf-Dieter Fessner and Thorleif Anthonsen

ISBN: 978-3-527-32071-4

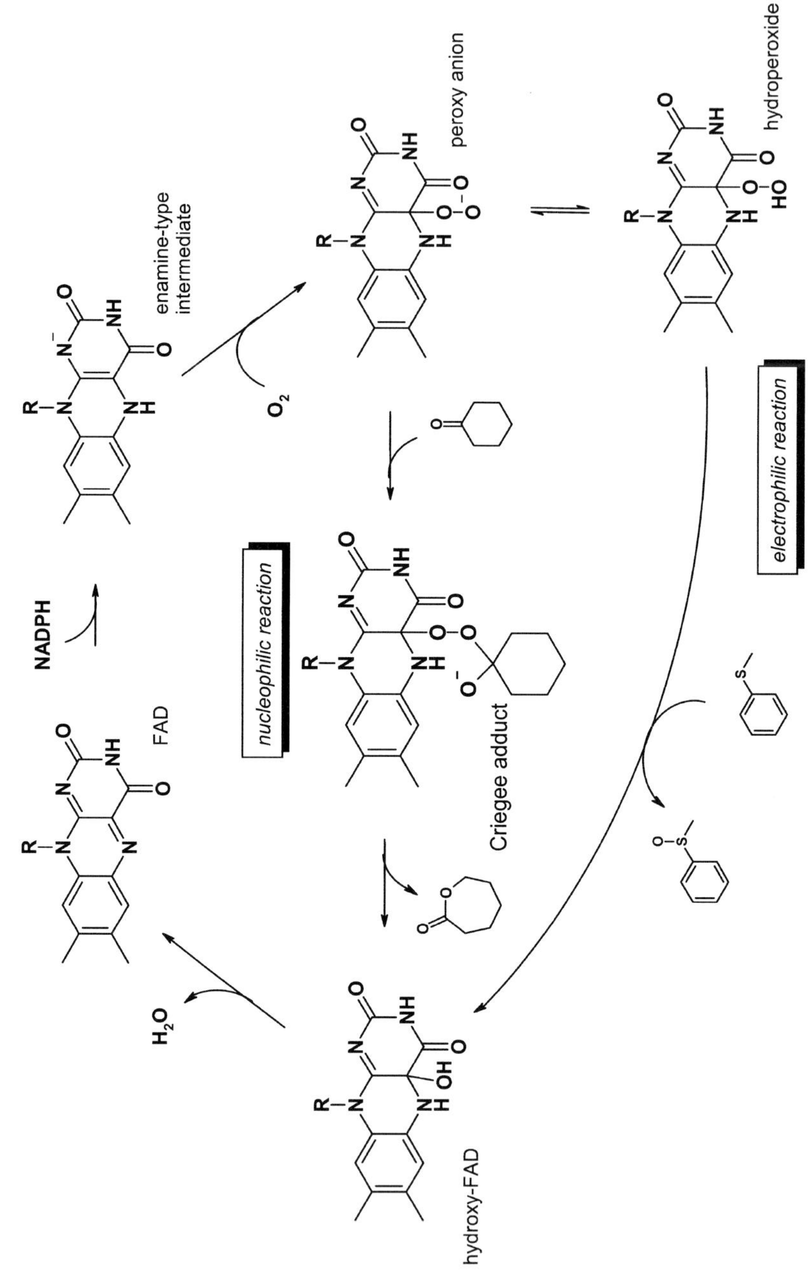

Figure 21.1 Mechanism and catalytic cycle of flavin-containing BVMOs.

This species reacts readily with molecular oxygen to yield the 4a-peroxyanion as equivalent to an organic peracid. Nucleophlic attack of the anion at a carbonyl group leads to formation of the Criegee intermediate analogous to the conventional reaction mechanism. Now, both stereoelectronic and conformational effects govern the rearrangement process: the spatial and electrostatic composition of the enzyme's active site determine which C—C bond can adopt the antiperiplanar conformation *vis-à-vis* the leaving group as a key aspect for migratory preference. The ester or lactone formed are released from the active site and the catalytic cycle is closed by elimination of water to regenerate the flavin co-factor.

The peroxyanion is also in equilibrium with the corresponding hydroperoxide and this protonation is believed to become prominent in the absence of a target for the nucleophilic attack. Consequently, an electrophilic reactive species is formed, which is then responsible for the oxygenation processes of heteroatoms.

Substantial progress in the understanding of this enzyme family was achieved by the recent determination of the three-dimensional structure of phenylacetone monooxygenase (PAMO) from the moderately thermophilic bacterium *Thermobifida fusca* [12], which exhibits a substrate preference for aryl containing ketones [13]. The enzyme displays a two-domain architecture: the FAD-binding domain is linked to the NADPH-binding domain and the active site is located in a cleft at the domain interface. The linker region contains a highly conserved sequence motif which represents an indicative 'fingerprint' sequence (FxGxxxHxxxW) for all BVMOs [14]. The two-domain set-up of PAMO suggests substantial conformational variations in the course of the biocatalytic oxygenation process.

BVMOs display a remarkable substrate promiscuity combined with usually high chemo-, regio-, and stereoselectivity, which make them appealing enzymes for applications in asymmetric synthetic chemistry. In recent years, a substantial number of newly identified enzymes have become available with complementary biocatalytic properties, in particular with respect to chemo- as well as stereoselectivity. While several detailed review articles have been published lately covering aspects of biochemistry [15, 16] and synthetic applications of BVMOs [17, 18] in a historical context, this chapter represents a highlight-type contribution focusing on recent developments in the field in order to promote the application of this enzyme group in organic synthesis further. The main emphasis will be put on modernized approaches for exploiting the well-established knowledge base in BVMO biocatalysis towards a 'second generation' of BVMO-type catalytic systems. The aspects discussed will mainly focus on progress resulting from the transnational cooperation by the research groups of the authors within a CERC3 research program in cooperation with the COST Action D25–Applied Biocatalysis.

21.2 BVMO Enzyme Platform

The majority of applications of enzyme-mediated Baeyer–Villiger oxidations have focused on the conversion of cyclic ketones to chiral lactones because of the importance of these building blocks in asymmetric synthesis. Many of the particular features of BVMOs were discovered with the cyclohexanone monooxygenase from *Acinetobacter* NCIMB 9871 ($CHMO_{Acineto}$), establishing this enzyme as the most prominent role model [19]. However, a large diversity of different BVMOs have become available within recent years. Various techniques in molecular biology were used for accessing novel genes, either by knowledge-based [14] or random strategies [20]. The applicability of these enzymes was largely improved by the generation of recombinant whole cells (*Saccharomyces cerevisiae* [21] and *Escherichia coli* [22] hosts), thereby providing simple solutions to the obstacles associated with protein isolation and co-factor regeneration applicable to synthetic chemists without special know-how in biochemistry.

Table 21.1 provides an overview of the most prominent enzymes together with information on representative substrate profiles in a historical context. The presently available BVMO platform includes enzymes that are highly stereoselective for the conversion of cyclic ketones in both desymmetrization reactions and kinetic resolutions capable of converting larger rings or sterically more demanding systems. Valuable extensions are BVMOs capable of selectively oxygenating aryl-containing substrates as well as linear ketones.

Based on a phylogenetic relationship, a certain clustering of those enzymes with a relevant number of successfully converted substrates can be observed based on their chemo- and stereoselectivity as outlined in Figure 21.2. Of particular relevance was the recent identification of two groups of cycloketone-converting BVMOs enabling access to enantiocomplementary lactones on a large number of substrates [32, 46, 47]. The enzymes included in this phylogenetic analysis represent only a very minor fraction of tentatively assigned BVMOs from recent genome sequencing programs. Well-defined substrate profiles, including structurally and electronically diverse ketone precursors, are available for many of these biocatalysts. Together with enzyme models derived from the structure of PAMO this may even allow careful predictions for novel substrates and/or modifications of the catalytic performance. However, a large variety of yet uncharacterized enzymes still awaits discovery for novel types of biotransformations.

21.3
Engineering of BVMOs

Altering the enzymatic properties of previously characterized enzymes by mutagenesis methodology using DNA technology is a well-accepted and often-used approach in the field of biocatalysis. By this approach, not only can substrate specificities be optimized, but the stability towards external factors such as temperature, organic solvents, salinity, and pH can also be improved. In recent years it has become popular to employ random mutagenesis combined with functionality screening ('directed evolution') for enzyme redesign. However, such an approach is often not very efficient for changing the substrate specificity of a certain biocatalyst as random mutations often occur far from the active site [48].

Table 21.1 BVMOs available in recombinant form with demonstrated potential for synthetic applications.

BVMO/Origin	Year		Representative
	Identical	Cloned	substrate profile
Baeyer–Villiger monooxygenase ($BVMO_{Mtb5}$) *Mycobacterium tuberculosis* H37Rv	2006 [23]	2006 [23]	[23, 24]
Cyclododecanone monooxygenase (CDMO) *Rhodococcus* SC1	2001 [25]	2001 [25]	[26]
Cyclohexanone monooxygenase ($CHMO_{Acineto}$) *Acinetobacer* NCIMB 9871	1976 [27]	1988 [28]	[29]
Cyclohexanone monooxygenase ($CHMO_{Arthro}$) *Arthrobacter* BP2	2003 [30]	2003 [31]	[26, 32]
Cyclohexanone monooxygenase ($CHMO_{Brachy}$) *Brachymonas petroleovorans*	2003 [33]	2003 [33]	[26, 32]
Cyclohexanone monooxygenase ($CHMO_{Brevi1\&2}$) *Brevibacterium* HCU	2000 [34]	2000 [34]	[26, 35]
Cyclohexanone monooxygenase ($CHMO_{Rhodo1\&2}$) *Rhodococcus* Phi1 and Phi2	2003 [30]	2003 [30]	[26, 32]
Cyclopentadecanone monooxygenase (CPDMO) *Pseudomonas* HI-70	2006 [36]	2006 [36]	[36]
Cyclopentanone monooxygenase ($CPMO_{Coma}$) *Comamonas* NCIMB 9872	1976 [37]	2002 [38]	[32, 38]
Hydroxyacetophenone monooxygenase (HAPMO) *Pseudomonas fluorescens* ACB	2001 [39]	2001 [39]	[40, 41]
Linear ketone monooxygenase ($BVMO_{Pflu}$) *Pseudomonas fluorescens* DSM 50106	2006 [42]	2006 [42]	[42]
Phenylacetone monooxygenase (PAMO) *Thermobifica fusca*	2004 [43]	2005 [44]	[45]

Such long-distance mutations typically do not strongly influence the substrate acceptance profile of enzymes. In fact, it appears that, for effectively altering the substrate specificity of an enzyme, one should preferentially target active site residues. Such a targeted mutagenesis of 'first shell' residues has been shown to result in dramatic changes in substrate specificity and/or enantioselectivity [49]. However,

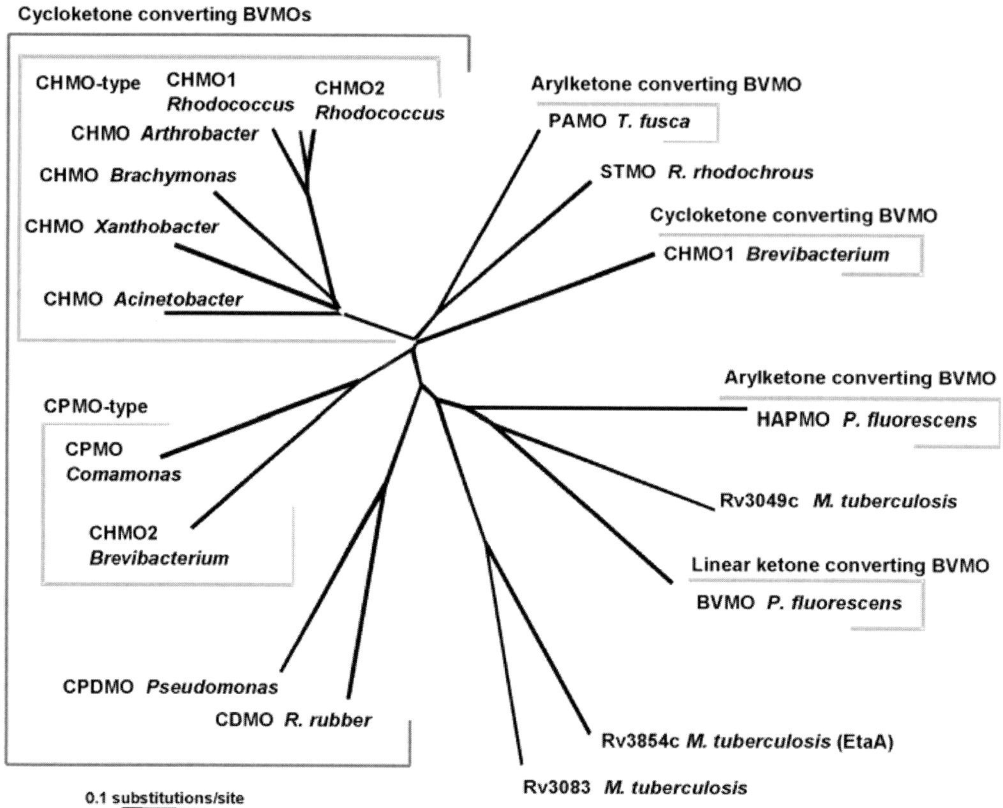

Figure 21.2 Phylogenetic relationships within BVMOs. (sequences of 18 proteins with confirmed BVMO activity were aligned and an unrooted phylogenetic tree was calculated using Clustal X v. 1.83 and TreeView v. 1.4).[1]

1) Abbreviations and GenBank accession numbers of protein sequences: CHMO *Acinetobacter*: CHMO *Acinetobacter* sp. NCIMB 9871: BAA86293; CHMO2 *Brevibacterium*: CHMO 2 *Brevibacterium* sp. HCU: AAG01290; CHMO *Arthrobacter*: CHMO *Arthrobacter* sp. BP2: AAN37479; CHMO *Brachymonas*: CHMO *Brachymonas petroleovorans*: AAR99068; CHMO1 *Brevibacterium*: CHMO 1 *Brevibacterium* sp. HCU: AAG01289; CHMO1 *Rhodococcus*: CHMO *Rhodococcus* sp. Phi1: AAN37494; CHMO2 *Rhodococcus*: CHMO *Rhodococcus* sp. Phi2: AAN37491; CHMO *Xanthobacter*: BVMO *Xanthobacter* sp. ZL5: CAD10801; CPMO *Comamonas*: cyclopentanone monooxygenase *Comamonas* sp. NCIMB 9872: BAC22652; PAMO *T. fusca*: phenylacetone monooxygenase *Thermobifida fusca*: 1W4X_A; HAPMO *P. fluorescens*: 4-hydroxyacetophenone monooxygenase *Pseudomonas fluorescens*: AAK54073; STMO *R. rhodochrous*: steroid monooxygenase *Rhodococcus rhodochrous*: BAA24454; CDMO *R. rubber*: cyclododecanone monooxygenase *Rhodococcus rubber*: AAL14233; Rv3854c *M. tuberculosis*: ETaA *Mycobacterium tuberculosis* H37Rv: CAB06212; CPDMO *Pseudomonas*: cyclopentadecanone monooxygenase *Pseudomonas* sp. HI-70: BAE93346; Rv3049c *M. tuberculosis:* BVMO from *Mycobacterium tuberculosis* H37Rv: CAA16134; Rv3083 *M. tuberculosis:* BVMO from *Mycobacterium tuberculosis* H37Rv: CAA16141; BVMO *P. fluorescens*: BVMO from *Pseudomonas fluorescens*: AAC36351.

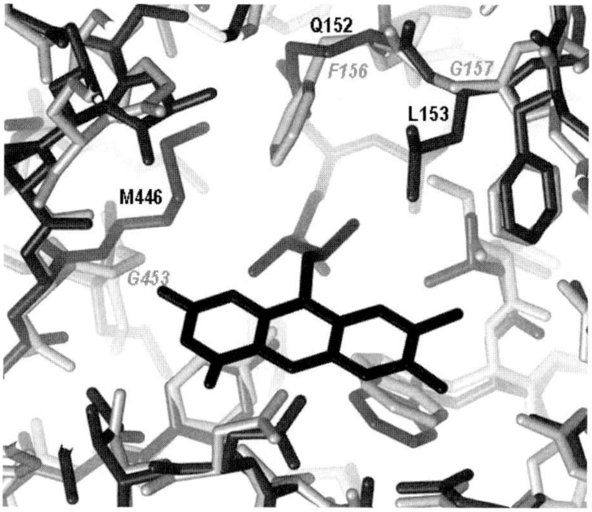

Figure 21.3 Overlay of the active site of PAMO (dark gray) and the model of CPMO (light gray). The FAD co-factor is in black. The three non-conserved residues that have been mutated in PAMO are indicated by the respective numbers for PAMO and CPMO.

such a structure-inspired (random) mutagenesis approach is only feasible when a structure or a structural model of the targeted biocatalyst is available.

In our recent work on BVMOs, we exploited the crystal structure of PAMO, which represents the only BVMO structure presently available [12, 13]. In a first study to alter substrate specificity, the PAMO structure and a CHMO model were compared in order to identify a bulge region in PAMO which appeared to be crucial for substrate specificity. Removing this bulge by deleting a few residues resulted in mutant PAMO enzymes that displayed activity with relatively bulky substrates, which are poorly or not accepted by wild-type PAMO [50]. A comparison of the PAMO structure and a homology-built structural model of the sequence-related CPMO revealed that both active sites are remarkably similar (Figure 21.3). In fact, the above-mentioned bulge region is also absent in CPMO, while it shows a similar substrate acceptance profile when compared with CHMO. Still, PAMO and CPMO show huge differences in their substrate profiles: while PAMO prefers aromatic substrates, CPMO is active with a wide range of (cyclic) aliphatic ketones. Most residues that surround the flavin co-factor in both BVMOs, thereby forming the active site, were found to be identical. Only three residues could be identified as significantly different: Q152, L153, and M446. It was postulated that these 'first shell' residues determine the molecular basis for the two widely different substrate specificities of PAMO and CPMO to some extent.

We replaced these residues in PAMO by the corresponding residues found in CPMO, thereby yielding three PAMO mutants [51]. Surprisingly, two of these mutants were found to be inactive (Q152F/L153G and Q152F/L153G/M446G).

These two mutant proteins were still able to bind the NADPH and FAD co-factors. Nevertheless, the mutants had lost the ability to perform oxygenation reactions. Apparently, the simultaneous mutation of Q152 and L153 resulted in an altered active site in which the organic substrates are not able to be oxidized by the peroxyflavin. In a recent mutagenesis study of CPMO these two homologous residues were randomly modified yielding several mutants with improved enantioselective behavior [51]. However, without exception, all reported CPMO mutants were less active and/or stable when compared with wild-type enzyme as evidenced by poor to moderate conversions using cells expressing these mutant enzymes. Cells containing wild-type CPMO resulted in full conversion of most tested substrates. This indicates that there is a subtle interplay between 'first shell' and 'second shell' residues resulting in a balanced and suitable substrate binding pocket.

Interestingly, the single mutant M446G PAMO was found to display a drastically altered substrate specificity and enantioselectivity. Several new compounds were identified as substrates for this PAMO variant (e.g. indole and benzaldehyde). The mutant displayed a higher specificity towards substrates containing the carbonyl group or the heteroatom in close proximity to the aromatic ring. A different positioning of the substrate's aromatic ring in the active site might cause this observed shift in regioselectivity. Furthermore, an altered substrate-binding pocket also explains the substantial changes in enantioselectivity observed towards sulfides and ketones. This confirms the role of this specific amino acid residue in modulating the substrate-binding pocket. Interestingly, the active site of mutant M446G is to some extent mimicking the effect of addition of methanol to wild-type PAMO. Recently we have shown that usage of up to 30% methanol is tolerated by PAMO and also affects the enantioselectivity [52]. Whether the organic solvent specifically binds in the active site or induces structural changes in wild-type PAMO, ultimately resulting in an altered substrate-binding pocket, is still disputable. The fact that replacement of M446 can result in dramatic changes in substrate specificity and enantioselectivity suggests that specific interaction of one or two methanol molecules in the active site may explain the observed solvent effect.

While the exact binding mode of substrates in PAMO and other BVMOs is uncertain due to the lack of a crystal structure containing a bound substrate or product, the mutagenesis studies described deliver clues about the residues involved in forming the substrate-binding pocket. In the last few years several random and directed mutagenesis studies have been performed with PAMO, CHMO and CPMO [53–56]. An inventory of all mutants that display an alteration of enantioselectivity shows that many targeted residues are part of the first shell of the proposed substrate-binding pocket of PAMO. As can be seen in Figure 21.4, most of these residues align a pocket at the *si*-face of the flavin co-factor. It is also striking to note that most of the observed mutations are located in the FAD binding domain and the neighboring helical subdomain, while no mutation occurs in the NADPH-binding domain. The identification of M446 in PAMO as being crucially involved in determining the substrate specificity and enantioselectivity complements the previously identified hotspots that influence BVMO specificity (Figure 21.4, Table 21.2).

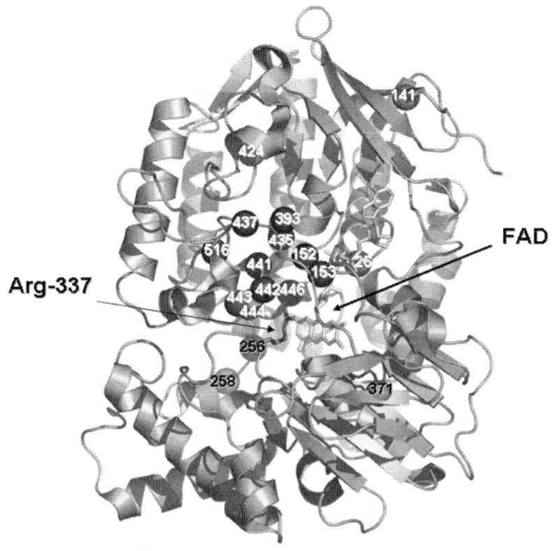

Figure 21.4 Cartoon picture of the PAMO crystal structure. The FAD co-factor is shown in sticks in the center together with the active site R337. Residues that are <15 Å from the isoalloxazin moiety of the FAD co-factor are shown in dark gray balls, while more distant residues are represented as light gray balls.

Future BVMO redesign studies can exploit the localization of these hotspots by simultaneously mutating the respective residues that determine the plasticity of the substrate-binding pocket of BVMOs. In this context it is noteworthy that, for several mutations in CHMO, an extension of substrate acceptance was observed with concomitant high turnover to allow for preparative exploitation [56]. By this strategy, novel valuable BVMOs can be created that extend the catalytic potential of the presently available BVMOs and may ultimately combine substrate promiscuity with other enzymatic properties, for example thermal stability in case of PAMO.

21.4 Baeyer–Villiger Biooxidation in Synthetic Chemistry

21.4.1 Chemoselectivity

Many molecules with different skeletons have been assayed with BVMOs, but only a few studies on substrates with more than one reactive group have been conducted. In the case of sulfoxidation reactions, 1,3-dithiane represents an interesting example for chemoselectivity: the initial biooxygenation product is a

Table 21.2 Inventory of observed mutations in BVMOs and the related improvements in enantioselectivity.

Effect on enantioselectivity[a]	Enzymes/Mutations[b]	Hotspots
CHMO		
9% *R* → 34% *R* [53]	**F432Y**, *K500R*	**L443**
9% *R* → 40% *R* [53]	**L143F**	**Q152**
9% *R* → 49% *R* [53]	**F432I**	**L443**
9% *R* → 54% *R* [53]	**L426P**, *A541V*	**A435**
9% *R* → 18% *S* [53]	*L220Q*, **P428S**, **T433A**	**P437**, **S444**
9% *R* → 46% *S* [53]	*D41N*, **F505Y**	**L516**
9% *R* → 78% *S* [53]	*K78E*, **F432S**	**L443**
9% *R* → 79% *S* [53]	**F432S**	**L443**
9% *R* → 90% *R* [53]	**L143F**, *E292G*, **L435Q**, *T464A*	**Q152**, **M446**
14% *R* → 99% *R* [57]	**D384H**	**T393**
14% *R* → 99% *R* [57]	**F432S**	**L443**
14% *R* → 98% *S* [57]	*K229I*, L248P	R258
14% *R* → 99% *S* [57]	Y132C, **F246I**, V361A, T415A	I141, **T256**, V371, S424
14% *R* → 95% *S* [57]	**F16L**, *F277S*	**F26**
CPMO		
5% *S* → 59% *R* [55]	**G449S**, **F450Y**	**A442**, **L443**
5% *S* → 90% *R* [55]	**F156N**, **G157Y**	**Q152**, **L153**
PAMO		
$E = 1 \rightarrow E = 100$ [50]	**ΔS441**, **ΔA442**	**S441**, **A442**
6% *S* → 95% *R* [51]	**M446G**	**M446**

a Improvements in the enantioselectivity of specific oxidation reactions are given for the indicated mutants.

b Residues that are located on the protein surface are shown in italic and residues that are <15 Å from the isoalloxazin moiety of the FAD co-factor are shown in bold.

The locations in the PAMO structure are shown in Figure 21.4.

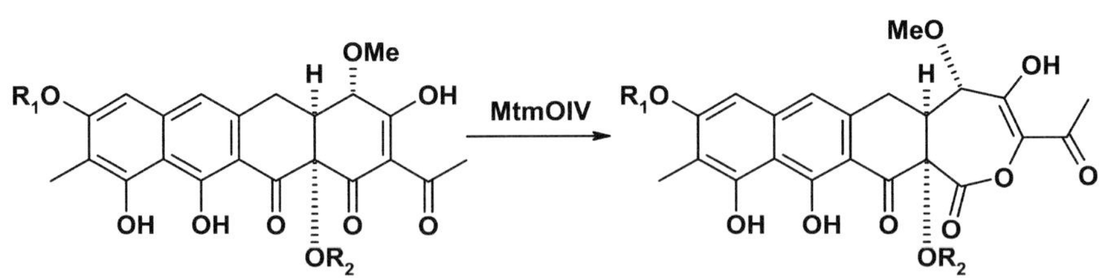

premithramycin B R_1 = disaccharide, R_2 =trisaccharide

Scheme 21.1 Site selectivity in the Baeyer–Villiger oxidation of premithramycin B by MtbOIV.

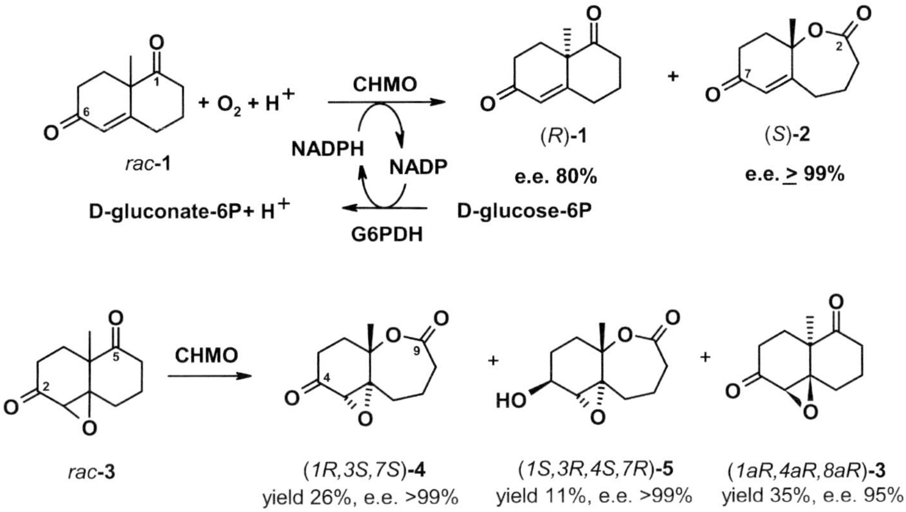

Scheme 21.2 Chemoselective oxygenation of dicarbonyl substrates using CHMO from *Acinetobacter*.

monosulfoxide, which is then preferably converted to the sulfone as opposed to the formation of disulfoxides [58].

Another impressive result for the selective recognition of a particular carbonyl group was reported for MtmOIV monooxygenase as the only BVMO for which the natural product has been clearly identified: this enzyme is able to insert an oxygen atom through a Baeyer–Villiger reaction in premithramycin B in high site specificity, discriminating olefinic, aromatic, cyclic, and non-cyclic carbonyl groups (Scheme 21.1) [59].

In a recent contribution by our groups we investigated the chemoselectivity of BVMOs acting on non-native substrate. The enzymatic Baeyer–Villiger oxidation of bicyclic diketones with the CHMO from *Acinetobacter* has been used to generate the corresponding keto lactones, structures frequently found in natural products or key intermediates in synthesis [60]. The enzymatic oxidation of racemic Wieland–Miescher ketone **1** with CHMO was fully selective giving the corresponding lactone **2** with the (*S*)-configuration and >99% enantiomeric excess (e.e.) (Scheme 21.2). Furthermore, the corresponding reduced compound, the racemic *cis*-8a-methyl-hexahydro-naphthalene-1,6-dione, was selectively oxidized by the CHMO only in position 1 to give the (5a*R*,9a*S*)-keto lactone with a very high e.e. (≥99). The epoxidation of the racemic Wieland–Miescher ketone gave epoxidione **3**, which gave the enantiopure lactone (1*R*,3*S*,7*S*)-**4** upon treatment with a crude preparation of CHMO plus the hydroxylactone (1*S*,3*R*,4*S*,7*R*)-**5** due to the presence of a contaminant dehydrogenase activity.

Where the selectivity of BVMOs regarding the Baeyer–Villiger process versus a heteroatom oxidation is concerned, usually the oxygenation at the carbonyl center is favored. This was demonstrated in a study of heterocyclic substrates of type **6** to give good yields of the corresponding lactone product **7** (Scheme 21.3) [61, 62].

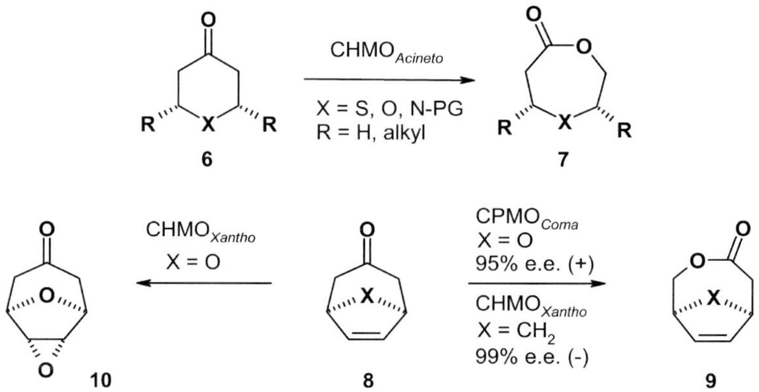

Scheme 21.3 Functional group selectivity of BVMOs.

Moreover, functional tolerance of these enzymes usually includes groups labile towards oxidation under conventional chemical conditions such as olefins (**7**: R = $CH{=}CH_2$) [63]. However, electron-deficient Michael-type acceptor alkenes have been reported to undergo oxygenation to epoxides in the case of alkenyl phosphonate precursors [64]. In a very recent contribution we could demonstrate that even non-activated C=C double bonds can react to epoxides and this biotransformation was clearly assigned to the activity of a BVMO [65]. However, such an epoxidation (**10**) has only been observed in a single case so far using a BVMO originating from a *Xanthobacter* sp.: it is noteworthy that a highly complementary behavior was found depending on the nature of the substrate **8** (carbocyclic versus heterocyclic) and transformation with CPMO from *Comamonas* selectively gave the Baeyer–Villiger product (+)-**9** (X = O).

While certain interesting aspects have been recently discovered, the potential of BVMO-mediated biooxygenation of multifunctional substrates and, in particular, of polyketone compounds, is yet to be fully investigated. Further studies in this area certainly contain the prospect of high impact contributions for the further improvement and increase in efficiency for future applications in single-operation multistep synthesis.

21.4.2
Dynamic Kinetic Resolutions

Besides the level of enantioselectivity, a main drawback of kinetic resolution is the well known limitation of product yield to 50%. Among several strategies elaborated to overcome this barrier [66] and improve kinetic resolution, dynamic kinetic resolution (DKR), coupling kinetic resolution and *in situ* racemization, was applied to Baeyer–Villiger biooxidations. Such an approach theoretically enables the synthesis of a enantiomerically pure compound in 100% yield from a racemate. The key point is the ability of a given substrate to be easily racemized under conditions compatible with the kinetic resolution. From this point of view, ketones

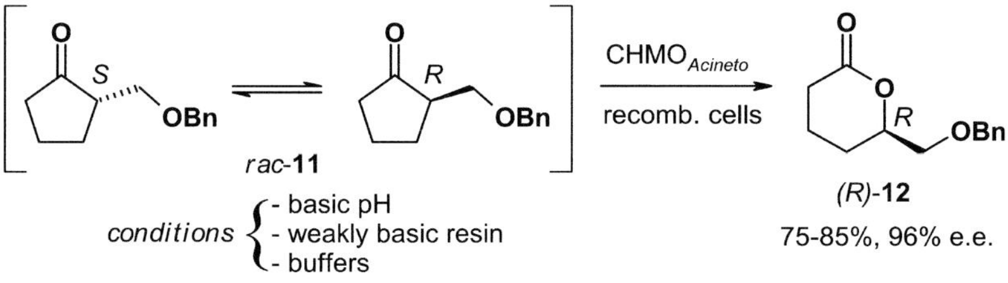

Scheme 21.4 Dynamic kinetic resolution by CHMO from *Acinetobacter*.

bearing a substituent at the α position are possible candidates since they are likely to undergo racemization via a keto-enol tautomerization under acidic or basic conditions. However, as the racemization rate is comparably slow relative to the biooxidation rate, the choice of highly enantioselective biotransformations is mandatory.

Following this line, several DKRs were described using 2-benzyloxycyclopentanone **11** as a model substrate and whole cells of *E. coli* TOP10[pQR239] overexpressing $CHMO_{Acineto}$ as the biocatalyst (Scheme 21.4). Preparative-scale biotransformations (1 g) were carried out directly in a benchtop fermenter. In the simplest case, racemization was catalyzed under basic conditions (pH 9) controlled by the addition of NaOH solution all over the biotransformation. The lactone product **12** was obtained in 85% yield and 96% e.e. for a substrate concentration below 0.3 g/l [67]. Milder conditions (pH 7) using anion-exchange resins as the 'racemizing' agent were employed with the same success in a higher concentration (1 g/l). Surprisingly, strongly basic resins bearing quaternary ammonium proved to be less effective compared to weakly basic resins bearing tertiary amino groups as Lewatit MP62 [68]. A model based on a pseudo-first-order kinetics was validated in a first approximation by a good correlation between experimental values and calculations of kinetic constants.

Efficient DKRs were also carried out in the presence of a high concentration of phosphate and imidazole buffers acting as only the 'racemizing' agents. Benzyloxycaprolactone was isolated in 75–80% yield and 97% e.e. Thus, at pH 7.2, racemization was determined as being approximatively 10-fold faster in imidazole or phosphate buffer than in classical biotransformation medium.

21.4.3
Regio- and Stereoselectivity

The identification of sub-clusters of BVMOs enabling access to antipodal lactones for an overlapping group of structurally diverse substrates has represented a key contribution to the field in recent years [32, 46, 47]. Cycloketone-converting BVMOs can be roughly sub-divided into 'CHMO-type' and 'CPMO-type' enzymes with $CHMO_{Acineto}$ and $CPMO_{Coma}$ as prototypes for the particular clusters. The two clades display distinctly diverse biotransformation characteristics: often enantiocomple-

Table 21.3 Desymmetrization of prochiral cycloketones to enantiocomplementary lactones by CHMO- ($CHMO_{Acineto}$ and $CHMO_{Xantho}$) and CPMO-type ($CHMO_{Brevi2}$ and $CPMO_{Coma}$) enzymes (representative examples).

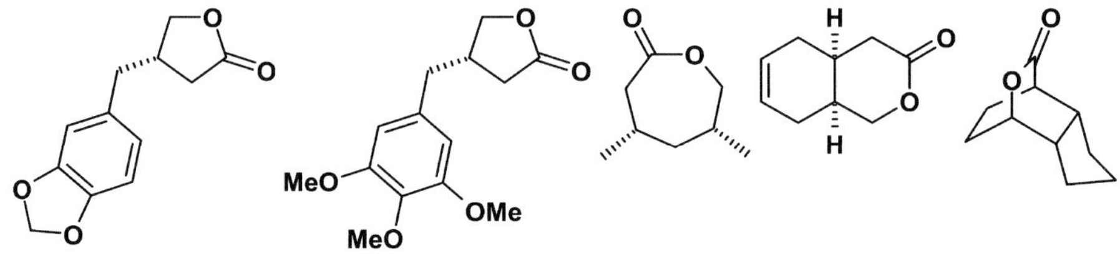

$CHMO_{Acineto}$	97% e.e. (−)	90% e.e. (−)	>99% e.e. (−)	5% e.e. (−)	97% e.e. (−)
$CHMO_{Xantho}$	99% e.e. (−)	95% e.e. (−)	97% e.e. (−)	>99% e.e. (−)	96% e.e. (−)
$CHMO_{Brevi1}$	75% e.e. (+)	79% e.e. (+)	97% e.e. (−)	71% e.e. (−)	94% e.e. (−)
$CHMO_{Brevi2}$	37% e.e. (−)	n.c.	99% e.e. (+)	94% e.e. (+)	92% e.e. (+)
$CPMO_{Coma}$	40% e.e. (−)	n.c.	91% e.e. (+)	>99% e.e. (+)	91% e.e. (+)

$CHMO_{Brevi1}$ adopts a borderline position $CHMO_{Brevi1}$ (n.c.: no conversion; the sign of specific rotation is indicated after the value for the e.e.).

mentary lactones are obtained, in some cases one group only exhibits high stereoselectivity, and in the minority of cases a difference in substrate acceptance is observed. The CHMO from *Brevibacterium* adopts a borderline position between these groups with some similarities to CHMO-type enzymes, albeit displaying a distinct stereospecificity with several substrates (Table 21.3; see also Figure 21.2 for phylogenetic analysis).

This suite of cycloketone-converting BVMOs has recently been expanded for enzymes displaying appealing properties in the kinetic resolution of linear ketones. A BVMO from *Pseudomonas fluorescens* turned out to be particularly useful for the conversion of terminal ketones to chiral diol derivatives [42, 69], while PAMO and HAPMO are suitable biocatalysts for the kinetic resolution of aromatic ketones and aldehydes [70].

The formation of BVMO clusters is also confirmed to a certain extent in regiodivergent biooxygenations by this enzyme group. Migratory preference of the more nucleophilic (higher substituted) center is observed generally for chemical oxidations and in the majority of enzyme-mediated transformations. This behavior is governed by the difference in electron density between carbons bearing an additional group and those with all hydrogen substituents. However, stereoelectronic effects leading to a preferred conformational arrangement to promote a different direction of the rearrangement process is particularly frequently encountered in biotransformations. In selected cases, racemic precursors are then oxidized by BVMOs into two types of regioisomeric lactones in a resolution process: migration of the more substituted carbon atom leads to formation of the expected 'normal' lactone, while the 'abnormal' lactone is produced upon migration of the less substituted carbon atom (Scheme 21.5).

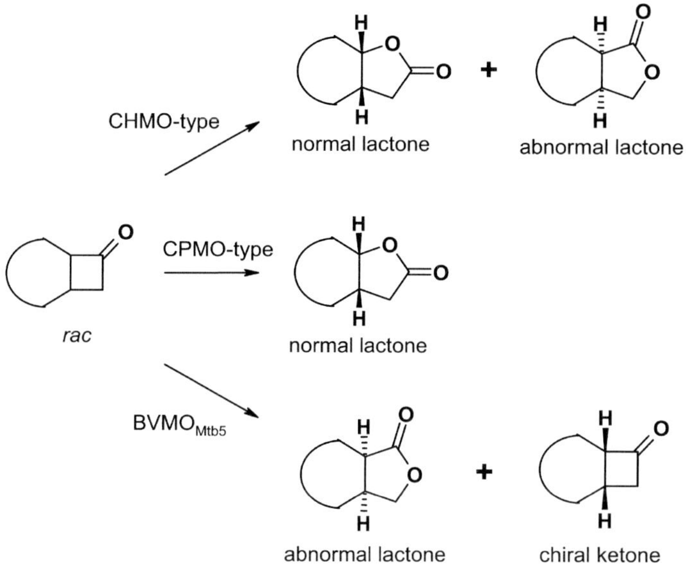

Scheme 21.5 BVMO portfolio for regiodivergent biooxidation of fused bicycloketones.

While this remarkable behavior was initially observed using CHMO$_{Acineto}$ with fused bicycloketones bearing a cyclobutanone structural motif [71] (see [72] for mechanistic considerations), a more general trend could be identified for CHMO- and CPMO-type enzymes: the prior cluster usually displays a clear regiodivergent biooxygenation and racemic precursors are converted into isomeric 'normal' and 'abnormal' lactones in approximately 1 : 1 ratios and in high optical purities, while representatives of the CPMO group preferentially yield 'normal' lactones, albeit in racemic form [73, 74]. The latter catalysts therefore enable chemoselective entry to chiral lactones when starting from optically pure ketone precursors.

Recently, screening of a small BVMO library originating from *Mycobacterium tuberculosis* led to the discovery of BVMO$_{Mtb5}$ as the selective biocatalyst for the predominant preparation of 'abnormal' lactones. This enzyme performs a kinetic resolution of racemic precursors to yield enantiomerically pure ketone and lactone, subsequently simplifying the separation of these products by chromatography. In addition, it nicely complements the portfolio of divergent biooxygenations with fused bicyclic ketones incorporating a cyclobutanone structural motif [75, 76].

In the case of larger ring systems, we could most recently apply the available collection of cycloketone-converting BVMOs to access all possible regio- and stereoisomeric lactones from terpenone precursors, as exemplified in Scheme 21.6 for biotransformations of dihydrocarvone [77]. Isolated cases, in which the formation of an unexpected regioisomer was observed, included CHMO$_{Acineto}$-mediated oxygenation of cyclohexanones bearing an electron-withdrawing cyano substituent in the 2-position [78] as well as the biotransformation of β-substituted cycloketones [79, 80]. Certainly, there remains ample opportunities for contributing to this

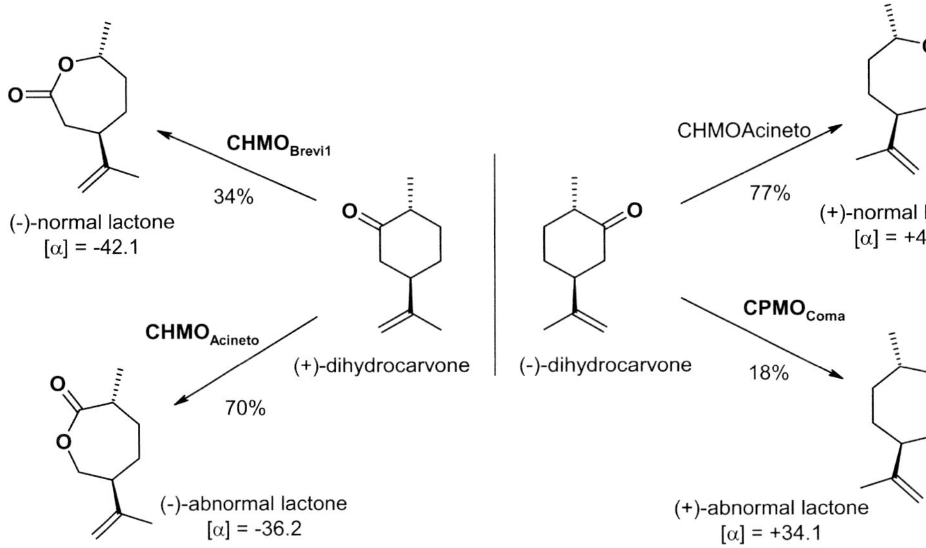

Scheme 21.6 Regiodivergent biooxidation of dihydrocarvone.

interesting sub-domain, in which enzymatic transformations enable rapid entry to diverse chiral building blocks that are otherwise difficult to obtain via conventional organic chemistry.

21.4.4 Natural Product and Bioactive Compound Synthesis

All of the strategies in the Baeyer–Villiger biooxygenations outlined above enable efficient access to optically pure building blocks of relevance for the synthesis of bioactive compounds or natural products. Several appealing applications have already been outlined in the past. Both chiral lactones or ketones obtained from kinetic resolutions of simple cycloketones have been utilized for the production of lipoic acid [81, 82] and various pheromones [83]. In addition, both products have been applied in synthetic routes to prostaglandins [84], algae pheromones [85], and cytostatics [86] in regiodivergent biooxygenations.

More recent applications in target-oriented synthesis took advantage of the BVMO platform for the generation of enantiocomplementary lactones. In this context, butyrolactones represent appealing intermediates due to the facile availability of prochiral ketone substrates for enzyme-mediated desymmetrizations; Scheme 21.7 indicates the potential in lignan total synthesis based on products obtained in a previous study [46]. The collection of BVMOs exploited by our groups also allowed efficient access to various indole alkaloids via enantiocomplementary lactones obtained in the desymmetrization of fused bicycloketones [35].

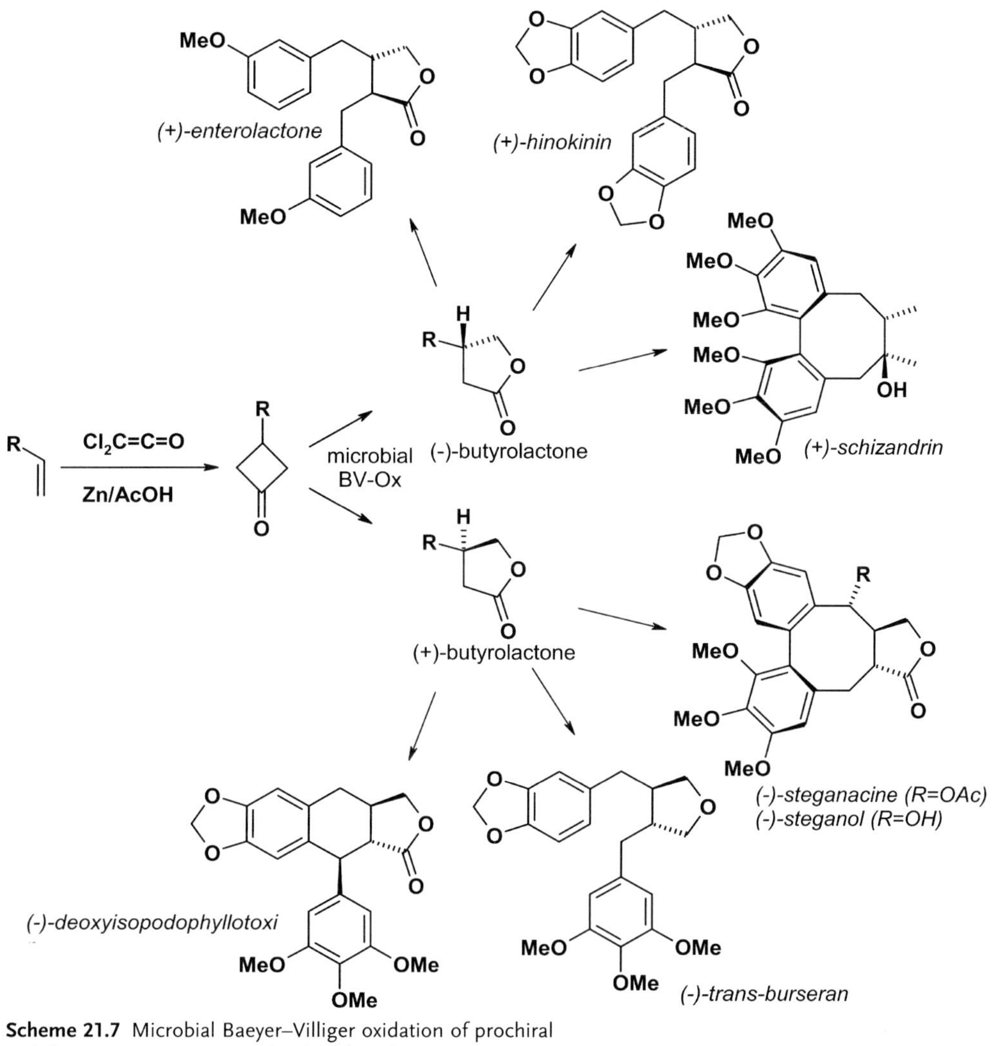

Scheme 21.7 Microbial Baeyer–Villiger oxidation of prochiral cyclobutanones to access antipodal butyrolactones as a platform for the synthesis of various lignans.

Bridged bicycloketones are attractive substrates due to the structural complexity of the scaffolds. The high chemoselectivity of the biotransformation for the carbonyl oxygenation allows for a functional decoration without the need for elaborate protection strategies. Subsequent chemical elaboration of the biooxidation products opens up rapid and efficient novel synthetic pathways towards various target compounds.

Biotransformation of hetero-bicyclic ketone **8** enabled access to unsaturated lactone **9** in a desymmetrization reaction without compromising the olefin func-

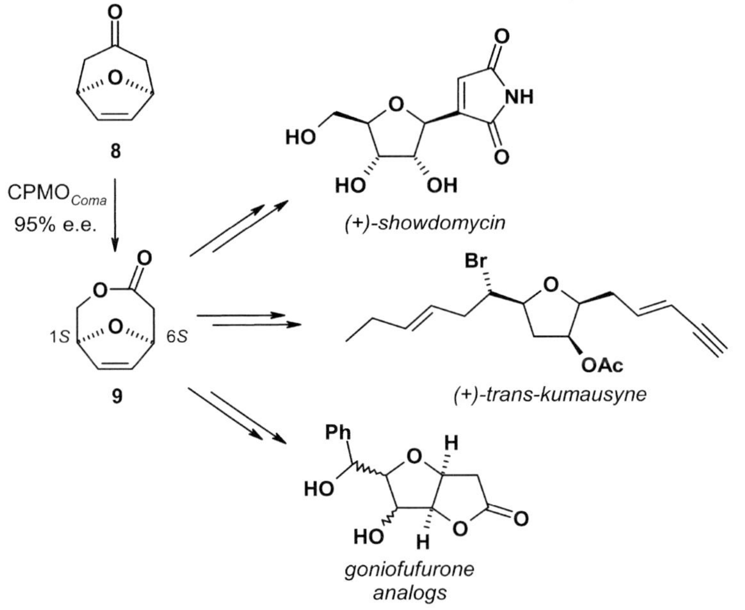

Scheme 21.8 BVMO-mediated desymmetrization of prochiral hetero-bicycloketone 8 and subsequent synthetic elaboration towards structurally diverse natural products.

tionality. This intermediate was then converted to showdomycin as a structural representative of the compound class of *C*-nucleoside antibiotics in order to establish the absolute configuration of this novel BVMO-derived metabolite. Alternative synthetic exploitation of the residual functionalities (alkene and lactone) opened up access to tetrahydrofuran natural products like kumausyne as well as goniofufurone analogs (Scheme 21.8) [87].

Green chemistry methods such as biocatalysis become particularly attractive when combined with additional sustainable strategies: very recently we proposed a combined photochemical and biocatalytic approach to access bicyclo[4.2.0] octanes via a novel strategy. Usually, the Cu-catalyzed [2 + 2] photocycloaddition of distantly located terminal diolefins is very difficult. However, when the double bonds are in close proximity the cyclization reaction proceeds smoothly (**11**). Consequently, incorporation of a cleavable bridge was expected to open a route to the above target compounds. In a first approach, a keto functionality was established at the bridge (**12**), which subsequently allowed for several cleavage protocols to be applied. The microbial Baeyer–Villiger oxidation again turned out to be the most appealing approach, as biotransformations proceeded in high yields to **13**. In addition, the biocatalytic route allowed access to antipodal lactones with full control over four stereogenic centers utilizing representatives of the CHMO and CPMO groups. Ultimately, the target bicyclo[4.2.0]octane system (**14**) was obtained after chemical lactone hydrolysis (Scheme 21.9) [88, 89].

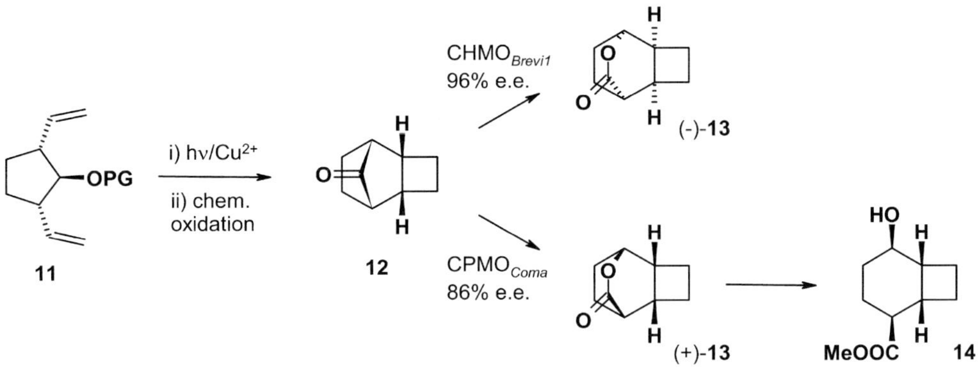

Scheme 21.9 Photochemical Cu-assisted [2 + 2] cycloaddition followed by BVMO-mediated desymmetrization as key steps in a novel synthetic approach to optically active bicyclo[4.2.0]octanes.

21.5 BVMOs in Stereoselective Sulfoxidations

The typical *S*-oxidation with BVMOs allows the formation of chiral sulfoxides from organic sulfides. This oxidation has received much interest in organic chemistry due to its use in the synthesis of enantiomerically enriched materials as chiral auxiliaries or directly as biologically active ingredients. This reaction has been studied extensively with CHMO from *Acinetobacter* showing high enantioselectivities in the sulfoxidation of alkyl aryl sulfides, disulfides, dialkyl sulfides, and cyclic and acyclic 1,3-dithioacetals [90]. CHMO also catalyzes the enantioselective oxidation of organic cyclic sulfites to sulfates [91].

Recently the sulfoxidation of some organic sulfides has been tested with two other enzymes: PAMO and HAPMO [92, 93]. The oxidation of sulfides by PAMO or HAPMO was coupled to a second enzymatic reaction to regenerate NADPH using glucose 6-phosphate dehydrogense (G6PDH). With regard to the stereoselectivity of enzymatic reactions, the data indicated that, for PAMO, it is highly dependent on substrate structure. Thus, the optical purities of the products range from 80% e.e. and (*R*)-configuration with methyl phenethyl sulfoxide to 98% e.e. and the (*S*)-configuration with benzyl ethyl sulfoxide. The alkyl benzyl sulfides possessing a small alkyl chain appeared to be very good PAMO substrates in terms of selectivity. It is quite interesting to point out that the best PAMO substrate is phenylacetone, which shares a similar structure to alkyl benzyl sulfides, suggesting that this structural motif is the preferred by this enzyme.

Moreover, the e.e. of the sulfoxide products changes with the progress of the reaction indicating that the PAMO was able to catalyze the kinetic resolution of the sulfoxide to the sulfone. The enantiomeric ratio measured with methyl phenethyl sulfoxide was 110. Previous studies carried out with CHMO revealed that the oxidation of the sulfoxide to sulfone was very slow, thereby preventing exploitation of this kinetic resolution [90].

With HAPMO the stereoselectivity was generally high: thioanisole oxidation was almost complete after 1 day, resulting in the formation of enantiopure (*S*)-methyl phenyl sulfoxide. A very high e.e. was also obtained in the sulfoxidation of thioanisole with CHMO (>99%), but with (*R*)-configuration. This finding was true for several aryl alkyl sulfides suggesting that these two enzymes might be enantiocomplementary in sulfoxidation reactions.

PAMO and HAPMO have been used in aqueous organic media in the sulfoxidation of several substrates. The effect of the solvent on the catalytic properties of the enzymes was quite remarkable, apart from a general decrease of enzyme activity. A high increase in enantioselelctivity was found with methanol and PAMO, which was one of the best solvents for this enzyme. However, for HAPMO the best solvent found was *i*-Pr_2O. A pH dependence of enantioselectivity has been observed with thioanisole and PAMO [94], showing that the e.e. increased from 10 to 45% by raising the pH from 6 to 10.

21.6 Towards a Technology Platform

21.6.1 Fermentation Up-Scaling

In addition to the high chemo- and stereoselectivity of BVMOs, the operational benefits of enzymatic oxygenation processes also include safety issues compared to chemical Baeyer–Villiger oxidations, since classical oxidants and flammable solvents are replaced, respectively, by air and water. However, contrary to expectation, no real breakthroughs have been reached at the industrial level. Even at the laboratory-scale, Baeyer–Villiger biocatalysts are under employed by synthetic chemistry [95]. Several explanations can be pointed out: the apparent complexity of the process [96] (e.g. compared to hydrolytic enzymes) with the need of co-factor recycling, but also the importance of providing a high level of oxygenation. Another bottleneck is the low productivity, which arises from a low working concentration in substrate and product due to frequent inhibition or toxicity phenomena. Nevertheless, various techniques were recently developed successfully in order to overcome, or at least minimize, these limitations.

21.6.1.1 Whole Cells

One of the easiest means of solving the co-factor need is the use of a whole cell-based process. NAD(P)H recycling is provided by the endogenous cellular tools themselves. The implementation of this process was largely facilitated by the use of recombinant cells overexpressing the target enzyme. The criteria for the host strain selection are simplicity of culture, fast growth rate, absence of natural side reactions, ease of genetic modification, and, of course, existence of an internal efficient co-factor regeneration system They were mostly fulfilled by *E. coli* species like *E. coli* TOP10 [97], *E. coli* BL21(DE3) [22], and *E. coli* DH5α [98]. Enzyme overexpression in such systems is under the control of a strong promoter and can be well controlled by induction using arabinose and isopropylthio galacto pyranoside (IPTG). Reducing equivalents, which are crucial for driving bioconversion to completion, are supplied from glucose [99] or glycerol [100] addition. However, whatever the strain, productivity is still rather low with respect to industrial demand. The standard bioconversion of bicyclo[3.2.0]hept-2-en-6-one barely reached a space–time yield of 0.87 g/l/h for an initial substrate concentration of 1 g/l [95c].

Although the parameters governing a whole cell-based process are numerous, the key limitation of the low productivity was clearly identified as substrate and product inhibition. Substrate inhibition could conveniently be overcome by continuous feeding. A pilot-scale fed-batch experiment (55 l) was carried out [101] in which the feeding rate was adjusted to maintain the ketone concentration below 0.7 g/l. A final concentration of about 4 g/l of lactones was reached within 4 h. After downstream product recovery on charcoal, over 200 g of combined lactones was obtained, corresponding to a 76% yield.

To improve productivity further, limitation by product inhibition has also to be circumvented, but that implies a deeper modification of the process. A classical method is the use of a two-liquid phase process. Only small-scale experiments were carried out with moderate success [100, 102, 103]. Moreover, given the strong oxygen requirements of the biotransformation (cf. below), the up-scaling of this method would probably be hampered.

A quantitative jump was finally brought by a 'two-in-one' resin-based *in situ* substrate feeding product removal method (SFPR) [104]. An adsorbent polymeric resin acts simultaneously as a reservoir for the substrate and as a trap for the product (Figure 21.5). The substrate, preloaded onto the resin, slowly diffuses into the broth and is transformed by cells while the formed product is readsorbed onto the solid. The system is conveniently tuned by a simple choice of substrate : resin ratio. Thus, both the substrate and product concentrations are controlled in order to maintain them below their inhibitory or toxic levels and allow a higher productivity. Additional benefits are also to be considered for practical large-scale applica-

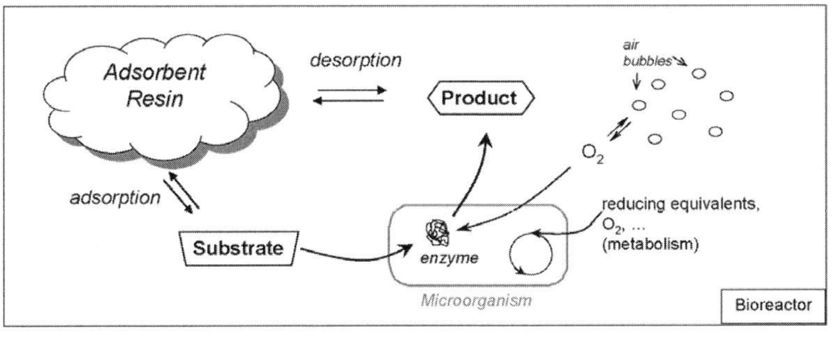

Figure 21.5 Principle of 'two-in-one' resin-based *in situ* SFPR.

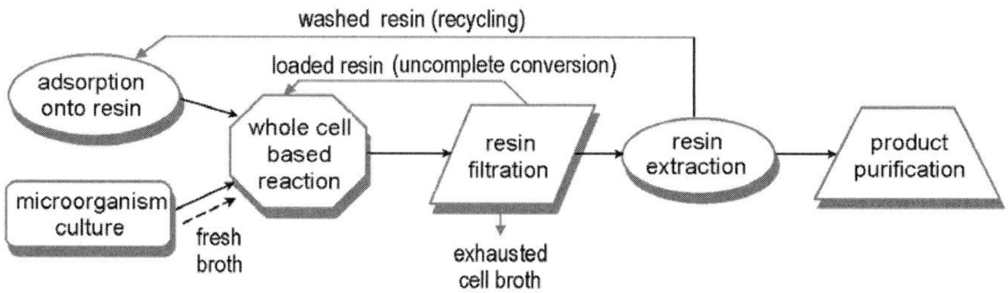

Figure 21.6 Flowsheet of the resin-based SFPR process.

tions such as the protection of sensitive compounds against evaporation or degradation, the simplified extraction protocol, and the reusability of the adsorbent resin. Product can be recovered from resin either by washings with a polar solvent or, to minimize the solvent amount, by a soxhlet extraction. The resin can also be reused after washing with methanol and water.

Several engineering designs (recycle reactor, conventional fermentor, and bubble column) bound for testing the SFPR concept at the laboratory scale were described [105]. At best, 25 g/l of bicyclo[3.2.0]hept-2-en-6-one were transformed by whole cells of CHMO-overexpressing *E. coli* TOP10[pQR239] in a space–time yield up to 1.2 g/l/h. Interestingly, exhausted cells could be replaced by fresh ones [106] by simple resin decantation when needed, which allowed the reaction to go to completion via a second cycle (Figure 21.6).

Oxygenation also proved to be crucial [107]: At high cell density, the oxygen requirements of Baeyer–Villiger oxidation compete with the needs of cell metabolism (respiration, co-factor recycling, etc.). The optimum biocatalyst concentration has to be set with respect to the oxygenation ability of a given reactor. Techniques for enhancing oxygen mass transfer such as a longer bubble residence time, the formation of smaller air bubbles (to increase the interface area) [107a], or the use of oxygen-enriched air have been tested [107b]. In the context of industrial applications, where the versatility of the equipment can be decisive for the choice of the process type, the best results were obtained by simply replacing the common fermenter ring sparger by a sintered-metal sparger. In collaboration with Sigma Aldrich, this simple modification was applied to a 50-l conventional industrial fermenter and enabled a 1-kg-scale operation in batch mode in less than 20 h [108]. An amount of 6 kg of adsorbent Dowex Optipore L-493 was used and regioisomeric lactones were directly separated by simulated moving bed chromatography after extraction [95b].

The resin-based SFPR concept is appearing more and more as a general tool for highly regio-, stereo-, or enantioselective Baeyer–Villiger biotransformations. Its scope has been successfully extended to various ketones [98] or sulfides [109] and different expression systems. Thus, typically 1 l of CPMO-overexpressing *E.coli* DH5α cells in the presence of Lewatit VPOC1163 transformed 5–15 g of ketones (3- and 4-methylcyclohexanone) into lactones in an approximately 80–95% yield.

21.6.1.2 Enzyme

To date, only a few examples of laboratory preparative-scale processes based on purified enzyme have been reported. Several studies have focused on the small-scale implementation of processes associating a new co-factor regenerating system, enzyme immobilization, membrane reactor, continuous substrate feeding, or resin-based SFPR with various results [110]. Using the outstanding stability of PAMO, a 200 ml biotransformation of 5 g/l phenyl cyclohexanone by an engineered mutant under two-liquid phase conditions using methyl tert-butyl ether as solvent was described [102].

21.6.2 Immobilization of BVMOs

$CHMO_{Acineto}$, the most used BVMO so far, is inherently unstable: therefore the operational stability is of high importance for any potential bioprocess. Certainly, a way of increasing the operational stability and also facilitating the work-up of the products is the immobilization of the biocatalysts, but there are only a few examples of immobilization of BVMOs. CHMO has been co-immobilized with alcohol dehydrogenase (ADH) or G6PDH as a recycling system for the regeneration of the co-factor (NADPH). CHMO and G6PDH have been co-immobilized with poly(acrylamide-*co*-*N*-acryloxysuccinimide) [111] and used for Baeyer–Villiger reactions lasting up to 10 days. More recently CHMO was co-immobilized with ADH on Eupergit (oxirane acrylic beads) [112], showing that the rate of substrate transformation was approximately 80% of that obtained with free enzyme and that the catalyst could be reused for up to 16 days with complete substrate conversion. The half-life of the immobilized CHMO was 2.5 days, which doubled the stability of the free enzyme. CHMO was also immobilized with a polyethylenimine-porous agarose polymeric composite and subjecting to a dose of γ-radiation [113]. After optimization 87% of the activity wais retained and this catalyst was reused for 16 cycles in the Baeyer–Villiger oxidation of a cyclohexanone derivative.

21.6.3 Self-sufficient Fusion Protein BVMOs

Several strategies of co-factor regeneration as one of the key obstacles in redox biocatalyst application have already been outlined in the previous section of this chapter. Most recently, we employed a novel approach to combine the catalytic activity of redox biocatalysts with concomitant co-enzyme recycling in a single fusion protein. During the last decade, a number of fusion protein tags have been developed that are used intensely in life science-related research and commercial activities. While the fusing of proteins is a widely applied strategy in, for example, enzyme purification protocols (e.g. GST-tag) [114] and subcellular visualization of target proteins (e.g. GFP-tag), [115] this concept is hardly encountered in the context of synthetic applications. Only a few isolated examples in the literature provide evidence that fusing separate enzymes can result in improved biocatalytic properties [116].

In this particular case, we decided to take advantage of the favorable thermodynamic equilibrium constant that drives the oxidation of phosphite to phosphate mediated by a recently described phosphite dehydrogenase (PTDH) [117] to a nearly irreversible process [118]. The exquisite selectivity of PTDH for phosphite also precludes any side reaction that can occur in case, for example, an ADH is used. These characteristics render PTDH as an ideal candidate for use as a co-enzyme regenerating enzyme (CRE) in combination with BVMOs or other NAD(P)H-dependent enzymes.

We created covalently connected fusion enzyme biocatalysts between PAMO, $CHMO_{Acineto}$, or $CPMO_{Coma}$ with PTDH (Figure 21.7) (CRE-PAMO, CRE-CHMO, and CRE-CPMO) within a proof-of-concept study [119]. Remarkably, we did not observe a significant change in the kinetic parameters of the bifunctional enzymes. Moreover, the substrate specificity as well as the stereoselectivity of the fusion BVMO–PTDH biocatalysts was hardly affected at all. Consequently, the fusion enzymes displayed a very similar biocatalytic behavior as compared to wild-type proteins. Based on the substrate acceptance profile of the three BVMOs, a first library of self-sufficient biocatalysts is now available for the conversion of aromatic and cycloaliphatic ketones for accessing even enantiocomplementary lactone products.

The novel biocatalysts were successfully employed in whole cell and purified protein biotransformations. However, a most appealing mode of operation is the utilization of a crude cell extract, which usually contains a very high amount of the CRE-BVMOs together with sufficient quantities of NADPH from the living host organism, ultimately making the addition of any co-factor recycling agent (additional enzyme, nicotinamide co-factor) obsolete apart from the auxiliary and cheap substrate phosphite.

At present, studies to extend this strategy to other enzymes are being conducted in our groups and we hope that this concept may become an appealing solution to the problem of co-factor regeneration in redox biocatalysis.

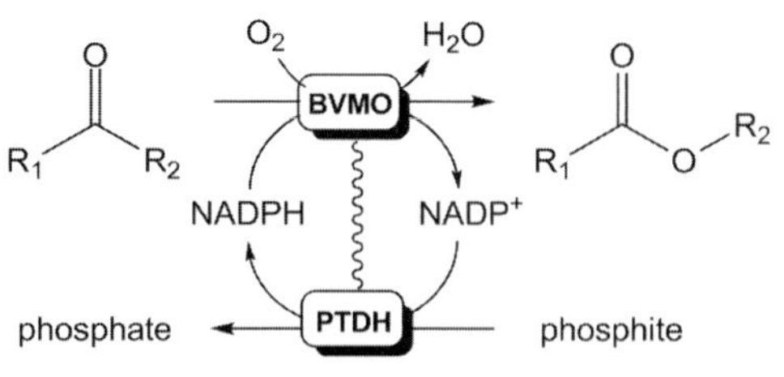

Figure 21.7 Schematic representation of co-enzyme regeneration by CRE BVMO fusion enzymes.

21.7 Outlook

Since the early days of Baeyer–Villiger biooxidations using a handful of crude biocatalysts originating from wild-type organisms with difficult to control enzyme production, the application of BVMOs in synthetic chemistry has matured to a platform technology. Today, a comparably large collection of diverse proteins with complementary substrate specificity and stereoselectivity has become available. In the light of the increasing number of genomes decoded every year and novel technologies for identifying potential BVMOs, there is a clear prospect of discovering new biocatalysts with novel properties and features. In addition, we are starting to understand the architecture of these enzymes based on the structure determination of at least one representative, subsequently allowing for random or knowledge-based optimization of these biocatalysts.

BVMOs open up a multitude of appealing transformations for accessing versatile building blocks for asymmetric synthesis. The number of applications in natural product and bioactive compound synthesis is constantly increasing. Still, considering the efficient and facile access to complex scaffolds incorporating several *de novo*-created stereogenic centers within the biooxygenation process, BVMOs deserve a larger community of users. In particular, in combination with novel fermentation strategies, these enzymes provide access to multigram-scale amounts of valuable intermediates.

We consider the recent design of the first self-sufficient fusion protein BVMOs as a very promising contribution by our groups for facilitating the use of this interesting enzyme family further. Due to the simplicity in applying these two-in-one second-generation BVMOs in synthetic chemistry, which can even be operated by scientists relatively inexperienced in biochemistry, we hope to proliferate the number of adventurous chemists eager to test some novel and exotic 'reagents' to ultimately add biocatalysis to their general portfolio of synthetic methodology.

References

1 Baeyer, A. and Villiger, V. (1899) *Chemische Berichte*, **32**, 3625–3633.
2 Krow, G.R. (1993) *Organic Reactions*, **43**, 251–798.
3 Renz, M. and Meunier, B. (1999) *European Journal of Organic Chemistry*, 737–750.
4 Mihovilovic, M.D., Rudroff, F. and Grötzl, B. (2004) *Current Organic Chemistry*, **8**, 1057–1069.
5 Turfitt, G.E. (1948) *Biochemical Journal*, **42**, 376–383.
6 Prairie, R.L. and Talalay, P. (1963) *Biochemistry*, **2**, 203–208.
7 Forney, F.W. and Markovetz, A.J. (1969) *Biochemical and Biophysical Research Communications*, **37**, 31–38.
8 Conrad, H.E., DuBus, R., Namtvedt, M.J. and Gunsalus, I.C. (1965) *Journal of Biological Chemistry*, **240**, 495–503.
9 Willetts, A. (1997) *Trends in Biotechnology*, **15**, 55–62.
10 Ryerson, C.C., Ballou, D.P. and Walsh, C. (1982) *Biochemistry*, **21**, 2644–2655.
11 Sheng, D., Ballou, D.P. and Massey, V. (2001) *Biochemistry*, **40**, 11156–11167.
12 Malito, E., Alfieri, A., Fraaije, M.W. and Mattevi, A. (2004) *Proceedings of the*

National Academy of Sciences of the United States of America, **101**, 13157–13162.

13 Fraaije, M.W., Kamerbeek, N.M., Heidekamp, A.J., Fortin, R. and Janssen, D.B. (2004) *Journal of Biological Chemistry*, **279**, 3354–3360.

14 Fraaije, M.W., Kamerbeek, N.M., van Berkel, W.J.H. and Janssen, D.B. (2002) *FEBS Letters*, **518**, 43–47.

15 Torres Pazmiño, D.E. and Fraaije, M.W. (2007) Future directions, in *Future Directions in Biocatalysis* (ed. T. Matsuda), Elsevier Science, Tokyo, pp. 107–128.

16 Kamerbeek, N.M., Janssen, D.B., van Berkel, W.J.H. and Fraaije, M.W. (2003) *Advanced Synthesis Catalysis*, **345**, 667–678.

17 Mihovilovic, M.D. (2006) *Current Organic Chemistry*, **10**, 1265–1287.

18 Mihovilovic, M.D., Müller, B. and Stanetty, P. (2002) *European Journal of Organic Chemistry*, 3711–3730.

19 Stewart, J.D. (1998) *Current Organic Chemistry*, **2**, 195–216.

20 Van Beilen, J.B., Mourlane, F., Seeger, M.A., Kovac, J., Li, Z., Smits, T.H.M., Fritsche, U. and Witholt, B. (2003) *Environmental Microbiology*, **5**, 174–182.

21 Stewart, J.D., Reed, K.W. and Kayser, M.M. (1996) *Journal of the Chemical Society–Perkin Transactions 1*, 755–757.

22 Chen, G., Kayser, M.M., Mihovilovic, M.D., Mrstik, M.E., Martinez, C.A. and Stewart, J.D. (1999) *New Journal of Chemistry*, **23**, 827–832.

23 Bonsor, D., Butz, S.F., Solomons, J., Grant, S., Fairlamb, I.J.S., Fogg, M.J. and Grogan, G. (2006) *Organic & Biomolecular Chemistry*, **4**, 1252–1260.

24 Snajdrova, R., Grogan, G. and Mihovilovic, M.D. (2006) *Bioorganic & Medicinal Chemistry Letters*, **16**, 4813–4817.

25 Kostichka, K., Thomas, S.M., Gibson, K.J., Nagarajan, V. and Cheng, Q. (2001) *Journal of Bacteriology*, **183**, 6478–6486.

26 Kyte, B.G., Rouviere, P., Cheng, Q. and Stewart, J.D. (2004) *Journal of Organic Chemistry*, **69**, 12–17.

27 Donoghue, N.A., Norris, D.B. and Trudgill, P.W. (1976) *European Journal of Biochemistry*, **63**, 175–192.

28 Chen, Y.-C.J., Peoples, O.P. and Walsh, C.T. (1988) *Journal of Bacteriology*, **170**, 781–789.

29 Stewart, J.D. (1998) *Current Organic Chemistry*, **2**, 195–216.

30 Brzostowicz, P., Walters, D.M., Thomas, S.M., Nagarajan, V. and Rouviere, P.E. (2003) *Applied and Environmental Microbiology*, **69**, 334–342.

31 Cheng, Q., Thomas, S.M., Kostichka, K., Valentine, J.R. and Nagarajan, V. (2000) *Journal of Bacteriology*, **182**, 4744–4751.

32 Mihovilovic, M.D., Rudroff, F., Grötzl, B., Kapitan, P., Snajdrova, R., Rydz, J. and Mach, R. (2005) *Angewandte Chemie–International Edition*, **44**, 3609–3613.

33 (a) Bramucci, M.G., Brzostowicz, P.C., Kostichka, K.N., Nagarajan, V., Rouviere, P.E., Thomas, S.M. (2003) PCT Int. Appl, WO 2003020890.
(b) Bramucci, M.G., Brzostowicz, P.C., Kostichka, K.N., Nagarajan, V., Rouviere, P.E., Thomas, S.M. (2003) *Chemical Abstracts*, **138**, 233997.

34 Brzostowicz, P.C., Gibson, K.L., Thomas, S.M., Blasko, M.S. and Rouviere, P.E. (2000) *Journal of Bacteriology*, **182**, 4241–4248.

35 Mihovilovic, M.D., Rudroff, F., Müller, B. and Stanetty, P. (2003) *Bioorganic & Medicinal Chemistry Letters*, **13**, 1479–1482.

36 Iwaki, H., Wang, S., Grosse, S., Bergeron, H., Nagahashi, A., Lertvorachon, J., Yang, J., Konishi, Y., Hasegawa, Y. and Lau, P.C.K. (2006) *Applied and Environmental Microbiology*, **72**, 2707–2720.

37 Griffin, M. and Trudgill, P.W. (1976) *European Journal of Biochemistry*, **63**, 199–209.

38 Iwaki, H., Hasegawa, Y., Lau, P.C.K., Wang, S. and Kayser, M.M. (2002) *Applied and Environmental Microbiology*, **68**, 5681–5684.

39 Kamerbeek, N.M., Moonen, M.J.H., van der Ven, J.G.M., van Berkel, W.J.H., Fraaije, M.W. and Janssen, D.B. (2001) *European Journal of Biochemistry*, **268**, 2547–2557.

40 Kamerbeek, N.M., Olsthorn, A.J.J., Fraaije, M.W. and Janssen, D.B. (2003) *Applied and Environmental Microbiology*, **69**, 419–426.

41 Mihovilovic, M.D., Kapitan, P., Rydz, J., Rudroff, F., Ogink, F.H. and Fraaije, M.W. (2005) *Journal of Molecular Catalysis B: Enzymatic*, **32**, 135–140.

42 Kirschner, A. and Bornscheuer, U.T. (2006) *Angewandte Chemie – International Edition*, **45**, 7004–7006.

43 Malito, E., Alfieri, A., Fraaije, M.W. and Mattevi, A. (2004) *Proceedings of the National Academy of Sciences of the United States of America*, **101**, 13157–13162.

44 Fraaije, M.W., Wu, J., Neuts, D.P.H., van Hellemond, E.W., Spelberg, J.H.L. and Janssen, D.B. (2005) *Applied Microbiology and Biotechnology*, **66**, 393–400.

45 Fraaije, M.W., Kamerbeek, N.M., Heidekamp, A.J., Fortin, R. and Janssen, D.B. (2004) *The Journal of Biological Chemistry*, **279**, 3354–3360.

46 Rudroff, F., Rydz, J., Ogink, F., Fink, M. and Mihovilovic, M.D. (2007) *Advanced Synthesis Catalysis*, **349**, 1436–1444.

47 Snajdrova, R., Braun, I., Bach, T., Mereiter, K. and Mihovilovic, M.D. (2007) *Journal of Organic Chemistry*, **72**, 9597–9603.

48 Morley, K.L. and Kazlauskas, R.J. (2005) *Trends in Biotechnology*, **23**, 231–237.

49 Fraaije, R.H.H., van den Heuvel, M.W., Ferrer, M., Mattevi, A. and van Berkel, W.J.H. (2000) *Proceedings of the National Academy of Sciences of the United States of America*, **97**, 9455–9460.

50 Bocola, M., Schulz, F., Leca, F., Vogel, A., Fraaije, M.W. and Reetz, M.T. (2005) *Advanced Synthesis Catalysis*, **347**, 979–986.

51 Torres Pazmino, D.E., Snajdrova, R., Rial, D.V., Mihovilovic, M.D. and Fraaije, M.W. (2007) *Advanced Synthesis Catalysis*, **349**, 1361–1368.

52 de Gonzalo, G., Ottolina, G., Zambianchi, F., Fraaije, M.W. and Carrea, G. (2006) *Journal of Molecular Catalysis B: Enzymatic*, **39**, 91–97.

53 Reetz, M.T., Brunner, B., Schneider, T., Schulz, F., Clouthier, C.M. and Kayser, M.M. (2004) *Angewandte Chemie – International Edition*, **43**, 4075–4078.

54 Clouthier, C.M., Kayser, M.M. and Reetz, M.T. (2006) *Journal of Organic Chemistry*, **71**, 8431–8437.

55 Clouthier, C.M. and Kayser, M.M. (2006) *Tetrahedron: Asymmetry*, **17**, 2649–2653.

56 Mihovilovic, M.D., Rudroff, F., Winninger, A., Schneider, T., Schulz, F. and Reetz, M.T. (2006) *Organic Letters*, **8**, 1221–1224.

57 Reetz, M.T., Daligault, F., Brunner, B., Hinrichs, H. and Deege, A. (2004) *Angewandte Chemie – International Edition*, **43**, 4078–4081.

58 Zambianchi, F., Raimondi, S., Pasta, P., Carrea, G., Gaggero, N. and Woodley, J.M. (2004) *Journal of Molecular Catalysis B: Enzymatic*, **31**, 165–171.

59 Gibson, M., Nur-e-alam, M., Lipata, F., Oliveira, M.A. and Rohr, J. (2005) *Journal of the American Chemical Society*, **127**, 17594–17595.

60 Ottolina, G., de Gonzalo, G., Carrea, G. and Danieli, B. (2005) *Advanced Synthesis Catalysis*, **347**, 1035–1040.

61 Latham, J.A. and Walsh, C. (1987) *Journal of the American Chemical Society*, **109**, 3421–3427.

62 Mihovilovic, M.D., Müller, B., Kayser, M.M., Stewart, J.D., Fröhlich, J., Stanetty, P. and Spreitzer, H. (2001) *Journal of Molecular Catalysis B: Enzymatic*, **11**, 349–353.

63 Mihovilovic, M.D., Grötzl, B., Kandioller, W., Muskotal, A., Snajdrova, R., Rudroff, F. and Spreitzer, H. (2008) *Chemistry & Biodiversity*, **5**, 490–498.

64 Colonna, S., Gaggero, N., Carrea, G., Ottolina, G., Pasta, P. and Zambianchi, F. (2002) *Tetrahedron Letters*, **43**, 1797–1799.

65 Rial, D.V., Bianchi, D.A., Kapitanova, P., Lengar, A., van Beilen, J.B. and Mihovilovic, M.D. (2008) *European Journal of Organic Chemistry*, 1203–1213.

66 Faber, K. (2001) *Chemistry – A European Journal*, **7**, 5004–5010.

67 Berezina, N., Alphand, V. and Furstoss, R. (2002) *Tetrahedron: Asymmetry*, **13**, 1953–1955.

68 Gutiérrez, M.C., Furstoss, R. and Alphand, V. (2005) *Advanced Synthesis Catalysis*, **347**, 1051–1059.

69 Geitner, K., Kirschner, A., Rehdorf, J., Schmidt, M., Mihovilovic, M.D. and

Bornscheuer, U.T. (2007) *Tetrahedron: Asymmetry*, **18**, 892–895.

70 Rodriguez, C., de Gonzalo, G., Fraaije, M.W. and Gotor, V. (2007) *Tetrahedron: Asymmetry*, **18**, 1338–1344.

71 Alphand, V. and Furstoss, R. (1992) *Journal of Organic Chemistry*, **57**, 1306–1309.

72 Kelly, D.R., Knowles, C.J., Mahdi, J.G., Taylor, I.N. and Wright, M.A. (1995) *Journal of the Chemical Society D – Chemical Communications*, 729–730.

73 Mihovilovic, M.D. and Kapitan, P. (2004) *Tetrahedron Letters*, **45**, 2751–2754.

74 Mihovilovic, M.D., Kapitan, P. and Kapitanova, P. (2008) *ChemSusChem*, **1**, 143–148.

75 Bonsor, D., Butz, S.F., Solomons, J., Grant, S., Fairlamb, I.J.S., Fogg, M.J. and Grogan, G. (2006) *Organic & Biomolecular Chemistry*, **4**, 1252–1260.

76 Snajdrova, R., Grogan, G. and Mihovilovic, M.D. (2006) *Bioorganic & Medicinal Chemistry Letters*, **16**, 4813–4817.

77 Cernuchova, P. and Mihovilovic, M.D. (2007) *Organic & Biomolecular Chemistry*, **5**, 1715–1719.

78 Berezina, N., Kozma, E., Furstoss, R. and Alphand, V. (2007) *Advanced Synthesis Catalysis*, **349**, 2049–2053.

79 Alphand, V. and Furstoss, R. (1992) *Tetrahedron: Asymmetry*, **3**, 379–382.

80 Kyte, B.G., Rouviere, P., Cheng, Q. and Stewart, J.D. (2004) *Journal of Organic Chemistry*, **69**, 12–17.

81 Adger, B., Bes, M.T., Grogan, G., McCague, R., Pedragosa-Moreau, S., Roberts, S.M., Villa, R., Wan, P.W.H. and Willetts, A.J. (1995) *Journal of the Chemical Society D – Chemical Communications*, 1563–1564.

82 Adger, B., Bes, M.T., Grogan, G., McCague, R., Podragosa-Moreau, S., Roberts, S.M., Villa, R., Wan, P.W.H. and Willetts, A.J. (1997) *Bioorganic and Medicinal Chemistry*, **5**, 253–261.

83 Alphand, V., Archelas, A. and Furstoss, R. (1990) *Journal of Organic Chemistry*, **55**, 347–350.

84 Alphand, V., Archelas, A. and Furstoss, R. (1989) *Tetrahedron Letters*, **30**, 3663–3664.

85 Lebreton, J., Alphand, V. and Furstoss, R. (1997) *Tetrahedron*, **53**, 145–160.

86 Andrau, L., Lebreton, J., Viazzo, P., Alphand, V. and Furstoss, R. (1997) *Tetrahedron Letters*, **38**, 825–826.

87 Mihovilovic, M.D., Bianchi, D.A. and Rudroff, F. (2006) *Chemical Communications*, 3214–3216.

88 Braun, I., Rudroff, F., Mihovilovic, M.D. and Bach, T. (2006) *Angewandte Chemie – International Edition*, **45**, 5541–5543.

89 Braun, I., Rudroff, F., Mihovilovic, M.D. and Bach, T. (2007) *Synthesis*, 3896–3906.

90 Colonna, S., Gaggero, N., Pasta, P. and Ottolina, G. (1996) *Chemical Communications*, 2303–2307.

91 Colonna, S., Gaggero, N., Carrea, G. and Pasta, P. (1998) *Chemical Communications*, 415–416.

92 de Gonzalo, G., Torres Pazmiño, D.E., Ottolina, G., Fraaije, M.W. and Carrea, G. (2005) *Tetrahedron: Asymmetry*, **16**, 3077–3083.

93 de Gonzalo, G., Torres Pazmiño, D.E., Ottolina, G., Carrea, M.W. and Fraaije, G. (2006) *Tetrahedron: Asymmetry*, **17**, 130–135.

94 Zambianchi, F., Fraaije, M.W., Carrea, G., de Gonzalo, G., Rodríguez, C., Gotor, V. and Ottolina, G. (2007) *Advanced Synthesis Catalysis*, **349**, 1327–1331.

95 (a) Alphand, V., Carrea, G., Wohlgemuth, R., Furstoss, R. and Woodley, J.M. (2003) *Trends in Biotechnology*, **21**, 318–323.
(b) Wohlgemuth, R. (2006) *Engineering in Life Sciences*, **6**, 577–583.
(c) Law, H.E.M., Baldwin, C.V.F., Chen, B.H. and Woodley, J.M. (2006) *Chemical Engineering Science*, **61**, 6646–6652.

96 van Beilen, J.B., Duetz, W.A., Schmid, A. and Witholt, B. (2003) *Trends in Biotechnology*, **21**, 170–177.

97 Doig, S.D., Simpson, H., Alphand, V., Furstoss, R. and Woodley, J.M. (2003) *Enzyme and Microbial Technology*, **32**, 347–355.

98 Rudroff, F., Alphand, V., Furstoss, R. and Mihovilovic, M.D. (2006) *Organic Process Research & Development*, **10**, 599–604.

99 (a) Walton, A.Z. and Stewart, J.D. (2002) *Biotechnology Progress*, **18**, 262–268.
(b) Walton, A.Z. and Stewart, J.D. (2004) *Biotechnology Progress*, **20**, 403–411.

100 Simpson, H., Alphand, V. and Furstoss, R. (2001) *Journal of Molecular Catalysis B: Enzymatic*, **16**, 101–108.

101 Doig, S.D., Avenell, P.J., Bird, P.A., Gallati, P., Lander, K.S., Lye, G.J., Wohlgemuth, R. and Woodley, J.M. (2002) *Biotechnology Progress*, **18**, 1039–1046.

102 Schulz, F., Leca, F., Hollmann, F., Reetz, M.T. and Beilst, J. (2005) *Organic Chemistry*, **1**, 10.

103 Brosa, C., Rodriguez-Santamarta, C., Salva, J. and Barbera, E. (1998) *Tetrahedron*, **54**, 5781–5788.

104 Sometimes called extractive biocatalysis. It was also successfully applied to ketone reduction (a) Vicenzi, J.T., Zmijewski, M.J., Reinhard, M.R., Landen, B.E., Muth, W.L. and Marier, P.G. (1997) *Enzyme and Microbial Technology*, **20**, 494–499.
(b) D'Arrigo, P., Lattanzio, M., Fantoni, G.P. and Servi, S. (1998) *Tetrahedron: Asymmetry*, **9**, 4021–4026.
(c) Shorrock, V.J., Chartrain, M. and Woodley, J.M. (2004) *Tetrahedron*, **60**, 781–788.

105 (a) Hilker, I., Alphand, V., Wohlgemuth, R. and Furstoss, R. (2004) *Advanced Synthesis Catalysis*, **346**, 203–214.
(b) Hilker, I., Gutierrez, M.C., Alphand, V., Wohlgemuth, R. and Furstoss, R. (2004) *Organic Letters*, **6**, 1955–1958.

106 Cells kept at +4 °C were still active after several weeks. Berezina, N., Kozma, E., Furstoss, R. and Alphand, V. (2007) *Advanced Synthesis Catalysis*, **439**, 2049–2053.

107 (a) Hilker, I., Baldwin, C., Alphand, V., Furstoss, R., Woodley, J. and Wohlgemuth, R. (2006) *Biotechnology and Bioengineering*, **93**, 1138–1144.
(b) Baldwin, C.V.F. and Woodley, J.M. (2006) *Biotechnology and Bioengineering*, **95**, 362–369.

108 Hilker, I., Wohlgemuth, R., Alphand, V. and Furstoss, R. (2005) *Biotechnology and Bioengineering*, **92**, 702–710.

109 Zambianchi, F., Raimondi, S., Pasta, P., Carrea, G., Gaggero, N. and Woodley, J.M. (2004) *Journal of Molecular Catalysis B: Enzymatic*, **31**, 165–171.

110 Zambianchi, F., Pasta, P., Carrea, G., Colonna, S., Gaggero, N. and Woodley, J. M. (2002) *Biotechnology and Bioengineering*, **78**, 489–496.

111 Abril, O., Ryerson, C.C., Walsh, C. and Whitesides, G.M. (1989) *Bioorganic Chemistry*, **17**, 41–52.

112 Zambianchi, F., Pasta, P., Carrea, G., Colonna, S., Gaggero, N. and Woodley, J. M. (2002) *Biotechnology and Bioengineering*, **78**, 489–496.

113 Atia, K.S. (2005) *Radiation Physics and Chemistry*, **73**, 91–99.

114 Recent reviews: (a) Stahl, S., Hober, S., Nilsson, J., Uhlen, M. and Hygren, P.-A. (2003) *Biotechnology and Bioengineering*, **27**, 95–129.
(b) Trejo, F., Gelpi, J.L., Busquets, M. and Cortes, A. (1999) *Current Topics in Peptide and Protein Research*, **3**, 173–180.

115 Examples from the recent literature: (a) Narita, J., Okano, K., Tateno, T., Tanino, T., Sewaki, T., Sung, M.-H., Fukuda, H. and Kondo, A. (2006) *Applied Microbiology and Biotechnology*, **70**, 564–572.
(b) Giridhar, P.-H., Wu, R. and Wu, W.-T. (2006) *Biotechnology and Bioengineering*, **95**, 1138–1147.
(c) Liu, D., Schmid, R.D. and Rusnak, M. (2006) *Applied Microbiology and Biotechnology*, **72**, 1024–1032.

116 (a) Zhan, Y., Li, S.-Z., Li, J., Pan, X., Cahoon, R.E., Jaworski, J.G., Wang, X., Jez, J.M., Chen, F. and Yu, O. (2006) *Journal of the American Chemical Society*, **128**, 13030–13031.
(b) Khang, Y.H., Kim, I.W., Hah, Y.R., Hwangbo, J.H. and Kang, K.K. (2003) *Biotechnology and Bioengineering*, **82**, 480–488.
(c) Chen, X., Liu, Z., Wang, J., Fang, J., Fan, H. and Wang, P.G. (2000) *Journal of Biological Chemistry*, **275**, 31594–31600.

117 (a) Metcalf, W.W. and Wolfe, R.S. (1998) *Journal of Bacteriology*, **180**, 5547–5558.
(b) Garcia Costas, A.M., White, A.K. and Metcalf, W.W. (2001) *Journal of Biological Chemistry*, **276**, 17429–17436.

(c) Vrtis, J.M., White, A.K., Metcalf, W.W. and van der Donk, W.A. (2002) *Angewandte Chemie*, **114**, 3391–3393.
(d) Vrtis, J.M., White, A.K., Metcalf, W. W. and van der Donk, W.A. (2002) *Angewandte Chemie – International Edition*, **41**, 3257–3259.

118 Woodyer, R.D., van der Donk, W.A. and Zhao, H. (2003) *Biochemistry*, **42**, 11604–11614.

119 Pazmino, D.E.T., Snajdrova, R., Baas, B.-J., Ghobrial, M., Mihovilovic, M.D. and Fraaije, M.W. (2008) *Angewandte Chemie – International Edition*, **47**, 2275–2278.

Index

Modern Biocatalysis: Stereoselective and Environmentally Friendly Reactions.
Edited by Wolf-Dieter Fessner and Thorleif Anthonsen

ISBN: 978-3-527-32071-4